T0202895

Lecture Notes in Computer Science 14634

Founding Editors

Gerhard Goos
Juris Hartmanis

The series Lecture Notes in Computer Science (LNCS), including its subseries Lecture Notes in Artificial Intelligence (LNAI) and Lecture Notes in Bioinformatics (LNBI), has established itself as a medium for the publication of new developments in computer science and information technology research, teaching, and education.

LNCS enjoys close cooperation with the computer science R & D community, the series counts many renowned academics among its volume editors and paper authors, and collaborates with prestigious societies. Its mission is to serve this international community by providing an invaluable service, mainly focused on the publication of conference and workshop proceedings and postproceedings. LNCS commenced publication in 1973.

Stephen Smith · João Correia · Christian Cintrano
Editors

Applications of Evolutionary Computation

27th European Conference, EvoApplications 2024
Held as Part of EvoStar 2024
Aberystwyth, UK, April 3–5, 2024
Proceedings, Part I

 Springer

Editors
Stephen Smith 🆔
University of York
York, UK

João Correia 🆔
University of Coimbra
Coimbra, Portugal

Christian Cintrano 🆔
University of Málaga
Málaga, Spain

ISSN 0302-9743   ISSN 1611-3349 (electronic)
Lecture Notes in Computer Science
ISBN 978-3-031-56851-0   ISBN 978-3-031-56852-7 (eBook)
https://doi.org/10.1007/978-3-031-56852-7

This Springer imprint is published by the registered company Springer Nature Switzerland AG
The registered company address is: Gewerbestrasse 11, 6330 Cham, Switzerland

Paper in this product is recyclable.

Preface

This volume contains the proceedings of EvoApplications 2024, the International Conference on the Applications of Evolutionary Computation. The conference was part of Evo*, the leading event on bio-inspired computation in Europe, and was held in Aberystwyth, UK, as a hybrid event, between Wednesday, April 3, and Friday, April 5, 2023.

EvoApplications, formerly known as EvoWorkshops, aims to bring together high-quality research focusing on applied domains of bio-inspired computing. At the same time, under the Evo* umbrella, EuroGP focused on the technique of genetic programming, EvoCOP targeted evolutionary computation in combinatorial optimization, and EvoMUSART was dedicated to evolved and bio-inspired music, sound, art, and design. The proceedings for these co-located events are available in the LNCS series.

EvoApplications 2024 received 77 high-quality submissions distributed among the main session on Applications of Evolutionary Computation and 10 additional special sessions chaired by leading experts on the different areas: Analysis of Evolutionary Computation Methods: Theory, Empirics, and Real-World Applications (Thomas Bartz-Beielstein, Carola Doerr, and Christine Zarges); Applications of Bio-inspired Techniques on Social Networks (Giovanni Iacca and Doina Bucur); Computational Intelligence for Sustainability (Valentino Santucci, Fabio Caraffini, and Jamal Toutouh); Evolutionary Computation in Edge, Fog, and Cloud Computing (Diego Oliva, Seyed Jalaleddin Mousavirad, and Mahshid Helali Moghadam); Evolutionary Computation in Image Analysis, Signal Processing, and Pattern Recognition (Pablo Mesejo and Harith Al-Sahaf); Machine Learning and AI in Digital Healthcare and Personalized Medicine (Stephen Smith and Marta Vallejo); Problem Landscape Analysis for Efficient Optimisation (Bogdan Filipič and Pavel Krömer); Resilient Bio-inspired Algorithms (Carlos Cotta and Gustavo Olague); Soft Computing Applied to Games (Alberto P. Tonda, Antonio M. Mora, and Pablo García-Sánchez); and Surrogate-Assisted Evolutionary Optimisation (Tinkle Chugh, Alma Rahat, and George De Ath). We selected 24 of these papers for full oral presentation, while 9 works were presented in short oral presentations and as posters. Moreover, these proceedings also include contributions from the Evolutionary Machine Learning (EML) joint track, a combined effort of the International Conference on the Applications of Evolutionary Computation (EvoAPPS) and European Conference on Genetic Programming (EuroGP), organized by Penousal Machado and Mengjie Zhang. EML received 28 high-quality submissions. After careful review, eleven were selected for oral presentations and six for short oral presentations and posters. Since EML is a joint track, the "Evolutionary Machine Learning" part of these proceedings contains 16 of these papers. The remaining one is published in the EuroGP proceedings. All accepted contributions, regardless of the presentation format, appear as full papers in this volume.

An event of this kind would not be possible without the contribution of a large number of people:

- We express our gratitude to the authors for submitting their works and to the members of the Program Committee for devoting selfless effort to the review process.
- We would also like to thank Nuno Lourenço (University of Coimbra, Portugal) for his dedicated work as Submission System Coordinator.
- We thank Evo* Graphic Identity Team, Sérgio Rebelo, Jéssica Parente, and João Correia (University of Coimbra, Portugal), for their dedication and excellence in graphic design.
- We are grateful to Zakaria Abdelmoiz (University of Málaga, Spain) and João Correia (University of Coimbra, Portugal) for their impressive work managing and maintaining the Evo* website and handling the publicity, respectively.
- We credit the invited keynote speakers, Jon Timmis (Aberystwyth University, UK) and Sabine Hauert (University of Bristol, UK), for their fascinating and inspiring presentations.
- We would like to express our gratitude to the Steering Committee of EvoApplications for helping organize the conference.
- Special thanks to Christine Zarges (Aberystwyth University, UK) as local organizer and to Aberystwyth University, UK, for organizing and providing an enriching conference venue.
- We are grateful to the support provided by SPECIES, the Society for the Promotion of Evolutionary Computation in Europe and its Surroundings, for the coordination and financial administration.

Finally, we express our continued appreciation to Anna I. Esparcia-Alcázar, from SPECIES, Europe, whose considerable efforts in managing and coordinating Evo* helped build a unique, vibrant, and friendly atmosphere.

April 2024

Stephen Smith
João Correia
Christian Cintrano

Organization

EvoApplications Conference Chair

Stephen Smith   University of York, UK

EvoApplications Conference Co-chair

João Correia   University of Coimbra, Portugal

EvoApplications Publication Chair

Christian Cintrano   University of Málaga, Spain

Analysis of Evolutionary Computation Methods: Theory, Empirics, and Real-World Applications Chairs

Thomas Bartz-Beielstein   TH Köln, Germany
Carola Doerr   CNRS and Sorbonne Université, France
Christine Zarges   Aberystwyth University, UK

Applications of Bio-inspired Techniques on Social Networks Chairs

Giovanni Iacca   Università di Trento, Italy
Doina Bucur   University of Twente, The Netherlands

Computational Intelligence for Sustainability Chairs

Valentino Santucci   Università per Stranieri di Perugia, Italy
Fabio Caraffini   Swansea University, UK
Jamal Toutouh   University of Málaga, Spain

Evolutionary Computation in Edge, Fog, and Cloud Computing Chairs

Diego Oliva   Universidad de Guadalajara, México
Seyed Jalaleddin Mousavirad   Hakim Sabzevari University, Iran
Mahshid Helali Moghadam   RISE Research Institutes of Sweden, Sweden

Evolutionary Computation in Image Analysis, Signal Processing, and Pattern Recognition Chairs

Pablo Mesejo   Universidad de Granada, Spain
Harith Al-Sahaf   Victoria University of Wellington, New Zealand

Machine Learning and AI in Digital Healthcare and Personalized Medicine Chairs

Stephen Smith   University of York, UK
Marta Vallejo   Heriot-Watt University, UK

Problem Landscape Analysis for Efficient Optimisation Chairs

Bogdan Filipič   INRAE, Jožef Stefan Institute, Slovenia
Pavel Krömer   Technical University of Ostrava, Czech Republic

Resilient Bio-inspired Algorithms Chairs

Carlos Cotta   University of Málaga, Spain
Gustavo Olaguer   CICESE, Mexico

Soft Computing Applied to Games Chairs

Alberto P. Tonda   INRAE, France
Antonio M. Mora   Universidad de Granada, Spain
Pablo García-Sánchez   Universidad de Granada, Spain

Surrogate-Assisted Evolutionary Optimisation Chairs

Tinkle Chugh   University of Exeter, UK
Alma Rahat   Swansea University, UK
George De Ath   University of Exeter, UK

Evolutionary Machine Learning Chairs

Penousal Machado   University of Coimbra, Portugal
Mengjie Zhang   Victoria University of Wellington, New Zealand

EvoApplications Steering Committee

Stefano Cagnoni   University of Parma, Italy
Pedro A. Castillo   University of Granada, Spain
Anna I. Esparcia-Alcázar   Universitat Politècnica de València, Spain
Mario Giacobini   University of Torino, Italy
Paul Kaufmann   University of Mainz, Germany
Antonio Mora   University of Granada, Spain
Günther Raidl   Vienna University of Technology, Austria
Franz Rothlauf   Johannes Gutenberg University Mainz, Germany
Kevin Sim   Edinburgh Napier University, UK
Giovanni Squillero   Politecnico di Torino, Italy
Cecilia di Chio   King's College London, UK
   (Honorary Member)

Program Committee

Jacopo Aleotti   University of Parma, Italy
Mohamad Alissa   Edinburgh Napier University, UK
Anca Andreica   Babes-Bolyai University, Romania
Claus Aranha   University of Tsukuba, Japan
Aladdin Ayesh   De Montfort University, UK
Kehinde Babaagba   Edinburgh Napier University, UK
Jaume Bacardit   Newcastle University, UK
Marco Baioletti   Università degli Studi di Perugia, Italy
Illya Bakurov   Universidade NOVA de Lisboa, Portugal
Wolfgang Banzhaf   Michigan State University, USA
Tiago Baptista   University of Coimbra, Portugal
Thomas Bartz-Beielstein   TH Köln, Germany

Giulio Biondi	University of Florence, Italy
Philip Bontrager	New York University, USA
János Botzheim	Eötvös Loránd University, Hungary
Jörg Bremer	University of Oldenburg, Germany
Will Browne	Queensland University of Technology, Australia
Doina Bucur	University of Twente, The Netherlands
Maxim Buzdalov	ITMO University, Russia
Stefano Cagnoni	University of Parma, Italy
Fabio Caraffini	De Montfort University, UK
Oscar Castillo	Tijuana Institute of Technology, Mexico
Pedro Castillo	University of Granada, Spain
Josu Ceberio	University of the Basque Country, Spain
Ying-Ping Chen	National Yang Ming Chiao Tung University, Taiwan
Francisco Chicano	University of Málaga, Spain
Anders Christensen	University of Southern Denmark, Denmark
Tinkle Chugh	University of Exeter, UK
Christian Cintrano	University of Málaga, Spain
Anthony Clark	Pomona College, USA
José Manuel Colmenar	Universidad Rey Juan Carlos, Spain
Feijoo Colomine	Universidad Nacional Experimental del Táchira, Venezuela
Stefano Coniglio	University of Southampton, UK
Antonio Cordoba	University of Seville, Spain
Oscar Cordon	University of Granada, Spain
João Correia	University of Coimbra, Portugal
Carlos Cotta	Universidad de Málaga, Spain
Fabio D'Andreagiovanni	CNRS, Sorbonne University – UTC, France
Gregoire Danoy	University of Luxembourg, Luxembourg
George De Ath	University of Exeter, UK
Amir Dehsarvi	University of York, UK
Antonio Della Cioppa	University of Salerno, Italy
Bilel Derbel	CRIStAL (Univ. Lille), France
Travis Desell	Rochester Institute of Technology, USA
Laura Dipietro	HI
Federico Divina	Pablo de Olavide University, Spain
Carola Doerr	Sorbonne University, CNRS, France
Bernabe Dorronsoro	University of Cadiz, Spain
Tome Eftimov	Jožef Stefan Institute, Slovenia
Abdelrahman Elsaid	Rocheser Institute of Technology, USA
Ahmed Elsaid	University of Puerto Rico at Mayagüez, Puerto Rico

Edoardo Fadda	Politecnico di Torino, Italy
Andres Faina	IT University of Copenhagen, Denmark
Thomas Farrenkopf	Technische Hochschule Mittelhessen, Germany
Francisco Fernandez De Vega	Universidad de Extremadura, Spain
Antonio J. Fernández Leiva	Universidad de Málaga, Spain
Bogdan Filipič	Jožef Stefan Institute, Slovenia
Francesco Fontanella	Università di Cassino e del Lazio meridionale, Italy
James Foster	University of Idaho, USA
Alex Freitas	University of Kent, UK
Marcus Gallagher	University of Queensland, Australia
Pablo García Sánchez	Universidad de Granada, Spain
Mario Giacobini	University of Torino, Italy
Tobias Glasmachers	Institut für Neuroinformatik, Switzerland
Kyrre Glette	University of Oslo, Norway
Guillermo Gomez-Trenado	Universidad de Granada, Spain
Antonio Gonzalez-Pardo	Universidad Rey Juan Carlos, Spain
Michael Guckert	THM, Germany
Heiko Hamann	University of Lübeck, Germany
Mahshid Helali Moghadam	RISE SICS, Sweden
Daniel Hernandez	Tecnológico Nacional de México/Instituto Tecnológico de Tijuana, Mexico
Ignacio Hidalgo	Universidad Complutense de Madrid, Spain
Rolf Hoffmann	TU Darmstadt, Germany
Giovanni Iacca	University of Trento, Italy
Anja Jankovic	Sorbonne University, France
Juan Luis Jiménez Laredo	LITIS – Université Le Havre Normandie, France
Yaochu Jin	Universität Bielefeld, Germany
Karlo Knezevic	University of Zagreb, Croatia
Anna Kononova	Leiden University, The Netherlands
Ana Kostovska	Jožef Stefan Institute, Slovenia
Pavel Krömer	Technical University of Ostrava, Czech Republic
Gurhan Kucuk	Yeditepe University, Turkey
Waclaw Kus	Silesian University of Technology, Poland
Yuri Lavinas	University of Tsukuba, Japan
Joel Lehman	University of Central Florida, USA
Kenji Leibnitz	National Institute of Information and Communications Technology, Japan
Fernando Lobo	University of Algarve, Portugal
Michael Lones	Heriot-Watt University, UK
Nuno Lourenço	University of Coimbra, Portugal
Francisco Luna	Universidad de Málaga, Spain

Gabriel Luque	University of Málaga, Spain
Evelyne Lutton	INRAE, France
João Macedo	University of Coimbra, Portugal
Penousal Machado	University of Coimbra, Portugal
Katherine Malan	University of South Africa, South Africa
Luca Mariot	Radboud University, The Netherlands
Jesús Mayor	Universidad Politécnica de Madrid, Spain
David Megias	Universitat Oberta de Catalunya, Spain
Paolo Mengoni	Hong Kong Baptist University, China
Pablo Mesejo	University of Granada, Spain
Krzysztof Michalak	Wroclaw University of Economics, Poland
Mohamed Wiem Mkaouer	Rochester Institute of Technology, USA
Mahshid Helali Moghadam	RISE Research Institutes of Sweden, Sweden
Salem Mohammed	Mustapha Stambouli University, Algeria
Antonio Mora	University of Granada, Spain
Seyed Jalaleddin Mousavirad	Hakim Sabzevari University, Iran
Mario Andres Muñoz Acosta	University of Melbourne, Australia
Ferrante Neri	University of Nottingham, UK
Geoff Nitschke	University of Cape Town, South Africa
Jorge Novo Buján	Universidade da Coruña, Spain
Gustavo Olague	CICESE, Mexico
Diego Oliva	Universidad de Guadalajara, Mexico
Carlotta Orsenigo	University of Milan, Italy
Marcos Ortega Hortas	University of A Coruña, Spain
Anna Paszynska	Jagiellonian University, Poland
David Pelta	University of Granada, Spain
Diego Perez Liebana	Queen Mary University of London, UK
Yoann Pigné	LITIS – Université Le Havre Normandie, France
Clara Pizzuti	National Research Council of Italy (CNR), Institute for High Performance Computing and Networking (ICAR), Italy
Arkadiusz Poteralski	Silesian University of Technology, Poland
Petr Pošík	Czech Technical University in Prague, Czechia
Raneem Qaddoura	Al Hussein Technical University, Jordan
Alma Rahat	Swansea University, UK
José Carlos Ribeiro	Polytechnic Institute of Leiria, Portugal
Guenter Rudolph	TU Dortmund University, Germany
Jose Santos	University of A Coruña, Spain
Valentino Santucci	University for Foreigners of Perugia, Italy
Lennart Schäpermeier	TU Dresden, Germany
Enrico Schumann	University of Basel, Switzerland
Sevil Sen	University of York, UK

Roman Senkerik	Tomas Bata University in Zlin, Czechia
Chien-Chung Shen	University of Delaware, USA
Sara Silva	Universidade de Lisboa, Portugal
Kevin Sim	Edinburgh Napier University, UK
Anabela Simões	Coimbra Institute of Engineering, Portugal
Stephen Smith	University of York, UK
Maciej Smołka	AGH University of Science and Technology, Poland
Yanan Sun	Sichuan University, China
Shamik Sural	Indian Institute of Technology Kharagpur, India
Ernesto Tarantino	ICAR-CNR, Italy
Andrea Tettamanzi	Université Côte d'Azur, France
Renato Tinós	USP, Brazil
Marco Tomassini	University of Lausanne, Switzerland
Alberto Tonda	INRAE, France
Jamal Toutouh	University of Málaga, Spain
Heike Trautmann	University of Münster, Germany
Marta Vallejo	Heriot-Watt University, UK
Koen van der Blom	Leiden Institute of Advanced Computer Science, The Netherlands
Frank Veenstra	IT-University of Copenhagen, Denmark
Sebastián Ventura	University of Cordoba, Spain
Diederick Vermetten	Leiden Institute for Advanced Computer Science, The Netherlands
Marco Villani	University of Modena and Reggio Emilia, Italy
Rafael Villanueva	Instituto Universitario de Matematica Multidisciplinar, Spain
Markus Wagner	University of Adelaide, Australia
Hao Wang	Leiden University, The Netherlands
Jaroslaw Was	AGH University of Science and Technology, Poland
Thomas Weise	University of Science and Technology of China, China
Simon Wells	Edinburgh Napier University, UK
Dennis Wilson	ISAE-Supaero, France
Anil Yaman	Korea Advanced Institute of Science and Technology, South Korea
Furong Ye	Leiden Institute of Advanced Computer Science, The Netherlands
Ales Zamuda	University of Maribor, Slovenia
Christine Zarges	Aberystwyth University, UK
Mengjie Zhang	Victoria University of Wellington, New Zealand

Contents – Part I

**Analysis of Evolutionary Computation Methods: Theory, Empirics,
and Real-World Applications**

Computational Intelligence for Sustainability

Evolutionary Computation in Edge, Fog, and Cloud Computing

Evolutionary Computation in Image Analysis, Signal Processing and Pattern Recognition

Contents – Part II

**Machine Learning and AI in Digital Healthcare and Personalized
Medicine**

Problem Landscape Analysis for Efficient Optimization

Soft Computing Applied to Games

Surrogate-Assisted Evolutionary Optimisation

Applications of Evolutionary Computation

Finding Near-Optimal Portfolios
with Quality-Diversity

Bruno Gašperov[✉], Marko Đurasević, and Domagoj Jakobovic

Faculty of Electrical Engineering and Computing, University of Zagreb,
Zagreb, Croatia
{bruno.gasperov,marko.durasevic,domagoj.jakobovic}@fer.hr

Abstract. The majority of standard approaches to financial portfolio optimization (PO) are based on the mean-variance (MV) framework. Given a risk aversion coefficient, the MV procedure yields a single portfolio that represents the optimal trade-off between risk and return. However, the resulting optimal portfolio is known to be highly sensitive to the input parameters, i.e., the estimates of the return covariance matrix and the mean return vector. It has been shown that a more robust and flexible alternative lies in determining the entire region of near-optimal portfolios. In this paper, we present a novel approach for finding a diverse set of such portfolios based on quality-diversity (QD) optimization. More specifically, we employ the CVT-MAP-Elites algorithm, which is scalable to high-dimensional settings with potentially hundreds of behavioral descriptors and/or assets. The results highlight the promising features of QD as a novel tool in PO.

Keywords: quality-diversity · illumination algorithm · MAP-Elites · portfolio optimization · near-optimal portfolios

1 Introduction

1.1 Portfolio Optimization and Near-Optimal Portfolios

Portfolio optimization (PO) entails finding the optimal allocation of limited capital across a range of available financial assets within a specified timeframe. The optimality is typically defined with respect to a risk-adjusted return metric, such as the Sharpe ratio [1], or other metrics that account for the investor's specific risk preferences like the CARA utility function [2]. Classical approaches to PO are deeply rooted in the mean-variance (MV) framework, also referred to as modern portfolio theory (MPT) [3]. MV optimization provides a systematic methodology for constructing optimal portfolios that leverage the principle of diversification. By distributing investments among assets with dissimilar risk-return profiles (e.g., assets with mutually uncorrelated or negatively correlated returns), individual risks offset each other, thereby reducing the total portfolio risk. Given a risk aversion coefficient, the MV optimization process diversifies

S. Smith et al. (Eds.): EvoApplications 2024, LNCS 14634, pp. 3–18, 2024.
https://doi.org/10.1007/978-3-031-56852-7_1

assets to produce a single optimal portfolio representing the best risk-return trade-off. The set of optimal portfolios for different risk preferences constitutes a Pareto front called the efficient frontier.

However, the MV framework is predicated upon a number of simplified and unrealistic assumptions, including the normality of returns and the stationarity of the return covariance matrix. It also tends to produce portfolios highly concentrated in only a few assets, jeopardizing diversification. Furthermore, MV optimizers are particularly sensitive to estimation errors, with small changes in input parameters, especially expected return estimates, noticeably affecting the resulting optimal portfolio weights [4]. Such notorious issues have spurred researchers to propose novel PO approaches focusing on improving robustness. Various attempts have been made in this direction, including methods based on shrinking the covariance matrix [5] or expected returns [6], introducing constraints to the original MV framework [7], resampling the efficient frontier [8], or directly applying techniques from the area of robust optimization [9].

Another alternative, the focus of our work, involves identifying portfolios that are not strictly optimal but rather near-optimal (with respect to MV optimality). In the first phase (the optimization process), an entire subspace of mutually diverse[1] near-optimal portfolios is determined without focusing on any specific portfolio. This offers enhanced robustness to estimation errors [10], as it sidesteps the complex issues stemming from the interplay between the objective and the inputs, which are inherent to the MV optimization. In the second phase (the *a posteriori* analysis), the investor selects the final portfolio from the subspace of near-optimal portfolios. This decision-making scheme allows investors to incorporate their expert opinions, subjective views, or any other soft factors that may be challenging to integrate directly into an optimization problem formulation. Simultaneously, it permits the *ad-hoc* consideration of market frictions (e.g., transaction costs) and other regulatory, liquidity, and risk concerns, which are crucial in real-life PO but were not necessarily applied as constraints during the first phase. For instance, when rebalancing portfolio weights, investors may prefer a near-optimal portfolio similar to the current one over the MV optimal portfolio due to the former's lower turnover, and hence also lower transaction costs. Therefore the consideration of near-optimal portfolios presents a fruitful opportunity for improving the process of PO.

1.2   Prior Research

Near-Optimal Portfolios. The problem of identifying and analyzing near-optimal portfolios has been scrutinized in several studies. Van der Schans and de Graaf [11] pioneered a novel methodology for constructing such portfolios, outlined as follows. Let $\mathcal{F} : \mathcal{W} \to \mathcal{Z}$ be the mapping from the set of admissible portfolio weights $\mathcal{W}$ to the risk-return space $\mathcal{Z}$. First, an optimal portfolio $\boldsymbol{w_0}$ (satisfying $\mathcal{F}(\boldsymbol{w_0}) \in \mathcal{E}$ where $\mathcal{E} \subset \mathcal{Z}$ is the efficient frontier) is selected as a

---

[1] In terms of their distance in the space of admissible portfolio weights or, more generally, some behavior space.

reference portfolio in accordance with the investor's risk preferences. A small region $\mathcal{R} \subset \mathcal{Z}$ around $\mathcal{F}(\boldsymbol{w_0})$ is then defined and the portfolio $\boldsymbol{w_1}$, s.t. $\mathcal{F}(\boldsymbol{w_1}) \in \mathcal{R}$, located furthest away[2] from $\boldsymbol{w_0}$, is found. Subsequently, the portfolio $\boldsymbol{w_2}$, again s.t. $\mathcal{F}(\boldsymbol{w_2}) \in \mathcal{R}$, positioned furthest away from the convex hull $\mathcal{H}$ spanned by previously found portfolios $\{\boldsymbol{w_0}, \boldsymbol{w_1}\}$ is itself added to $\mathcal{H}$. Generally, in step i:

$$\boldsymbol{w_i} = \underset{\mathcal{F}(w) \in \mathcal{R}}{\arg \max} \; d\left(\boldsymbol{w}, \mathcal{H}\right), \tag{1}$$

with $d\left(\boldsymbol{w}, \mathcal{H}\right) = \min_{w' \in \mathcal{H}} \|\boldsymbol{w} - \boldsymbol{w'}\|$. This process iterates until $\mathcal{S} \subset \mathcal{Z}$, the convex hull in the risk-return space corresponding to the portfolios in $\mathcal{H}$, covers $\mathcal{R}$ to the required level of precision ϵ. Finally, a set of K diverse near-optimal portfolios $\{\boldsymbol{w_0}, \boldsymbol{w_1}, \ldots, \boldsymbol{w_{K-1}}\}$ is obtained. Since a convex combination of near-optimal portfolios is also near-optimal [10], any portfolio in $\mathcal{H}$ acts as a viable option for the investor. The authors finally show that this region of near-optimal portfolios is more robust to input estimate uncertainties than a single optimal portfolio. However, this approach exhibits several limitations. Firstly, it involves solving a difficult non-convex optimization problem of finding the portfolio furthest from the convex hull, leading the authors to use a somewhat *ad hoc* combination of a support vector machine [12] and a basin-hopping algorithm [13]. Secondly, it becomes unfeasible in high-dimensionality settings, i.e., when faced with a large number of assets (N), which is typically the case in modern PO[3]. Thirdly, it is restricted to finding near-optimal portfolios directly in the space of portfolio weights and is not easily applicable to other possible behavior (feature) spaces, which may comprise variables such as fundamentals, technicals, and risk factors. Put differently, it does not tackle the generalized variant where in step i:

$$\boldsymbol{w_i} = \underset{\mathcal{F}(w) \in \mathcal{R}}{\arg \max} \; d\left(\phi(\boldsymbol{w}), \mathcal{H}\right) \tag{2}$$

where $\mathcal{H} = \{\phi(\boldsymbol{w_0}), \ldots, \phi(\boldsymbol{w_{i-1}})\}$ for some function ϕ. To ameliorate some of these problems, Cajas [15] proposed the near-optimal centering (NOC) method, based on finding the analytic centers of near-optimal regions. The method can be used in conjunction with any convex risk measure and has been empirically demonstrated to lead to improved diversification and robustness when compared to traditional PO methods. Van Eeghen [14] expands on [11] by using polytope theory to inspect the structure and robustness of near-optimal regions. Moreover, the author proposes a new implementation of the method from [11] that reduces computation time while maintaining accuracy.

The topic of near-optimal portfolios has been investigated or touched upon in several other works as well. Chopra [16] uses a grid search to discover a subset of near-optimal portfolios and analyzes their composition. Benita *et al.* [17] emphasize that near-optimal portfolios might provide a higher degree of robustness to various scenarios due to significant differences in their makeups and provide an

---

[2] In the sense of the Euclidean distance between the portfolio weight vectors $\|\boldsymbol{w_1} - \boldsymbol{w_0}\|$.

[3] Van Eeghen [14] reports computation times of around 2 hours and more per run already for $N > 20$.

illustrative example. Lastly, Fagerström and Oddshammar demonstrate that the Conditional Value-at-Risk (CVaR) optimization model tends to produce portfolios that are near-optimal under the MV framework [18], i.e., located very close to the efficient frontier.

Evolutionary Computation and Quality-Diversity Optimization. Evolutionary computation (EC) methods have a rich history of successful applications in PO [19], partly owing to their ability to handle non-convex search spaces that arise when real-world constraints (e.g., buy-in thresholds, turnover constraints, other regulatory and risk constraints) are imposed into the problem setting [20,21]. However, there is a significant lack of research regarding the applications of exploration algorithms, such as quality-diversity (QD) [22] and novelty search (NS) [23] in PO and quantitative finance generally, despite their significant potential and their successes in other domains [24,25]. The links between portfolio diversification and the divergent search paradigm, while arguably natural, remain understudied. Several somewhat related works nevertheless exist. Zhang *et al.* [26] address the specific problem of finding formulaic alpha factors that can predict and explain asset returns, making them suitable for use with multi-factor asset pricing models. To efficiently explore the space of formulaic alphas, with a focus on less frequently visited regions, they propose a search that combines QD and principal component analysis (PCA). This PCA-enhanced search is then used as an integral part of their `AutoAlpha` hierarchical evolutionary algorithm. Another approach is put forth by Yuksel [27], in which meta-learning QD optimization is employed to tackle the problem of large-scale sparse index tracking. It is concluded that the proposed method can be utilized in other scenarios where diversity among co-optimized solutions is needed, as well as in the presence of noisy reinforcement learning rewards.

1.3   Objectives and Contributions

In this paper, we tackle the problem of identifying a diverse set of near-optimal portfolios (i.e., portfolios with risk-return profiles located close to that of the reference optimal portfolio), as part of the broader PO problem. Specifically, we aim to answer the following research question (**Q**): *How to obtain a wide range of portfolios that are all near-optimal but mutually diverse in the portfolio weight space or the otherwise defined behavior space (BS)?* While previous research has found that, in some tasks, elite solutions are concentrated within a small part of the genotypic space ("the elite hypervolume") [28], our task runs in the opposite direction, as it involves finding genotypically different solutions (portfolios) that are all elite. We set out to test the following hypothesis (**H**): *The combination of convergent and divergent search provided through QD algorithms can be leveraged to obtain a set of diverse near-optimal portfolios.* The hypothesis stems from the observation that QD algorithms provide a natural choice for the underlying problem due to their ability to yield a range of diverse yet high-performing solutions. While some related approaches exist [26,27], to the best of our knowledge, this paper is the first to approach MV-based PO via QD optimization. To ensure the scalability to high-dimensional behavioral and/or asset spaces, which are ubiquitous in modern finance, the approach is powered by the `CVT-MAP-Elites` algorithm, as vanilla `MAP-Elites` faces the curse of dimensionality. We first use a toy example with only three assets to show that the approach is competitive against a similar approach based on the construction of convex hulls [11], and later extend our investigation to a higher-dimensional setting. As will be shown, the experimental results collectively clearly point to the promising capabilities of QD as a novel tool in the arsenal of modern PO practitioners.

2 Methodology

2.1 Problem Formulation

Let $\mathcal{W}$ be the set of all admissible portfolio weights:

$$\mathcal{W} = \left\{ (w_1, \ldots, w_i, \ldots, w_N) \mid w_i \geq 0 \quad \forall i, \quad \sum_{i=1}^{N} w_i = 1 \right\}, \tag{3}$$

where $N \geq 2$ is the total number of assets and w_i denotes the portfolio weight in the i-th asset. For simplicity, short selling is not permitted, but it can be easily integrated into the framework if needed, by relaxing the $w_i \geq 0$ constraint. Also, let $b : \mathcal{W} \to \mathcal{B}$ be a behavior function, mapping $\mathcal{W}$ to a BS denoted by $\mathcal{B}$. Assume that $\mathcal{B}$ is split into M niches (regions), i.e., $\mathcal{B} = \mathcal{N}_1 \cup \mathcal{N}_2 \cup \ldots \cup \mathcal{N}_M$. Finally, let $f : \mathcal{W} \to \mathbb{R}$ be a fitness function. Each candidate portfolio w is then associated with its behavioral descriptor (BD) $b_w \in \mathcal{B}$ and fitness value $f(w) \in \mathbb{R}$. The goal is to find:

$$\forall \mathcal{N}_i \quad w^* = \underset{w, \, b_w \in \mathcal{N}_i}{\arg \max} \; f(w). \tag{4}$$

2.2 Behavior Function and Space

We explore two different behavior functions and BS designs. In both cases, centroidal Voronoi tessellation (CVT) is used to partition the BS into niches.

Portfolio Weights. In the simplest variant, the behavior function b_1 is set as an identity function, i.e., $\forall w$, $b_1(w) = w$, with $\mathcal{B}_1 = \mathcal{W}$. Portfolio weight vectors w simultaneously serve as both genotypes and phenotypes. Similar can be seen in some of the approaches used to tackle the Rastrigin function benchmark [29] through QD algorithms [30, 31].

Asset's Fundamentals. In another variant, we separate the two spaces (genotypic and phenotypic) and set $b_2(w) = p_w$, where $p_w \in \mathcal{B}_2$ is a vector that describes the fundamental properties of the assets that dominate w weight-wise. More specifically:

$$\mathcal{B}_2 = \left\{ (s_1, \ldots, s_i, \ldots, s_L, c) \mid s_i \geq 0 \quad \forall i, \quad c > 0, \quad \sum_{i=1}^{L} s_i = 1 \right\}, \tag{5}$$

where s_i denotes the sector exposure of a portfolio to sector i, and L is the number of sectors. Sector exposure is defined as the sum of portfolio weights assigned to assets belonging to the respective sector, i.e., $s_i = \sum_{j \in S_i} w_j$, with S_i denoting the set of indices of assets belonging to sector i. The variable c denotes normalized market capitalization. Note that, unlike in the previous case, it is not possible to uniformly sample from the BS directly. Also, any other asset characteristics or factors of importance to the investor (e.g., ESG[4] factors, geographical diversification, etc.) might be used instead.

2.3 Fitness Functions

Fitness1 - The fitness (quality) function can be given by the negative distance between the risk profile of the candidate portfolio w and that of the reference optimal portfolio w_0 (obtained via MV optimization):

$$f_1(w) = -||\mathcal{F}(w_0) - \mathcal{F}(w)|| = -||(\mu_0, \sigma_0) - (\mu, \sigma)||, \qquad (6)$$

where $(\mu, \sigma) \in \mathcal{Z}$ is a vector consisting of the expected return and the volatility of the portfolio w (and equivalently for w_0). This fitness function will be used to take the convex hull approximation method to the near-optimal region.
Fitness2 - Alternatively, a modification of **Fitness1** is proposed:

$$f_2(w) = \begin{cases} -||(\mu_0, \sigma_0) - (\mu, \sigma)||, & \text{if } (\mu, \sigma) \text{ not in } \mathcal{R} \\ ||w - w_0||, & \text{otherwise} \end{cases} \qquad (7)$$

where $\mathcal{R}$ is the region of near-optimality around (μ_0, σ_0), $\mathcal{R} \subset \mathcal{Z}$, defined as:

$$\mathcal{R} = \{(\mu, \sigma) \mid \mu \geq (1 - c)\mu_0, \ \sigma \leq (1 + c)\sigma_0\} \qquad (8)$$

for some constant $c \in (0, 1)$. **Fitness2** promotes solutions in each niche whose risk profiles are as close as possible to the risk profile of w_0 until close proximity is reached. Once inside the near-optimality region $\mathcal{R}$, solutions that are furthest weight-wise from w_0 are preferred to ensure more genotypic diversity through an additional mechanism. Ideally, the distance from the convex hull of the entire archive of near-optimal solutions (instead of only from w_0) should be used as the measure. However, for computational efficiency, we employ this simple heuristic which will be shown to demonstrate strong performance in lower-dimensional settings.

2.4 Recombination Operator

Given the conditions in Eq. 3, a suitable constraint-preserving reproduction operator is required. To this end, a recombination operator (with mutation) that also includes clipping and normalization is used. It is described in great detail in Algorithm 1. Additional constraints, such as cardinality restrictions or buy-in thresholds, can be incorporated as needed.

---

[4] Environment, social and governance.

Algorithm 1: Recombination operator

Inputs: Parent portfolios w_1, w_2, mutation rate m
Result: Child portfolio w_3

1 $\lambda \leftarrow$ UniformRandom$[0, 1]$;   // Randomly generated weight parameter
2 $\delta \leftarrow$ UniformRandom$[-m, m, \text{len}(w_1)]$;   // Mutation vector
3 $w_3 \leftarrow \lambda w_1 + (1 - \lambda)w_2 + \delta$;   // Child portfolio
4 $w_3 \leftarrow$ Clip$(w_3, 0, 1)$;   // Clipping to ensure non-negative weights
5 $w_3 \leftarrow \dfrac{w_3}{\sum_{i=1}^{\text{len}(w_3)} w_{3,i}}$;   // Normalization of the child portfolio
6 **return** w_3

2.5 Algorithm

Due to the continual increase in the number of investable securities in financial markets, modern PO typically operates in high-dimensional settings, potentially encompassing hundreds or even thousands of financial assets. The ensuing high-dimensionality can be tackled with the CVT-MAP-Elites algorithm, where the number of niches is constant and independent of the dimensionality of the BS [32]. It is a variant of the vanilla MAP-Elites algorithm, which is itself based on maintaining an archive of elite solutions, one for each niche in the BS. The underlying idea is to find a plethora of behaviorally diverse yet high-performing solutions. Unless noted otherwise, unstructured archive $\mathcal{A}$ is assumed. For smaller BS dimensionalities (typically 2D to 6D [32]), the basic MAP-Elites algorithm [33] can be used instead. We use CVT-MAP-Elites in all of the performed experiments. The used hyperparameter values are provided separately for each of the experiments.

3 Experimental Results

3.1 Toy Example

We begin by considering the toy example first introduced in [16] and later revisited in [10], in which only three ($N = 3$) asset classes - stocks, bonds, and treasury bills are considered[5]. The mean return, standard deviation, and correlation estimates for the three asset classes (given in Table 1) are used to construct the expected return vector $\hat{\mu}$ and the return covariance matrix $\hat{S}$ estimates. The resulting MV efficient frontier is shown in Fig. 1, accompanied by additional elements, including the near-optimality region $\mathcal{R}$. To ensure consistency with [10], the same reference portfolio w_0 is adopted as well as the identical value of $c = 0.1$. After running the QD algorithm $R = 100$ times[6] for each

---

[5] More precisely, the assets include the S&P 500 market index, Lehman Brothers Long Term Government Bond Index, and one-month Treasury bills. The original data is presented monthly and spans the period from 1980 to 1990, but the estimates are transformed into annual values in our work.

[6] At the start of each QD run, niches are recalculated, discarding the old CVT results.

fitness function, while using $\mathcal{B}_1$ and with $M = 200$, the convex hulls depicting regions of near-optimality in the portfolio weights space are obtained. Each run takes less than 100 s even without any parallelization. The maximum number of evaluations is set to $N_{max} = 250\ 000$ and the number of CVT samples equals $N_{CVT} = 10\ 000$. Random initialization is done until $P_{init} = 10\%$ of niches are filled with solutions. Figure 2 presents the results for four different methods (**M1** - Chopra [16], **M2** - van der Schans and de Graaf [10,11], **M3** - QD with `Fitness1`, and **M4** - QD with `Fitness2`), with more details given in Table 2. Larger hull surfaces (volumes or hypervolumes in higher-dimensional spaces) are desirable, as they indicate greater compositional diversity within the set of found near-optimal portfolios, allowing investors more freedom to accommodate their specific preferences. The best results overall are obtained by **M4**, the modified version of **M3**, as it benefits from the additional diversification mechanism. The dominance of **M2** over **M3** with respect to the convex hull surface is not a matter of concern but rather an anticipated outcome. Namely, **M2** selects portfolios (exclusively from $\mathcal{R}$) by directly maximizing distances from the convex hull of previously found solutions, whereas **M3** seeks solutions within each niche whose risk profiles are strictly closest to $\mathcal{F}(\boldsymbol{w_0})$. As a result, **M3** pushes strongly towards $\mathcal{F}(\boldsymbol{w_0})$ (lying on the efficient frontier) regardless of whether exploring inside or outside of $\mathcal{R}$ and hence shrinks $\mathcal{S}$, the corresponding region in the risk-return space. We therefore suspect **M3** to yield portfolios with better risk-reward profiles (measured by the Sharpe ratio) than **M2** and also a lower $\mathcal{S}$ surface, all of which is indeed confirmed by the results laid out in Table 1. It is also noteworthy that **M4** achieves an even higher value of the Sharpe ratio than **M3**, despite its emphasis on genotypic diversity. The likely reason is that, given that $\boldsymbol{w_0}$ is not the maximum Sharpe portfolio, it is possible for the neighboring risk profiles in $\mathcal{R}$ to dominate over it.

Table 1. The estimates from historical data (annual)

	Mean (%)	Std (%)	Corr. stocks	Corr. bonds	Corr. T-Bills	Optimal weights (%) (moderate risk aversion)
Stocks	15.876	16.603	1.000	–	–	58.1
Bonds	12.324	13.801	0.341	1.000	–	22.8
T-Bills	8.748	0.759	−0.081	0.050	1.000	19.1

3.2   Higher-Dimensional Setting

In this section, encouraged by the positive results from the toy example, we apply the proposed method to a higher-dimensional setting comprised of a large number of assets. Under such conditions, QD is expected to offer heuristic solutions in a reasonable time. More specifically, we consider the universe of $N = 105$ different equity assets covering $L = 11$ different sectors. The Sharpe optimal portfolio is used as the reference portfolio. Moreover, the Ledoit-Wolf shrinkage

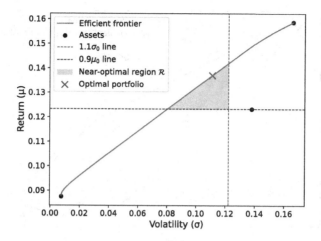

Fig. 1. The MV efficient frontier with the chosen optimal portfolio and its region of near-optimality.

Table 2. Performance evaluation of the four tested methods

Method	Convex hull $\mathcal{H}$ surface	Risk-return subspace $\mathcal{S}$ surface ($\times 10^3$)	Sharpe ratio of portfolios	% of niches with opt. port
M1[a]	0.0948	0.1973	1.1368 ± 0.0959	–
M2[b]	0.1724	0.3712	1.1762 ± 0.1299	–
M3	0.1529 ± 0.0015	0.3652 ± 0.0076	1.2696 ± 0.0009	48.980 ± 0.003
M4	$\mathbf{0.1746 \pm 0.0002}$	$\mathbf{0.3882 \pm 0.0008}$	$\mathbf{1.2824 \pm 0.0009}$	$\mathbf{49.070 \pm 0.003}$

[a]Results from [16] (deterministic).
[b]Results from [10] (deterministic).

[5] is used[7] for estimating the covariance matrix $\hat{S}$, while CAPM returns [34] are utilized to derive the mean return estimate $\hat{\mu}$. The look-back window used for parameter estimation encompasses the period from January 2nd, 2020, to September 1st, 2023, corresponding to exactly $T = 924$ trading days. We again set $c = 0.01$, leading to a very strict criterion of near-optimality (i.e., a small surface of $\mathcal{R}$), and also select $M = 5000$. Due to a plethora of assets, investors might prefer to first perform asset allocation at the sector/industry or some other macro level[8] before selecting individual constituent securities. Consequently, we use the BS design $\mathcal{B}_2$. As the fitness function, f_1 is selected. The algorithm ultimately returns an archive of solutions $\mathcal{A}$ comprised of both near-optimal portfolios and non-near-optimal portfolios (for niches in which no near-optimal portfolios have been found).

---

[7] With constant variance set as the shrinkage target.
[8] Similar approaches are employed in top-down investment strategies such as Tactical Asset Allocation (TAA) [35].

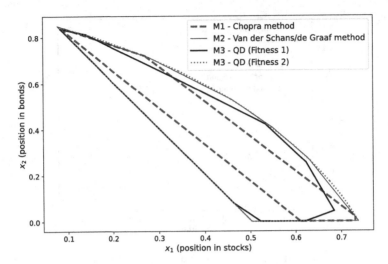

Fig. 2. Convex hulls in the genotypic/phenotypic space for different methods for a single QD run. The constraint $\sum x_i = 1$ introduces coplanarity in the otherwise three-dimensional space.

Considering both the absence of suitable benchmarks and the computational challenges in calculating convex hull volumes in spaces with higher dimensionalities[9], we draw inspiration from previous research on benchmarking QD algorithms [37] and use it as a starting point in creating appropriate performance metrics. All of the used metrics are described in Table 3. Remark that metrics $\mathcal{C}'$, QDScore2 and AP2 only consider niches in which near-optimal portfolios have been found, unlike QDScore1 and AP1, which (also) require information on other niches. Following the notation from Table 3, the number of niches with and without near-optimal portfolios is $\mathcal{C}'M$ and $|\mathcal{A}| - \mathcal{C}'M$, respectively.

Experimental Results. Figure 3 presents the main experimental results. The upper subplots show the first three metrics ($\mathcal{C}'$, QDScore1, QDScoreMOD) plotted against the number of QD evaluations, as well as the archive profiles (AP1 and AP2). The mean values as averaged over $R = 20$ runs are shown, together with empirical percentiles (the 5-th and the 95-th p^{th}). As before, niches are recalculated from scratch at the beginning of each QD run. To ensure the high quality of tessellation, the maximum number of QD evaluations is set to $N_{max} = 2.2 \times 10^6$, $P_{init} = 10\%$, and the number of CVT samples equals $N_{CVT} = 50\,000$. In terms of the modified coverage, it is evident that the percentage of niches with near-optimal solutions sharply rises, reaching 60% already at a bit over 500 000 evaluations, and finally getting to approximately 75%. Such high percentages clearly indicate that, despite the use of a stringent criterion for near-optimality, the method succeeds in finding a wide range of near-optimal

---

[9] With the QuickHull algorithm [36], the execution time grows by $n^{\lfloor \frac{d}{2} \rfloor}$, where n is the input size and d the dimensionality.

Table 3. List of metrics

Metric	Description	Expression
Modified coverage ($\mathcal{C}'$)	The proportion of niches with near-optimal portfolios	$\frac{\text{No. of niches with n.o. portfolios}}{\text{No. of niches}}$
QD-score (QDScore1)	The cumulative normalized fitness of all portfolios in $\mathcal{A}$	$\sum_{i=1}^{\lvert\mathcal{A}\rvert} \frac{f_i - \min_j\left(f_j\right)}{\max_j\left(f_j\right) - \min_j\left(f_j\right)}$
Modified QD-score (QDScoreMOD)	The cumulative normalized fitness of all found near-optimal portfolios	$\sum_{i=1}^{\mathcal{C}'M} \frac{f_i - \min_j\left(f_j\right)}{\max_j\left(f_j\right) - \min_j\left(f_j\right)}$
Modified archive profile 1 (AP1)	The proportion of found non-near-optimal portfolios exceeding some threshold value	$\sum_{i=1}^{\lvert\mathcal{A}\rvert - \mathcal{C}'M} \mathbb{1}\left(f_i \geq f_{\text{threshold}}\right)$
Modified archive profile 2 (AP2)	The proportion of found near-optimal portfolios exceeding some threshold value	$\sum_{i=1}^{\mathcal{C}'M} \mathbb{1}\left(f_i \geq f_{\text{threshold}}\right)$

portfolios encompassing various industry compositions and market capitalization values. This is a fortunate conclusion, especially for investors who prefer human-in-the-loop approaches, enabling them to incorporate their own preferences that might be hard to formalize, into the decision-making process. As for the QD-score metrics, we observe that QDScore1 steadily increases over time, indicating an improvement in diversity/performance among the found near-optimal solutions (portfolios). Likewise, the QDScore1 curve shows that the majority of niches are populated relatively quickly (in under 500 000 evaluations). Also note that QDScore1 expectedly converges faster than QDScoreMOD, as filling niches with near-optimal solutions takes more time compared to arbitrary solutions. Modified archive profiles, calculated after performing the maximum number of QD evaluations, are provided at the bottom of Fig. 3. While AP1 and AP2 have similar shapes (under different threshold scales), a small right tail can be seen with AP2, showing a number of "super" near-optimal portfolios with risk profiles extremely similar to that of the reference portfolio. The significant compositional diversity of the obtained portfolios is depicted in Fig. 5, which displays the Sharpe optimal portfolio alongside two mutually diverse near-optimal portfolios generated by the method in the feature (behavior) space. Despite the multitude of potential solutions, in order to finalize the investment decision-making process, it is necessary to select a single portfolio from the set of obtained near-optimal portfolios. To this end, Algorithm 2 delineates the entire end-to-end investment decision-making process, incorporating the proposed QD method.

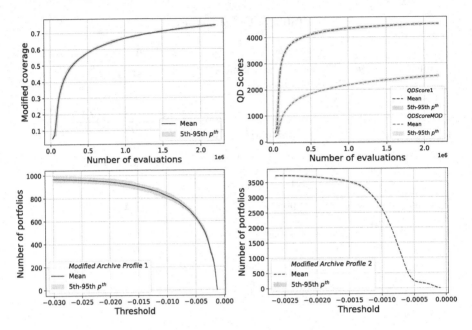

Fig. 3. The mean modified coverage $\mathcal{C}'$ (the upper left subplot), QDScore1, and QDScore2 (the upper right subplot) plotted against the number of evaluations, alongside empirical percentiles. The modified archive profiles are displayed at the bottom, with the set including non-near-optimal portfolios on the left and the set including near-optimal portfolios on the right.

Robustness. Generally, the evaluation of QD solutions is exacerbated by the presence of stochasticity in the underlying environment [29]. In our case, the evaluation is fully deterministic once the estimates $(\hat{\boldsymbol{\mu}}, \hat{\boldsymbol{S}})$ are fixed. However, there is stochasticity involved due to the very fact that "true" parameter values $(\boldsymbol{\mu}, \boldsymbol{S})$ are hidden, whereas the estimates (which can be derived in multiple ways) represent random variables. With this in mind, we investigate the robustness of the generated portfolios to a certain type of change in the estimates $(\hat{\boldsymbol{\mu}}, \hat{\boldsymbol{S}})$. More specifically, we study whether previously found near-optimal portfolios generally remain near-optimal when re-estimating under different estimation window[10] sizes T. The results are shown in Fig. 4. As expected, the mean modified coverage $\mathcal{C}'$ remains robust to changes in the estimation window size T when larger threshold constants c are employed. On the other hand, the sensitivity of $\mathcal{C}'$ to changes in T, in particular to its reduction, is much more emphasized for c values in the range $[0.5\%, 2.5\%]$, i.e., for stricter near-optimality criteria. For example, with $c = 0.5\%$ and $T = 824$, on average only 136.15 niches

---

[10] The choice of the estimation window size is a non-trivial issue that has been studied before [38–40], with larger sizes leading to reduced estimation errors at the price of assuming unrealistically long stationarity periods.

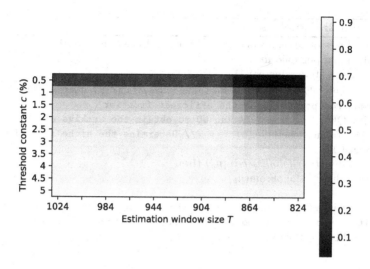

Fig. 4. The mean modified coverage $\mathcal{C}'$ for different threshold constants c and estimation window sizes T. Observe the relatively high sensitivity of $\mathcal{C}'$ to the shortening of T, especially for stricter near-optimality criteria (i.e., for smaller c values).

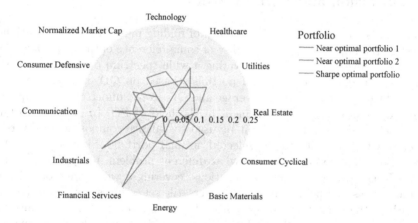

Fig. 5. Two of the obtained near-optimal portfolios juxtaposed against the Sharpe optimal portfolio in the BS. The first (second) near-optimal portfolio is highly concentrated in industrials (financial services) stocks and low (high) market cap stocks, while the Sharpe optimal portfolio remains more balanced.

(or 2.7%) contain near-optimal solutions. We leave for further research the study of whether solutions that remain near-optimal under a wider range of estimates present a superior investment choice.

Algorithm 2: Portfolio selection process

Inputs: Investor's risk aversion γ and preferred BD b, historical data $\mathcal{D}$
Result: Final portfolio w

1 $\hat{S}, \hat{\mu} \leftarrow$ EstimateParameters($\mathcal{D}$)
2 $w_0 \leftarrow$ CalculateEfficientFrontierPortfolio(γ, $\hat{S}$, $\hat{\mu}$) ; // MV step - calculating the required portfolio on the efficient frontier
3 $\mathcal{A} \leftarrow$ RunQD ;   // Run QD to obtain the archive of portfolios
4 $n_b \leftarrow$ DetermineNicheIndex(b) ;   // Determine the niche index for the preferred BD
5 **if** *NearOptimalPortfolioExistsIn(n_b)* **then**
6 | $w \leftarrow$ ElitePortfolioFrom(n_b)
7 **else**
8 | $n_b' \leftarrow$ ClosestNicheWithNearOptimalPortfolio(n_b) ;   // Among all niches with a near-optimal portfolio, select the one closest to n_b
9 | $w \leftarrow$ ElitePortfolioFrom(n_b')
10 **end**
11 **return** w

---

4 Conclusion and Further Work

This paper is concerned with the problem of finding mutually diverse portfolios located in the region of near-optimality of some reference optimal portfolio. We introduce a novel method for discovering a wide spectrum of mutually diverse (in a predefined BS) near-optimal portfolios based on QD optimization. The main findings, pointing to high coverage and QD-scores, underscore the capacity of QD to serve as a novel instrument in the field of PO. In addition to QD, the knobelty algorithm [41] might be used to balance compositional diversity (*novelty*) with proximity to the selected optimal portfolio. Similar approaches might also be employed for a somewhat different problem; to approximate the entire efficient frontier of optimal solutions, covering a wide range of different risk preferences and hence catering to a versatile set of investors. These include QD with a fitness function that penalizes distances from the efficient frontier in each niche, as well as vanilla NS formulations in which the risk-reward space is used as the phenotypic space. Other objectives besides MV may be explored in the future as well, together with different BS designs (e.g. factor-based, distance from the equally weighted or currently selected portfolio) and definitions of near-optimality (e.g. those employing soft constraints). Links with sparse PO should also be investigated. Lastly, we anticipate further work to leverage the power of QD in visualizing and illuminating the portfolio search space. More broadly, we hope to see future approaches harnessing the potential of open-endedness-based approaches, including QD and NS, in computational finance.

References

1. Sharpe, W.F.: The sharpe ratio. Streetwise-the Best J. Portfolio Manag. **3**, 169–185 (1998)
2. Babcock, B.A., Choi, E.K., Feinerman, E.: Risk and probability premiums for cara utility functions. J. Agricult. Resource Econ. **22**, 17–24 (1993)
3. Markowitz, H.M., Todd, G.P.: Mean-variance analysis in portfolio choice and capital markets, vol. 66. John Wiley & Sons (2000)
4. Best, M.J., Grauer, R.R.: On the sensitivity of mean-variance-efficient portfolios to changes in asset means: some analytical and computational results. Rev. Financial Stud. **4**(2), 315–342 (1991)
5. Ledoit, O., Wolf, M.: Honey, i shrunk the sample covariance matrix. UPF economics and business working paper, vol. (691) (2003)
6. Black, F., Litterman, R.: Asset allocation: combining investor views with market equilibrium. Goldman Sachs Fixed Income Res. **115**(1), 7–18 (1990)
7. DeMiguel, V., Garlappi, L., Nogales, F.J., Uppal, R.: A generalized approach to portfolio optimization: improving performance by constraining portfolio norms. Manag. Sci. **55**(5), 798–812 (2009)
8. Michaud, R.O., Michaud, R.O.: Efficient asset management: a practical guide to stock portfolio optimization and asset allocation. Oxford University Press (2008)
9. Yin, C., Perchet, R., Soupé, F.: A practical guide to robust portfolio optimization. Quantitative Finance **21**(6), 911–928 (2021)
10. de Graaf, T.: Robust Mean-Variance Optimization. PhD thesis, Master Thesis, Leiden University & Ortec Finance (2016)
11. van der Schans, M., de Graaf, T.: Robust optimization by constructing near-optimal portfolios. Available at SSRN 3057258 (2017)
12. Wang, L.: Support vector machines: theory and applications, vol. 177. Springer Science & Business Media (2005). https://doi.org/10.1007/b95439
13. Wales, D.J., Doye, J.P.K.: Global optimization by basin-hopping and the lowest energy structures of lennard-jones clusters containing up to 110 atoms. J. Phys. Chem. A **101**(28), 5111–5116 (1997)
14. van Eeghen, W.J.B., van Gaans, O.W., van der Schans, M.: Analysis of near-optimal portfolio regions and polytope theory (2018)
15. Cajas, D.: Robust portfolio selection with near optimal centering. Available at SSRN 3572435(2019)
16. Vijay Kumar Chopra: Improving optimization. J. Invest. **2**(3), 51–59 (1993)
17. Benita, G., Baudot-Trajtenberg, N., Friedman, A.: The challenges of managing large fx reserves: the case of israel. BIS Paper, (104m) (2019)
18. Fagerström, S., Oddshammar, G.: Portfolio optimization-the mean-variance and cvar approach (2010)
19. Brabazon, A., O'Neill, M., Dempsey, I.: An introduction to evolutionary computation in finance. IEEE Comput. Intell. Mag. **3**(4), 42–55 (2008)
20. Branke, J., Scheckenbach, B., Stein, M., Deb, K., Schmeck, H.: Portfolio optimization with an envelope-based multi-objective evolutionary algorithm. Eur. J. Oper. Res. **199**(3), 684–693 (2009)
21. Qi, R., Yen, G.G.: Hybrid bi-objective portfolio optimization with pre-selection strategy. Inform. Sci. **417**, 401–419 (2017)
22. Chatzilygeroudis, K., Cully, A., Vassiliades, V., Mouret, J.-B.: Quality-diversity optimization: a novel branch of stochastic optimization. In: Pardalos, P.M., Rasskazova, V., Vrahatis, M.N. (eds.) Black Box Optimization, Machine Learning, and

No-Free Lunch Theorems. SOIA, vol. 170, pp. 109–135. Springer, Cham (2021). https://doi.org/10.1007/978-3-030-66515-9_4

23. Lehman, J., Stanley, K.O.: Novelty search and the problem with objectives. Genetic programming theory and practice IX, pp. 37–56 (2011)

24. Gomes, J., Urbano, P., Christensen, A.L.: Evolution of swarm robotics systems with novelty search. Swarm Intell. **7**, 115–144 (2013)

25. Pugh, J.K., Soros, L.B., Stanley, K.O.: Quality diversity: a new frontier for evolutionary computation. Front. Robot. AI **3**, 40 (2016)

26. Zhang, T., Li, Y., Jin, Y., Li, J.: Autoalpha: an efficient hierarchical evolutionary algorithm for mining alpha factors in quantitative investment. arXiv preprint arXiv:2002.08245 (2020)

27. Yuksel, K.A.: Generative meta-learning robust quality-diversity portfolio. In: Proceedings of the Companion Conference on Genetic and Evolutionary Computation, pp. 787–790 (2023)

28. Vassiliades, V., Mouret, J.-P.: Discovering the elite hypervolume by leveraging interspecies correlation. In: Proceedings of the Genetic and Evolutionary Computation Conference, pp. 149–156 (2018)

29. Digalakis, J.G., Margaritis, K.G.: On benchmarking functions for genetic algorithms. Inter. J. Comput. Math. **77**(4), 481–506 (2001)

30. Bossens, D.M., Tarapore, D.: Quality-diversity meta-evolution: customising behaviour spaces to a meta-objective. arXiv preprint arXiv:2109.03918 (2021)

31. Sfikas, K., Liapis, A., Yannakakis, G.N.: Monte carlo elites: Quality-diversity selection as a multi-armed bandit problem. In: Proceedings of the Genetic and Evolutionary Computation Conference, pp. 180–188 (2021)

32. Vassiliades, V., Chatzilygeroudis, K., Mouret, J.-B.: Using centroidal voronoi tessellations to scale up the multidimensional archive of phenotypic elites algorithm. IEEE Trans. Evol. Comput. **22**(4), 623–630 (2017)

33. Mouret, J.-B., Clune, J.: Illuminating search spaces by mapping elites. arXiv preprint arXiv:1504.04909 (2015)

34. Fama, E.F., French, K.R.: The capital asset pricing model: theory and evidence. J. Econ. Perspect. **18**(3), 25–46 (2004)

35. Faber, M.: A quantitative approach to tactical asset allocation. J. Wealth Manag. Spring (2007)

36. Barber, C.B., Dobkin, D.P., Huhdanpaa, H.: Qhull: Quickhull algorithm for computing the convex hull. Astrophysics Source Code Library, pp. ascl-1304 (2013)

37. Flageat, M., Lim, B., Grillotti, L., Allard, M., Smith, S.C., Cully, A.: Benchmarking quality-diversity algorithms on neuroevolution for reinforcement learning. arXiv preprint arXiv:2211.02193 (2022)

38. Gašperov, B., Šarić, F., Begušić, S., Kostanjčar, Z.: Adaptive rolling window selection for minimum variance portfolio estimation based on reinforcement learning. In: 2020 43rd International Convention on Information, Communication and Electronic Technology (MIPRO), pp. 1098–1102. IEEE (2020)

39. Wang, P.-T., Hsieh, C.-H.: On data-driven log-optimal portfolio: a sliding window approach. IFAC-PapersOnLine **55**(30), 474–479 (2022)

40. Chuanzhen, W.: Window effect with markov-switching garch model in cryptocurrency market. Chaos, Solitons Fractals **146**, 110902 (2021)

41. Kelly, J., Hemberg, E., O'Reilly, U.-M.: Improving genetic programming with novel exploration - exploitation control. In: Sekanina, L., Hu, T., Lourenço, N., Richter, H., García-Sánchez, P. (eds.) EuroGP 2019. LNCS, vol. 11451, pp. 64–80. Springer, Cham (2019). https://doi.org/10.1007/978-3-030-16670-0_5

Improving Image Filter Efficiency: A Multi-objective Genetic Algorithm Approach to Optimize Computing Efficiency

Julien Biau[1] , Sylvain Cussat-Blanc[2](✉) , and Hervé Luga[2]

[1] INIA SAS, Toulouse, France
julien.biau@gmail.com
[2] University of Toulouse, Toulouse, France
sylvain.cussat-blanc@ut-capitole.fr, herve.luga@irit.fr

Abstract. For real-time applications in embedded systems, an efficient image filter is not defined solely by its accuracy but by the delicate balance it strikes between precision and computational cost. While one approach to manage an algorithm's computing demands involves evaluating its complexity, an alternative strategy employs a multi-objective algorithm to optimize both precision and computational cost.

In this paper, we introduce a multi-objective adaptation of Cartesian Genetic Programming aimed at enhancing image filter performance. We refine the existing Cartesian Genetic Programming framework for image processing by integrating the elite Non-dominated Sorting Genetic Algorithm into the evolutionary process, thus enabling the generation of a set of Pareto front solutions that cater to multiple objectives.

To assess the effectiveness of our framework, we conduct a study using a Urban Traffic dataset and compare our results with those obtained using the standard framework employing a mono-objective evolutionary strategy. Our findings reveal two key advantages of this adaptation. Firstly, it generates individuals with nearly identical precision in one objective while achieving a substantial enhancement in the other objective. Secondly, the use of the Pareto front during the evolution process expands the research space, yielding individuals with improved fitness.

Keywords: Genetic Programming · Cartesian Genetic Programming · Multi-Objective · Genetic Improvement · Image processing · Real Time Applications · Embedded Systems

1 Introduction

When employing image filters on embedded systems with limited computing power, the challenge extends beyond precision; one must also consider the constraints of computational capacity. In the realm of real-time applications on embedded systems, an efficient image filter is one that strikes an optimal balance between fitness and computational time.

© The Author(s), under exclusive license to Springer Nature Switzerland AG 2024
S. Smith et al. (Eds.): EvoApplications 2024, LNCS 14634, pp. 19–34, 2024.
https://doi.org/10.1007/978-3-031-56852-7_2

While assessing an algorithm's complexity is a common practice, involving measurements of time and estimations of worst-case scenarios, controlling the equilibrium between precision and computation time remains a nuanced challenge. An alternative approach involves the use of multi-objective algorithms to concurrently optimize precision and computation.

In this work, we introduce a multi-objective evolutionary strategy within the framework of Cartesian Genetic Programming for Image Processing for Genetic Improvement (CGP-IP-GI). Our approach accounts for both filter precision and execution time, yielding a range of high-performing solutions. The choice of solution depends on the desired fitness level or the available computing power. Additionally, our multi-objective adaptation results in solutions with improved fitness when compared to previous mono-objective studies.

This paper is organized as follows: Sect. 2 presents the state-of-the-art in multi-objective genetic algorithms, while Sect. 3 outlines the adaptation of CGP-IP-GI, incorporating the elite non-dominated sorting genetic algorithm to achieve multi-objective optimization. Section 4 details the experiments underpinning this research, and Sect. 5 offers a comprehensive presentation of the results. Finally, in Sect. 6, we present our preliminary conclusions.

2 Related Works

This section discusses prior research in the realm of multiple objective problems, with a specific focus on the field of genetic algorithms. It also introduces Cartesian Genetic Programming for Image Processing (CGP-IP) and its recent advancement for genetic enhancement (CGP-IP-GI).

2.1 Multi-objective Evolution Algorithms

In contrast to a single-objective problem, which seeks to find an optimal solution, a multi-objective problem entails the simultaneous optimization of multiple objective functions. Enhancing one aspect in a multi-objective problem may have detrimental effects on the outcomes of other objectives. As a result, multi-objective genetic algorithms aim to generate a collection of efficient solutions that belong to the Pareto-optimal front.

Mathematically, the concept of Pareto optimality can be formally defined as follows (Eq. 1). Assuming, without loss of generality, a maximization problem, and given two decision vectors a and b belonging to the decision space X, vector a is said to *dominate* vector b if and only if:

$$\begin{cases} \forall i \in [1, 2, ..., n] : f_i(a) >= f_i(b) \\ \exists j \in [1, 2, ..., n] : f_j(a) > f_j(b) \end{cases} \tag{1}$$

All decision vectors which are not dominated by any other decision vector of a given set are called *nondominated* regarding this set. If it is clear from the context which set is meant, we simply leave it out. The decision vectors that are nondominated within the entire search space are denoted as Pareto optimal and constitute the *Pareto-optimal set* or *Pareto-optimal front*.

Non-dominated Sorting Genetic Algorithm

The Non-dominated Sorting Genetic Algorithm (NSGA), originally developed by Srivas and Deb [28], follows a methodology that starts by randomly sorting the obtained solutions based on their dominance. Within each subpopulation, the algorithm computes the proximity between solutions, and selection is carried out using the roulette method. This method gives a higher probability to solutions from the first non-dominated subpopulation. New solutions are generated through a combination of crossover and mutation applied to the selected solutions, and the algorithm continues its search for solutions until a predefined stop criterion is met. NSGA ranks solutions based on their dominance and assigns precision values.

NSGA has also been employed in a study focused on optimizing Stirling thermal engines, with the goal of achieving maximum power, thermal efficiency, and minimum pressure drop [1].

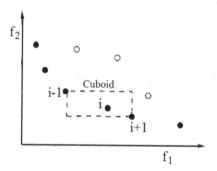

Fig. 1. The crowding distance is computed, with points identified by filled circles representing solutions belonging to the same nondominated front. (Source [6])

Elitist Non-dominated Sorting Genetic Algorithm

The elite non-dominated sorting genetic algorithm (NSGA-II), introduced by Deb and Goel [7,8], bears resemblance to NSGA. However, NSGA-II employs the crowding distance to select the most isolated results (as depicted in Figs. 1 and 2), eliminating the need to calculate the precision parameter (σ_{share}). Consequently, the algorithm ensures that the optimal Pareto solution discovered up to the current step is retained. The solution selection mechanism is employed to control population size, although this approach may diminish proximity to the optimal solution. As long as the number of solutions with the first non-dominated solutions does not exceed the number of primary populations, all solutions in this set are selected.

2.2   Cartesian Genetic Programming

Cartesian Genetic Programming (CGP) is a form of Genetic Programming (GP) in which programs are represented as directed, often acyclic graphs indexed by

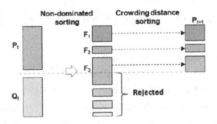

Fig. 2. Basics of NSGA-II procedure. (Source [14])

Cartesian coordinates (Fig. 3). CGP was invented by Miller and Thomson [20], [22] for use in evolving digital circuits, although it has been applied in a number of domains [23]. CGP is used in [16] to evolve neural networks, in [5, 10] for object detection in image processing, and in [15] for image noise reduction. Its benefits include node neutrality, being the encoded parts of the genome that don't contribute to the interpreted program, node reuse, and a fixed representation that reduces program bloat [21]. A recent review of CGP is given in [24].

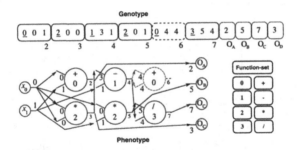

Fig. 3. A genotype in the form of a CGP (Cartesian Genetic Programming) and its associated schematic phenotype are presented for a group of four mathematical equations. The genes highlighted in the genotype dictate the function of individual nodes. The addresses for each program input and node in both the genotype and phenotype are displayed below. Inactive regions of both the genotype and phenotype are depicted as grey dashes, specifically in the case of node 6. (Source [25]). (Color figure online)

In CGP, functional nodes, defined as a set of evolved genes, connect to program inputs and to other functional nodes via their Cartesian coordinates. The outputs of the program are taken from any internal node or program input based on evolved output coordinates. The CGP nodes are organized in a rectangular grid with R rows and C columns. Nodes have the flexibility to establish connections with any node from preceding columns, and this connectivity is governed by a parameter L, representing the number of columns to which a node can connect. In this investigation, consistent with previous studies such as [24], the value of R is set to 1, indicating that all nodes reside in a single row.

The CGP genotype consists of a list of node genes; each node in the genome encodes the node function, the coordinates of the function inputs (here referred to as Connection 0 and Connection 1), and optionally, the parameters for the node function. Finally, the end of the genome encodes the nodes that give the final program output. By tracing back from these output nodes, a single function can be derived for each program output by offering a concise and legible program representation.

The genes in CGP are optimized through using the 1+λ algorithm. A population of λ individuals are randomly generated and evaluated on a test problem. The evaluation is performed by decoding the program graph from the individual genotype and applying the program to a specific problem such as image masking, as in this study. The best individual based on this evaluation is retained for the next generation. A mutation operator is applied to this individual to create λ new individuals; in CGP, the mutation operator randomly samples a subset of new genes from a uniform distribution. This new population is evaluated, and the best individual is retained for the next generation; this iterative process continues until a configured stopping criterion.

2.3   Cartesian Genetic Programming for Image Processing

An important choice to make when using CGP is the set of possible node functions. In the original circuit design, the node functions are logic gates such as AND and NOR. Applications of CGP in game playing and data analysis use a standard set of mathematical functions such as $x + y$, $x * y$, and $cos(x)$ for a node with inputs x and y. Function sets must be defined such that outputs of any node will be valid for another node; in mathematical functions, this is often guaranteed by limiting the domain and range of the functions between -1 and 1.

CGP-IP is an adaption of CGP that uses image processing functions and applies programs directly to images [10]. The inputs and outputs of the evolved functions are images that allows for consistency between node functions; each node function is defined to input an image of a fixed size and output an image of the same size. CGP-IP has previously used a set of 60 functions [10] from OpenCV, a standard open-source image processing library.

In a previous work [10], CGP-IP has used an island population distribution algorithm. In this method, multiple populations compete inside "islands" that are independent $1 + \lambda$ evolutionary algorithms. A migration interval parameter dictates the frequency of expert sharing between islands, allowing for the synchronization of the best individual across islands. Island models have been shown to be good alternatives to the genetic algorithm, as they help preserve genetic diversity [30]. Their use in CGP-IP has demonstrated an improvement compared to the $1 + \lambda$ algorithm.

CGP-IP individuals are evaluated by applying the evolved filter to a set of images, comparing them to target images, and computing a difference metric between the output image from the evolved filter and the target such as the mean error or Matthews Correlation Coefficient (MCC) [19]. In this paper, we use MCC, which measures the quality of binary classification and has been showed particularly adapted to classification tasks using CGP [9]. Calculations are based on the confusion matrix, which is the count of the true positives (TP), false positives (FP), true negatives (TN) and false negatives (FN).

$$mcc = \frac{TPxTN - FP*FN}{\sqrt{(TP+FP)(TP+FN)(TN+FP)(TN+FN)}} \qquad (2)$$

A MCC with a score of 1 corresponds to a perfect classification, 0 to a random classifier, and −1 to a fully inverted classification. Therefore, our fitness function for evolution is defined as follows:

$$fitness = 1 - mcc \qquad (3)$$

The closer our fitness value is to 0, the more accurate our phenotype is.

2.4 Genetic Improvement

Genetic Improvement (GI) is a relatively recent field of software engineering research that uses search to improve existing software [26]. Using handwritten code as a starting point, GI searches the space of program variants created by applying mutation operators. The richness of this space depends on the power and expressivity of the mutation operators, which can modify existing code by changing functions or parameters, add new code, and, in some cases, remove parts of a program. Over the past decade, the GI field has greatly expanded, and current research on GI has demonstrated its many potential applications. GI has been used to fix software bugs [2, 18], drastically speed up software systems [18, 29], port a software system between different platforms [17], transplant code features between multiple versions of a system [27], grow new functionalities [12], and more recently to improve memory [31] and energy usage [4].

The majority of GI work uses GP to improve the programs under optimization [2, 17, 18, 27, 29]. In most methods, applying GI on a existing program is achieved by encoding the program within a GP tree and then computing the corresponding genome. GP mutation operators are applied to the encoded program to generate adjacent programs. To this end, both the program encoding and the operators must be defined to suit the initial program that is to be enhanced with additional functions to improve the functional graph. The fitness used during the evolutionary optimization of the program can be based on various metrics such as program length, efficiency, or relevance to given test cases [2, 17, 29].

2.5 Genetic Improvement in Cartesian Genetic Programming for Image Processing

In this section, we describe the node insertion and deletion operators designed for GI with CGP-IP [3]. In general, CGP can create a graph through the random mutation of node gene parameters such as connections or functions. However, this can be destructive when improving a given genome, because the modifications to active nodes can remove parts of the function graph. Previous research has proposed self-modifying genomes [11] which use functions which can add or remove nodes upon function execution. CGP-IP-GI propose node insertion and deletion operators during the mutation process instead of during the execution of the program. These operators are designed to maintain the active subgraph of a program, that is, they are not destructive.

The mutation operator comprises three possible operators: node insertion, node deletion, and standard parameter modification using a uniform distribution. The node operators have configurable mutation rates: r_{ins} and r_{del}. If one of these structural operators is chosen, it will be the only mutation performed; otherwise, standard mutation occurs. In this study, we use $r_{ins} = 0.1$ and $r_{del} = 0.1$ for all experiments.

3   Multi-objective Implementation in Cartesian Genetic Programming for Image Processing

To accomplish multi-objective optimization, we will employ the NSGA-II algorithm. In order to do so, it is imperative to adapt CGP-IP-GI by making adjustments to its evolutionary algorithm and integrating a Pareto front as the evaluation function.

3.1   Evolutionary Algorithm

In the typical scenario, CGP-IP-GI employs an evolutionary algorithm with a $\mu + \lambda$ strategy, where $\mu = 1$ parent generates λ children. However, to ensure that the Pareto front is adequately representative, it becomes essential to consider $\mu > 1$. Accordingly, the evolution phase is adjusted to enable the generation of λ children from μ parents.

3.2   Adapting NSGA-II for CGP-IP-GI

In order to maintain multiple candidates following each selection phase, it becomes necessary to replace the evaluation function that previously considered only one objective with the NSGA-II algorithm, which takes into account two objectives and enables the selection and retention of a set of solutions.

The NSGA-II algorithm can be divided into two phases: the first phase involves the elimination of dominated solutions, while the second phase focuses on preserving the most isolated solutions by sorting them.

Selection of Non-dominated Solutions

Solutions not belonging to the Pareto front or dominated solutions are deleted. Only non-dominated solutions are retained (Algorithm 1). A solution x $\in$ E dominates if x' $\in$ E if:

$$\forall i, f_i(x) \leq f_i(x') \text{ and } \exists i, f_i(x) < f_i(x') \tag{4}$$

with $f_i(x)$ the function of objective **i**

Sorting by Isolation Distance

If the number of non-dominated solutions is greater than the number of parents μ, it is necessary to sort and retain only μ solutions. The extreme points must be kept in each generation and, therefore, be assigned an infinite crowding distance. For each solution between the extreme points, the crowding distance between the previous solution and the next solution is calculated. Solutions are sorted by crowding distance and the μ solutions with the greatest distance are kept (Algorithm 2).

Algorithm 1: Removal of dominated solutions

Data: solutions is an array containing all solutions
Result: frontpareto contains the solutions on the Pareto front
solutions.sort([fitness,duration]);
for $i \leftarrow 0$ **to** *solutions.length* $- 1$ **do**
    front = True;
    for $j \leftarrow i + 1$ **to** *solutions.length* **do**
        if *solutions[i].duration* $\geq$ *solutions[j].duration* **then**
            front = False;
        end
    end
    if *front* **then**
        frontpareto.push(solutions[i])
    end
end
frontpareto.push(solutions[solutions.length-1]);

Algorithm 2: Sorting solutions by crowding distance and selecting μ solutions

---

Data: solutions is an array containing all solutions
Result: parents contains μ solutions to use as parents
for $i \leftarrow 1$ **to** *solutions.length* $- 1$ **do**
    solutions[i].crowding = abs(solutions[i-1].fitness - solutions[i+1].fitness) +
    abs(solutions[i-1].duration - solutions[i+1].duration)
end
parents.push(solutions[0]);
parents.push(solutions[-1]);
solutions.sort(crowding);
for $i \leftarrow 0$ **to** $\mu - 2$ **do**
    parents.push(solutions[i]);
end

3.3   Synchronization of Islands

CGP-IP-GI uses a distribution of individuals who evolve on different islands. The islands are synchronized at a fixed interval. Synchronization involves taking the best of individuals from all the islands and implanting them as a parent individual on all the islands. The synchronization process is adapted to use the NSGA-II algorithm. During synchronization, all the individuals of each island are grouped together, and only the individuals belonging to the Pareto front are kept. If the number of individuals retained on the Pareto font is greater than μ (number of parents), a sorting by isolation distance is applied to keep only the most isolated individuals. The remaining μ individuals are implanted in the place of the previous parents on each island.

3.4   Objectives Functions

In the study presented in this paper, we will be optimizing two objectives simultaneously. The first objective pertains to result precision, and the objective function for precision is defined as: $1 - MCC$. Details on how to calculate this objective function can be found in Sect. 2.3 above. A lower value of the objective function implies better precision, and thus, the goal is to minimize it.

The second objective focuses on the execution time of the computation. To measure computation time, we employ the Python function **process_time**[1]. This function corresponds to the time required for executing the precision objective function. A smaller result of this objective function indicates a shorter computation time, and the objective here is also to minimize it.

4   Experiments

To assess this multi-objective adaptation in the context of Genetic Improvement (GI), we will apply the same experimental conditions as those outlined in a prior publication on image filter optimization using CGP-IP-GI [3]. This approach enables us to evaluate and, subsequently, compare the outcomes of the multi-objective adaptation with the results obtained in the single-objective version.

4.1   Urban Traffic Dataset

A: Input   B: Output

Fig. 4. Example from the urban traffic image dataset

In this study, we employed CGP-IP for the purpose of object identification within a cityscape. The dataset is sourced from urban traffic livestream cameras[2], and we generated output masks using Mask-RCNN [13]. These masks are used to isolate and retain only the relevant objects within the scene. The primary objective of this dataset is to create a filter capable of extracting and tracking specific objects in the video stream. To achieve this, the filter must discern the moving objects from one frame to the next

---

[1] https://docs.python.org/3/library/time.html.

[2] https://camstreamer.com/live/streams/14-traffic.

in a sequence of five-minute videos. The videos are captured in 16-bit RGB color with a resolution of 1024×576 pixels. Prior to input, we convert the images into grayscale (as depicted in Fig. 4.A). The target classification (as shown in Fig. 4.B) is aimed at identifying significant objects such as pedestrians and vehicles.

The expert filter used in this dataset was designed by engineers. It functions by subtracting the previous image from the current image and applying erode and dilate function to remove the noise. For this dataset, evolution was run for 5000 generations over forty independent trials.

4.2 CGP-IP Parameters

In this work, we have used the following parameters for CGP-IP:

- R: number of rows in CGP is 1
- C: number of columns in CGP is set to 50 but can change with the node addition and deletion operators
- r_{mut}: mutation rate for each gene is 0.25
- r_{ins}: node insertion mutation operator rate is 0.1
- r_{del}: node deletion mutation operator rate is 0.1
- number of islands: number of parallel $1 + \lambda$ evolutions is 4
- μ: number of parents on each island is 4
- λ: population size on each island is 8
- Synchronisation interval between islands: number of generations before islands that synchronize with the chromosome with the best fitness is 20
- number of generations is 5000

Each node within the graph is encoded with eight parameters. The first parameter signifies an index within the list of image processing functions. The second parameter, "Connection 0", represents a connection with a preceding node, where the output of the preceding node serves as the input for the function. The third parameter, "Connection 1", is also a connection with a previous node, using its output as input for the function (although not all functions make use of "Connection 1"). The fourth, fifth, and sixth parameters, labeled "Parameters 0", "Parameters 1", and "Parameters 2", are real numbers corresponding to the first, second, and third parameters of the function. These parameters are not necessarily used because not all functions require three parameters. For instance, Gabor Filter parameters are only utilized in conjunction with Gabor filter functions. Throughout the evolution process, mutations can occur on the function index, connections, or parameters.

4.3 Image Processing Functions

The function set employed in this study is primarily derived from the CGP-IP function set [10]. However, it's important to note that new functions have been introduced into the OpenCV library since the publication of the previous work. In addition to the pre-existing list of image processing functions [10] integrated into CGP-IP, we have incorporated two additional OpenCV functions: "watershed" and "distance transform".

Table 1. Significant values of the Pareto front at generation 5000 over 40 runs

Fitness	1.0	0.70	0.59	0.55	0.52	0.50	0.49	0.49	**0.39**	0.38	0.36	**0.35**
Duration	0.09	0.12	0.18	0.33	0.66	1.05	1.79	2.58	**2.63**	3.13	7.65	**11.7**

5 Results

In this section, we present the results obtained from our dataset over the course of 5000 generations and across 40 independent runs. The application of a multi-objective algorithm enabled us to procure a diverse range of efficient solutions catering to both objectives.

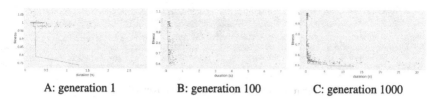

A: generation 1   B: generation 100   C: generation 1000

Fig. 5. All individuals over 40 runs for generation 1 (A), 100 (B) and 1000 (C). The line corresponds to Pareto front.

Figure 5 depicts the Pareto front at generation 1 in blue, generation 100 in green, and generation 1000 in red over the 40 runs. During this evolutionary progression, we observe a significant enhancement in precision from generation 1 to 1000, coinciding with a stabilization in execution time. Beyond generation 1000 and up to 5000, only four solutions exhibit noticeable precision improvements.

Figure 6.A showcases the evolution of the mean and standard deviation of the best precision achieved in each of the 40 runs throughout 5000 generations. This graph highlights a pronounced surge in precision up to generation 500, followed by a gradual but continuous improvement, with the standard deviation progressively increasing from generation 3500 and doubling by generation 5000.

Figure 6.B illustrates the evolution of the mean and standard deviation of the shortest execution duration in each of the 40 runs over 5000 generations. Here, we observe an increase in both mean and standard deviation up to generation 250, after which they remain at consistent levels until generation 5000.

A close examination of Figs. 6.A and 6.B uncovers the intertwined nature of precision and execution time, which stabilizes after generation 300.

Figure 7 showcases the solutions comprising the Pareto front at generation 5000, encompassing a wide array of solutions to accommodate diverse objectives. Table 1 provides a summary of representative values from this Pareto front.

Figure 8.A presents the evolution of the mean and standard deviation of the number of active nodes for the best precision achieved in each of the 40 runs throughout 5000

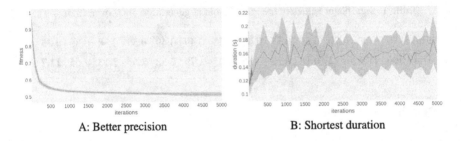

A: Better precision   B: Shortest duration

Fig. 6. Average and standard deviation of the best precision and shortest duration over 40 runs

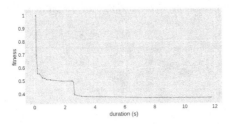

Fig. 7. Pareto front at generation 5000 over 40 runs

generations. Notably, there is a consistent rise in both the mean and standard deviation of active nodes during the entire 5000-generation span.

Figure 8.B mirrors this trend, capturing the evolution of the mean and standard deviation of the number of active nodes for the shortest execution duration in each of the 40 runs over 5000 generations. We observe an increase in both mean and standard deviation up to generation 200, followed by a stable trajectory up to generation 5000. The comparative analysis of Figs. 6 and 8 underscores the connection between improved objectives and an increased number of active nodes.

The Pareto front established after 5000 generations provides a broad spectrum of solutions, with precision ranging from 0.36 to 1.0 and execution times spanning from 90 milliseconds to 11.7 s (see Table 1). This diversity enables the selection of solutions based on the interplay of these two objectives. For instance, even when the highest precision achieved is **0.35**, there exists a solution offering slightly less precision (**0.39**, i.e., a 10% reduction) but with a significantly shorter execution time (**2.63** s instead of **11.7** s, representing a computational cost reduction of 78%).

5.1 Comparing Multi-objective to Mono-objective Results

The CGP-IP-GI framework has previously been examined in a single-objective setup using the same dataset and CGP-IP parameters [3], and Fig. 9 is extracted from this prior publication.

In Fig. 9.A, we observe the evolution of both the mean and standard deviation of precision over 40 runs spanning 5000 generations. When compared to Fig. 6.A, both graphs exhibit a consistent and swift enhancement up to generation 500, followed by a

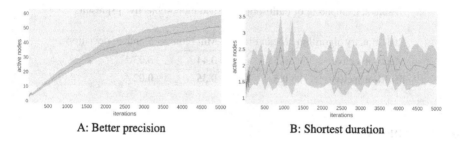

A: Better precision   B: Shortest duration

Fig. 8. Average and standard deviation of the number of actives nodes of the best precision and the shortest duration over 40 runs

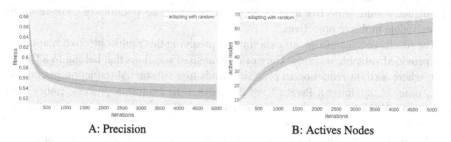

A: Precision   B: Actives Nodes

Fig. 9. Average and standard deviation of the precision and the number of actives nodes over 40 runs from [3]

more gradual improvement. Notably, the standard deviations in both figures are quite similar. The multi-objective adaptation yields a slightly improved precision (2%) at the 5000th generation.

Moving to Fig. 9.B, it illustrates the progression of the mean and standard deviation of the number of active nodes over 40 runs during 5000 generations. In comparison with Fig. 8.A, both graphs display a parallel evolution characterized by a continuous increase in the number of active nodes and a comparable standard deviation. At the 5000th generation in Fig. 9.A, the mean value stands at 0.53, with a minimum of 0.44, a maximum of 0.59, and a standard deviation of 0.03 (as outlined in Table 2). While the mean precision in Figs. 9.A and 6.A displays a slight difference at the 5000th generation, comparing their best fitness demonstrates that the multi-objective adaptation enables the attainment of individuals with superior fitness. Significantly, the T-test p-value between Figs. 6.A and 9.A falls below $1e^{-5}$.

The results from this experiment, conducted over 40 runs and spanning 5000 generations, reveal a consistent improvement in accuracy, as well as an ongoing increase in the number of active nodes. Moreover, when compared to the outcomes of previous studies [3], our findings demonstrate that the use of a multi-objective algorithm does not compromise precision over the course of 5000 generations. On the contrary, maintaining a Pareto front during the evolution process, as opposed to a single individual, results in an expanded search space and leads to more optimal solutions.

Table 2. Comparing of multi-objective and mono-objective [3] results

Evolutionary algorithm	Mean Fitness	Min Fitness	Standard deviation
Mono-objective	0.53	0.44	0.03
Multi-objective	**0.5**	**0.35**	**0.02**

6   Conclusion

In this study, our primary objective was to investigate the feasibility of effectively managing the computational demands inherent in the genetic evolution of image filters. We pursued a multi-objective approach by enhancing the adaptability of CGP-IP-GI to accommodate multiple objectives.

The results obtained from our experiments involving this multi-objective adaptation have provided valuable insights. They offer a range of solutions that belong to a Pareto front, where a slight reduction in precision leads to a substantial reduction in computation time. Maintaining a Pareto front throughout the evolution process, rather than relying on a single individual, significantly expands the search space, thus facilitating the discovery of efficient solutions.

This study, in conjunction with comparisons to prior publications, reinforces the notion that it is indeed possible to reduce the computational time required for image filters while preserving precision. This outcome holds significant promise for the utilization of GP-based algorithms in embedded applications, freeing up computational resources for other essential tasks. Additionally, our findings underscore the efficiency of adapting the NSGA-II algorithm within genetic algorithms for genetic improvement.

In practical terms, the evolution of the CGP-IP-GI framework now enables the execution of multi-objective genetic improvement, offering a wide array of effective solutions for various objectives. The selection of the most suitable solution remains a human-driven process, aligning with the specific constraints and requirements of the application. For systems with limited computing power, the dynamic adaptation to the available computational resources from this pool of solutions holds substantial potential.

One noteworthy avenue that merits exploration is the use of multi-objective algorithms to expand the search space of single-objective algorithms. While our study primarily aimed to reduce computational time as the second objective, an intriguing prospect exists in controlling the number of active nodes. This approach could involve either minimizing or maximizing them. The selection of the most suitable secondary objective to maximize the search space warrants further investigation.

Furthermore, our findings open the door to the exploration of divergent search approaches, offering the possibility of combining directed search, computational cost optimization, and divergent search. This avenue holds promise for future research to unlock new levels of efficiency and effectiveness in the field of genetic improvement.

References

1. Ahmadi, M.H., Hosseinzade, H., Sayyaadi, H., Mohammadi, A.H., Kimiaghalam, F.: Application of the multi-objective optimization method for designing a powered stirling heat engine: Design with maximized power, thermal efficiency and minimized pressure loss. Renewable Energy **60**, 313–322 (2013) . https://doi.org/10.1016/j.renene.2013.05.005,https://www.sciencedirect.com/science/article/pii/S0960148113002504

2. Arcuri, A., Yao, X.: A novel co-evolutionary approach to automatic software bug fixing. In: 2008 IEEE Congress on Evolutionary Computation (IEEE World Congress on Computational Intelligence), pp. 162–168 (2008). https://doi.org/10.1109/CEC.2008.4630793

3. Biau, J., Wilson, D., Cussat-Blanc, S., Luga, H.: Improving image filters with cartesian genetic programming. In: Proceedings of the 13th International Joint Conference on Computational Intelligence (IJCCI 2021), ECTA, vol. 1, pp. 17–27. INSTICC, SciTePress (2021). https://doi.org/10.5220/0010640000003063

4. Bruce, B.R., Petke, J., Harman, M.: Reducing energy consumption using genetic improvement. In: Proceedings of the 2015 Annual Conference on Genetic and Evolutionary Computation, GECCO 2015, pp. 1327–1334. Association for Computing Machinery, New York (2015). https://doi.org/10.1145/2739480.2754752,https://doi.org/10.1145/2739480.2754752

5. Cortacero, K., et al.: Evolutionary design of explainable algorithms for biomedical image segmentation. Nat. Commun. (2023)

6. Deb, K., Agrawal, S., Pratap, A., Meyarivan, T.: A fast and elitist multiobjective genetic algorithm: Nsga-ii. IEEE Trans. Evol. Comput. **6**, 182–197 (2002)

7. Deb, K., Goel, T.: Controlled elitist non-dominated sorting genetic algorithms for better convergence. In: Eckart, Z., Lothar, T., Kalyanmoy, D., Artemio, C.C., David, C. (eds.) Evolutionary Multi-Criterion Optimization, pp. 67–81. Springer, Berlin (2001). https://doi.org/10.1007/3-540-44719-9_5

8. Deb, K., Goel, T.: A hybrid multi-objective evolutionary approach to engineering shape design. In: Eckart, Z., Lothar, T., Kalyanmoy, D., Artemio, C.C., David, C. (eds.) Evolutionary Multi-Criterion Optimization, pp. 385–399. Springer, Berlin (2001). https://doi.org/10.1007/3-540-44719-9_27

9. Harding, S., Graziano, V., Leitner, J., Schmidhuber, J.: Mt-cgp: Mixed type cartesian genetic programming. In: Proceedings of the 14th Annual Conference on Genetic and Evolutionary Computation, GECCO 2012, pp. 751–758. Association for Computing Machinery, New York (2012). https://doi.org/10.1145/2330163.2330268,https://doi.org/10.1145/2330163.2330268

10. Harding, S., Leitner, J., Schmidhuber, J.: Cartesian Genetic Programming for Image Processing, pp. 31–44. Springer, New York (2013). https://doi.org/10.1007/978-1-4614-6846-2_3

11. Harding, S.L., Miller, J.F., Banzhaf, W.: Self-Modifying Cartesian Genetic Programming, pp. 101–124. Springer, Berlin (2011). https://doi.org/10.1007/978-3-642-17310-3_4

12. Harman, M., Jia, Y., Langdon, W.B.: Babel pidgin: Sbse can grow and graft entirely new functionality into a real world system. In: Le Goues, C., Yoo, S. (eds.) Search-Based Software Engineering, pp. 247–252. Springer International Publishing, Cham (2014). https://doi.org/10.1007/978-3-319-09940-8_20

13. He, K., Gkioxari, G., Dollár, P., Girshick, R.: Mask r-cnn (2018)

14. Jafarian, F., Amirabadi, H., Sadri, J.: Application of multi-objective optimization algorithm and artificial neural networks at machining process (March 2013)

15. Kalkreuth, R., Rudolph, G., Krone, J.: More efficient evolution of small genetic programs in Cartesian Genetic Programming by using genotypie age. In: 2016 IEEE Congress on Evolutionary Computation (CEC), pp. 5052–5059. IEEE (Jul 2016). https://doi.org/10.1109/CEC.2016.7748330,https://ieeexplore.ieee.org/document/7748330/

16. Khan, G.M., Miller, J.F., Halliday, D.M.: Evolution of cartesian genetic programs for development of learning neural architecture. Evol. Comput. **19**(3), 469–523 (2011) https://doi.org/10.1162/EVCO_00043

17. Langdon, W.B., Harman, M.: Evolving a cuda kernel from an nvidia template. In: IEEE Congress on Evolutionary Computation, pp. 1–8 (2010). https://doi.org/10.1109/CEC.2010.5585922

18. Langdon, W.B., Harman, M.: Optimizing existing software with genetic programming. IEEE Trans. Evol. Comput. **19**(1), 118–135 (2015). https://doi.org/10.1109/TEVC.2013.2281544

19. Matthews, B.: Comparison of the predicted and observed secondary structure of t4 phage lysozyme. Biochimica et Biophysica Acta (BBA) - Protein Structure **405**(2), 442–451 (1975). https://doi.org/10.1016/0005-2795(75)90109-9,https://www.sciencedirect.com/science/article/pii/0005279575901099

20. Miller, J.F.: An empirical study of the efficiency of learning boolean functions using a cartesian genetic programming approach. In: Proceedings of the 1st Annual Conference on Genetic and Evolutionary Computation, GECCO 1999, vol. 2, pp. 1135–1142. Morgan Kaufmann Publishers Inc., San Francisco, CA, USA (1999)

21. Miller, J.F.: What bloat? cartesian genetic programming on boolean problems. In: 2001 Genetic and Evolutionary Computation Conference Late Breaking Papers, pp. 295–302 (2001). https://www.elec.york.ac.uk/intsys/users/jfm7/gecco2001Late.pdf

22. Miller, J.F., Thomson, P.: Cartesian genetic programming. In: Genetic Programming. pp. 121–132. Springer, Berlin Heidelberg (2000). doi: https://doi.org/10.1007/978-3-642-17310-3_2

23. Miller, J.F.: Cartesian genetic programming. Springer (2011)

24. Miller, J.F.: Cartesian genetic programming: its status and future. Genetic Program. Evolvable Mach. 1–40 (2019)

25. Miragaia, R., Fernández, F., Reis, G., Inácio, T.: Evolving a multi-classifier system for multi-pitch estimation of piano music and beyond: an application of cartesian genetic programming. Appl. Sci. **11**(7), 2902 (2021)

26. Petke, J., Haraldsson, S.O., Harman, M., Langdon, W.B., White, D.R., Woodward, J.R.: Genetic improvement of software: a comprehensive survey. IEEE Trans. Evol. Comput. **22**(3), 415–432 (2018). https://doi.org/10.1109/TEVC.2017.2693219

27. Petke, J., Harman, M., Langdon, W.B., Weimer, W.: Using genetic improvement and code transplants to specialise a c++ program to a problem class. In: Nicolau, M., et al. (eds.) Genetic Programming, pp. 137–149. Springer, Berlin Heidelberg, Berlin, Heidelberg (2014). https://doi.org/10.1007/978-3-662-44303-3_12

28. Srinivas, N., Deb, K.: Muiltiobjective optimization using nondominated sorting in genetic algorithms. Evol. Comput. **2**(3), 221–248 (1994). https://doi.org/10.1162/evco.1994.2.3.221

29. White, D.R., Arcuri, A., Clark, J.A.: Evolutionary improvement of programs. IEEE Trans. Evol. Comput. **15**(4), 515–538 (2011). https://doi.org/10.1109/TEVC.2010.2083669

30. Whitley, D., Rana, S., Heckendorn, R.: The island model genetic algorithm: On separability, population size and convergence. J. Comput. Inform. Technol. **7** (1998)

31. Wu, F., Weimer, W., Harman, M., Jia, Y., Krinke, J.: Deep parameter optimisation. In: Proceedings of the 2015 Annual Conference on Genetic and Evolutionary Computation, GECCO 2015, pp. 1375–1382. Association for Computing Machinery, New York (2015). https://doi.org/10.1145/2739480.2754648

Low-Memory Matrix Adaptation Evolution Strategies Exploiting Gradient Information and Lévy Flight

Riccardo Lunelli and Giovanni Iacca$^{(\boxtimes)}$ (iD)

Department of Information Engineering and Computer Science, University of Trento,
Trento, Italy
riccardo.lunelli-1@studenti.unitn.it, giovanni.iacca@unitn.it

Abstract. The Low-Memory Matrix Adaptation Evolution Strategy is a recent variant of CMA-ES that is specifically meant for large-scale numerical optimization. In this paper, we investigate if and how gradient information can be included in this algorithm, in order to enhance its performance. Furthermore, we consider the incorporation of Lévy flight to alleviate stability issues due to possibly unreliably gradient estimation as well as promote better exploration. In total, we propose four new variants of LMMA-ES, making use of real and estimated gradient, with and without Lévy flight. We test the proposed variants on two neural network training tasks, one for image classification through the newly introduced Forward-Forward paradigm, and one for a Reinforcement Learning problem, as well as five benchmark functions for numerical optimization.

Keywords: Low-Memory Matrix Adaptation Evolution Strategy · Lévy Flight · Forward-Forward Algorithm · Reinforcement Learning · Gradient

1 Introduction

The Covariance Matrix Adaptation Evolution Strategies (CMA-ES) [1] is a well-known evolutionary algorithm that, to date, is considered among the state-of-the-art evolutionary optimizers for real-valued, box-constrained optimization. The main strength of this algorithm consists of the use of a Covariance Matrix, calculated over a subset of the best individuals in each generation of the evolutionary process, for building a multivariate Gaussian distribution model. This model is employed, at each generation, to sample the directions for the mutation step of the ES algorithm. Then, the covariance matrix is adapted at each generation in order to increase the probability of mutations along the most promising mutation directions. As such, CMA-ES performs an inherent dimensionality reduction, aimed at identifying the principal directions of improvement over the fitness landscape at hand. On top of this Covariance Matrix Adaptation (CMA) mechanism, CMA-ES also uses the so-called Cumulative Step-size Adaptation

S. Smith et al. (Eds.): EvoApplications 2024, LNCS 14634, pp. 35–50, 2024.
https://doi.org/10.1007/978-3-031-56852-7_3

(CSA) [2], which is a method that complements the effect of CMA (that provides promising mutation directions) by adapting the step-size in a way that keeps track of a history of previous step-sizes. Overall, the two mechanisms together (CMA and CSA) allow the algorithm to conduct an efficient search even in complex, multi-modal fitness landscapes.

Despite its well-established effectiveness in terms of optimization, which was demonstrated not only in single-objective, box-constrained problems [3,4] but also in the context of multi-objective [5] and constrained optimization [6–8], CMA-ES has however one main drawback, which is its heavy computational complexity. In fact, the original version of CMA-ES is much more computationally expensive than traditional Evolution Strategies, as it needs to store in memory the whole covariance adaptation (of size n^2, where n is the problem size) and adapt it at each generation (which requires several steps whose computational complexity is $\mathcal{O}(n^2)$ [9]). Therefore, since its introduction several research works have investigated various ways to simplify CMA-ES by either revisiting the CMA process [10,11], introducing active covariance updates [12], or even removing the covariance matrix altogether [13]. Other methods constraint the covariance matrix to be diagonal [9], hence trading off the possibility of capturing pairwise variable dependencies (which is instead possible in a full matrix) with a reduced (i.e., linear) time and space complexity. Surprisingly, the authors in [9] show that even though this variant (called sep-CMA-ES) is designed to handle essentially separable functions, where it significantly outperforms the original CMA-ES, for some large-scale non-separable problems, such as the Rosenbrock function, it converges faster, and to comparable results, than CMA-ES.

Among these modified variants of CMA-ES, the recently proposed Low-Memory Matrix Adaptation Evolution Strategies (LMMA-ES) algorithm [14] differs from the original CMA-ES algorithm in that it builds an approximation of the covariance matrix based on an iteratively updated transformation matrix that takes into account only $m << n$ dimensions (where m is a hyperparameter that can in principle be as small as 1). As such, LMMA-ES dramatically reduces the memory complexity of the original CMA-ES, hence resulting particularly suitable for solving large-scale optimization problems.

While ES-based algorithms are mostly meant for zeroth order (and, in the specific case of LMMA-ES, large-scale) optimization, hence without making use of gradient information, this kind of algorithm can be an effective tool also in those contexts where instead gradients are available, such as in neural network training. For instance, in the context of Reinforcement Learning (RL), it has been shown that ES can provide better scalability to traditional RL techniques such as Q-learning and policy gradients [15].

In this work, we therefore address the question as to whether gradient information can be incorporated into the LMMA-ES algorithm. In particular, we present two variants of LMMA-ES, one that incorporates real gradient information and another that uses an *estimation* of the gradient. Moreover, for the two proposed variants of LMMA-ES, we also consider the incorporation of the Lévy flight mechanism (see [16]), which is an alternative to random walk that allows occasional large perturbations hence facilitating escaping from possible local minima, as a way to better balance the algorithm explorability with the typical exploitativity of gradient-based search.

We test the proposed approaches on three classes of problems. First, we consider the case of training of a neural network for image classification, using the recently proposed Forward-Forward (FF) algorithm [17]. Subsequently, we consider a classic control task, namely Cart Pole, as an instance of an RL problem. Finally, we complete our experimentation by testing the algorithms on five well-known benchmark functions, to investigate the scalability of our approaches.

The rest of the paper is structured as follows. The next section presents the methods. Then, Sect. 3 and Sect. 4 present, respectively, the experimental setup and the numerical results. Finally, Sect. 5 concludes this work.

2   Methods

LMMA-ES (Vanilla). Our baseline is the vanilla version of the LMMA-ES algorithm, as proposed in [14]. As discussed in the introduction, LMMA-ES is essentially a variant of CMA-ES that reduces the complexity of the covariance matrix adaptation. Similarly to CMA-ES, at each generation, the algorithm generates each new offspring x_i by applying a Gaussian mutation, as follows:

$$x_i = y + \sigma \cdot \mathcal{N}(0, C) \tag{1}$$

where y is a weighted mean (i.e., the centroid) of μ best individuals in the previous generation, σ is the step-size and C is the covariance matrix. As in CMA-ES, the step-size is adapted by means of the CMA mechanism [2], which allows the algorithm to learn the best step-size σ increasing it when recent steps are in the same direction and decreasing it when steps tend to have different directions. As said earlier, the main difference between CMA-ES and LMMA-ES is instead in the way C is built and updated, since LMMA-ES takes into account only $m << n$ dimensions to estimate C. The other aspects of the algorithm are in principle the same as in CMA-ES. We refer the reader to [14] for further details about the structure of the LMMA-ES algorithm.

LMMA-ES (Real Gradient). In the original paper proposing LMMA-ES [14], the authors suggested as a possible improvement of the algorithm using momentum information based on the gradient, to enhance results. Here, we incorporate this information by calculating the gradient of the centroid vector y, i.e., its delta at every update step of the original LMMA-ES algorithm. The gradient is used to compute the **momentum** vector, using either the AMSGrad [18] variant of the Adam algorithm, or the legacy momentum[1], depending on the problem at hand (see Sect. 3). Following this, the momentum vector is scaled by a constant factor and added to a noise vector, and then normalized. Note that

---

[1] https://boostedml.com/2020/07/gradient-descent-and-momentum-the-heavy-ball-method.html.

each element of the noise vector is drawn from a Gaussian distribution $\mathcal{N}(0,1)$. The procedure for creating each i-th individual is then modified as follows:

$$\mathbf{z_i} = \mathbf{momentum} \cdot scale + \mathbf{noise_i} \tag{2}$$

$$\mathbf{z_i} = \mathbf{z_i}/std(\mathbf{z_i}) \tag{3}$$

where the *scale* factor is used as a hyperparameter to regulate how much information of the momentum we want to use in our algorithm. Normalization is necessary to maintain the std. dev. of the resulting vector $\mathbf{z_i}$ at $\mathbf{1}$. It should be noted that $\mathbf{z_i}$ does not have zero mean, as it is biased in the direction of the momentum. In other words, using the momentum it could be possible to bias the creation of new individuals towards a better point in space. Thus, the formula for the creation of new individuals, shown in Eq. (1), in this case becomes:

$$\mathbf{x_i} = \mathbf{y} + \sigma \cdot \mathcal{N}(\mathbf{momentum} \cdot scale, \mathbf{C}) \tag{4}$$

where:

$$\mathcal{N}(\mathbf{momentum} \cdot scale, \mathbf{C}) = \mathbf{z_i} \times L \tag{5}$$

and $\mathbf{L}$ is the Cholesky decomposition of the covariance matrix: $\mathbf{C} = \mathbf{L} \times \mathbf{L}^T$. The reason for using the Cholesky decomposition is that it provides a lower triangular matrix $\mathbf{L}$ which, when multiplied by a vector sampled from a multivariate Gaussian distribution, results in a vector that follows the desired multivariate Gaussian distribution with covariance matrix $\mathbf{C}$.

LMMA-ES (Estimated Gradient). This variant generalizes the previous one for problems where gradient information is not readily available. When direct gradient calculation is unfeasible, strategies for gradient estimation can in fact be used. Although an estimated gradient might lack the precision of a directly calculated one, it could still improve upon the vanilla LMMA-ES. The chosen strategy for gradient estimation leverages information provided by the individuals, necessitating only the calculation of the fitness value of $\mathbf{y}$. Therefore, only the top μ individuals are considered for the gradient estimation. The first step involves calculating the direction from $\mathbf{y}$ to the focal i-th individual:

$$\mathbf{dir_i} = \mathbf{y} - \mathbf{x_i}. \tag{6}$$

Considering only the best μ individuals is beneficial as it provides more insights into the optimal direction, especially in high-dimensional spaces. Next, the difference in fitness values between $\mathbf{y}$ and the i-th individual is calculated:

$$\Delta f_i = f(\mathbf{y}) - f(\mathbf{x_i}). \tag{7}$$

The direction $\mathbf{dir_i}$ is then multiplied by this difference in fitness values. To penalize distant individuals that could result in poor gradient estimation, the result is divided by the squared norm of the direction:

$$\mathbf{grad_i} = (\mathbf{dir_i} \cdot \Delta f_i)/||\mathbf{dir_i}||^2. \tag{8}$$

The final step is to calculate the mean of all the estimated gradients:

$$\textbf{gradient} = 1/\mu \cdot \sum_{i}^{\mu} \textbf{grad}_i. \tag{9}$$

This **gradient** vector is then used to calculate the **momentum** for Eq. (2) and Eq. (4), according to the selected algorithm (AMSGrad or legacy momentum).

Lévy Flight. The variant based on the estimated gradient sometimes exhibits stability issues due to a bias induced by an inaccurate estimation of the gradient. To avoid these issues and promote exploration, we consider the incorporation of the Lévy flight (see [16]). This mechanism allows random weights to be assigned to the momentum vector, with the weights being drawn from a Lévy distribution. In this way, the weights are mostly situated in the range $[0, 1]$, but occasionally they can attain very high values[2]. Conceptually, this permits the exploitation of gradient information when this offers a reliable estimation. In this case, individuals with higher weights will be favored. Conversely, if the estimation is poor, the algorithm will select individuals with lower weights for the gradient, effectively reverting to the vanilla implementation.

The Lévy flight is applied (with $\beta = 1$) separately for each i-th individual, prior to the addition of noise to the gradient, as follows:

$$\textbf{momentum}_i = \textbf{momentum} \cdot |levy\_flight()| \tag{10}$$

Subsequently, $\textbf{z}_i$ is calculated using the i-th momentum:

$$\textbf{z}_i = \textbf{momentum}_i + \textbf{noise}_i \tag{11}$$

Finally, $\textbf{z}_i$ is normalized as in Eq. (3) and is used as in Eq. (4) and Eq. (5) where in this case **momentum** is replaced by **momentum**$_i$.

3 Experimental Setup

As detailed in the following subsections, we evaluated our approaches on three classes of problems, namely: (1) the FF algorithm [17] for image classification; (2) a classic RL problem, i.e., Cart Pole; (3) five well-known benchmark functions for numerical optimization. Note that, for the momentum calculation, in the case of the FF algorithm we used AMSGrad [18]; otherwise, we used legacy momentum.

All the experiments have been carried out on an Apple M2 PRO machine, comprising of 12 CPU cores and 19 GPU cores with 32 GB of shared memory. Given that the FF algorithm requires numerous matrix operations, this was implemented using PyTorch, which allows to naturally leverage the GPUs efficiency for this kind of tasks. Our code is available at: https://github.com/luna97/matrix-adaptation-exploiting-gradient-evolution-strategy.

---

[2] Note that this method can be useful also with real gradients. In fact, in certain circumstances, even real gradients may not help find better points in the space.

3.1 Forward-Forward Algorithm

The FF algorithm [17], recently introduced as an alternative to the widely used Back-Propagation, proposes a novel paradigm for neural network training. The concept behind this approach is to simulate the learning process of the brain more closely than what happens with the Back-Propagation, as the learning procedure is composed of a Forward-Backward mechanism. In other words, the information is passed only forward, so that the data are processed one layer at a time, hence reducing the computational demands compared to traditional methods based on Back-Propagation. In order to be applied, the FF algorithm necessitates the creation of both *positive* and *negative* samples, with the labels embedded within the data themselves using one-hot encoding. For instance, in the case of the MNIST dataset, the original paper [17] suggests replacing the first ten pixels of the image with the one-hot encoding label (for positive data) or with a random one-hot encoding label (for negative data).

Starting from the implementation available in [19], we modified the loss and the method for label information injection, as proposed in [17], replacing them with the approaches suggested in the SymBa algorithm [20] to expedite convergence. Accordingly, the loss function used in our experiments was:

$$L_{SymBa} = log(1 + e^{-\alpha\Delta}) \tag{12}$$

where $\alpha = 4$ (as in [20]) and $\Delta = G_{pos} - G_{neg}$, with $G_{pos} = \sum_j y^2_{pos,j}$ and $G_{neg} = \sum_j y^2_{neg,j}$, where y_{pos} and y_{neg} are the activations of the neurons within the positive and the negative samples respectively.

Furthermore, rather than substituting the first 10 pixel-values of the image with the one-hot encoding representing the class of that image, this information was concatenated. In [20], the authors proposed a similar approach where an entire channel containing a unique mask with the same size as the image representing the class was concatenated. However, this method would significantly expand the search space. To be precise, the search space would increase by 28×28 (784) variables in the case of MNIST[3], and 32×32 (1024) variables for CIFAR10[4], which are the two datasets used in our experiments. By utilizing the one-hot encoding instead, the search space is augmented by only 10 variables.

From the optimization perspective, the various LMMA-ES variants were run separately for one layer at a time. In each layer, an individual from the population represented a possible set of weights from the current layer to the next one. The loss function in Eq. (12) was used as the fitness function to be minimized, and no bounds on the weights were employed.

In the case of MNIST, we used a fully connected network with three linear layers. Specifically, the first layer consisted of 794 neurons, accounting for the 784 (28×28) pixel values of the image and 10 additional values for the one-hot encoding of the corresponding class. The second and the third layer were respectively composed of 512 and 316 neurons. For the case of CIFAR10, we utilized

---

[3] http://yann.lecun.com/exdb/mnist/.

[4] https://www.cs.toronto.edu/~kriz/cifar.html.

again a fully connected network, with four layers. The first layer contained 3082 neurons, with 3072 values derived from the image (3 channels with 1024 values each) and 10 added values from the one-hot encoding. The next three layers consisted of 1024, 500, and 500 neurons, respectively. For all the layers on both datasets, the ReLU activation function was used.

3.2   Cart Pole

The Cart Pole task [21], often referred to as the "inverted pendulum" problem, is a classic benchmark in the field of RL. The objective of the task is to balance a pole, which is hinged to a cart, in an upright position by only moving the cart horizontally on a track. The system is inherently unstable, and without intervention, the pole will fall. The challenge is to determine the right actions (moving the cart left or right) to keep the pole balanced for as long as possible.

To benchmark this problem, the Python environment provided by the Gymnasium library was used[5]. A simple multi-layer perceptron with two layers was employed. Specifically, the first layer consisted of 64 neurons, each receiving 4 input values corresponding to the cart position, cart velocity, pole angle, and pole angular velocity, respectively. The second layer had 2 neurons, representing the actions of moving the cart to the left and to the right, respectively. The ReLU activation function was used, and a softmax layer was added at the end of the network to determine which action to take. In this problem, the entire network was considered as an individual. The fitness function (to be maximized) is the mean of the sum of the discounted rewards over 3 runs of the individual:

$$f(\mathbf{x_i}) = \frac{1}{3} \sum_{j=1}^{3} \sum_{t}^{N_j} discounted\_reward_j(\mathbf{x_i}) \tag{13}$$

where N_j is the number of total time steps for the j-th run. Note that N_j is not fixed since a run can be shorter if the cart pole fails its task. Also in this case, no bounds on the weights were employed.

3.3   Benchmark Functions

Finally, we considered five benchmark functions that are commonly used in numerical optimization, namely the Sphere, Rastrigin, Rosenbrock, Ackley, and Griewank functions, defined respectively from Eq. (14) to Eq. (18). These functions have been chosen due to their different characteristics in terms of fitness landscape. For all the functions, we considered both the case of 100 and 1000 variables, to verify the scalability of the proposed approaches. The following bonds were employed: $[-5.12, 5.12]^n$ for the Sphere and the Rastrigin functions; $[-5, 10]^n$ for the Rosenbrock function; $[-32.768, 32.768]^n$ for the Ackley function; and $[-600, 600]^n$ for the Griewank function.

---

[5] https://gymnasium.farama.org/environments/classic_control/cart_pole/.

$$f(\mathbf{x}) = \sum_{i=1}^{n} x_i^2 \tag{14}$$

$$f(\mathbf{x}) = 10d + \sum_{i=1}^{n} [x_i^2 - 10cos(2\pi x_i)] \tag{15}$$

$$f(\mathbf{x}) = \sum_{i=1}^{n-1} [100(x_{i+1} - x_i^2)^2 + (x_i - 1)^2] \tag{16}$$

$$f(\mathbf{x}) = -20\exp\left(-0.2\sqrt{\frac{1}{n}\sum_{i=1}^{n} x_i^2}\right) - \exp\left(\frac{1}{n}\sum_{i=1}^{n}\cos(cx_i)\right) + 20 + e \tag{17}$$

$$f(\mathbf{x}) = \frac{1}{4000}\sum_{i=1}^{n} x_i^2 - \prod_{i=1}^{n}\cos\left(\frac{x_i}{\sqrt{i}}\right) + 1. \tag{18}$$

4   Experimental Results

All the LMMA-ES variants were run with the default parametrization regarding the population size $n_{pop} = 4 + \lfloor 3 \cdot \ln n \rfloor$, with n being the problem size. The number of generations was set to 500 in the case of the FF algorithm, 30 for Cart Pole, while for the benchmark functions, it was set to 400 and 1500, respectively for the case of 100 and 1000 variables. All the other parameters were kept at their default values as in [14]. In all the experiments, the $\mathbf{y}$ vector was initialized to a random vector (with each variable sampled from $\mathcal{N}(0,1)$). We assess the statistical significance of our results by applying, for each problem, the Wilcoxon Rank-Sum test ($\alpha = 0.05$) pairwise to each comparison between the vanilla algorithm and each of the compared algorithms.

4.1   Forward-Forward Algorithm

MNIST Dataset. In the case of MNIST, the network was not trained using batches, therefore we set the β_1 hyperparameter for the AMSGrad algorithm to 0.6 (while keeping all the other hyperparameters at their default values defined in PyTorch), because we expected the gradient not to be excessively noisy. The *scale* factor was set to 1. In total, 10 independent runs (i.e., evolutionary processes for training) were executed.

As depicted in Fig. 1, the most effective algorithms are, somehow expectably, the ones exploiting the real gradient. In this case, there is no significant performance difference observed when using Lévy flight. The variants using estimated gradient demonstrate faster convergence than the vanilla algorithm, moreover, the use of the Lévy flight enhances the convergence speed. This finding is of significant interest because it suggests the application to other scenarios where the gradient calculation is not possible but a gradient estimation is available.

CIFAR10 Dataset. In the case of CIFAR10, the network was trained with batches of size 4096. We decided to set the value of β_1 to 0.9, in order to have a more stable gradient. The *scale* factor was set to 1. Also in this case 10 independent runs were executed.

As shown in Fig. 2, the results align closely with those obtained using the MNIST dataset. The variants employing the real gradient clearly outperform the others. Here, it becomes more evident that Lévy flight does not enhance the algorithm using real gradient, but it does improve again the convergence in the case of estimated gradients. This finding further substantiates that Lévy flight can enhance the estimated gradient when its accuracy is lacking. An interesting discrepancy with the MNIST dataset is that in this case the variant utilizing only estimated gradient (without Lévy flight) and the vanilla algorithm seem to have similar behavior. This might be due to the fact that the estimated gradient is very small, rendering the contribution to the creation of z_i nearly negligible.

Accuracy. Table 1 presents the overall test accuracy and F1 score on the two datasets, where FF indicates the original algorithm [17], which was executed for 500 epochs, while the other entries are based on LMMA-ES.

Apart from the case of vanilla vs. estimated gradient (without Lévy) on CIFAR10, where the two algorithms are equivalent, each difference is statistically significant. Overall, the test results confirm that the variants utilizing the real gradient consistently deliver the highest accuracy. It is also interesting to note that these variants outperform the original FF algorithm.

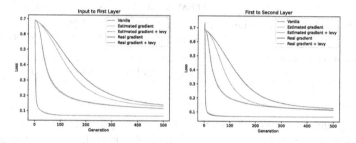

Fig. 1. Results on the MNIST dataset for the vanilla LMMA-ES and the proposed variants used for optimizing the weights between the input and the first hidden layer (left) and between the first and the second hidden layer (right).

4.2 Cart Pole

Since there is no real gradient information available for this problem, we only tested the variants that estimate the gradient, along with the vanilla algorithm. Also in this case, 10 runs were run. The *scale* factor was set to 0.1. For the momentum calculation, we used the legacy momentum algorithm, with $\beta = 0.6$.

Upon examining Fig. 3, it appears that the two variants with estimated gradient show a very similar trend w.r.t. the one shown by the vanilla algorithm.

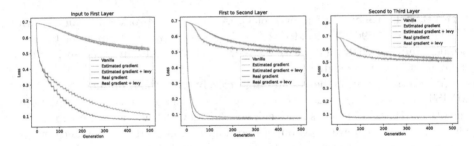

Fig. 2. Results on the CIFAR10 dataset for the vanilla LMMA-ES and the proposed variants used for optimizing the weights between the input and the first hidden layer (left), between the first and the second hidden layer (center), and between the second and the third hidden layer (right).

Table 1. Accuracy and F1 score (in parenthesis) on the test dataset for MNIST and CIFAR10. Underlined values indicate the highest accuracy obtained on each dataset.

	MNIST	CIFAR10
Vanilla	88.15% ± 0.21 (0.8791 ± 0.0021)	31.75% ± 0.77 (0.2990 ± 0.0089)
Real Gradient	95.20% ± 0.21 (0.9515 ± 0.0022)	<u>47.69%</u> ± 0.34 (<u>0.4757</u> ± 0.0037)
Real Gradient + Lévy	<u>95.24%</u> ± 0.25 (<u>0.9518</u> ± 0.0025)	46.84% ± 0.22 (0.4669 ± 0.0026)
Estimated Gradient	89.01% ± 0.16 (0.8885 ± 0.0017)	32.27% ± 0.47 (0.3033 ± 0.0054)
Estimated Gradient + Lévy	86.61% ± 0.53 (0.8644 ± 0.0054)	33.52% ± 0.55 (0.3274 ± 0.0048)
Forward-Forward	93.96% ± 0.61 (0.9390 ± 0.0045)	45.61% ± 0.26 (0.4557 ± 0.0024)

In this case, the three algorithms are statistically equivalent in terms of performance. However, given the relative simplicity of this task, further research into more complex RL tasks is needed to better assess whether gradient estimation can effectively aid convergence.

4.3   Benchmark Functions

It is essential to highlight that highly multimodal functions can have very high gradients. This can cause offspring generation to lean heavily in one direction. A larger gradient will have a more pronounced influence on offspring creation, and if this influence becomes overwhelming, all the offspring might cluster too closely together. The *scale* factor therefore plays a crucial role in this context, as it helps modulate the gradient's contribution. In other words, given the nature of these benchmark problems, sometimes the gradient, and especially its estimation, may

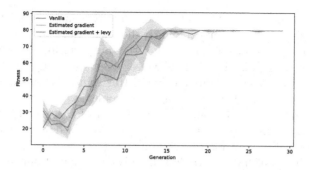

Fig. 3. Fitness trend (mean ± std. dev. across 10 runs) on the Cart Pole task.

be too high, causing numerical problems or simply impeding the algorithm to converge. To solve this issue, relatively small *scale* factors were used in all of our benchmark experiments, especially in the case of estimated gradients, as we detail below for each function.

The experiments were conducted on each benchmark function with two different dimensionalities, namely $n = 100$ and $n = 1000$. As done on Cart Pole, we used legacy momentum with $\beta = 0.6$. For each benchmark function, 10 runs were conducted. The results, in terms of mean and std. dev. of the best function values across the 10 runs found by each algorithm on each function, are reported in Tables 2 and 3, respectively for the 100- and the 1000-dimensional cases.

Table 2. Mean ± std. dev. of the best function values found in 10 runs on the benchmark functions in 100 variables. Underlined values indicate the lowest mean value achieved on each function.

	Sphere	Rastrigin	Rosenbrock	Ackley	Griewank
Vanilla	0.000005 ± 0.000002	1118.8 ± 21.9	25.220 ± 2.78	0.2639 ± 0.503	0.00013 ± 0.0005
Real Gradient	0.000980 ± 0.000600	1098.0 ± 16.9	31.824 ± 2.62	0.0163 ± 0.007	0.00180 ± 0.0025
Real Gradient + Lévy	0.000650 ± 0.000600	1098.1 ± 17.9	31.664 ± 1.39	0.0076 ± 0.004	0.00050 ± 0.0001
Estimated Gradient	0.000950 ± 0.000600	1108.6 ± 12.8	32.475 ± 1.55	0.0135 ± 0.004	0.00210 ± 0.0012
Estimated Gradient + Lévy	0.001060 ± 0.000600	1111.8 ± 19.6	31.675 ± 1.66	0.0184 ± 0.005	0.00160 ± 0.0010

Table 3. Mean ± std. dev. of the best function values found in 10 runs on the benchmark functions in 1000 variables. Underlined values indicate the lowest mean value achieved on each function.

	Sphere	Rastrigin	Rosenbrock	Ackley	Griewank
Vanilla	0.0097 ± 0.0011	11831.7 ± 62.6	330.42 ± 16.11	1.7154 ± 0.092	0.0172 ± 0.0009
Real Gradient	0.0028 ± 0.0003	11613.2 ± 109.8	309.07 ± 8.58	1.5935 ± 0.091	0.0179 ± 0.0040
Real Gradient + Lévy	0.0001 ± 0.0001	11612.5 ± 92.4	272.38 ± 7.08	1.0397 ± 0.454	0.0068 ± 0.0021
Estimated Gradient	0.0088 ± 0.0007	11763.6 ± 59.5	325.89 ± 9.08	1.6172 ± 0.068	0.0183 ± 0.0015
Estimated Gradient + Lévy	0.0083 ± 0.0006	11822.1 ± 129.1	326.69 ± 8.78	1.5651 ± 0.111	0.0189 ± 0.0011

Sphere Function. For the Sphere function, the *scale* factor was set at 0.01 for all the proposed variants. When examining the results for the problem with 100 variables, it is evident that the vanilla algorithm outperforms our proposed variants. Although Fig. 4 (left) does not offer clear evidence to support this, Table 2 shows that the vanilla algorithm can find better solutions (this is statistically confirmed). Conversely, when analyzing the problem with 1000 variables, the advantages of our methods become more evident. In Fig. 4 (right) the improvement (in terms of time to convergence) of the proposed LMMA-ES variants becomes clearer, especially for the variant using the real gradient with Lévy flight. This is also confirmed in Table 3. In this case, apart from the case of the vanilla algorithm vs. estimated gradient, all the other pairwise differences are statistically significant.

Rastrigin Function. For this problem, a smaller *scale* factor of 0.001 was chosen for all the proposed variants. Echoing our earlier observations on the Sphere function, also in this case the larger the problem dimension, the more pronounced the improvement yielded by our proposed variants seems over the vanilla LMMA-ES algorithm, especially when the actual gradient is involved. Figure 5 (left) shows that, for the 100-dimensional case, the vanilla algorithm seems to reach an optimal solution faster. Nonetheless, all variants ultimately succeed in finding better solutions than the vanilla algorithm, as shown in Table 2 (the difference is statistically significant for the comparison between the vanilla algorithm and real gradient variants). Figure 5 (right) distinctly showcases the enhancement compared to the vanilla LMMA-ES algorithm in the 1000-dimensional case (the difference is statistically significant in all cases apart from vanilla vs. estimated gradient with Lévy flight).

Rosenbrock Function. For this problem, we set the *scale* factor to 0.0001 for all the proposed variants. Regarding the fitness trends, shown in Fig. 6, our findings here are similar to what we observed on the previous benchmark functions. The vanilla algorithm seems to find better solutions for $n = 100$, compared to our proposed variants (all differences are statistically significant). When looking at the 1000-dimensional case, the results match what was observed on the other benchmark functions, where all of our methods were able to find better solutions on average, especially when the real gradient is involved (the difference is statistically significant for the comparisons between vanilla algorithm and real gradient variants).

Ackley Function. For this problem, a *scale* factor of 0.1 was chosen for the variants with real gradient, and 0.01 for those with estimated gradient. The fitness trends shown in Fig. 7 indicate that the variant with real gradient and Lévy flight outperforms, in terms of time to convergence, all the other algorithms. However, in the 100-dimensional case all the algorithms are statistically equivalent. Instead, in the 1000-dimensional case all differences are statistically significant. Furthermore, in this case we can see that the worst curve is the one relative to the vanilla algorithm. As for the other problems, we can conclude that for high-dimensional problems our proposed variants can expedite convergence and find, in general, better solutions.

Griewank Function. For this problem, a *scale* factor of 1 was chosen for all the proposed variants. In Fig. 8 (left), it is evident that in this case the vanilla algorithm converges more rapidly. In the 100-dimensional case, all differences are statistically significant. For the 1000-dimensional case, all variants appear quite close in terms of fitness trend, although when looking at Table 3, it can be seen that all of our proposed variants outperform the vanilla algorithm on average. In this case, however, the difference is statistically significant only for the comparisons between the vanilla algorithms and the variants with real gradient and Lévy flight.

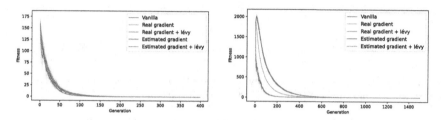

Fig. 4. Fitness trend (mean ± std. dev. of the best function values across 10 runs) on the Sphere function: 100 (left) and 1000 variables (right).

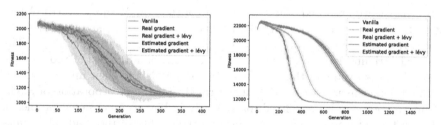

Fig. 5. Fitness trend (mean ± std. dev. of the best function values across 10 runs) on the Rastrigin function: 100 (left) and 1000 variables (right).

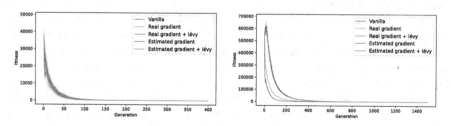

Fig. 6. Fitness trend (mean ± std. dev. of the best function values across 10 runs) on the Rosenbrock function: 100 (left) and 1000 variables (right).

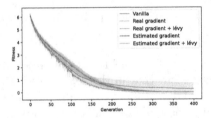
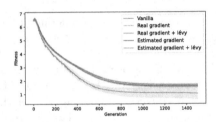

Fig. 7. Fitness trend (mean ± std. dev. of the best function values across 10 runs) on the Ackley function: 100 (left) and 1000 variables (right).

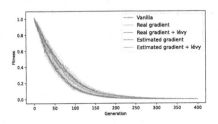
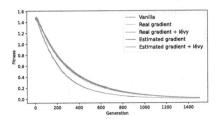

Fig. 8. Fitness trend (mean ± std. dev. of the best function values across 10 runs) on the Griewank function: 100 (left) and 1000 variables (right).

5   Conclusion

In this paper, we proposed four new variants of the Low-Memory Matrix Adaptation Evolution Strategies (LMMA-ES) algorithm, which make use of gradient information (either based on real or estimated gradient) and Lévy flight. The proposed variants underwent extensive testing across a number of problems of various kinds, to validate the efficacy of our proposals and their applicability.

In our experiments, we consistently verified that the incorporation of gradient information can be beneficial, particularly in higher-dimensional problems, although this advantage heavily depends on the quality of the gradient estimation. Furthermore, we have seen that the Lévy flight can be an effective way to improve the convergence speed and reduce some stability issues that may derive from the gradient estimation.

We believe that this study can be expanded in various directions. For instance, better gradient estimators could be studied in order to further optimize the algorithm's performance. As evidenced by the results obtained by the algorithms using real gradients, better estimations can indeed lead to faster convergence and better results.

Another possible future research direction should focus on evaluating the proposed variants on a broader set of problems. In this sense, RL emerges as a promising domain for the application of the proposed approaches, and further investigations in this field are needed.

Finally, investigating ways to improve or consolidate the proposed variants could be beneficial, such as exploring strategies allowing online learning of the *scale* factor. At the moment, setting the *scale* factor, a pivotal hyperparameter, requires a foundational understanding of the problem at hand to strike an optimal balance.

References

1. Hansen, N., Müller, S.D., Koumoutsakos, P.: Reducing the time complexity of the derandomized evolution strategy with covariance matrix adaptation (CMA-ES). Evol. Comput. **11**(1), 1–18 (2003)
2. Ostermeier, A., Gawelczyk, A., Hansen, N.: Step-size adaptation based on non-local use of selection information. In: Davidor, Y., Schwefel, H.-P., Männer, R. (eds.) PPSN 1994. LNCS, vol. 866, pp. 189–198. Springer, Heidelberg (1994). https://doi.org/10.1007/3-540-58484-6_263
3. Caraffini, F., Iacca, G., Neri, F., Picinali, L., Mininno, E.: A CMA-ES super-fit scheme for the re-sampled inheritance search. In: IEEE Congress on Evolutionary Computation, pp. 1123–1130. IEEE (2013)
4. Caraffini, F., Iacca, G., Yaman, A.: Improving (1+1) covariance matrix adaptation evolution strategy: a simple yet efficient approach. In: International Global Optimization Workshop (2019)
5. Igel, C., Hansen, N., Roth, S.: Covariance matrix adaptation for multi-objective optimization. Evol. Comput. **15**(1), 1–28 (2007)
6. Arnold, D.V., Hansen, N.: A (1+ 1)-CMA-ES for constrained optimisation. In: Genetic and Evolutionary Computation Conference, pp. 297–304 (2012)
7. de Melo, V.V., Iacca, G.: A CMA-ES-based 2-stage memetic framework for solving constrained optimization problems. In: IEEE Symposium on Foundations of Computational Intelligence, pp. 143–150. IEEE (2014)
8. de Melo, V.V., Iacca, G.: A modified covariance matrix adaptation evolution strategy with adaptive penalty function and restart for constrained optimization. Expert Syst. Appl. **41**(16), 7077–7094 (2014)
9. Ros, R., Hansen, N.: A simple modification in CMA-ES achieving linear time and space complexity. In: Rudolph, G., Jansen, T., Beume, N., Lucas, S., Poloni, C. (eds.) PPSN 2008. LNCS, vol. 5199, pp. 296–305. Springer, Heidelberg (2008). https://doi.org/10.1007/978-3-540-87700-4_30
10. Beyer, H.-G., Sendhoff, B.: Covariance matrix adaptation revisited – the CMSA evolution strategy –. In: Rudolph, G., Jansen, T., Beume, N., Lucas, S., Poloni, C. (eds.) PPSN 2008. LNCS, vol. 5199, pp. 123–132. Springer, Heidelberg (2008). https://doi.org/10.1007/978-3-540-87700-4_13
11. Beyer, H.-G., Sendhoff, B.: Simplify your covariance matrix adaptation evolution strategy. IEEE Trans. Evol. Comput. **21**(5), 746–759 (2017)
12. Jastrebski, G.A., Arnold, D.V.: Improving evolution strategies through active covariance matrix adaptation. In: IEEE Congress on Evolutionary Computation, pp. 2814–2821. IEEE (2006)
13. Arabas, J., Jagodziński, D.: Toward a matrix-free covariance matrix adaptation evolution strategy. IEEE Trans. Evol. Comput. **24**(1), 84–98 (2019)
14. Loshchilov, I., Glasmachers, T., Beyer, H.-G.: Large scale black-box optimization by limited-memory matrix adaptation. IEEE Trans. Evol. Comput. **23**(2), 353–358 (2019)
15. Salimans, T., Ho, J., Chen, X., Sidor, S., Sutskever, I.: Evolution strategies as a scalable alternative to reinforcement learning. arXiv preprint arXiv:1703.03864 (2017)

16. Iacca, G., dos Santos Junior, V.C., de Melo, V.V.: An improved Jaya optimization algorithm with Lévy flight. Expert Syst. Appl. **165**, 113902 (2020)
17. Hinton, G.: The forward-forward algorithm: some preliminary investigations (2022)
18. Reddi, S.J., Kale, S., Kumar, S.: On the convergence of adam and beyond. arXiv preprint arXiv:1904.09237 (2019)
19. Pezeshki, M., Rahman, H., Yun, J.: pytorch_forward_forward. https://github.com/mohammadpz/pytorch_forward_forward
20. Lee, H.-C., Song, J.: SymBa: symmetric backpropagation-free contrastive learning with forward-forward algorithm for optimizing convergence (2023)
21. Barto, A.G., Sutton, R.S., Anderson, C.W.: Neuronlike adaptive elements that can solve difficult learning control problems. IEEE Trans. Syst. Man Cybern. **13**(5), 834–846 (1983)

Memory Based Evolutionary Algorithm for Dynamic Aircraft Conflict Resolution

Sarah Degaugue[1,2]([✉]), Nicolas Durand[1], and Jean-Baptiste Gotteland[1]

[1] Fédération ENAC ISAE-SUPAERO ONERA, Université de Toulouse,
Toulouse, France
`{sarah.degaugue,nicolas.durand,jean-baptiste.gotteland}@enac.fr`
[2] DGAC-DSNA-DTI, Toulouse, France

Abstract. In this article, we focus on a dynamic aircraft conflict resolution problem. The objective of an algorithm dedicated to dynamic
problems shifts from finding the global optimum to detecting changes
and monitoring the evolution of the optima over time. In the air traffic
control domain, there is added value in dealing quickly with the dynamic
nature of the environment and providing the controller with solutions
that are stable over time. In this article, we compare two approaches
of an evolutionary algorithm for the management of aircraft in a control sector at a given flight level: one is naive, i.e. the resolution of the
current situation is reset to zero at each time step, and the other is
memory-based, where the last population of the optimisation is stored
to initiate the resolution at the next time step. Both approaches are evaluated with basic and optimised operators and settings. The results are in
favour of the optimised version with explicit memory, where conflict-free
solutions are found quicker and the solutions are more stable over time.
Furthermore in the case of an external action, although the diversity
of the population could be lower with the memory-based approach, the
presence of memory does not appear to be a hindrance and, on average,
improves the solver's responsiveness.

Keywords: Evolutionary Algorithm · Dynamic Optimisation
Problem · Aircraft Conflict Resolution

1 Introduction

Aircraft conflict resolution is operated by air traffic controllers based on a two-
dimensional representation of aircraft on a screen. The underlying problem has
been modelled in many different ways allowing various metaheuristics to give
efficient solutions, such as Evolutionary Algorithm (EA) [11], Ant Colony Optimisation [10], Particle Swarm Optimisation or Differential Evolution [28]. Mathematical models were also used to address this problem. In such models, the
hypotheses made on trajectory predictions were generally very restrictive in
order to allow mathematical resolution. For example, Pallottino's approach [22]
used Mixed Integer Linear Programming and relied on constant speed trajectories that are changed all at once. This is also the case in more recent papers

from Vela, Escudero or Rey [4, 24, 29]. Constraint Programming methods [2] and Maximum Clique Search in a graph [18] have also been explored in the last decade, they can handle realistic models and perform well on small instances on which the optimality can be proved. The most realistic models take into account uncertainties and operational constraints and are thus good candidates for metaheuristics, because the trajectories need to be simulated to evaluate a set of manoeuvres. Furthermore population based metaheuristics have the great advantage to return a population of solutions instead of a single option. This gives an opportunity to imagine an intelligent support tool for air traffic controllers who could pick manoeuvre options in a pool of good solutions. Most of the research on aircraft conflict resolution studied the static problem without taking into account its dynamic over time. Indeed, aircraft constantly move in a control sector. The conflict resolution problem is thus not static but must take into account its dynamic aspect. Some changes can be modelled as continuous such as the aircraft positions or the trajectory prediction evolution over time. Other changes are discrete, such as the entrance of an aircraft in the control sector or the exit of an aircraft from the sector. An air traffic controller can suddenly decide to manoeuvre an aircraft. Because of these changes over time, conflict resolution is a dynamic problem. Besides eliminating all conflicts, the aim of conflict resolution is also to minimise delays and the number of aircraft manoeuvred. For this reason, conflict resolution is a dynamic optimisation problem (DOP).

A first approach [8] studied the operational benefits of using a memory-based EA (with basic crossover and mutation operators) and its impact on an air traffic control point of view. This article focuses on the behaviour of an EA in such a dynamic environment and introduces an optimised algorithm modifying the crossover and mutation operators. We compare the basic and optimised versions with or without a memory process on different levels of traffic densities. We also address in this article the robustness of an automatic solver with external actions on aircraft.

In Sect. 2, we discuss the state of the art on the use of memory in EAs for DOPs. Section 3 introduces the problem modelling. Section 4 details two algorithmic adaptations of an EA for our problem. In Sect. 5, two versions of the memory management are described. Section 6 applies the different algorithmic versions on three levels of traffic densities and discusses the effect of external actions on the dynamic simulations.

2   EA for Dynamic Optimisation Problems

DOPs are most commonly described in two different ways: either by a succession of static problems [5, 21, 30], or by a mathematical expression with time-dependant parameters [6, 31]. The simple search for an optimal solution is no longer sufficient in a Evolutionary Dynamic Problem (EDP). Detecting changes in the environment of the current problem and consequently in the objective function are important points in the solution search.

2.1   Naive Approach

The most naive approach to solve Dynamic Optimisation Problems (DOPs) is to reset the evolutionary process when a change occurs. A similar approach restarts the population based on the convergence of EAs [17]. Hu and Eberhart describe in [16] Rerandommisation PSO (RPSO) where all or part of the swarm is randomly moved around in the search space when a change occurs. The shortcoming of these approaches is that the past evolutionary material is not always used, although this could accelerate convergence.

2.2   Implicit Memory

In implicit memory approaches, a local memory is added to each chromosome (e.g. adding characteristics specific to genes (redundant, recessive, etc.)). These characteristics can be used, for example, to re-introduce past good solutions. Diploid scheme [20] and triallelic scheme [13] have been introduced, respectively, by Ng and Wong in 1995 and by Goldberg and Smith in 1987. The specific nature of our problem, dealing with constantly moving aircraft, aircraft entrances and exits, makes the use of an explicit memory more adapted for this problem.

2.3   Explicit Memory

Explicit memory approaches save information, either in the structure of individuals or in a memory external to the population, to preserve old elements of the population. Several uses of this memory are possible. Unlike implicit memory, the use of explicit memory is more controlled because it keeps information on when and which memory elements are reused. A first intuitive approach was introduced by Louis and Xu in 1996 [19] and consists of reusing old population elements when a change of environment occurs, while initialising certain chromosomes. In the same vein, in [26] authors introduce a short-term memory of ancestors present in previous generations, and in [27] authors add some of their ancestors locally to chromosomes when they are assessed as good. However, the effectiveness of these approaches is highly problem-dependent. Limits appear particularly for significant changes in the environment, as in the classic task planning problem where adding or removing a task considerably modifies the location of the optimum.

In [23], good population elements and their associated environment are stored periodically, then re-inserted when the environment changes. Instead of storing good whole individuals, and potentially their associated environment, it is also possible to store an abstract form of these individuals by recording their position in the search space. To do this, they first need to partition the search space and then define a representation matrix for this space. Following this, a spatial distribution of good individuals is obtained to guide the initialisation of a future optimisation if a change occurs [25]. This approach seems to be more useful for chaotic changes in the environment, whereas the crude safeguarding of individuals appears to be more effective for regular or cyclical changes.

3   Problem Modelling

This section introduces the environment and the decision variables of our problem.

3.1   Environment

Let us consider a situation with $n (\in \mathbb{N})$ aircraft flying in an en-route sector. The required separation between two aircraft is 5 nautical miles (NM) in the horizontal plane or 1000 feet in the vertical plane. Two aircraft are in conflict when there is a loss of separation in their predicted trajectories. Let us suppose that all aircraft fly with constant speeds at the same flight level. Adding different flight levels generally eases the problem because it gives more options to solve conflicts. It also partitions the problem in different sub-problems that can often be solved independently. For the sake of simplicity, we do not investigate this aspect in this article. In order to comply with air traffic controllers behaviour, our model takes into account uncertainties in the trajectory prediction. Many realistic uncertainty models have been presented in previous work [1,3]. Here we use a simplified version of uncertainty described in [12] that increases the size of the horizontal separation standard linearly with time in the prediction. With such an uncertainty model, when looking too far ahead, many conflicts are predicted but may never occur. In order to limit this phenomenon, we limit the uncertainty growth to a time window: if t is the current time, uncertainties will grow according to our model in the time window $[t, t + T]$ ($T = 6$ min in the experiments) and will then be capped at their value at time $t + T$ for subsequent predictions. This helps focusing on short term conflicts while keeping long term detection.

3.2   Decision Variables

Our model considers heading change manoeuvres to solve conflicts. Here, a manoeuvre is a α degree heading change, starting at time t_0 and ending at time t_1. α is relative to the current heading. Once a manoeuvre is finished, the aircraft heads toward its destination (see Fig. 1).

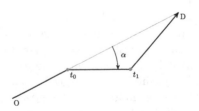

Fig. 1. Manoeuvre model for an aircraft flying from O to D.

For each aircraft i, $(t_{0_i}, t_{1_i}, \alpha_i)$ are bounded because of aerodynamic, sector boundaries or time constraints. For instance, t_{0_i} can take values in $[\underline{t_{0_i}}, \overline{t_{0_i}}]$. α_i is discretised with a step of 5°C in the range $[-45°, 45°]$, which corresponds to air traffic controllers practices.

Air traffic controllers only give one manoeuvre at a time to each aircraft. Once the manoeuvre has started, only t_1 can be modified (delayed or advanced to the current time plus 60 s (S) at the earliest). In our model, a manoeuvred aircraft can be manoeuvred again only once it is heading back to its destination (between t_1 and D on Fig. 1) but a second manoeuvre cannot be predicted before the aircraft has finished its current one. This model complies to air traffic controllers habits and favours a better balance of manoeuvres between aircraft.

4   Algorithm Versions

As first described by [15], the principle of an EA can be summarised as follows: given an evaluation function (fitness) to maximise, we initially randomly create a population of candidate solutions, and apply the fitness function as a measure of quality. At each generation, a selection is performed on the population, followed by crossovers and mutations between some individuals leading to a new population composed of some good old elements and new ones. This process is iterated until a good enough solution is found or a time limit is reached.

Two versions (basic BV, and optimised OV) of the algorithm are introduced in this section. Basic BV uses the operators introduced in [8]. In the OV version, we optimised the algorithm by defining new crossover and mutation operators, favouring diversity and by applying a local optimisation at the end of the algorithm in order to reduce as much as possible the current manoeuvres.

4.1   Population Element Structure

For a traffic situation including n aircraft, a population element e is a potential solution of the problem composed of n genes where each gene i (g_i^e) is the ith aircraft manoeuvre $(t_{0_i}^e, t_{1_i}^e, \alpha_i^e)$ with:

- $t_{0_i}^e$, the manoeuvre start time chosen for aircraft i;
- $t_{1_i}^e$, the manoeuvre ending time chosen for aircraft i;
- α_i^e, the deviation angle chosen for aircraft i. If $\alpha_i^e = 0$, the aircraft i is not manoeuvred.

4.2   Population Element Fitness

In this work, we have several metrics that can be sorted by priority:

1. Solve all the conflicts;
2. Minimise the number of manoeuvres;
3. Minimise the delay due to the manoeuvres;

4. Start the manoeuvres as late as possible in order to avoid useless manoeuvres, given that detected conflicts can disappear over time with uncertainty reduction.

As these different metrics are sorted, and given that the EA will provide us with different solutions of the problem, we decided to stay in a mono-objective definition of the problem, by defining a single fitness balancing the different criteria.

Let us introduce for gene i of a population element e:

- d_i^e the delay of the aircraft i due to the manoeuvre;
- $l_i^e = \overline{t_{0_i}} - t_{0_i}^e$ where $\overline{t_{0_i}}$ is the maximum time allowed to start a manoeuvre. $l_i^e = 0$ if the aircraft i is not manoeuvred;
- S^e the set of remaining conflicts in the population element e;
- S_i^e the set of remaining conflicts involving aircraft i in element e;
- d_c the conflict duration for a conflict $c \in S^e$.

We introduce a local fitness as presented by Durand et al. in [9]. Each gene i of a population element e has a local fitness f_i^e, which is useful to improve the crossover and mutation steps. f_i^e is expressed as follows:

$$f_i^e = \begin{cases} \dfrac{1}{1+ \sum\limits_{c \in S_i^e} d_c} & \text{if } S^e \neq \emptyset \\[2ex] 1 + \dfrac{1}{1+2\times d_i^e + l_i^e} & \text{else} \end{cases}$$

Let us define ts_c the starting time of conflict $c \in S^e$. If a conflict remains, it should start as late as possible. The global fitness of a population element e is defined as follows:

$$F^e = \begin{cases} \dfrac{1}{2} - \dfrac{1}{2\,(1+ \min\limits_{c \in S^e} ts_c)} + \dfrac{1}{2\,(1+ \sum\limits_{c \in S^e} d_c)} & \text{if } S^e \neq \emptyset \\[2ex] \dfrac{1}{n} \sum\limits_{i=0}^{n} f_i^e & \text{else} \end{cases}$$

4.3   EA Operators

Crossover: The crossover operator creates two new elements from two parent elements. In the BV version, one is created by mixing the genes of the two parents and the other by an arithmetic operation on the genes of the two parents. Let us consider two population elements e_1 and e_2.

For the first child e_a, we use the strategy described in [9] which consists in using the partial separability of a population element's fitness. For all $i \in [|1, n|]$, if $f_i^{e_1} > f_i^{e_2}$ then $g_i^{e_a} = g_i^{e_1}$ else $g_i^{e_a} = g_i^{e_2}$.

The second child e_b is created by applying barycentres. For all $i \in [|1, n|]$, let us choose randomly $\lambda_{t_{0_i}}, \lambda_{t_{1_i}}, \lambda_{\alpha_i} \in [-50, 100]$ and define:

$$t_{0_i}^{e_b} = (\lambda_{t_{0_i}} \times t_{0_i}^{e_1} + (100 - \lambda_{t_{0_i}}) \times t_{0_i}^{e_2}) \div 100$$
$$t_{1_i}^{e_b} = (\lambda_{t_{1_i}} \times t_{1_i}^{e_1} + (100 - \lambda_{t_{1_i}}) \times t_{1_i}^{e_2}) \div 100$$
$$\alpha_i^{e_b} = (\lambda_{\alpha_i} \times \alpha_i^{e_1} + (100 - \lambda_{\alpha_i}) \times \alpha_i^{e_2}) \div 100$$

The OV version emphasises the diversification role of the crossover operator by randomly creating two new random population elements one time out of three.

In both BV and OV versions, the population crossover rate $p_{cross} = 30\%$ was chosen following a former dedicated study [7].

Mutation: The mutation operator locally modifies one of the genes of a population element. The gene is chosen according to the value of its local fitness in order to focus on the worst values. A roulette wheel draw is performed on the set M of genes that can still be modified. Let us consider a population element e. If $S^e \neq \emptyset$, the probability for gene $i \in M$ to be chosen is proportional to $1/min(1, f_i^e)$. When no conflict remains the probability for gene $i \in M$ to be chosen is proportional to $1/(f_i^e - 1)$.

Once a gene i has been selected, we randomly modify $t_{0_i}^e, t_{1_i}^e$ or α_i^e if the aircraft does not yet have a manoeuvre in progress, and $t_{1_i}^e$ if the aircraft is already manoeuvred and can still be adjusted.

In the BV algorithm, we randomly change one of the three parameters of the manoeuvre of aircraft i. In the OV algorithm, if the current element e has no conflict, we first try to remove the manoeuvre of aircraft i (set $\alpha_i^e = 0$). If it creates a conflict we apply the former process.

In both BV and OV versions, the population mutation rate $p_{mut} = 40\%$ was chosen following a former dedicated study [7].

4.4 Sharing Process

Keeping a diverse population is essential in a dynamic environment, and also very useful to avoid premature local convergence of the EA. We use a cluster based sharing process as described in [11,32].

The sharing process requires to define a distance between population elements. Therefore we consider three different manoeuvre directions (turn right, turn left or do not turn) in order to match the way air traffic controllers understand different resolutions. We define the distance between two population elements as the number of aircraft that are not manoeuvred in the same direction. Two population elements are zero-distant if all of their aircraft are manoeuvred in the same direction and thus belong to the same cluster. This defines the notion of cluster, grouping all the individuals giving the same manoeuvre directions to all their aircraft. There can be potentially up to 3^n clusters. In the experiments, considering the exponential growth of this number, and the constraints on manoeuvres in progress, we will rarely find the maximum number of clusters in a population. The sharing process helps control the diversity of the population. Let C be the set of all clusters and f_c the fitness of the best individual in a cluster $c \in C$. Let c_{best} be the cluster in which the best individual in the population is found and f_{best} its fitness. Our sharing process works in two steps. First, we define a sharing rate $s_r \in [0; 1]$. The best elements of all the clusters c which respect Eq. 1 are automatically selected in the new population.

$$f_c \geq s_r \times f_{best} \qquad (1) \qquad\qquad F_s^e = \frac{F^e}{card(c)} \qquad (2)$$

Adjusting s_r is challenging. When s_r is high, only the best clusters are protected in the selection process, whereas when s_r is low, the algorithm is more conservative.

The second step applies the selection process described in 4.5 to modified fitness of the population elements. The modified fitness of an element of a cluster c is given by Eq. 2: the fitness of a population element is divided by the cardinal of its cluster.

4.5 Selection

The selection step is based on the stochastic remainder without replacement method [14], taking into account the fitness F_s calculated during sharing.

4.6 Population Size and Ending Criterion

After a large number of experiments to check the quality of the convergence of the EA (as explained in [7]), we selected the following parameters: the population size is fixed to 200 and the algorithm is stopped after 200 generations, or when the best fitness corresponds to a non-conflicting solution and has not been improved during the last 20 generations. The EA can thus converge before reaching 200 generations.

4.7 Final Optimisation

Some useless non-zero manoeuvres can appear and survive with the previous operators (especially the crossover and sharing operators which preserve a large population diversity). To eliminate these non-zero manoeuvres in the last generation, a final local optimisation is performed in the OV version on all individuals. $|\alpha|$ and then t_1 are decreased as much as possible (without creating new conflicts).

5 Memory Management

The traffic continuously evolves over time. Aircraft enter, leave and fly through the sector. The goal is to avoid conflicts while minimising aircraft deviations. At each time step (every 30 s in the experiments), every manoeuvre started or starting in less than 1 min is updated or applied. Manoeuvres starting in more than 1 min are ignored because they can be recalculated later.

5.1 Naive Approach (NA)

In a classical EA initialisation, each population element is randomly created. This strategy is used in the naive approach, NA in the following. The solver only receives information about the current environment.

5.2   Explicit Memory Approach (EMA)

The explicit memory approach stores all the population elements of the last generation and reintroduces them to initialise the next resolution. Some previous solutions (respecting decision variable bounds at the previous step) may no longer be correct. An aircraft may have been manoeuvred and cannot change direction anymore, another may have finished a manoeuvre and is free to start a new one. Furthermore, the number of decision variables may have changed if some aircraft have left or entered the sector. In these cases, the genes representing the outgoing aircraft are deleted from the population elements and new genes corresponding to incoming aircraft are randomly created.

5.3   Summary of the Tested Algorithms

In the experiments, we call BVNA and BVEMA the EA versions associated with basic operators (crossover and mutation), and executed with either the NA or EMA approaches. Similarly, we call OVNA and OVEMA the EA versions associated with adapted operators, final optimisation and executed with either NA or EMA. s_r was experimentally adjusted at $s_r = 0.1$, a low value favouring a large diversity in the population.

6   Experimental Results

6.1   Exercises

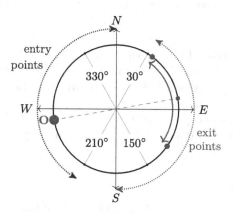

Fig. 2. Geometry of conflict scenario generation.

Creation: We create 3 scenario sizes, involving on average 35 aircraft (Baseline), 50 aircraft (Baseline+50%) and 70 aircraft (Baseline+100%) for around one hour of simulation. The Baseline scenario is already much denser than the current traffic encountered by air traffic controllers in real life. They can generally handle up to 30 aircraft, but on several independent flight levels. Each aircraft is assigned a random speed between 385 kts and 550 kts, a random entry point O taken on a circle of 90 NM radius on the angle $a_{in} \in [210°; 360°]$ (see Fig. 2), and an exit point D on the angle $a_{out} = a_{in} - 180° \pm Random(45°)$. We impose $a_{out} \in [30°; 180°]$ to avoid interactions between entering and exiting aircraft. Air traffic control generally uses an analogue semi-circular rule for the same purpose.

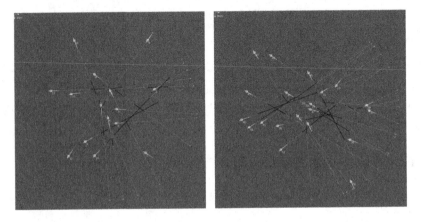

Fig. 3. Extracted situations from the Baseline (left) and the Baseline+100% (right) scenarios.

Preliminary Analysis: Figure 3 shows two extracted situations simulated without resolution of the Baseline, on the left, and the Baseline+100%, on the right. The aircraft are represented by the white squares, their velocity vectors by the white lines and their previous positions by comets. The grey lines show the remaining trajectories and potential conflict zones are coloured in black. Current conflicts are coloured in red. A higher traffic density increases the number of potential conflicts.

6.2 First Results

In a first experiment, we evaluate the combinations of the EA versions with and without memory (BVNA, BVEMA, OVNA, OVEMA) on the three exercises previously described. Each combination is run twenty times on each exercise, using different random seeds to ensure the statistical validity of the results. We apply a Wilcoxon Rank Sum Test with continuity correction using R to compare the results between two different versions. The returned statistical elements are denoted by W and p.

Algorithm Evaluation: The different simulations are compared using the following criteria:

- F: the fitness of the best element found at each resolution time step;
- $|C|$: the number of clusters at the end of each resolution;
- $|C^0|$: the number of clusters without conflict at the end of each resolution;
- G_{tot}: the total number of generations for each resolution;
- G_c: the minimal number of generations before a non-conflicting solution is found, at each resolution;
- $|S|$: the number of remaining conflicts at the end of the exercise.

Ideally, the higher F, $|C|$ and $|C^0|$ are, and the lower G_{tot}, G_c and $|S|$ remain, the better the EA behaves.

Table 1. Algorithmic results

Version	BVNA	BVEMA	OVNA	OVEMA
Criteria	**Baseline**			
$\mu(F)$	1.77	1.79	1.78	1.79
$\mu(|C|)$	84.7	82.1	120	120
$\mu(|C^0|)$	59.9	54.2	64.0	59.8
$\mu(G_{tot})$	131	51.9	129	51.0
$\mu(G_c)$	3.04	0.04	3.19	0.05
$\sum |S|$	0	0	0	0
Criteria	**Baseline+50%**			
$\mu(F)$	1.76	1.72	1.77	1.76
$\mu(|C|)$	80.8	79.2	112	112
$\mu(|C^0|)$	55.8	49.8	55.1	50.4
$\mu(G_{tot})$	135	66.7	136	61.1
$\mu(G_c)$	4.86	0.71	4.22	0.68
$\sum |S|$	0	0	0	0
Criteria	**Baseline+100%**			
$\mu(F)$	1.63	1.59	1.65	1.66
$\mu(|C|)$	82.0	82.9	115	115
$\mu(|C^0|)$	51.0	45.1	53.1	50.4
$\mu(G_{tot})$	167	100	167	86.2
$\mu(G_c)$	18.6	19.3	20.8	2.95
$\sum |S|$	0	4	0	0

Table 1 shows for the three traffic densities, and the four combinations of the algorithm (BVNA, BVEMA, OVNA, OVEMA) the mean values on the 20 runs of the mean values of criteria F, $|C|$, $|C^0|$, G_{tot} and G_c on all the time steps of the simulation where a manoeuvre occurs. The sum of remaining conflicts over the 20 runs of each exercise is shown in the last row $\sum |S|$.

The criteria between BVNA and OVNA, and between BVEMA and OVEMA improve in a majority of cases. For example in the Baseline+100% exercise, all criteria improve between BVEMA and OVEMA ($\mu(F) : 1.59 \to 1.66$ (W > 10^7, p < 10^{-3}), $\mu(|C|) : 82.9 \to 115$ (W > 10^6, p < 10^{-15}), $\mu(|C^0|) : 45.1 \to 50.4$ (W > 10^7, p < 10^{-14}), $\mu(G_{tot}) : 100 \to 86.2$ (W > 10^7, p < 10^{-14}), $\mu(G_c) : 19.3 \to 2.95$ (W > 10^7, p < 10^{-13}) and $\sum |S| : 4 \to 0$). The increase of the total number of clusters is notable. The new crossover operator creating new individuals one time out of three in the OV version could explain this phenomenon. The number of remaining conflicts drops to zero in the OV version.

Averages of $\mu(F)$ are similar for OVNA and OVEMA but differences appear when looking at the number of generation and cluster criteria. Using memory does not seem to penalise the diversity of the whole population. The constant reuse of former population elements does not impact the total number of clusters. However, even if the number of clusters $|C|$ are similar, the number of clusters without conflict remains slightly higher without memory, showing that starting with a random population remains better for the diversity of acceptable solutions. Keeping a high diversity is important in the eventuality of external actions on aircraft. The main advantage of using memory is shown with the optimised version for which $\mu(G_{tot})$ and $\mu(G_c)$ decrease a lot with the use of memory. For example, on the Baseline+100% exercise, the mean number of generations necessary to find a conflict-free solution is divided by almost ten, dropping from 20.8 to 2.95 (W > 10^7, p < 10^{-15}). This last point could be very advantageous if the EA was used to help an air traffic controller making decision in real time.

Aeronautical Performances: Table 2 evaluates the following criteria on the same simulations. The lower the criteria are, the better the aeronautical performances are.

- M: the number of manoeuvres divided by the number of aircraft at the end of each simulation;
- D: the average percentage of additional flight time per aircraft;
- V: the percentage of aircraft with varying manoeuvres planned. A manoeuvre is varying if it has at least been planned in the opposite direction (turn right then left or turn left then right) between two successive resolutions.

In the most of the cases, especially when the traffic density increases, OVNA and OVEMA have better results than BVNA and BVEMA. When memory is used, the OV version tends to minimise the number of manoeuvres performed, but the manoeuvre durations are higher. The balance between these two criteria is modified with the use of memory. This can be adjusted in the fitness definition.

The major observation here is the improved stability of the manoeuvres planned by the conflict solver when memory is used. Indeed, for the most dense exercise, criterion V drops from 33% (without memory) to 9% (with memory) (W = 0, p < 10^{-7}), showing that with memory, only 9% of the aircraft are planned

Table 2. Aeronautical results

Version	BVNA	BVEMA	OVNA	OVEMA
Criteria	**Baseline**			
M	0.28	0.28	0.30	0.28
D	1.06	1.18	1.18	1.11
V	14.1	9.11	12.5	7.14
Criteria	**Baseline+50%**			
M	0.28	0.32	0.27	0.27
D	1.23	2.10	1.30	1.55
V	17.7	7.82	14.1	6.54
Criteria	**Baseline+100%**			
M	0.64	0.62	0.59	0.54
D	3.08	3.91	2.84	3.55
V	42.1	20.8	33.3	9.04

an opposite direction between two time steps. This could be a real advantage if our EA had to help an air traffic controller make decisions over time.

6.3   External Action Impacts

In this section, we focus on the effect of external actions on EA resolutions. At some time steps (every 300 s if possible, or more if not), a random manoeuvre is associated with an aircraft (in the aircraft manoeuvring bounds) and both the state of the environment and the resolution at the next time step are saved. The main simulation is run using the OVEMA version. When an external manoeuvre is applied, we compare two resolutions, OVNA and OVEMA at this specific time step.

We randomly apply a manoeuvre to one aircraft that can still be manoeuvred. If this manoeuvre does not cause a conflict within the next three minutes, it is added to the current solution, otherwise the process is repeated until an acceptable manoeuvre is found. This random action is initiated if the EA has at least one conflict-free solution at the current time and if at least one aircraft can be manoeuvred.

In Table 3, for each exercise, we evaluate the mean values of F, $|C|$, $|C^0|$, G_{tot} and G_c with the OVNA and OVEMA solvers. The percentage of times for which one version is strictly better than the other is shown in the brackets next to the mean value of the criteria.

The mean fitness is generally better with memory than without, and the fitness is strictly better without memory only 11%, 17% and 27% of the time. As previously, the major result concerns the number of generations necessary to optimise a solution or to find a conflict-free solution. For the most dense scenario (Baseline+100%), the total number of generations G_{tot} drops from 154 to 104

Table 3. Effect of external actions

Criteria	Baseline		Baseline+50%		Baseline+100%	
	OVNA	OVEMA	OVNA	OVEMA	OVNA	OVEMA
$\mu(F)$	1.76 (11%)	1.84 (80%)	1.77 (17%)	1.76 (36%)	1.57 (27%)	1.6 (54%)
$\mu(|C|)$	93 (40%)	99 (53%)	102 (40%)	101 (40%)	111 (46%)	111 (43%)
$\mu(|C^0|)$	57 (48%)	61 (42%)	57 (53%)	54 (30%)	49 (57%)	48 (26%)
$\mu(G_{tot})$	92 (5%)	47 (54%)	108 (24%)	70 (56%)	154 (10%)	104 (64%)
$\mu(G_c)$	6 (6%)	2.4 (21%)	7.6 (3%)	5.4 (33%)	39 (5%)	21 (61%)

(W > 10^5, p < 10^{-15}) with memory and the number of necessary generations to find a conflict-free solution drops from 39 to 21 (W > 10^5, p < 10^{-15}) with memory. Using a memory approach has not prevented the EA to adjust to an unexpected event.

7   Conclusion

In this article, we introduce an optimised version of an EA combined with a memory process and compare it with a basic version on a very dense dynamic conflict resolution optimisation problem. We define a new crossover operator to help the EA keep a diverse population even when a memory process is used. We also define a new mutation operator to keep reduced manoeuvres allowing aircraft to remain free for future manoeuvres. We add a final optimisation process to help remove useless manoeuvres and reduce delays. We use a cluster based sharing process with a low sharing rate to make sure that the population covers the search space as much as possible through the simulations.

Using an explicit memory process drastically reduces the total number of generations and the number of generations necessary to find a conflict-free solution. It also reduces the variation of manoeuvres planned over time. These two results are essential if such an algorithm was used to help air traffic controllers make real time decisions in a dynamic environment.

We show that the memory based approach can handle external actions and still quickly find good solutions despite the fact that old options are kept in the population. This is an important point if such an algorithm should interact with an air traffic controller making decisions that are not present in the manoeuvres covered by the population.

For future research, it could be wise, at every time step, to run in parallel the EA with several random seeds and keep the best solution found. This could be tested and compared with or without memory. We could also imagine a hybrid version of the naive and explicit memory approaches, where only the best element of each cluster at the last generation would be reintroduced in the next population initialisation. The rest of the population could then be randomly created which would favour some diversity in the population.

This work will be tested with air traffic controllers to measure the capacity for collaboration between humans and a population based automatic solver and the quality of interactions with such a memory based algorithm.

References

1. Allignol, C., Barnier, N., Durand, N., Alliot, J.M.: A new framework for solving en-routes conflicts. In: 10th USA/Europe Air Traffic Management Research and Developpment Seminar (2013)
2. Allignol, C., Barnier, N., Durand, N., Gondran, A., Wang, R.: Large scale 3D en-route conflict resolution. In: ATM Seminar, 12th USA/Europe Air Traffic Management R&D Seminar, Seattle (2017). https://enac.hal.science/hal-01592235
3. Allignol, C., Barnier, N., Durand, N., Gondran, A., Wang, R.: Large scale 3D en-route conflict resolution. In: ATM Seminar, 12th USA/Europe Air Traffic Management R&D Seminar, Seattle (2017). https://hal-enac.archives-ouvertes.fr/hal-01592235
4. Alonso-Ayuso, A., Escudero, L., Martin-Campo, F.: Collision avoidance in air traffic management: a mixed-integer linear optimization approach. IEEE Trans. Intell. Transp. Syst. **12**(1), 47–57 (2011)
5. Aragon, V.S., Esquivel, S.C.: A evolutionary algorithm to track changes of optimum value locations in dynamic environments. J. Comput. Sci. Technol. **4**(3), 127–134 (2004)
6. Bosman, P.A. N..: Learning and anticipation in online dynamic optimization. In: Yang, S., Ong, Y.-S., Jin, Y. (eds.) Evolutionary Computation in Dynamic and Uncertain Environments, pp. 129–152. Springer, Heidelberg (2007). https://doi.org/10.1007/978-3-540-49774-5_6
7. Degaugue, S., Gotteland, J., Durand, N.: Algorithme évolutionnaire pour la résolution, en continu, de conflits aériens. In: ROADEF (2023)
8. Degaugue, S., Durand, N., Gotteland, J.B.: Impact of explicit memory on dynamic conflict resolution. In: 10th International Conference on Research in Air Transportation (ICRAT 2022), Tampa, p. 53. (2022). https://hal.science/hal-03878000
9. Durand, N., Alliot, J.M.: Genetic crossover operator for partially separable functions. In: GP 1998, 3rd Annual Conference on Genetic Programming, Madison (1998). https://enac.hal.science/hal-00937718
10. Durand, N., Alliot, J.M.: Ant colony optimization for air traffic conflict resolution. In: ATM Seminar 2009, 8th USA/Europe Air Traffic Management Research and Development Seminar, Napa (2009). https://enac.hal.science/hal-01293554
11. Durand, N., Alliot, J.M., Noailles, J.: Automatic aircraft conflict resolution using genetic algorithms. In: Proceedings of the Symposium on Applied Computing, Philadelphia. ACM (1996)
12. Durand, N., Gotteland, J.-B., Matton, N.: Visualizing complexities: the human limits of air traffic control. Cognit. Technol. Work **20**(2), 233–244 (2018). https://doi.org/10.1007/s10111-018-0468-0
13. Goldberg, D.E., Smith, R.E.: Nonstationary function optimization using genetic algorithms with dominance and diploidy. In: ICGA (1987)
14. Goldberg, D.: Genetic algorithms in search. In: Optimization and Machine Learning. Addison Wesley, Reading (1989)
15. Holland, J.: Adaptation in Natural and Artificial Systems. University of Michigan Press (1975)

16. Hu, X., Eberhart, R.: Adaptive particle swarm optimization: detection and response to dynamic systems. In: Proceedings of the 2002 Congress on Evolutionary Computation. CEC 2002 (Cat. No.02TH8600), vol. 2, pp. 1666–1670 (2002). https://doi.org/10.1109/CEC.2002.1004492

17. Krishnakumar, K.: Micro-genetic algorithms for stationary and non-stationary function optimization. In: Rodriguez, G. (ed.) Intelligent Control and Adaptive Systems, vol. 1196, pp. 289–296. International Society for Optics and Photonics, SPIE (1990). https://doi.org/10.1117/12.969927

18. Lehouillier, T., Omer, J., Soumis, F., Desaulniers, G.: A flexible framework for solving the air conflict detection and resolution problem using maximum cliques in a graph (2015)

19. Louis, S., Xu, Z.: Genetic algorithms for open shop scheduling and re-scheduling. In: ISCA 11th International Conference on Computers and their Applications, pp. 99–102 (1996)

20. Ng, K.P., Wong, K.C.: A new diploid scheme and dominance change mechanism for non-stationary function optimization (1995). https://cir.nii.ac.jp/crid/1373949023319151361

21. Rohlfshagen, P., Yao, P.K.L.: Dynamic evolutionary optimisation: an analysis of frequency and magnitude of change. In: Proceedings of the 2009 Genetic and Evolutionary Computation Conference GECCO 2009, pp. 1713–1720 (2009)

22. Pallottino, L., Féron, E., Bicchi, A.: Conflict resolution problems for air traffic management systems solved with mixed integer programming. IEEE Trans. Intell. Transp. Syst. 3(1), 3–11 (2002)

23. Ramsey, C.L., Grefenstette, J.J.: Case-based initialization of genetic algorithms. In: Proceedings of the 5th International Conference on Genetic Algorithms, pp. 84–91. Morgan Kaufmann Publishers Inc., San Francisco (1993)

24. Rey, D., Rapine, C., Fondacci, R., Faouzi, N.E.: Minimization of potential air conflicts through speed regulation. Transp. Res. Record: J. Transp. Res. Board 2300, 59–67 (2012)

25. Richter, H., Yang, S.: Memory based on abstraction for dynamic fitness functions. In: Giacobini, M., et al. (eds.) Applications of Evolutionary Computing, pp. 596–605. Springer, Heidelberg (2008). https://doi.org/10.1007/978-3-540-78761-7_65

26. Trojanowski, K., Michalewicz, Z.: Searching for optima in non-stationary environments. In: Proceedings of the 1999 Congress on Evolutionary Computation-CEC99 (Cat. No. 99TH8406), vol. 3, pp. 1843–1850 (1999). https://doi.org/10.1109/CEC.1999.785498

27. Trojanowski, K., Michalewicz, Z., Xiao, J.: Adding memory to the evolutionary planner/navigator. In: Proceedings of 1997 IEEE International Conference on Evolutionary Computation (ICEC 1997), pp. 483–487 (1997). https://doi.org/10.1109/ICEC.1997.592359

28. Vanaret, C., Gianazza, D., Durand, N., Gotteland, J.B.: Benchmarking conflict resolution algorithms. In: Proceedings of the 5th International Conference on Research in Air Transportation, Berkeley (ICRAT 2012). (2012). https://hal.science/hal-00863090

29. Vela, A., Solak, S., Singhose, W., Clarke, J.: A mixed integer program for flight-level assignment and speed control for conflict resolution. In: Proceedings of the Joint 48th IEEE Conference on Decision and Control and 28th Chinese Control Conference. IEEE (2009)

30. Weicker, K.: An analysis of dynamic severity and population size. In: Parallel Problem Solving from Nature VI (2002)

31. Woldesenbet, Y.G., Yen, G.G.: Dynamic evolutionary algorithm with variable relocation. IEEE Trans. Evol. Comput. **13**(3), 500–513 (2009)
32. Yin, X., Germay, N.: A fast genetic algorithm with sharing scheme using cluster analysis methods in multimodal function optimization. In: Albrecht, R.F., Reeves, C.R., Steele, N.C. (eds.) Artificial Neural Nets and Genetic Algorithms, pp. 450–457. Springer, Vienna (1993). https://doi.org/10.1007/978-3-7091-7533-0_65

GM4OS: An Evolutionary Oversampling Approach for Imbalanced Binary Classification Tasks

Davide Farinati[✉] and Leonardo Vanneschi

NOVA Information Management School (NOVA IMS), Universidade Nova de Lisboa, Campus de Campolide, 1070-312 Lisboa, Portugal
dfarinati@novaims.unl.pt

Abstract. Imbalanced datasets pose a significant and longstanding challenge to machine learning algorithms, particularly in binary classification tasks. Over the past few years, various solutions have emerged, with a substantial focus on the automated generation of synthetic observations for the minority class, a technique known as oversampling. Among the various oversampling approaches, the Synthetic Minority Oversampling Technique (SMOTE) has recently garnered considerable attention as a highly promising method. SMOTE achieves this by generating new observations through the creation of points along the line segment connecting two existing minority class observations. Nevertheless, the performance of SMOTE frequently hinges upon the specific selection of these observation pairs for resampling. This research introduces the Genetic Methods for OverSampling (GM4OS), a novel oversampling technique that addresses this challenge. In GM4OS, individuals are represented as pairs of objects. The first object assumes the form of a GP-like function, operating on vectors, while the second object adopts a GA-like genome structure containing pairs of minority class observations. By co-evolving these two elements, GM4OS conducts a simultaneous search for the most suitable resampling pair and the most effective oversampling function. Experimental results, obtained on ten imbalanced binary classification problems, demonstrate that GM4OS consistently outperforms or yields results that are at least comparable to those achieved through linear regression and linear regression when combined with SMOTE.

Keywords: Oversampling · Imbalanced Data · Binary Classification · Genetic Programming · Genetic Algorithms

1 Introduction

In real-world classification tasks, it is common to encounter datasets where the proportion of labels is not homogeneous among the different classes. This problem is commonly referred to as imbalance. Classification models tend to exhibit higher accuracy when handling observations from the class that is more prevalent, often termed as the majority class, as opposed to the less frequently occurring class, referred to as the minority class. A common strategy is to add new observations, typically created artificially, to the minority class, to improve the balancing of the dataset. The approach is called oversampling [1,2]. One of the most used oversampling approaches is the Synthetic Minority Oversampling Technique (SMOTE) [3]. This oversampling algorithm works by selecting a random pair of neighboring observations, drawing a straight line segment between the two, and randomly sampling a new observation along that segment. One of its notable drawbacks lies in the sensitivity of its performance to the choice of the set of data points used for resampling. The quality and relevance of the selected points play a crucial role in determining the effectiveness of SMOTE in generating synthetic samples that accurately represent the minority class. In this paper, we introduce the Genetic Methods for OverSampling (GM4OS), a novel oversampling method that joins the representation power of two evolutionary algorithms, Genetic Algorithms (GAs) [4] and Genetic Programming (GP) [5], to overcome some of the disadvantages of SMOTE. In GM4OS, individuals are represented as pairs of objects. The first one is a function, represented as a standard GP individual. The other one is a string, as a traditional GA individual. The GA part controls which existing observations from the minority class will belong to the resampling set. The GP part evolves an oversampling function that will combine points that belong to the resampling set, and create the new synthetic points needed to balance the dataset. The fitness of a GM4OS individual is given by the performance of a model trained by a previously chosen target machine learning algorithm on the newly balanced training set. The objective of the evolutionary process is to look for the best resampling set and the most effective oversampling function at the same time.

The paper is organized as follows: in Sect. 2 we revise previous and related work. In Sect. 3 we present GM4OS. Section 4 delves into the employed experimental framework, including the set of parameters, the datasets chosen as test cases, the models utilized as baseline for comparison with GM4OS and the used metrics. Section 5 presents and discusses the obtained experimental results. Finally, Sect. 6 concludes the work and suggests ideas for future research.

2   Literature Review

Imbalanced datasets represent a recurrent challenge in real-world classification tasks. In the literature, both internal and external approaches have been used in an attempt to obviate the impact of imbalanced data on the model's performance. External approaches (data level) rebalance the dataset by either removing observations from the majority class (undersampling) or by adding observations to the minority class (oversampling) [1,2]. In internal approaches (algorithm level), the algorithm is adapted to handle automatically imbalanced observations. This can be done by either assigning a different weight to each class [6] or by using multiple classifiers simultaneously, also referred to as ensemble learning [7]. A combination of both internal and external approaches can be implemented as well. The Synthetic Minority Oversampling

Technique (SMOTE) is an external approach that was designed by Chawla *et al.* in 2002 [3]. SMOTE generates new observations belonging to the minority class as a combination of existing points of the same class. One of the drawbacks of SMOTE is the over generalization of the minority class, leading to a possible overlap between classes [8]. Borderline observations are more important for classification task, being the ones more prone to misclassification. For this reason a variant of SMOTE, named borderline-SMOTE [9], uses as resampling set only the borderline and nearby observations. Another known oversampling technique is the Adaptive Synthetic Sampling Approach (Adasyn) [10]. In Adasyn, minority data samples are generated according to their distribution. To mitigate the learning bias present in the initial dataset, this approach employs a strategy where a greater number of synthetic observations is generated from those minority observations that happen to be more challenging to learn. This is done by assigning weights to the various minority class samples.

The GM4OS method presented in this study is an evolutionary algorithm that joins the representation power of GP and GAs. For this reason, particularly interesting for this literature review are some existing oversampling methodologies grounded in the use of evolutionary algorithms. GP has been previously applied to imbalanced classification tasks. For instance, standard GP and some of its variants have been successfully applied to several classification tasks in [11]. Also, recent research contributions by Pei *et al.* [12] as well as Kumar [13] have introduced novel fitness functions tailored specifically for GP. These innovative fitness functions have been designed to enhance GP's performance when addressing the intricacies of imbalanced classification scenarios. The GenSample algorithm introduced by Karia *et al.* in 2019 [14], implements an oversampling technique based on GAs. GenSample iteratively learns which minority samples are best suited for resampling and the authors reported on promising results compared to a set of state-of-the-art models.

The oversampling function evolved by the GP component of GM4OS combines two vectors (existing observations) to create a single one (a new synthetic observation). This process resembles the vectorial GP approach that was presented by Azzali *et al.* in [15,16]. For instance, similarly to the GP component of a GM4OS individual, vectorial GP allows aggregate functions and vectorial operations in the primitive GP set and vectorial variables as terminals.

3   Genetic Methods for Oversampling

As stated above, the Genetic Methods for OverSampling (GM4OS) is a novel oversampling approach for binary classification problems, based on a combination of GP and GAs representations. GM4OS employs the conventional workflow of evolutionary algorithms. Its peculiarity resides in the representation of the evolving solutions. In GM4OS, in fact, individuals are represented as pairs of objects, one resembling a standard GP individual and the other one a string-like GA individual. The GP part is a function, that can be represented for instance as a tree, that is able to take as input two vectors (existing observations) and combine them to create a new observation (synthetic observation).

The GA part is a string of length $n = size\_majclass - size\_minclass$, where $size\_majclass$ and $size\_minclass$ are, respectively, the number of observations belonging to the majority and minority class in the training set. Each allele of the GA part of the individual contains a pair of observations of the training set belonging to the minority class. Figure 1 provides a visual representation of an arbitrary GM4OS individual.

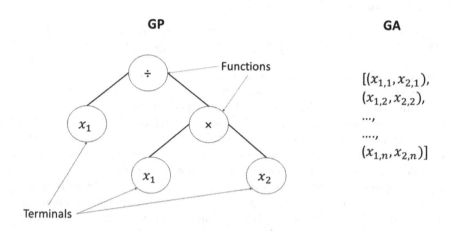

GP   **GA**

Fig. 1. Visual representation of a GM4OS individual. The GP function reported on the left side takes as input two vectors, x_1 and x_2 (existing observations) and combines them to create a new observation, given by its output vector (synthetic observation). The GA genome reported on the right side contains pairs of minority class observations. By giving each one of these pairs as input to the GP part, n synthetic observations can be created.

At the beginning of the evolution, the classification dataset is partitioned into 3 subsets: the training, validation and test sets. Then, by applying the function represented in the GP part to each one of the n pairs of observations of the GA part, n new synthetic observations are produced. The new observations are labeled as minority class and added to the training set, making it balanced.

Example. The GP part of the individual in Fig. 1, shown in the left side of the figure, represents the function:

$$P(x_1, x_2) = x_1/(x_1 \cdot x_2)$$

where x_1 and x_2 are two vectors of the same dimension as the feature space, and the operators / and · represent the element-by-element vectorial division and multiplication, respectively. Now, let us assume, for the sake of simplicity, that $n = 2$ (i.e., the majority class has only two observations more than the minority class). In such a case, the GA part would be represented as a vector of two pairs of observations, chosen randomly from the minority class. In the simplistic hypothesis that the feature space has a dimension equal to three, and using arbitrary numbers as data, let us assume that the GA part is:

$$[([2,\ 1,\ 4],[1,\ 2,\ 0.5]),([1,\ 3,\ 2],[2,\ 1,\ 0.5])]$$

where $[2,\ 1,\ 4],[1,\ 2,\ 0.5],[1,\ 3,\ 2]$ and $[2,\ 1,\ 0.5]$ are existing training observations, labeled as minority class. In such a simple example, GM4OS would create the two following synthetic observations:

- $\mathcal{P}([2,\ 1,\ 4],[1,\ 2,\ 0.5]) = [2/(2\cdot1),\ 1/(1\cdot2),\ 4/(4\cdot0.5)] = [1,\ 0.5,\ 2]$
- $\mathcal{P}([1,\ 3,\ 2],[2,\ 1,\ 0.5]) = [1/(1\cdot2),\ 3/(3\cdot1),\ 2/(2\cdot0.5)] = [0.5,\ 1,\ 2]$

These two newly created observations would now be inserted in the training set and labeled as minority class. In this way, the training set is now balanced.

At this point, a classification model is fitted on the newly balanced training set, and it is then used to make predictions on the validation set. The loss between expected and calculated outputs on the validation set is finally used as fitness for the GM4OS individual. In this work, the classification model chosen is Logistic Regression [17], given its simplicity and training efficiency. Figure 2 shows a flowchart representing the fitness evaluation process of a GM4OS individual.

During the evolution, distinct genetic operators are applied independently to each one of the two parts of a GM4OS individual: when crossover or mutation need to be applied to a GM4OS individual (according to the probabilities presented in Sect. 4), the same operator type (mutation or crossover) is applied simultaneously to both its GP and GA parts. The genetic operators used in this work for the GP and GA parts are also specified in Sect. 4. Traditionally, GA individuals are represented as vectors of scalar values. However, the fact that in GM4OS the GA part is a vector of pairs (the minority class observations that will be used by the GP part) does not represent an issue: it is still possible to use standard GA operators, treating each pair as a single and indivisible piece of information. So, for instance, one-point crossover will exchange substrings of pairs between parents, while one-point mutation will replace an existing pair with a new pair of minority class observations, generated at random. Note that when one-point crossover is applied the crossover point is restricted to fall in the boundaries between observations within a pair.

4   Experimental Settings and Test Problems

Table 1 reports all the GM4OS parameters employed in this experimental study. To assess the performance of GM4OS, we have employed two baseline methods for comparative evaluation. The first one is a simple Logistic Regression (denoted as LR in the continuation) [17,18] fitted on the imbalanced training dataset. The second one is still a Logistic Regression, but this time fitted on the training dataset re-balanced using the traditional SMOTE algorithm (denoted as SMOTE+LR in the continuation) [3,19]. Table 2 presents the binary classification datasets used as test problems in our experimental study. The imbalance

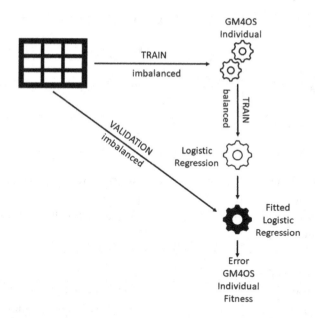

Fig. 2. Flowchart of the evaluation of a GM4OS individual. The initial dataset is split into training, test and validation. Then a GM4OS individual is used to produce enough new synthetic observations to balance the training set. A model, Logistic Regression in our experiments, is fitted on the balanced training set. Then, the fitted model is used to make predictions on the validation set. Finally, the loss between predictions and true labels on the validation set is used as fitness for the GM4OS individual.

ratio is calculated as the ratio between the number of observations of the majority class and the number of observations of the minority class. In this way, the imbalance ratio is, by definition, a positive number and, given that we are using imbalanced datasets, it is strictly larger than one in our test problems. All these datasets belong to the Penn Machine Learning Benchmarks (PMLB) library [20]. In general, several different performance measures that can be used to evaluate classification algorithms; among others, one may mention for instance precision, recall, accuracy, and F1-score. The choice of which metric to prioritize depends largely on the specific problem and its inherent characteristics. In our case, since we are addressing imbalanced classification problems, the F1-score of the minority class emerges as a particularly pertinent fitness metric for GM4OS. In fact, the F1-score provides a balanced measure of both precision and recall. So, it is well-suited for situations where the imbalanced distribution of classes demands a focus on the accurate identification of the minority class instances, striking a balance between minimizing false positives and false negatives [21].

Table 1. GM4OS parameters.

Parameter	Value
Population size	50
Generations	50
Mutation probability	0.2
Crossover probability	0.8
Elitism	True
Elite size	1
Selection algorithm	Tournament
Tournament size	2
GP initialization	Ramped-half-half
GA initialization	Random
GP constant set	$\{-1, 2, 3, 4, 5\}$
GP constant probability	0.3
GP function set	{add, sub, mul, div, mean}
GP crossover	Subtree single point swap crossover
GP mutation	Single node mutation
GA crossover	One point crossover
GA mutation	One point mutation
Maximum GP tree depth for initialization	8
Maximum GP tree depth for evolution	8

Table 2. Specifications of the datasets used for model comparison.

Dataset	Number of observations	Number of features	Imbalance ratio
flare	1066	10	4.8
haberman	306	3	2.7
spect	267	22	3.8
spectf	349	44	2.7
ionosphere	351	34	1.8
hungarian	294	13	1.77
diabetes	768	8	1.8
hepatitis	155	19	3.84
appendicitis	106	7	4.04
analcatdata	264	4	21.8

5   Experimental Results

All the results reported in this section are medians over 30 independent runs. At each run, the datasets are split randomly with uniform distribution into training, validation and test partitions, composed of 60%, 20% and 20% of the data observations, respectively. The proportion between majority and minority class observations is kept constant across the different splits. The same training/validation/test partition has been used for all the studied methods at each particular run. For LR and SMOTE+LR, the training and validation sets were joined and both were used as training set. The obtained experimental results are reported in Fig. 3. More particularly, Plot 3a (3b, 3c, 3d, 3e, 3f, 3g, 3h, 3i and 3j, respectively) reports the results for the flare (haberman, spect, spectf, ionosphere, hungarian, diabetes, hepatitis, appendicitis and analcatdata, respectively) test case. Each one of these plots represents a comparison between GM4OS, SMOTE+LR and LR. The comparison is done using box-plots of the F1-score of the minority class on the 30 different test sets. Table 3 reports the p-values of the Mann-Whitney U test for pairwise comparison of the methods, for five different metrics. The p-values are in bold when they indicate a statistically significant difference, using a significance level $\alpha = 0.05$, with the Bonferroni correction [22]. Also, the presence of the symbol ^{g} indicates that GM4OS outperforms the baseline, while the symbol ^{b} indicates the opposite. The last two rows of the table are a summary of the results for each measure(column). The first number, always positive, it indicates how many times GM4OS significantly outperforms LR+SMOTE or LR. While the second number, always negative, it indicates how many times GM4OS is significantly outperformed by LR+SMOTE or LR. For GM4OS, in both Fig. 3 and Table 3, the performance of the best individual at the last generation is used. Observing the F1-score metric for the minority class, we can notice that GM4OS is never significantly outperformed by the baselines. This outcome is corroborated by the box-plots presented in Fig. 3 and the p-values of the third column of Table 3. It is also possible to see that in four test cases out of ten GM4OS significantly outperforms LR, and in three test cases out of ten GM4OS significantly outperforms SMOTE+LR. The F1-score is the measure that we have used as fitness to guide the evolution of GM4OS. However, it is also interesting to show how GM4OS performs in terms of other metrics. Looking at the p-values of the accuracy, shown in Table 3, it is possible to observe that LR significantly outperforms GM4OS in four test cases out of ten, while having comparable performance in the remaining six cases. This is probably due to the fact that accuracy is a measure that can be misleading when working with imbalanced datasets. A model that predicts the majority class for every instance can be a very poor quality model, but still achieve a high accuracy when data is imbalanced. For the other studied metrics, it is difficult to identify a specific pattern. Depending on the test case, GM4OS can significantly outperform, be significantly outperformed or have comparable performance to the baseline models, as shown by the p-values presented in Table 3. Finally, Fig. 4 presents the evolution of the test fitness of the best individual of GM4OS along generations, compared to the one of the two baseline algorithms, namely LR and

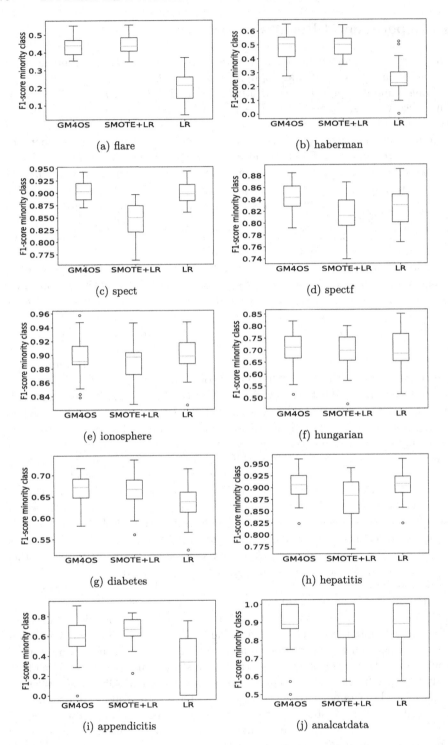

Fig. 3. Box-plots of the F1-score of the minority class of GM4OS against the baselines for all the datasets.

Table 3. p-values of the Mann-Whitney U test [23] for different evaluation metrics of GM4OS against the baselines. In bold when the p-value indicates a statistically significant difference. Symbol ^{g} indicates that GM4OS outperforms the baseline, while symbol ^{b} indicates the opposite. The last two rows are a summary of the results for each measure(column). The first number, always positive, it indicates how many times GM4OS significantly outperforms LR+SMOTE or LR. While the second number, always negative, it indicates how many times GM4OS is significantly outperformed by LR+SMOTE or LR.

Problem	Baseline	F1-score minority class	recall minority class	precision minority class	recall majority class	precision majority class	accuracy
flare	LR+ SMOTE	0.695	0.21	0.137	**0.002**^{b}	0.663	0.041
	LR	**3.29e − 11**^{g}	**4.65e − 6**^{g}	**2.88e − 04**^{b}	**1.55e − 8**^{b}	**4.05e − 5**^{g}	**5.99e − 7**^{b}
haberman	LR+ SMOTE	0.888	0.243	0.025	**0.005**^{b}	0.739	0.023
	LR	**8e − 9**^{g}	**1.11e − 5**^{g}	**0.001**^{b}	**5.62e − 08**^{b}	**8.14e − 4**^{g}	**1.84e − 4**^{b}
spect	LR+ SMOTE	**1.61e − 8**^{g}	**5.72e − 18**^{b}	**4.77e − 7**^{g}	**2.97e − 18**^{b}	**4.89e − 12**^{b}	**1.67e − 4**^{g}
	LR	0.437	**1.24e − 12**^{b}	**1.46e − 4**^{g}	**5.01e − 12**^{g}	**8.89e − 8**^{b}	**0.011**^{b}
spectf	LR+ SMOTE	**8.99e − 4**^{g}	**9.88e − 9**^{b}	**0.015**^{g}	**1.9e − 9**^{g}	**4.03**^{b}	0.222
	LR	0.45	**6.52e − 7**^{b}	**0.015**^{g}	**1.57e − 8**^{g}	**2.93e − 05**^{b}	0.128
ionosphere	LR+ SMOTE	0.662	**0.002**^{b}	**2.63e − 6**^{g}	**3.4e − 7**^{g}	**0.006**^{b}	0.599
	LR	0.558	**0.006**^{b}	**1.79e − 4**^{g}	**3.22e − 5**^{g}	**0.1**^{b}	0.208
hungarian	LR+ SMOTE	0.721	0.92	0.894	0.815	0.947	0.699
	LR	0.693	0.811	0.524	0.55	0.781	0.234
diabetes	LR+ SMOTE	0.673	0.191	**0.008**^{b}	0.192	**0.008**^{b}	0.15
	LR	**0.001**^{g}	**2.63e − 7**^{b}	0.041	**2.63e − 7**^{b}	**0.041**^{b}	**1.16e − 5**^{b}
hepatitis	LR+ SMOTE	**0.012**^{g}	**1.47e − 5**^{b}	**9.22e − 4**^{b}	**1.1e − 6**^{g}	**3.72e − 4**^{b}	**0.016**^{g}
	LR	0.898	**0.001**^{b}	**0.015**^{b}	**1.5e − 4**^{g}	**0.008**^{b}	0.092
appendicitis	LR+ SMOTE	0.039	**0.01**^{b}	0.344	0.036	0.022	0.634
	LR	**1.11e − 4**^{g}	0.2	0.838	0.112	0.313	0.586
analcatdata	LR+ SMOTE	0.895	0.322	0.443	0.402	0.34	0.856
	LR	0.315	0.787	0.676	0.779	0.767	0.99
summary	LR+ SMOTE	+3, 0	+2, −5	+3, −2	+4, −2	0, −5	+2, 0
	LR	+4, 0	0, −5	+3, −3	+4, −3	+2, −5	0, −4

SMOTE+LR, which are presented as horizontal straight lines in the figure. More specifically, Plot 4a (4b, 4c, 4d, 4e, 4f, 4g, 4h, 4i and 4j, respectively) reports the results for the flare (haberman, spect, spectf, ionosphere, hungarian, diabetes, hepatitis, appendicitis and analcatdata, respectively) test case.

From these plots, it is possible to observe that GM4OS outperforms LR in the entire evolution process in six out of ten total test cases. Similarly GM4OS

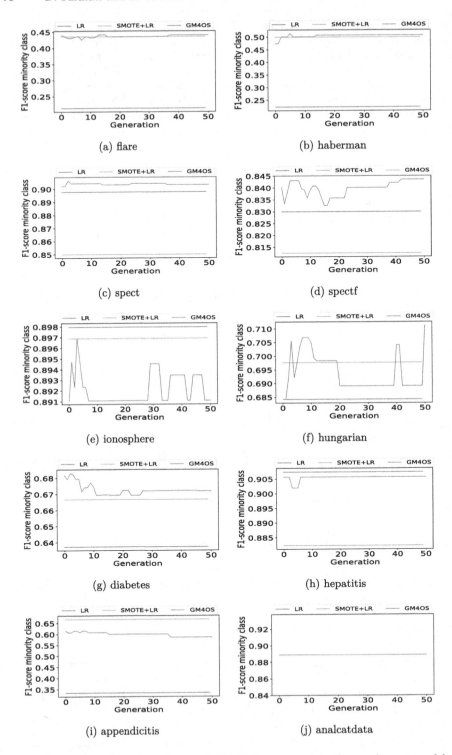

Fig. 4. Evolution of the test fitness of GM4OS, compared to the results returned by LR and SMOTE+LR, over the studied datasets.

outperforms SMOTE+LR over all the generations in four out of ten total test cases. In the case of Plot 4j, reporting results on the analcatdata test problem, all three models have the same performance, around 0.89, and GM4OS has it starting from the beginning of the evolution (and this is why the figure looks like a unique horizontal line). This result is coherent with the box-plot shown in Fig. 3j, where the median of all the models is around 0.89 as well.

6   Conclusions and Future Work

This paper introduced the Genetic Methods for OverSampling (GM4OS), an evolutionary oversampling approach for imbalanced binary classification problems. In real-world binary classification tasks, imbalancing between the classes represents a recurrent issue. In those cases, in fact, classification models typically struggle to classify correctly the observations belonging to the minority class. A popular approach to tackle this issue is to add synthetic observations to the minority class, commonly referred to as oversampling. One of the most known oversampling approaches is the Synthetic Minority Oversampling Technique (SMOTE) [3]. SMOTE creates new observations by selecting two existing minority class observations, and sampling a new synthetic point from the straight line segment that connects them. However, this approach relies heavily on the set of points that are used for resampling. GM4OS integrates the representation power of Genetic Algorithms (GAs) and Genetic Programming (GP), to look for the most appropriate resampling set and resampling function at the same time. GM4OS individuals, in fact, are represented as pairs of objects, one of which is a GP-like function, while the other one is a GA-like string. The GP part strongly resembles the recently introduced vectorial GP approach [15,16] and combines two vectors (existing observations) to generate a new single vector (synthetic observation), while the GA part controls which existing minority class observations will constitute the resampling set. GM4OS was experimentally compared with a simple Logistic Regression (LR) [17] and SMOTE combined with LR, on ten imbalanced binary classification test problems, taken from the Penn Machine Learning Benchmarks library [20]. The experimental results show that, on all the studied test problems, GM4OS is able to find models that have an F1-score on the minority class that is better, or at least comparable, to the baseline models.

Despite the positive experimental outcomes achieved, a significant scope remains open to future research. One such avenue involves the adaptation of the GM4OS framework to address multi-class classification problems. Another area of investigation pertains to the exploration of alternative fitness metrics for GM4OS, distinct from the F1-score utilized in this study. Specifically, the pursuit of metrics that have demonstrated efficacy for GP applied to imbalanced classification tasks, as evidenced for instance in [12,13], holds significant promise. Multi-objective optimization, using several fitness functions at the same time, also deserves investigation. An additional idea for future research is inspired by the observation that the GP component within GM4OS is susceptible to the issue of bloat, a common occurrence in GP. While the present work implemented a strategy to restrict tree depth to a maximum of 8, a myriad of other strategies have been advanced to mitigate bloat. Notably, existing references suggest that the introduction of a dynamic GP population size effectively alleviates bloat, thereby reducing computational overhead while maintaining excellent performance levels, as documented for instance in [24–26]. In light of these findings, the incorporation of dynamic population sizing into GM4OS presents itself as an intriguing avenue for future development. Furthermore, as a forthcoming research endeavor, it is imperative to acknowledge that the present study employed a Logistic Regression model to evaluate the oversampling function/resampling set pair, due to its computational efficiency [17]. Nevertheless, the framework remains adaptable to experimentation with alternative models, such as for instance Decision Tree classifiers [27] or many others. In this study, we conducted a comparative analysis to assess the performance of GM4OS against an alternative oversampling technique, namely SMOTE (Synthetic Minority Over-sampling Technique). Our investigation aimed to elucidate the effectiveness of these oversampling approaches in addressing the challenge of class imbalance within classification tasks. Additionally, our future research endeavors will encompass further comparisons with alternative oversampling methodologies, including Adasyn [10] and borderline-SMOTE [9]. Lastly, an interesting prospect for future research entails the extension of GM4OS to encompass synthetic data generation. This extension can be realized through the modification of the resampling set governed by the GA component of GM4OS, extending its influence to the entire dataset.

Acknowledgments. This work was supported by national funds through FCT (Fundação para a Ciência e a Tecnologia), under the project - UIDB/04152/2020 - Centro de Investigação em Gestão de Informação (MagIC)/NOVA IMS.

References

1. Batista, G.E.A.P.A., Prati, R.C., Monard, M.C.: A study of the behavior of several methods for balancing machine learning training data. SIGKDD Explor. **6**, 20–29 (2004)
2. Chawla, N., Japkowicz, N., Kołcz, A.: Editorial: special issue on learning from imbalanced data sets. SIGKDD Explor. **6** , 1–6 (2004). https://doi.org/10.1145/1007730.1007733
3. Chawla, N.V., Bowyer, K.W., Hall, L.O., Kegelmeyer, W.P.: Smote: synthetic minority over-sampling technique. J. Artif. Int. Res. **16**(1), 321–357 (2002)
4. Mitchell, M.: An Introduction to Genetic Algorithms. MIT Press (1996)
5. Poli, R., Langdon, W.B., McPhee, N.F.: A Field Guide to Genetic Programming. Lulu Enterprises, UK Ltd (2008)
6. Ali, A., Shamsuddin, S.M., Ralescu, A.: Classification with class imbalance problem: a review **7**, 176–204 (2015)
7. Huang, J., Ling, C.: Using AUC and accuracy in evaluating learning algorithms. IEEE Trans. Knowl. Data Eng. **17**(3), 299–310 (2005). https://doi.org/10.1109/TKDE.2005.50

8. Gosain, A., Sardana, S.: Handling class imbalance problem using oversampling techniques: a review. In: 2017 International Conference on Advances in Computing, Communications and Informatics (ICACCI), pp. 79–85 (2017). https://doi.org/10.1109/ICACCI.2017.8125820

9. Han, H., Wang, W.-Y., Mao, B.-H.: Borderline-smote: a new over-sampling method in imbalanced data sets learning. In: Proceedings of the 2005 International Conference on Advances in Intelligent Computing - Volume Part I (ICIC 2005), Springer, Heidelberg (2005), pp. 878–887. https://doi.org/10.1007/11538059_91

10. He, H., Bai, Y., Garcia, E.A., Li, S.: Adasyn: adaptive synthetic sampling approach for imbalanced learning. In: 2008 IEEE International Joint Conference on Neural Networks (IEEE World Congress on Computational Intelligence), pp. 1322–1328 (2008). https://api.semanticscholar.org/CorpusID:1438164

11. Frank, F., Bacao, F.: Advanced genetic programming vs. state-of-the-art automl in imbalanced binary classification. Emerg. Sci. J. 7(4), 1349–1363 (2023). https://doi.org/10.28991/ESJ-2023-07-04-021

12. Pei, W., Xue, B., Shang, L., Zhang, M.: New fitness functions in genetic programming for classification with high-dimensional unbalanced data. In: 2019 IEEE Congress on Evolutionary Computation (CEC), pp. 2779–2786 (2019). https://doi.org/10.1109/CEC.2019.8789974

13. Kumar, A.: A new fitness function in genetic programming for classification of imbalanced data. J. Exp. Theor. Artif. Intell. 1–13 (2022). https://doi.org/10.1080/0952813X.2022.2120087

14. Karia, V., Zhang, W., Naeim, A., Ramezani, R., Gensample: a genetic algorithm for oversampling in imbalanced datasets. arXiv preprint arXiv:1910.10806 (2019)

15. Azzali, I., Vanneschi, L., Silva, S., Bakurov, I., Giacobini, M.: A vectorial approach to genetic programming. In: Sekanina, L., Hu, T., Lourenço, N., Richter, H., García-Sánchez, P. (eds.) Genetic Programming: 22nd European Conference, EuroGP 2019, Held as Part of EvoStar 2019, Leipzig 24–26 April 2019, Proceedings, pp. 213–227. Springer, Cham (2019). https://doi.org/10.1007/978-3-030-16670-0_14

16. Azzali, I., Vanneschi, L., Bakurov, I., Silva, S., Ivaldi, M., Giacobini, M.: Towards the use of vector based GP to predict physiological time series. Appl. Soft Comput. 89, 106097 (2020). https://doi.org/10.1016/j.asoc.2020.106097

17. Cox, D.R.: The regression analysis of binary sequences. J. Roy. Stat. Soc.: Ser. B (Methodol.) 20(2), 215–232 (1958)

18. Pedregosa, F., et al.: Scikit-learn: machine learning in Python. J. Mach. Learn. Res. 12, 2825–2830 (2011)

19. Lemaître, G., Nogueira, F., Aridas, C. K.: Imbalanced-learn: a python toolbox to tackle the curse of imbalanced datasets in machine learning. J. Mach. Learn. Res. 18(17), 1–5 (2017)

20. Romano, J.D., et al.: Pmlb v1.0: an open source dataset collection for benchmarking machine learning methods. arXiv preprint arXiv:2012.00058v2 (2021)

21. Ferrer, L.: Analysis and comparison of classification metrics. arXiv preprint arXiv:2209.05355 (2023)

22. Bonferroni, C.: Teoria statistica delle classi e calcolo delle probabilità, Pubblicazioni del R. Istituto superiore di scienze economiche e commerciali di Firenze, Seeber (1936)

23. Mann, H.B., Whitney, D.R.: On a test of whether one of two random variables is stochastically larger than the other. Annal. Math. Statist. 18(1), 50–60 (1947). https://doi.org/10.1214/aoms/1177730491

24. Fernandez, F., Vanneschi, L., Tomassini, M.: The effect of plagues in genetic programming: a study of variable-size populations. In: Ryan, C., Soule, T., Keijzer, M., Tsang, E., Poli, R., Costa, E. (eds.) Genetic Programming, pp. 317–326. Springer, Heidelberg (2003). https://doi.org/10.1007/3-540-36599-0_29

25. Rochat, D., Tomassini, M., Vanneschi, L.: Dynamic size populations in distributed genetic programming. In: Keijzer, M., Tettamanzi, A., Collet, P., van Hemert, J., Tomassini, M. (eds.) Genetic Programming: 8th European Conference, EuroGP 2005, pp. 50–61. Springer, Heidelberg (2005). https://doi.org/10.1007/978-3-540-31989-4_5

26. Farinati, D., Bakurov, I., Vanneschi, L.: A study of dynamic populations in geometric semantic genetic programming. Inf. Sci. **648**, 119513 (2023). https://doi.org/10.1016/j.ins.2023.119513

27. Breiman, L., Friedman, J.H., Olshen, R.A., Stone, C.J.: Classification and regression trees. Biometrics **40**, 874 (1984). https://api.semanticscholar.org/CorpusID:29458883

Evolving Staff Training Schedules Using an Extensible Fitness Function and a Domain Specific Language

Neil Urquhart[✉] and Kelly Hunter

School of Computing, Edinburgh Napier University, Edinburgh, UK
{n.urquhart,k.hunter}@napier.ac.uk

Abstract. When using a meta-heuristic based optimiser in some industrial scenarios, there may be a need to amend the objective function as time progresses to encompass constraints that did not exist during the development phase of the software. We propose a means by which a Domain Specific Language (DSL) can be used to allow constraints to be expressed in language familiar to a domain expert, allowing additional constraints to be added to the objective function without the need to recompile the solver. To illustrate the approach, we consider the construction of staff training schedules within an organisation where staff are already managed within highly constrained schedules. A set of constraints are hard-coded into the objective function in a conventional manner as part of a Java application. A custom built domain specific language (named Basil) was developed by the authors which is used to specify additional constraints affecting individual members of staff or groups. We demonstrate the use of Basil and show how it allows the specification of additional constraints that enable the software to meet the requirements of the user without any technical knowledge.

1 Introduction and Motivation

Evolutionary Algorithms and related meta-heuristics have been developed to solve a range of real-world problems. However, it is inevitable that an organisation's needs change over time: as a result, the constraints within the original system may no longer meet their needs. While the users of a system may have problem domain knowledge, they may not have any software engineering experience. Updating constraints within the system may pose difficulties, requiring a software-engineering specialist to alter the system. This paper proposes a mechanism by which a user with knowledge of the problem domain can add additional constraints to the objective function, in a manner that does not require specialist software engineering skills, using a custom built domain-specific language (DSL) designed and implemented by the authors.

We consider the specific case of an industrial partner within the public transportation sector, who has limited software development expertise. The partner has a requirement to provide staff training schedules within a heavily constrained

© The Author(s), under exclusive license to Springer Nature Switzerland AG 2024
S. Smith et al. (Eds.): EvoApplications 2024, LNCS 14634, pp. 83–97, 2024.
https://doi.org/10.1007/978-3-031-56852-7_6

environment. Staff must be to be allocated to training slots in a manner that causes least disruption to the existing schedules. As with many problems that involve the scheduling of people, many constraints exist based around the specific requirements of individuals that are not necessarily known in the original design phase.

The contribution of this paper is to address the following research questions:

1. To what extent can constraints be expressed in a custom built DSL a manner that is achievable by a domain expert?
2. What mechanisms could be used to evaluate constraints specified using a DSL against candidate solutions at run time?
3. How does the evaluation time scale as more constraints are added at run-time?

The principle contribution of this paper is the development of the DSL and the use of pattern matching to evaluate constraints.

This paper is organised as follows, Sect. 2 describes related work in the field of optimisation. The problem domain is described in Sect. 3 and the evolutionary algorithm used to produce solutions is described in Sect. 4. The development of a domain specific language (DSL) specifically for this application is described in Sect. 5 along with the mechanism by which Basil statements are compiled into regular expressions which are then matched against the solution being evaluated. Finally, conclusions and future work are described in Sect. 6.1.

2   Related Work

Evolutionary Algorithms and other meta-heuristics have been applied to problems related to a number of industrial sectors including timetabling [1,11], staff scheduling, [2,3,10], vehicle routing and logistics [4,6] and job shop/factory scheduling [5]. The domain of staff scheduling, and in particular nurse scheduling has received a great deal of attention, for a recent survey of this domain the reader is directed to [7]. Whilst some industrial scheduling problems map closely to traditional benchmark problem types (e.g., the Travelling Salesman Problem, Vehicle Routing Problem or Flow Shop Problems) many incorporate constraints that are specific to the organisation who own the problem.

There exists the issue of how to specify these organisation specific constraints in a manner that is suitable for organisations who do not have the capability to modify the underlying software. One option is the use of a domain specific language (DSL) to specify constraints. Regenell and Kuchcinski [8] describe the use of an embedded DSL for combinatorial optimisation. The approach taken is based on the Scala platform, the resultant DSL making use of the Scala syntax. Whilst this approach has much to recommend it, not least the ability to integrate the DSL compilation with that of the main Scala application. Constraints can also be specified in a constraint modelling language such as MiniZinc [12].

The DSL-based approaches outlined above have the disadvantages that the DSLs are difficult to use by domain experts who are not software engineers. In this work, the DSL presented (Basil) is designed specifically around entities within the problem domain in order to make it usable by problem domain experts within an organisation that does not have software development expertise.

3   Problem Domain

3.1   Problem Definition

A major Scottish public transport provider employs over 2,000 drivers. It is a requirement under current UK/EU legislation that professional drivers must undertake mandatory Certificate of Professional Competence (CPC) training [9]. Within a five-year period each driver must undertake 35 h of training: failure to complete the required amount of training results in the drivers' license expiring, losing their right to drive on a commercial basis.

An existing proprietary software package is used to schedule drivers to their routine duties, but this system does not schedule the time required for CPC training. The policy of the organisation is that each driver is allocated one CPC training day per year - this ensures that they will have undertaken the required 35 h within the five-year period. Each driver has a specific license expiry date based on when they completed their initial training which specifies the deadline by which their CPC training must have been completed within the fifth year.

A total of 2014 drivers employed are split into groups, representing the area of the organisation that they work for (see Table 1), each driver must have one CPC training day per year. Each training day can accommodate 12 trainees, training takes place for 40 weeks per year for 5 days per week, creating 2400 training places. Assuming that each member of staff attends on the day that they are scheduled then there is a 16% spare capacity. In practice this capacity is required to cover situations such as non-attendance due to illness or where the staff member cannot be released for training due operational requirements.

Table 1. The employee group sizes with the problem being considered. The maximum number of employees which may be allocated to training from each group on the same day is shown.

Group Size	50	24	750	400	450	140	140	60
Max. Trainees per Day	1	1	5	3	3	1	1	1

There are a number of constraints that govern the CPC training schedule:

1. Any driver whose license expires in the current year must have their training day, for that year, prior to their license expiring.
2. Each driver may only attend one CPC training day each year.
3. Each training day can only accommodate 12 trainees.
4. The number of trainees on each day from a specific group must not exceed the limit set for that group (see Table 1).

5. In their normal duties, each week drivers are allocated to duties that are classified as either 'early' or 'late', if possible, CPC training days should be scheduled for drivers when they are already allocated to early duties, this makes it easier to release them for the training.

3.2 Problem Instances

The problem instances used in this paper are generated randomly, based on statistics supplied by the partner. This avoids having to share commercially sensitive data during the development stage.

Table 2. The parameters used when generating the test instances.

Parameter	Value
Class_Size	12
Early_Shift_Probability	0.5
Probability_license_expires	0.2
Training_Weeks	40
Training_Days_Week	5

4 Solving Using an Evolutionary Algorithm

4.1 Algorithm Description

The algorithm used within this paper is named *CPC-EA* and is described in Algorithm 1, the parameters used are given in Table 3. A steady state population is employed: within the generational loop (Lines 4–3) one new *child* solution is created by either recombination of two parents (Lines 6–8) or by cloning a single parent (Line 10). The child then replaces the loser of a tournament (Line 15) providing the child fitness is an improvement on the loser (Lines 16–8).

Most academic use of EAs described in studies execute the EA for a fixed number of evaluations (referred to as an *evaluation budget*). In this application our aim is to produce a usable schedule for the business, so there is no requirement to limit the evaluations to a specific time frame, but instead the algorithm can execute until it cannot find any more improvements to the solution. A parameter *MAX_EVALS* is used to specify the maximum number of evaluations that will be carried out in order to prevent excessively long execution times.

Algorithm 1. CPC-EA

```
 1: pop = initialise(POP_SIZE)
 2: bestSol = findBest(pop)
 3: evalLeft = TIMEOUT
 4: while evalsLeft > 0 do
 5:     if random() < XOVERRATE then
 6:         p1 = tournament(pop, TOURSIZE)
 7:         p2 = tournament(pop, TOURSIZE)
 8:         child = recombine(p1, p2)
 9:     else if random() >= XOVERRATE then
10:         child = tournament(pop, TOURSIZE)
11:     mutate(child)
12:     evalute(child)
13:     evals + +
14:     evalsLeft − −
15:     rip = tournamentLoose(pop, TOURSIZE)
16:     if child.fitness < rip.fitness then
17:         pop.remove(rip)
18:         pop.add(child)
19:         if child.fitness < bestSol.fitness then
20:             bestSol = child
21:             evalLefts = TIMEOUT
22:     if evals > MAX_EVALS then
23:         return
```

Table 3. Parameters used within CPC-EA in this paper.

Parameter	Value
POP_SIZE	100
TIME_OUT	25,000
XOVER_RATE	0.5
TOUR_SIZE	2
MAX_EVALS	250,000

Representation. Each solution comprises a list of training slots, the total number of slots being calculated as $Class\_Size \times Training\_Days\_Week \times Training\_Weeks$ (see Table 2), for the instances under consideration this equates to 2400 training slots. Each slot may be empty or have a driver allocated to it. All the drivers in the problem must be allocated to a slot for a valid solution to exist (in our definition a valid solution is one in which all drivers are allocated a training day regardless of any other constraints). Table 4 gives an example of the representation used.

Table 4. A truncated example of a solution, one entry exists for each driver which has a training slot associated with it (labelled as <week>.<day>.<no>). As there are more slots than drivers, a list of unused slots is also maintained.

Driver	Training Slot
00001	2.1.1
00002	40.4.4
00003	34.3.5
00004	21.2.1
00005	16.1.12
00006	14.5.5
00007	21.2.3
00008	22.3.4
00009	35.2.2
00010	35.4.7
00011	2.1.2
00012	9.3.12
...	...

Unused Slots
17.1.1
40.4.12
8.5.11
16.2.8
21.5.4
24.5.12
33.1.11
35.3.1
26.2.12

Initialisation. When initialising the population, each solution has a random unused training slot allocated to each driver. Algorithm 2 illustrates the means by which each individual is initialised. The MAX_TRIES variable was set to 150 in order to ensure that a reasonable proportion of training slots were potentially tried for each driver. For drivers whose license is due to expire (Line 6), the algorithm is biased towards finding a slot that allows training before their expiry date (Line 7). For the remaining drivers (Line 11) the selection is biased towards finding a training slot that coincides with an early shift pattern.

Operators. When a new *child* solution is created, initially no training slots are allocated. Each driver d is then considered in turn, a parent is selected at random, and an attempt is made to allocate the slot associated with d in that parent. If the slot cannot be used (as it has already been allocated within the child) then the slot associated with d on the other parent is tried. If a slot from neither parent can be used, the d is allocated a slot chosen at random from the list of unused slots.

Two mutation operators are used:

– Select two drivers d_1 and d_2 at random from within the chromosome, swap the training slots allocated to d_1 and d_2.
– Select a driver d at random, move the training slot allocated to d to unused list. Select a slot from the unused list at random and allocate that slot to d.

Algorithm 2. Initialise Individual

1: $d = 0$
2: **while** d < drivers.length **do**
3:     $tries = 0$
4:     **while** $tries < MAX\_TRIES$ **do**
5:         $slot = getRandFreeSlot()$
6:         **if** $driver.expiresCurrentYear$ **then**
7:             **if** $slot.week < driver.expiry()$ **then**
8:                 $driver.trainingSlot = s$
9:                 $freeSlots.remove(s)$
10:                $tries = MAX\_TRIES$
11:         **else**
12:             **if** $driver.getShift(slot.week) == early$ **then**
13:                 $driver.trainingSlot = s$
14:                 $freeSlots.remove(s)$
15:                 $tries = MAX\_TRIES$

Objective Function. A penalty-based fitness function is used. The penalty weights used can be found in Table 5 (these values were determined by empirical investigation). The fitness function may be divided into two sub functions: *fitness-base* and *fitness-dsl* (see Sect. 5).

Fitness-base is hard-coded at design time and incorporates those constraints identified during the analysis and development stage undertaken with the partner. The base fitness function examines the solution for violations of constraints 1–3 (see Table 5). *Fitness-dsl* is used to allow the end-user to specify additional constraints using a custom DSL (see Sect. 5).

Table 5. The penalties used within the fitness function. Note that constraints 1–3 are evaluated by the base fitness function and 4–6 are evaluated by the extended fitness function using. Each Basil statement is compiled into a custom constraint.

	Constraint	Penalty
1	Final training day after license has expired	15
2	Unbalanced training group (see Table 1)	5
3	Training scheduled during late shift	5
4	Custom Constraint (low priority)	1
5	Custom Constraint (medium priority)	5
6	Custom Constraint (high priority)	10

4.2   Initial Results

Table 6 shows the results obtained with *CPC-EA* on the five test instances. In each case the *CPC-EA* was run 10 times and the best result shown (best being defined as lowest fitness). Note that in these instances no extended fitness function was specified.

A small number of training slots violated the late shift constraint (Table 5 Item 3). Examination of the solutions suggested that in most cases this occurred as the driver affected had a license that expired early in the year and so the training had to take place within the first few weeks, even if that meant violating the late shift constraint. Figure 2 shows the total number of late shift violations by week. Note that the violations all occur before week 12 and that 60% of the violations occur in the first two weeks. It should also be noted that every one of the drivers whose training week clashed with a late shift had a license due to expire in the current year.

As we are carrying out EA runs that do not have a fixed number of evaluations, we should examine the relationship between performance (fitness) and the number of evaluations used. Figure 1 plots the fitness and total evaluations for all 40 initial runs of *CPC-EA*. The results of the Pearson coefficient suggest that there is a significant small relationship, examination of the plot shows that there are a number of runs where a smaller number of evaluations has been accompanied by a low fitness, thus justifying the practice of executing the algorithm 10 times and selecting the best result achieved.

Table 6. The initial results obtained when using *fitness-base* only on the test instances. Results shown are based on the best of 10 runs - the average being shown in parenthesis.

Data Set	Fitness	Constraint Violations		
		Expired License	Imbalanced Groups	Late Shifts
801	60 (83.5)	0	0	6 (8.1)
480	50(71.5)	0	0	10(14.3)
665	40 (58.5)	0	0	8 (11.7)
135	30 (40.5)	0	0	6 (8.1)

5   Extending the Fitness Function

5.1   Introduction

When using an Evolutionary Algorithm within an industrial environment, a hard-coded fitness function can present a major disadvantage. As business and operational needs change, the constraints on the problem under consideration may change, requiring the fitness function to be modified. Modifying the fitness function is difficult and potentially expensive, to address this, as discussed in Sect. 4.1 we divide the fitness function into two sub functions:

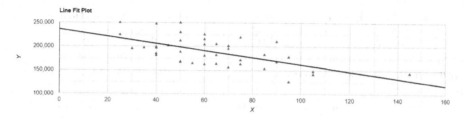

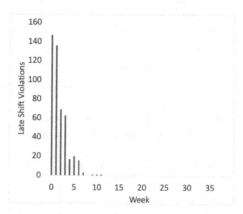

Fig. 1. A line fit plot for the fitness (x axis) versus evaluations y axis. The Pearson correlation coefficient returns a result of $r = -0.6058, p = 0.00003447$, which suggests a significant very small relationship between x and y.

Fig. 2. The total number of late shift constraint violations found within the 400 solutions summarised in Table 6. In every case the driver with the constraint violation also had a license due to expire in the current year.

– *fitness-base*: This function evaluates a set of constraints that are hard-coded in Java, it is not intended to be modified by the end-user.
– *fitness-dsl*: This function evaluates a set of constraints specified using the DSL by the end-user. The constraints are compiled and evaluated at run-time.

The fitness value assigned to a candidate solution is the sum of the penalty values assigned by *fitness-base* and *fitness-dsl*. As the end users do not have software engineering experience it is not desirable to use an existing scripting language, we investigate the development of a domain specific language (DSL) named Basil that allows constraints to be specified. The DSL is based around entities within the problem domain that will be familiar to the end user, making it easier for them to use, it is only intended for the specification of constraints for this problem domain. Each constraint within Basil specifies whether a particular characteristic should not appear in the solution, in this manner the constraints specified using Basil are binary.

5.2   The Basil Language Syntax, Compilation and Evaluation

The Basil language is a DSL used to specify constraints which the user wishes to apply to the solution, as it is only used to specify constraints based around entities in the problem domain, it is not Turing complete (it has no branch or jump constructs).

A Basil script comprises a list of constraints and the priorities associated with these constraints. A constraint specified in Basil takes the basic form:

`<entity> <condition> <time> <priority>`

Each constraint specifies that a particular entity (driver or group of drivers) should be placed (or not placed) before, after or in a specific time (training week). Optionally, a priority may be assigned to the constraint, each priority level has a penalty value assigned to it (see Table 5 Items 4–6).

- **entity** The entities upon which constraints may be imposed are Drivers or Groups. These are denoted by use of the keywords *driver* or *group*, followed by the appropriate driver or group identifier.
- **condition** Each condition begins with the phrase *must be* (which can be negated with the phrase **must not be** followed by one of the keywords *before*, *after* or *in*
- **time** Times are specified using the keyword *week* followed by the week number.
- **priority** The optional priority may be set using the *priority* keyword followed by *high*, *medium* or *low*. Where a priority is not specified, the constraint is allocated a medium priority.

An example Basil script may be seen Algorithm 3.

Algorithm 3. An example of a Basil script. Line 1 is a comment, lines 2-5 describe constraints to be applied to the problem being solved.

```
1: #A set of test constraints
2: driver 123 must not be before week 23 with high priority
3: driver 226 must not be after week 23 with low priority
4: driver 1500 must not be in week 12 with medium priority
5: group depot1 must not be in week 35
```

5.3   Basil Execution

Basil is based around the concept of regular expression-based pattern matching, each Basil constraint being compiled into a regular expression. In order to evaluate a solution against a regular expression each solution is converted into what is termed an *intermediate* format (Fig. 3) describing the allocation of training slots to drivers. Table 7 shows examples of Basil statements (constraints) and their resulting Regex expressions.

```
...
:ID: 1987:GR:DP1:WK:31:DY:2:DT:2:XW:33:FY:0
:ID: 1988:GR:DP1:WK:28:DY:2:DT:2:XW:06:FY:0
:ID: 1989:GR:DP1:WK:21:DY:1:DT:2:XW:37:FY:0
...
```

Fig. 3. An extract from the intermediate format. This format presents the solution in a manner that supports pattern matching via regular expressions. Each line describes the assignment of one driver to a training slot.

Table 7. Statements written in Basil are parsed and compiled into regular expressions which are then evaluated against a solution presented in the intermediate format (Fig. 3. The appears flag specifies if the regex expression must appear in the solution or not. The priority weight field specifies the penalty weight to be associated with a violation.

Basil Statement	Regex	Appears Flag	Priority Weight
driver 123 must not be before week 23 with low priority	:ID:123:GR:...:WK:23:DY:.:DT:..:XW:...:FY:	false	low
group DP1 must not be in week 35	:ID:...:GR:DP1:WK:35:DY:..:DT:..:XW:...:FY:	false	medium (default)
driver 456 must be before week 23 with high priority	:ID:456:GR:...:WK:23:DY:..:DT:..:XW:...:FY:	true	high
driver 567 must be after week 33 with high priority	:ID:567:GR:...:WK:33:DY:..:DT:..:XW:...:FY:	true	high

5.4 Results with Basil

Basil scripts containing 10, 25, 50 and 100 constraints were generated at random. The generation of constraints at random simulates the arbitrary constraints which might stem from individual staff requests and organisational constraints. Some of these random constraints will conflict with each other or with base constraints (Table 5 Items 1–3). Our interest is not in avoiding this conflict but in managing it.

The results obtained may be seen in Table 8. The reader should note that the best solutions never break the license expiry constraint: this is very desirable given the importance to the business of ensuring that drivers' licenses are not allowed to expire. If we explore the relationship between the fitness of the best solution found (over 10 runs) and the number of custom constraints, we find that there exists a significant large positive relationship (calculated using a Pearson Correlation Coefficient where $r = 0.5651$ and $p = 0009$). This is as we might expect, adding more constraints results in a reduction in solution quality.

As we are examining an industrial application, we should examine the effects of adding the additional constraints and the overhead of evaluating them. We are not concerned with the overall run-time required, the user can adjust the *TIME_OUT* property to find an appropriate balance between the time they are willing to wait and the quality of the resulting solution. In this section we are concerned with the general effect of adding numbers of additional constraints into the fitness function via Basil and the increased time taken to evaluate these constraints. Figure 4 shows the average time ev (milliseconds) to evaluate 2,000 individuals we are concerned with the trend in ev, as the number constraints is increased. Figure 4 suggests that the increase in evaluation time is super linear.

The initial system was implemented in Java and executed on a MacBook based round the Apple M1 CPU.

Table 8. Results obtained using up to 100 randomly specified Basil constraints.

DataSet	Custom Constraints	Fitness		Constraints			Custom Constraints		
				Shift Violation (late shift)	Expired License	Unbalanced Groups	Low	Medium	High
135	0	Best	55	8	0	0			
		Avg	88.5	14.1	0	0			
	5	Best	2934	13	0	109	4	0	234
		Avg	2663.4	15.9	0.1	109.4	4.3	0	230.7
	10	Best	943	11	0	0	878	0	0
		Avg	962.3	10.9	0	0	880.8	0	0
	25	Best	449	9	0	0	3	70	0
		Avg	485.2	14.5	0	1	3.4	77.7	0
	50	Best	3815	6	0	113	566	60	218
		Avg	3844.7	10.7	0	126.2	575.7	61.9	225.9
	100	Best	7311	19	0	145	445	775	204
		Avg	7353.4	23.1	0.2	138.3	448.4	778.2	210.9
801	0	Best	105	17	0	0			
		Avg	127.5	20.7	0	0			
	5	Best	125	16	0	0	35	0	0
		Avg	145.5	16.4	0	0	36.5	0	0
	10	Best	976	16	0	0	610	50	1
		Avg	973.5	18.6	0	0	611.2	53.2	1
	25	Best	880	17	0	17	25	16	50
		Avg	913.7	21.9	1.3	21.1	26.2	17.8	53.8
	50	Best	5896	20	0	177	5	144	412
		Avg	5914.7	22.8	0.4	179.4	6.2	145	414
	100	Best	1918	19	0	22	448	28	105
		Avg	1956.9	21.4	0.1	27.7	451.4	29.5	108.7
665	0	Best	65	10	0	0			
		Avg	84.5	13.6	0	0			
	5	Best	80	13	0	0	0	0	3
		Avg	100.6	14.3	0	0	0.1	3.1	0
	10	Best	131	12	0	0	1	12	0
		Avg	156.3	15.7	0	0	1.8	12.1	0
	25	Best	954	11	0	61	22	2	54
		Avg	977	16	0	61.4	23.5	2	54.1
	50	Best	239	11	0	0	39	19	2
		Avg	261.6	13.4	0.1	0	44.6	20.1	2.8
	100	Best	8234	13	0	0	506	23	752
		Avg	8250.9	15.8	0	0	507.4	23.9	753.2
480	0	Best	75	11	0	0			
		Avg	98.5	15.3	0	0			
	5	Best	644	10	0	3	19	107	1
		Avg	670.6	10	0	1.7	20.1	108.3	1
	10	Best	535	11	0	0	455	1	1
		Avg	564.8	14.7	0	0	0	1.1	1.2
	25	Best	565	15	0	8	35	1	33
		Avg	581.1	18.9	0	10.5	38.1	1	37.1
	50	Best	1332	15	0	0	966	7	21
		Avg	1355.3	19.8	0.5	0	966.8	7.6	21.5
	100	Best	3807	15	0	38	10	392	147
		Avg	3838.2	18.9	0	44.7	10.7	392.4	152.7

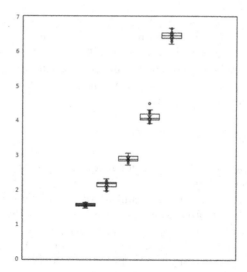

Fig. 4. Milliseconds per evaluation on the results obtained in Table 8 the box plots show evaluation times for 0, 10, 25, 50 and 100 custom constraints respectively. Note the lack of overlap between the box plots, also note that the trend in the evaluation time versus quantities of custom constraint is super-linear.

6 Discussion and Future Work

6.1 Conclusions

The principle contribution presented in this paper is the development of the Basil DSL and the mechanism by which constraints are evaluated at run time. The basic problem is not novel, nor is the algorithm used to solve it. The contribution of this paper lies the development and the use of the Basil DSL and the use of the intermediate representation and pattern matching (see Sect. 5) to allow evaluation of constraints at run-time.

In addressing the first research question stated in the introduction, the results presented in Table 8 suggest that can be implemented within a DSL and evaluated at run time. Assessing whether the DSL is usable by a domain expert is more difficult proposition, informal discussions suggest that domain experts can utilise the BASIL language, future work will include a more formal evaluation of BASIL with regards to usability through a user study.

The mechanism described in Sect. 5 highlights the use of the intermediate representation and pattern matching as the means of addressing the second research question. The use of regular expressions allows and existing well-proven mechanism to be used, one which is implemented in most commonly used programming environments. This avoids the use of more complex solutions such as full compilation of the custom constraints into the main code base, a procedure which also has potential security and integrity issues.

If we consider the last question, Fig. 4 shows the increase in evaluation time as the number of custom constraints increases. We note that further work may be required on the execution mechanism for Basil as scripts of more than 100 constraints are not viable in terms of execution. This could be partially negated by executing the software on more powerful hardware and examining the implementation of the regex pattern matcher.

6.2   Future Work

This paper has laid a solid foundation for future work on the problem of incorporating, custom constraints into a solver in a manner that is appropriate for non-technical users. As well as more technical work in he implementation, future work will also include the integration of Natural Language Processing into Basil to allow constraints to be expressed in natural language. In a problem such as this where there are many stakeholders (e.g., over 2,000 drivers) the ability for them to articulate their constraints directly to the system would be a very powerful feature.

Acknowledgements. The authors are indebted to management of the industrial partner for their time in explaining the problem and the feedback given on the work undertaken.

References

1. A tabu search algorithm with controlled randomization for constructing feasible university course timetables. Comput. Oper. Res. **123**, 105007 (2020). https://doi.org/10.1016/j.cor.2020.105007
2. Abdelghany, M., Yahia, Z., Eltawil, A.B.: A new two-stage variable neighborhood search algorithm for the nurse rostering problem. RAIRO - Oper. Res. **55**(2), 673–687 (2021). https://doi.org/10.1051/ro/2021027
3. Burke, E.K., Curtois, T., Qu, R., Vanden-Berghe, G.: A time predefined variable depth search for nurse rostering. INFORMS J. Comput. **25**(3), 411–419 (2013). https://doi.org/10.1287/ijoc.1120.0510
4. Kent, E., Atkin, J.A.D., Qu, R.: Vehicle routing in a forestry commissioning operation using ant colony optimisation. In: Dediu, A.-H., Lozano, M., Martín-Vide, C. (eds.) TPNC 2014. LNCS, vol. 8890, pp. 95–106. Springer, Cham (2014). https://doi.org/10.1007/978-3-319-13749-0_9
5. Kittel, F., Enenkel, J., Guckert, M., Holznigenkemper, J., Urquhart, N.: Optimisation algorithms for parallel machine scheduling problems with setup times. In: Proceedings of the Genetic and Evolutionary Computation Conference Companion. GECCO '21, New York, NY, USA, pp. 131–132. Association for Computing Machinery (2021). https://doi.org/10.1145/3449726.3459487
6. Kondratenko, Y., Kondratenko, G., Sidenko, I., Taranov, M.: Fuzzy and evolutionary algorithms for transport logistics under uncertainty. In: Kahraman, C., Cevik Onar, S., Oztaysi, B., Sari, I.U., Cebi, S., Tolga, A.C. (eds.) INFUS 2020. AISC, vol. 1197, pp. 1456–1463. Springer, Cham (2021). https://doi.org/10.1007/978-3-030-51156-2_169

7. Ngoo, C.M., Goh, S.L., Sze, S.N., Sabar, N.R., Abdullah, S., Kendall, G.: A survey of the nurse rostering solution methodologies: the state-of-the-art and emerging trends. IEEE Access **10**, 56504–56524 (2022). https://doi.org/10.1109/access.2022. 3177280

8. Regnell, B., Kuchcinski, K.: A scala embedded DSL for combinatorial optimization in software requirements engineering. In: First Workshop on Domain Specific Languages in Combinatorial Optimization, pp. 19–34 (2013)

9. Service, G.D.: Driver CPC training for qualified drivers (2021). https://www.gov.uk/driver-cpc-training

10. Si Ying, P., Mohd-Yusoh, Z.I.: Staff scheduling for a courier distribution centre using evolutionary algorithm. Indonesian J. Electric. Eng. Comput. Sci. **27**(2), 1043 (2022). https://doi.org/10.11591/ijeecs.v27.i2.pp1043-1050

11. Siddiqui, A.W., Arshad Raza, S.: A general ontological timetabling-model driven metaheuristics approach based on elite solutions. Expert Syst. Appl. **170**, 114268 (2021). https://doi.org/10.1016/j.eswa.2020.114268. https://www.sciencedirect.com/science/article/pii/S0957417420309799

12. University, M.: Minizinc constraint modelling language (2020). https://www.minizinc.org/

On the Utility of Probing Trajectories for Algorithm-Selection

Quentin Renau$^{(\boxtimes)}$ (ID) and Emma Hart (ID)

Edinburgh Napier University, Edinburgh, Scotland, UK
{q.renau,e.hart}@napier.ac.uk

Abstract. Machine-learning approaches to algorithm-selection typically take data describing an instance as input. Input data can take the form of features derived from the instance description or fitness landscape, or can be a direct representation of the instance itself, i.e. an image or textual description. Regardless of the choice of input, there is an implicit assumption that instances that are similar will elicit similar performance from algorithm, and that a model is capable of learning this relationship. We argue that viewing algorithm-selection purely from an instance perspective can be misleading as it fails to account for how an algorithm 'views' similarity between instances. We propose a novel 'algorithm-centric' method for describing instances that can be used to train models for algorithm-selection: specifically, we use short *probing trajectories* calculated by applying a solver to an instance for a very short period of time. The approach is demonstrated to be promising, providing comparable or better results to computationally expensive landscape-based feature-based approaches. Furthermore, projecting the trajectories into a 2-dimensional space illustrates that functions that are similar from an algorithm-perspective do not necessarily correspond to the accepted categorisation of these functions from a human perspective.

Keywords: Algorithm Selection · Black-Box Optimisation · Algorithm Trajectory

1 Introduction

We are motivated by the future goal of designing optimisation systems that are capable of learning from past experience. For example, algorithm-selection (AS) methods such as machine-learning based classifiers [35] learn by being trained on results obtained from solving a large set of instances, while transfer-learning methods [3,39] reuse information extracted from one instance in solving a new instance (e.g. warm-starting with a previous solution). In both of these scenarios, there is an implicit assumption that if two instances are similar to each other, then they might elicit similar performance from the same solver or that information can be transferred from one instance to another, for example to seed or warm-start an optimiser [20].

Supplementary Information The online version contains supplementary material available at https://doi.org/10.1007/978-3-031-56852-7_7.

Typical approaches to AS train a machine-learning (ML) model to predict either algorithm-performance or best-solver based on a description of an instance as input to the model in some form. There have been many advances made in recent AS literature regarding the choice of model and in defining appropriate model inputs [1,16]. In particular, a lot of attention has been directed towards defining features to train a model. Human-designed features can be used, but have the potential disadvantage of being domain-specific and often costly to compute [21], while it has also been noted that instances that are close in human-designed feature-spaces are not necessarily close in the performance space of a given algorithm [33]. Defining features via Exploratory Landscape Analysis (ELA) [22] has become a popular alternative, particularly in the continuous optimisation domain, creating a feature-vector describing the fitness landscape of an instance. However, there is significant overhead cost induced by the ELA feature computation: furthermore, the sample points used to compute features are usually discarded, hence wasting computational budget. More recent 'feature-free' methods avoid calculating features altogether by directly using a description of the instance as input, e.g. using text-based descriptions [2,35] or images [32]. However, we argue that all of the above approaches are potentially flawed in that the model is trained on data that takes only an 'instance perspective' of the data (through human-designed features, landscape features or a description of the instance data directly): instances are described by features that are calculated independently of the execution data obtained by any algorithm.

Some recent research [13,14,19] has begun to address this, training selection models whose input data includes information derived from running a solver, in addition to (or instead of) using purely instance-centric data. We continue to push in this direction in proposing a novel method for training a selector that uses *only* time-series information obtained from a *probing-trajectory*. A probing trajectory is defined by either the best or current performance of a meta-heuristic solver over its first n function evaluations on an instance, where n is deliberately very short. We propose that an algorithm that produces similar trajectories on two instances 'sees' some commonality between those instances (and vice versa). Each trajectory (a time-series) can be used as input to an AS classification method, either directly or using time-series features derived from the trajectory. This use of probing-trajectories has the following benefits:

- It completely removes the need to either define or calculate features of any type in order to create training data.
- The trajectory provides an 'algorithm-perspective' of an instance, in contrast to feature-based approaches which only describe the instance (or its associated landscape) in isolation from any solver. We hypothesise that taking the 'algorithm perspective' might make it easier for an AS approach to learn as the trajectory is a very close proxy to true algorithm behaviour.
- The probing-trajectory used to get a prediction from the model can be re-used to warm-start a selected solver, hence saving budget.

We evaluate the approach using the BBOB functions [9] as a test-bed. We show that trajectory-based algorithm-selection can outperform the classical landscape-aware approach in continuous optimisation. In this scenario we also show that for sampling budgets where a landscape-aware approach cannot be applied (i.e., when sampling budgets are too small), trajectory-based algorithm-selection still performs well, making it a good low-budget alternative to ELA features.

The outline of this paper is as follows. Section 2 gives an overview of the background and related work. Section 3 describes the data used, the methods for obtaining probing-trajectories, and describes the experiments conducted in this paper. Section 4 describes the results obtained with the probing-trajectories on an algorithm selection task. Section 5 provides insights into trajectory similarities, while Sect. 6 exposes the pros and cons of using the probing-trajectories Finally, Sect. 7 highlights concluding remarks and future work.

2   Background and Related Work

The majority of previous work in algorithm selection is performed using information describing an *instance*, for example extracting features at the instance level; extracting features depicting the landscape of the objective function at hand; using feature-free Deep Learning techniques. The use of *Instance features* is most common in combinatorial optimisation domains. They rely intrinsically on the problem domain and are usually manually designed [12], differing from one domain to the other, i.e., Travelling Salesperson Problem (TSP) features and Knapsack features cannot be interchanged. However, the main drawback of instance features is that they often do not correlate well with algorithm performance data. For instance, Sim *et al.* [33] demonstrate in the TSP domain that instances that are close in the feature-space can be very distant in the performance space (i.e., the Euclidean distance between their feature-vectors is small while the distance between the performance of two algorithms on the instances is very large). Furthermore, they can also be computationally expensive [21].

On the other hand, the use of *landscape-features* is common in numerical black-box optimisation, typically via Exploratory Landscape Analysis (ELA) [22] which calculates landscape features. ELA has grown over the years with a gradual introduction of new features [8,17,23]. Features are numerical values obtained by sampling m points, $x_1, \ldots, x_m \in \mathbb{R}^d$ in a d-dimensional search space, and computing the associated objective function $f(x_1), \ldots, f(x_m)$. The features are then approximated given the pairs $(x_1, f(x_1)), \ldots, (x_n, f(x_m))$. ELA has been successfully applied to both algorithm configuration [4] and algorithm selection [16] on benchmark data as well as real-world optimisation problems [29]. A definition of the most used features and their properties can be found in [27]. The main drawback of ELA however is the overhead cost induced by the feature computation as the sample points that are usually used to compute features are then discarded. Other work provides evidence in some domains that ELA features need be used with care [27].

Feature free techniques avoid the problem of human-designed features by relying on Deep Learning to extract patterns and be able to perform algorithm selection.

Feature free approaches have been successfully applied both in combinatorial optimisation [1] using the instance definition as input and in continuous optimisation [32] using sample points in the search space. While the latter removes the human bias from the design of features, the overhead cost of sampling the search space on top of running the algorithm is still present. Moreover, Deep Learning approaches reduce the understanding of the instance space and make it difficult to truly understand algorithm behaviour.

All of the approaches just described take an instance-centric view: that is, the input to a selector is independent of the execution of any algorithm. We suggest this is problematic, given that previous work has suggested that there is not necessarily a strong correlation between the distance of two instances in a feature-space and the distance in the performance space according to a chosen portfolio of solvers [33]. Some recent work has begun to address this, using information derived from running a solver as input to a selector. For example, recent work from Jankovic *et al.* [13,14] proposes extracting ELA features from the search trajectory of an algorithm. In [13], they obtain a trajectory by using half the available budget to run an algorithm (250 function evaluations, corresponding to the ELA features recommended budget [18]) and combine this with the state variables of the algorithm to predict the algorithm performance. Overall, their approach gave encouraging results but was outperformed by classical ELA features computed on the full search space. In [14], they use trajectories of $30d$ points to compute ELA features to train a performance predictor in order to choose which algorithm to warm-start. They successfully compute features during the search of one algorithm to select the appropriate algorithm to switch to finish the run. This $30d$ points budget is slightly lower than the recommended ELA features budget of $50d$ points.

The work of [7] also obtain algorithm trajectories but then construct time-series based on concatenation of statistics derived from the population and fitness values (mean, standard deviation, minimum, maximum) at each generation to use as input to a classifier that predicts which of the 24 BBOB function the trajectory belongs to. The length of the trajectories obtained is 900 points on functions of dimension $d = 3$. This budget is very large as it is 6 times more than the recommended ELA features budget of $50d = 150$. Their approach successfully outperforms ELA features extracted from algorithm trajectories but is not compared to ELA features extracted from sample points in the full search space. Another approach combines ELA features with time-series features extracted from state variables of CMA-ES [19] to perform a per-run algorithm selection with warm-starting. The results of this work shows performance on par with ELA features on the per-run algorithm selection task. Although not specifically concerned with algorithm-selection, the work of [26] is also worthy of mention in taking an algorithm perspective by utilising information incorporated in CMA-ES *state variables* to train a surrogate model to predict performance.

We build on the nascent line of work in also proposing a trajectory-based approach to algorithm-selection. Specifically, we attempt to use short trajectories that only consume a small fraction of the available computational budget, and that can be re-used to warm-start an algorithm predicted by a selector. Unlike the work described in [13], we do not compute landscape features from search-trajectories and make use of the probing trajectories both directly and indirectly.

3   Methods

3.1   Data

We consider the first 5 instances of the 24 noiseless Black-Box Optimisation Benchmark (BBOB) functions from the COCO platform [10] as a test-bed. In BBOB, instances are transformations of the original function such as rotations, translations or scaling.

For each instance, we collect data from running three algorithms: CMA-ES [11], Particle Swarm Optimisation (PSO) [15], and Differential Evolution (DE) [36]. Each algorithm is run 5 times per instance. Thus, we have $24 \times 5 \times 5 = 600$ trajectories. Our data is obtained directly from [37] which records search-trajectories per run. Note that some automated algorithm configuration was performed by the authors before they collected this data and that population sizes are different for each algorithm, see [38] for further details.

We use data obtained from [30] on the BBOB suite to calculate ELA features. For each feature, 100 independent values are available per function instance. The sampling strategy used to sample points is the Sobol′ low-discrepancy sequence. However, we use only a fraction of the available data, i.e., 10-dimensional functions, a feature computation budget of $30d$ and the general recommendation for feature computation $50d$ [18]. We tried to compute ELA features at budgets lower than $30d$ but as the budget decreases, Nearest Better Clustering features start to output *Not a Number* values and some Dispersion features are not computed. We select 10 cheap features based on their expressiveness and invariance to transformations. The 10 features selected are the same as in [28].

3.2   Algorithm Selection Inputs

Using the data collected in Sect. 3.1, we create three types of inputs for the algorithm selection procedure: raw-trajectories, features extracted from the time-series formed by the raw-trajectories and ELA features.

Raw Probing-Trajectories. A probing-trajectory consists of a time-series of values o obtained from the first n iterations of an algorithm, where o is either the current best fitness (coined 'current' in the rest of the paper) or the best-so-far fitness values (coined 'best' in the rest of the paper). Points are added to the trajectory in the order in which they are sampled[1].

We evaluate three approaches to using raw probing-trajectories as input to an algorithm-selector: (1) using two different types of probing-trajectory (best-so-far, current); (2) using concatenated probing-trajectories from multiple algorithms as input, i.e., concatenate trajectories from two or more algorithms; (3)

---

[1] We evaluate the effect of this choice later in Sect. 4.3.

input trajectories of different lengths. Hence, the input of the algorithm selection procedure is a time-series with its length depending on the number of concatenated algorithms and its number of generations, representing one run on one function instance.

We test the impact of the length of the trajectories on the ability to select the best algorithm using two settings for the number of generations $g \in \{2, 7\}$. The former enables us to evaluate really short trajectories obtained using only 2 generations, while the latter results in trajectories using a similar computational budget to that used to obtain ELA features (only one setting described below requires more evaluations than ELA features). The function evaluations budget depends on the population size of the algorithm according to the dataset used, i.e., 10 for CMA-ES, 30 for DE, and 40 for PSO. For a single trajectory input this results in a budget of $20, 60, 80$ for CMA-ES, DE, PSO respectively at generations $= 2$, and $70, 210, 280$ at generations $= 7$. Note that all these budgets are less than the minimum budget of 300 at which it is feasible to compute ELA features. For concatenated trajectories, the maximum budget is 560 when three trajectories are joined $(7 \times (10 + 30 + 40))$, therefore comparable to the higher $50d$ budget used to compute ELA features.

Time-Series Features. We extract time-series features from the probing-trajectories described above as in [24], including tuning the process by using a feature selection technique. The time-series features are extracted on the probing-trajectories using the *tsfresh* Python package (version 0.20.1), while the feature selection is performed using the *Boruta* Python package (version 0.3).

Time-series features are extracted on the same settings as probing-trajectories, i.e., on the 'current' or 'best' trajectories for a single or concatenated trajectory given 2 or 7 generations. The input of the algorithm selection procedure is thus a vector of features where the dimension depends on the number of features extracted or selected, representing one run on one function instance.

ELA Features. As mentioned above (in Sect. 3.1), 100 $10-$dimensional feature vectors are available for each BBOB instance. These features are computed with $30d = 300$ and $50d = 500$ sample points. In order to have a fair comparison between ELA and probing-trajectories, we randomly sample 5 feature vectors out of 100 to match the number of available trajectories by instances, i.e, 5 runs by instances.

The input of the algorithm selection procedure is a $10-$dimensional vector of features, representing one function instance.

3.3 Algorithm Selection Methods

The algorithm selection is treated as a classification task, i.e, given an input representing an instance, the output is the algorithm to use on that particular instance. To train the classification models, both the trajectories and feature-vectors are labelled with the best performing algorithm, defined as the one having the best median target value after 100,000 function evaluations. Note that no single algorithm outperforms the others on all 24 functions: CMA-ES is the best performing algorithm for 11 functions, DE for 7, and PSO for 6.

In Sect. 3.2, we described two types of input (trajectories and feature vectors), hence two types of classifiers are used depending on the data type.

Classification with Features. Features inputs can be derived from calculating either ELA features or time-series features extracted from the probing-trajectories. As used in [5,29], we train a default Random Forests [6] from the *scikit-learn* package [25] (version 1.1.3). Separate models are trained using different input type, i.e., models using ELA features and time-series features are different.

Classification with Raw Trajectories. The raw probing trajectories form a time-series. For this type of input, a specialised time-series classifier is used, specifically the default Rotation Forests [31] from the *sktime* package (version 0.16.1). This choice of classifier is motivated by its closeness to the classifier used for features while accounting for the data being time-series. We train classifiers using single or concatenated trajectories to predict which algorithm of the portfolio to use, e.g., a classifier trained with 'CMA-ES' trajectories can predict which of the three algorithms from the portfolio to use.

Validation Procedure. Both type of inputs have the same validation procedure. As performed in [19], we perform a *leave-one-instance-out (LOIO) cross-validation* and we compute *the overall accuracy*. Classifiers are trained using data from all runs of the 24 functions on all except one instance. The data of the left out instance is used as the validation set, i.e, all the runs from the 24 functions on the ID of the left out instance. Overall, 480 inputs are used to train the model while the remaining 120 inputs are used for validation.

4   Results

In this section, we first present the results obtained with single trajectories (Sect. 4.1), followed by results obtained from using input to a classifier that is formed by concatenating trajectories from multiple algorithms (Sect. 4.2). In each case, we compare the classification results using three types of input: the raw-trajectory; a feature-vector derived from the time-series; an ELA feature-vector. We discuss the results in Sect. 4.3.

4.1   ELA Features vs Single Trajectories

In this section, we compare algorithm-selection performed using ELA features as input to trajectory-based inputs obtained from running a single algorithm. For the latter, we experiment with using inputs from (1) the raw trajectory data and (2) time-series features derived from the trajectory. We train three separate classifiers, using trajectories from each of the algorithms in turn[2].

Algorithm trajectories are computed for 2 and 7 generations. As explained previously, the minimum budget is 20 function evaluations for CMA-ES (2 * population size 10) and the largest is 280 function evaluations for PSO (7 * population size 40). As mentioned in Sect. 3.1, ELA features are extracted using $30d = 300$ and $50d = 500$ points to provide fair comparisons.

We observe that models trained with best-so-far trajectories outperforms models trained with current trajectories with 2 generations and vice versa with 7 generations. Due to space limitations, we will only present best-so-far trajectories for 2 generations and current trajectories for 7 generations. The other plots can be found in the supplementary materials.

Figure 1 compares the classification accuracy of the classifiers trained using probing-trajectories, time series features extracted from three trajectories and time series feature selection.

Classifiers trained using only 2 generations (using a maximum budget of 80 evaluations) exhibit a poor accuracy (Fig. 1a). Classifiers trained on ELA features perform better but recall that these require a minimum budget of 300 evaluations, almost 4 times the budget for the trajectories. Even with this additional budget, the median accuracy reached is relatively poor: 90% and 90.8% for 300 and 500 function evaluations respectively. This is not unexpected and consistent with previous literature, given that is known that ELA features are not all invariant to function transformations [27,34]. One of the trajectory-based classifier models exceeds the performance of the ELA trained classifier at a budget of 300 samples and matches the ELA trained classifier with 500 samples, demonstrating 90.8% median accuracy: PSO with raw probing-trajectories. In this context, the use of a PSO probing-trajectory is clearly an asset as it requires less than a sixth of ELA features budget to achieve the same level of performance.

When the number of generations increases to 7 (Fig. 1b), all but one of the classifiers trained on input obtained from a trajectory outperform both the ELA trained classifiers (budget 300, 500). A considerable increase in performance accuracy is obtained in most cases. Once again, the PSO trajectories reach the best performances with a peak for raw probing-trajectories at a median accuracy of 100%. In this case, almost perfect classification is achieved with substantially less function evaluations than needed to compute ELA features.

PSO trajectories may be more informative than other algorithms trajectories because PSO has the largest population size and evaluates more points. As a comparison, CMA-ES has the smallest population size and is in most cases the algorithm providing the worse accuracies. Hence, the information contained in the trajectories may be a matter of the number of generations and the number of points evaluated.

---

[2] A classifier trained only on e.g. CMA-ES trajectories can predict any of the three solvers, etc.

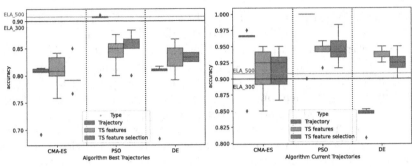

(a) Best trajectory for 2 generations   (b) Current trajectory for 7 generations

Fig. 1. Accuracy of classification on the LOIO cross-validation for best-so-far and current probing-trajectories, time series features and time series feature selection for 2 and 7 generations. Median ELA feature accuracy is represented by lines for 300 and 500 function evaluations.

4.2   ELA Features vs Multiple Trajectories

In this section, we compare algorithm-selection performed with ELA and time series features with classifiers using a concatenation of trajectories as input (i.e., the trajectories obtained from more than one algorithm are joined and used as input to the classifier to predict the best algorithm). By concatenating trajectories, we train four models named: $C - P$ (for the concatenation of CMA-ES and PSO trajectories), $C - D$ (for CMA-ES and DE), $D - P$ (for DE and PSO) and, ALL (for the concatenation of the three algorithms). These models are also trained for 2 or 7 generations.

As in Sect. 4.1, we observe that models trained with best-so-far trajectories outperforms models trained with current trajectories with 2 generations and vice versa with 7 generations. Due to space limitations, we will only present best-so-far trajectories for 2 generations and current trajectories for 7 generations. The other plots can be found in the supplementary materials.

Figure 2 compares the classification accuracy of the concatenation of trajectories on the LOIO cross-validation. As observed previously in Sect. 4.1, classifiers trained using the raw trajectories outperform those trained using the time-series features and ELA feature selection.

At 2 generations (Fig. 2a), classifiers trained on raw trajectories outperform those trained on ELA features except in a single case $(C - D)$. The classifiers trained with time-series features are all outperformed by classifiers trained on ELA features when using this low budget of 2 generations.

When the number of generations increases to 7 (Fig. 2b), the picture changes dramatically. Classification accuracies increase and all trajectory-based classifiers outperform classifiers using ELA features. Recall that the maximum budget for the trajectory based input here is 560 (ALL) and the minimum 280 $(C - P)$ compared to the ELA budgets of $(300, 500)$. While accuracies of time series-based classifiers are comparable, the median accuracy of raw trajectories increases to 100% (with one run around 90% for all classifiers).

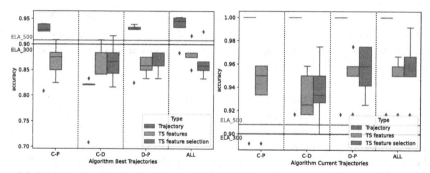

(a) Best trajectory for 2 generations   (b) Current trajectory for 7 generations

Fig. 2. Accuracy of classification on the LOIO cross-validation for best-so-far and current probing-trajectories, time series features and time series feature selection for 2 and 7 generations. Median ELA feature accuracy is represented by lines for 300 and 500 function evaluations.

4.3 Additional Insights into Trajectory Performance

Randomness in the 'Current' Trajectory. As noted in Sect. 3.2, the 'current' trajectory is obtained by adding each point sampled by an algorithm to a trajectory in the order in which they are evaluated. However, it should be clear that for a generational algorithm, the order that solutions are evaluated in is irrelevant. Therefore to satisfy ourselves that the particular ordering used had no influence of the results, we perform an additional experiment in which the points within each trajectory are randomly shuffled within a generation to obtain a new trajectory. We consider trajectories of 7 generations for the three algorithms in the portfolio. This process is repeated 5 times.

We repeat previous experiments using the raw trajectory, and time-series features as input to a classifier. We performed a Kolmogorov-Smirnov statistical test between the initial run and the 5 shuffled trajectories. The p-values for the five tests performed are between 0.84 and 1. These values do not enable the null hypothesis to be rejected at a confidence level of 0.05, thus we conclude that the order the points are evaluated in the 'current' trajectories does not impact the classification accuracy.

Order in the ALL Trajectory. In the same manner, the concatenated *ALL* trajectories used in the experiments used a fixed ordering to obtain the concatenation, i.e., CMA-ES, PSO, and DE $(C - P - D)$. Therefore, we also investigate the impact of the order used to construct the *ALL* trajectory to determine if it has an impact on the classification accuracy. We evaluate the classification accuracy of the six possible combinations for the *ALL* trajectory. We use the raw trajectory, compute time series features and perform feature selection on the LOIO cross-validation using the 'current' trajectory on 7 generations as before. We performed a Kolmogorov-Smirnov statistical test between the combination used in previous sections $(C - P - D)$ and all other combinations. The p-values for the five tests performed are between 0.87 and 1. These values do not enable the null hypothesis to be rejected at a confidence level of 0.05, thus we can conclude that the order used to create the *ALL* trajectory does not impact the classification accuracy.

Best-so-far vs Current Trajectories. We mention in Sect. 4.1 and Sect. 4.2, that the 'current' trajectory outperforms the 'best' for larger number of generation and vice versa. In Fig. 3, we display accuracy of classification of the *ALL* trajectories for different generations from 2 to 7.

We observe that the crossing point between the 'best' and 'current' trajectories happens with 4 generations. With fewer generations, the 'best' trajectory outperform the 'current' until they reach similar accuracies with 4 generations. Interestingly, we observe that only 3 generations are necessary for the ALL_{best} to reach a median accuracy of 99.8% (with one run at 89.2%) which is 9% better than ELA features for half their computation budget.

An explanation of the difference in performance of the two trajectories resides in the fact that the 'best' trajectories can be seen as elitist (i.e., keeping only the best value) and 'current' trajectories can be seen as non elitist (i.e., accepting lower fitness values). It has been seen that elitist algorithms are outperforming non elitist ones for small budgets and vice versa when the budget increases (an example of this behavior can be found in [29]). Hence the 'best' trajectory may contain more useful information when the budget is low whereas more generations may be needed for the 'current' trajectory to reach the same level of information.

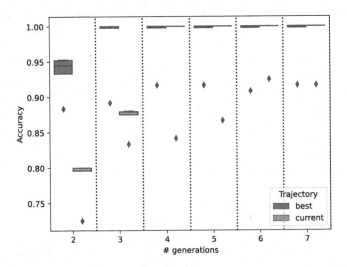

Fig. 3. Boxplot of accuracy of classification of the 'best' and 'current' trajectory-based classifiers on *ALL* trajectories from 2 to 7 generations on LOIO cross-validations.

5   Insights into Trajectory Similarity

To understand whether similarity between trajectories correlates with similarity in terms of solver performance, we project the time-series obtained into 2 dimensions using UMAP from the *umap-learn* Python library (version 0.5.3) with default parameters in an unsupervised setting.

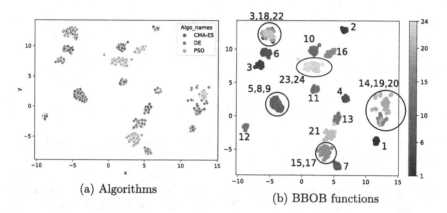

(a) Algorithms   (b) BBOB functions

Fig. 4. UMAP 2d projection of the $ALL_{current}$ time series of runs on function instances for 7 generations for each algorithm and BBOB functions (560 function evaluations).

Figure 4 presents projections for the $ALL_{current}$ time series using a budget of 560 function evaluations to create the trajectories, i.e. 7 generations. Each point represents a trajectory for one run on an instance. Colour labels in Fig. 4a represent the best algorithm for the function the trajectory was obtained from, i.e., the label used during classification for a given trajectory as in Sect. 3.3.

For a given algorithm, multiple clusters can be seen throughout the space. This may imply that for a given algorithm, distinct types of trajectories lead to the same choice of algorithm. However, we also observe some clusters of points where several algorithms are grouped in the same cluster. This may affect the accuracy of classification: in particular the figure suggests it may induce errors between CMA-ES and DE, and between PSO and DE.

The number of clusters is lower than the number of BBOB functions, implying that from the algorithm perspective, different functions may be similar. In order to verify the similarity of functions from the algorithm perspective, we coloured the instance runs by function. The result is shown in Fig. 4b. From the algorithm perspective, there are clear groups of functions won by the same algorithm. Trajectories from only one function (F3) are found in different groups. F3 belongs to two groups that are close in the space and hence may actually be one larger group.

From a trajectory perspective, ten functions have their own group (F1, F2, F4, F6, F10, F11, F12, F13, F16, and F21) while the others are distributed in groups of at least two functions. The largest groups are composed of three functions that are similar from a trajectory point of view. For two of these groups, all functions are won by the same algorithm: DE for F3, F18, and F22 and CMA-ES for F5, F8, and F9. Interestingly, these functions belong to different categories (BBOB functions are divided into five categories representing function properties) in the BBOB test-bed [10]: three different categories for F3, F18, and F22 and two for 5, F8, and F9. Recall that these categories are human-designed: however, the results imply that the properties of the functions that are observable by a human do not completely reflect the algorithm perspective. This reinforces the point made in Sect. 1 that determining instance similarity from an algorithm perspective might prove useful in future attempts to build optimisation systems that are capable of learning across instance sets.

6   Pros and Cons of Trajectories

The results in the previous sections show that trajectory-based algorithm selection is a good low-budget counterpart to classical approaches that tend to use ELA features. At very low budget (2 generations, 80 evaluations), a classifier trained on raw PSO trajectories performs on par with a classifier based on ELA features obtained using 500 samples. Three classifiers trained with concatenated trajectories ($C - P, D - P, ALL$) obtained from 2 generations with total budgets of $100, 140, 160$ respectively also outperform the ELA classifiers. Using a larger number of generations (7), 8 out of 9 classifiers trained on single raw trajectories or information derived from the trajectory outperforms the ELA based classifier using 500 samples. Using concatenated trajectories from 7 generations, all classifiers outperform the ELA classifiers, with the classifiers trained on concatenated raw, 'current' trajectories obtaining 100% median accuracy (with a maximum budget of 560 evaluations).

Given that computation of ELA features usually requires approximately $30d$ samples and that the computation is known to fail for some features at budgets less than this, using trajectories to train selectors is a promising way forward, providing reasonable results even at very low budget. In addition, the trajectory obtained to provide input to the selector can also be re-used in some cases: for example, if a classifier trained on 'CMA-ES' trajectories predicts CMA-ES for an instance, then the run can simply be continued from where it was stopped. Furthermore, if using the ALL trajectory which contains partial runs of all algorithms in the portfolio, then the budget is 'virtually' reduced by a third as any selected algorithm can simply continue its run from the already computed trajectory.

Comparing the results of the ALL classifiers that use three concatenated trajectories with results from trajectories formed from pairs of classifiers raises the question of how the method might scale with the number of algorithms in the portfolio. For example, for a portfolio of N solvers, it is possible that only a subset of trajectories is sufficient to train a classifier. This necessitates consideration of how to select the most appropriate combination of trajectories, i.e., $M < N$. As the fixed evaluation budget b must be divided between N trajectories (such that each trajectory is allocated $o = b/N$ function evaluations), then the influence of settings of o and N for a fixed value b should also be studied further. Finally, further investigation of how the approach scales as the N increases is also required.

7   Conclusion

The paper address the issue of algorithm-selection in a continuous optimisation, specifically in a setting whether there is a fixed budget of evaluations that must be shared between any computation required to derive input to an algorithm-selector and in running the selected algorithm. The goal is to find a form of input to a classifier that requires minimal computational budget, delivers high accuracy, and views similarity from an algorithm perspective. We hypothesised that short probing-trajectories obtained by running an algorithm for a small number of generations could be used to train a classifier, with the added benefit that the run used to obtain the trajectory could simply be continued for the chosen algorithm.

We demonstrated that a time-series classifier trained on raw trajectories and classifiers trained on features extracted from the trajectories can outperform classifiers trained on ELA features at considerably lower budget, in the best case using six times fewer sample points. Moreover, unlike using ELA features where sample points used for feature computation are often discarded, points from trajectories can be re-used if the algorithm that a trajectory belongs to is selected or points can be re-used for warm-starting another optimisation algorithm.

Obvious next steps include testing the approach on a more complicated task with larger algorithm portfolios as well as on different data-sets. A more thorough evaluation of state-of-the-art time-series classifiers (and tuning of their hyper-parameters) is also likely to improve results. Another key question lies in the scaling of the *ALL* concatenation of trajectories approach: does the number of trajectories needed depend linearly on the size of the portfolio or can judicious selection of specific trajectories suffice? Another interesting direction for future work is to use evaluate the use of the approach in combinatorial optimisation domains where there is much less work in defining appropriate ELA features and where calculation of domain-specific features is often expensive (particularly in TSP). Furthermore, in some combinatorial domains, hand-designed features have also been criticised for not correlating well with performance [33]. A trajectory-based approach could therefore be a promising avenue for research.

Acknowledgements. The authors are funded by the EPSRC 'Keep-Learning' project: EP/V026534/1 and EP/V027182/1

References

1. Alissa, M., Sim, K., Hart, E.: Automated algorithm selection: from feature-based to feature-free approaches. J. Heuristics **29**(1), 1–38 (2023). https://doi.org/10.1007/s10732-022-09505-4

2. Alissa, M., Sim, K., Hart, E.: Algorithm selection using deep learning without feature extraction. In: Proceedings of the Genetic and Evolutionary Computation Conference, pp. 198–206 (2019)

3. Ardeh, M., Mei, Y., Zhang, M.: Genetic programming hyper-heuristics with probabilistic prototype tree knowledge transfer for uncertain capacitated arc routing problems. In: 2020 IEEE Congress on Evolutionary Computation (CEC), pp. 1–8 (2020). https://doi.org/10.1109/CEC48606.2020.9185714

4. Belkhir, N.: Per Instance Algorithm Configuration for Continuous Black Box Optimization. phdthesis, Université Paris-Saclay (2017). https://hal.inria.fr/tel-01669527/document

5. Belkhir, N., Dréo, J., Savéant, P., Schoenauer, M.: Per instance algorithm configuration of CMA-ES with limited budget. In: Proceedings of the Genetic and Evolutionary Computation Conference, GECCO 2017, pp. 681–688. ACM (2017). https://doi.org/10.1145/3071178.3071343

6. Breiman, L.: Random forests. Mach. Learn. **45**(1), 5–32 (2001). https://doi.org/10.1023/A:1010933404324

7. Cenikj, G., Petelin, G., Doerr, C., Korosec, P., Eftimov, T.: Dynamorep: trajectory-based population dynamics for classification of black-box optimization problems. In: Proceedings of the Genetic and Evolutionary Computation Conference, GECCO 2023, Lisbon, Portugal, July 15–19, 2023, pp. 813–821. ACM (2023). https://doi.org/10.1145/3583131.3590401

8. Derbel, B., Liefooghe, A., Vérel, S., Aguirre, H., Tanaka, K.: New features for continuous exploratory landscape analysis based on the SOO tree. In: Proceedings of Foundations of Genetic Algorithms (FOGA) 2019, pp. 72–86. ACM (2019). https://doi.org/10.1145/3299904.3340308

9. Finck, S., Hansen, N., Ros, R., Auger, A.: Real-parameter black-box optimization benchmarking 2010: presentation of the noiseless functions (2010). http://coco.gforge.inria.fr/downloads/download16.00/bbobdocfunctions.pdf

10. Hansen, N., Auger, A., Ros, R., Mersmann, O., Tusar, T., Brockhoff, D.: COCO: a platform for comparing continuous optimizers in a black-box setting. Opt. Meth. Software **36**(1), 114–144 (2021). https://doi.org/10.1080/10556788.2020.1808977

11. Hansen, N., Ostermeier, A.: Completely derandomized self-adaptation in evolution strategies. Evol. Comput. **9**(2), 159–195 (2001). https://doi.org/10.1162/106365601750190398

12. Heins, J., Bossek, J., Pohl, J., Seiler, M., Trautmann, H., Kerschke, P.: A study on the effects of normalized TSP features for automated algorithm selection. Theor. Comput. Sci. **940**(Part), 123–145 (2023). https://doi.org/10.1016/j.tcs.2022.10.019

13. Jankovic, A., Eftimov, T., Doerr, C.: Towards feature-based performance regression using trajectory data. In: Castillo, P.A., Jiménez Laredo, J.L. (eds.) EvoApplications 2021. LNCS, vol. 12694, pp. 601–617. Springer, Cham (2021). https://doi.org/10.1007/978-3-030-72699-7_38

14. Jankovic, A., Vermetten, D., Kostovska, A., de Nobel, J., Eftimov, T., Doerr, C.: Trajectory-based algorithm selection with warm-starting. In: IEEE Congress on Evolutionary Computation, CEC 2022, Padua, Italy, July 18–23, 2022, pp. 1–8. IEEE (2022). https://doi.org/10.1109/CEC55065.2022.9870222

15. Kennedy, J., Eberhart, R.: Particle swarm optimization. In: Proceedings of ICNN 1995 - International Conference on Neural Networks, vol. 4, pp. 1942–1948 (1995). https://doi.org/10.1109/ICNN.1995.488968
16. Kerschke, P., Hoos, H., Neumann, F., Trautmann, H.: Automated algorithm selection: survey and perspectives. Evol. Comput. **27**(1), 3–45 (2019)
17. Kerschke, P., Preuss, M., Wessing, S., Trautmann, H.: Detecting funnel structures by means of exploratory landscape analysis. In: Proceedings of the Genetic and Evolutionary Computation Conference, GECCO 2015, pp. 265–272. ACM (2015). https://doi.org/10.1145/2739480.2754642,http://dl.acm.org/citation.cfm?doid=2739480.2754642
18. Kerschke, P., Preuss, M., Wessing, S., Trautmann, H.: Low-budget exploratory landscape analysis on multiple peaks models. In: Proceedings of the Genetic and Evolutionary Computation Conference, GECCO 2016, pp. 229–236. ACM (2016). https://doi.org/10.1145/2908812.2908845
19. Kostovska, A., et al.: Per-run algorithm selection with warm-starting using trajectory-based features. In: Rudolph, G., Kononova, A.V., Aguirre, H., Kerschke, P., Ochoa, G., Tusar, T. (eds.) PPSN 2022, Part I. LNCS, vol. 13398, pp. 46–60. Springer, Cham (2022). https://doi.org/10.1007/978-3-031-14714-2_4
20. Kostovska, A., et al.: Per-run algorithm selection with warm-starting using trajectory-based features. In: Rudolph, G., Kononova, A.V., Aguirre, H., Kerschke, P., Ochoa, G., Tusar, T. (eds.) PPSN 2022, pp. 46–60. Springer, Cham (2022)
21. Kotthoff, L., Kerschke, P., Hoos, H., Trautmann, H.: Improving the state of the art in inexact TSP solving using per-instance algorithm selection. In: Rudolph, G., et al. (eds.) PPSN 2022. LNCS, vol. 8994, pp. 202–217. Springer, Cham (2015). https://doi.org/10.1007/978-3-319-19084-6_18
22. Mersmann, O., Bischl, B., Trautmann, H., Preuss, M., Weihs, C., Rudolph, G.: Exploratory landscape analysis. In: Proceedings of the Genetic and Evolutionary Computation Conference, GECCO 2011, pp. 829–836. ACM (2011). https://doi.org/10.1145/2001576.2001690
23. Muñoz, M., Kirley, M., Halgamuge, S.: Exploratory landscape analysis of continuous space optimization problems using information content. IEEE Trans. Evol. Comput. **19**(1), 74–87 (2015). https://doi.org/10.1109/TEVC.2014.2302006
24. de Nobel, J., Wang, H., Bäck, T.: Explorative data analysis of time series based algorithm features of CMA-ES variants. In: GECCO 2021: Genetic and Evolutionary Computation Conference, Lille, France, July 10–14, 2021, pp. 510–518. ACM (2021). https://doi.org/10.1145/3449639.3459399
25. Pedregosa, F., et al.: Scikit-learn: machine learning in Python. J. Mach. Learn. Res. **12**, 2825–2830 (2011)
26. Pitra, Z., Repický, J., Holena, M.: Landscape analysis of Gaussian process surrogates for the covariance matrix adaptation evolution strategy. In: Proceedings of the Genetic and Evolutionary Computation Conference, GECCO 2019, pp. 691–699 (2019). https://doi.org/10.1145/3321707.3321861
27. Renau, Q.: Landscape-Aware Selection of Metaheuristics for the Optimization of Radar Networks. Ph.D. thesis, Polytechnic Institute of Paris, Palaiseau, France (2022). https://tel.archives-ouvertes.fr/tel-03593606
28. Renau, Q., Dreo, J., Doerr, C., Doerr, B.: Towards explainable exploratory landscape analysis: extreme feature selection for classifying BBOB functions. In: Castillo, P.A., Jiménez Laredo, J.L. (eds.) EvoApplications 2021. LNCS, vol. 12694, pp. 17–33. Springer, Cham (2021). https://doi.org/10.1007/978-3-030-72699-7_2

29. Renau, Q., Dréo, J., Peres, A., Semet, Y., Doerr, C., Doerr, B.: Automated algorithm selection for radar network configuration. In: Fieldsend, J.E., Wagner, M. (eds.) Proceedings of the Genetic and Evolutionary Computation Conference, GECCO 2022, pp. 1263–1271. ACM (2022). https://doi.org/10.1145/3512290.3528825

30. Renau, Q., Dreo, J., Doerr, C., Doerr, B.: Exploratory Landscape Analysis Feature Values for the 24 Noiseless BBOB Functions (2021). https://doi.org/10.5281/zenodo.4449934

31. Rodríguez, J., Kuncheva, L., Alonso, C.: Rotation forest: a new classifier ensemble method. IEEE Trans. Pattern Anal. Mach. Intell. **28**(10), 1619–1630 (2006). https://doi.org/10.1109/TPAMI.2006.211

32. Seiler, M.V., Prager, R.P., Kerschke, P., Trautmann, H.: A collection of deep learning-based feature-free approaches for characterizing single-objective continuous fitness landscapes. In: GECCO 2022: Genetic and Evolutionary Computation Conference, Boston, Massachusetts, USA, July 9–13, 2022, pp. 657–665. ACM (2022). https://doi.org/10.1145/3512290.3528834

33. Sim, K., Hart, E.: Evolutionary approaches to improving the layouts of instance-spaces. In: Rudolph, G., et al. (eds.) PPSN 2022, Part I. LNCS, vol. 13398, pp. 207–219. Springer, Cham (2022). https://doi.org/10.1007/978-3-031-14714-2_15

34. Skvorc, U., Eftimov, T., Korosec, P.: A comprehensive analysis of the invariance of exploratory landscape analysis features to function transformations. In: IEEE Congress on Evolutionary Computation, CEC 2022, Padua, Italy, July 18–23, 2022, pp. 1–8. IEEE (2022). https://doi.org/10.1109/CEC55065.2022.9870313

35. Song, Y., Bliek, L., Zhang, Y.: Revisit the algorithm selection problem for tsp with spatial information enhanced graph neural networks (2023)

36. Storn, R., Price, K.: Differential evolution - a simple and efficient heuristic for global optimization over continuous spaces. J. Global Optim. **11**(4), 341–359 (1997). https://doi.org/10.1023/A:1008202821328

37. Vermetten, D., Hao, W., Sim, K., Hart, E.: To Switch or not to Switch: Predicting the Benefit of Switching between Algorithms based on Trajectory Features - Dataset (2022). https://doi.org/10.5281/zenodo.7249389

38. Vermetten, D., Wang, H., Sim, K., Hart, E.: To switch or not to switch: predicting the benefit of switching between algorithms based on trajectory features. In: Correia, J., Smith, S., Qaddoura, R. (eds.) Applications of Evolutionary Computation. LNCS, vol. 13989, pp. 335–350. Springer, Cham (2023). https://doi.org/10.1007/978-3-031-30229-9_22

39. Zhuang, F., et al.: A comprehensive survey on transfer learning. Proc. IEEE **109**(1), 43–76 (2021). https://doi.org/10.1109/JPROC.2020.3004555

Nature-Inspired Portfolio Diversification Using Ant Brood Clustering

Ashish Lakhmani[(✉)] [iD], Ruppa K. Thulasiram[iD], and Parimala Thulasiraman[iD]

Department of Computer Science, University of Manitoba, Winnipeg, Canada
lakhmana@myumanitoba.ca,
{tulsi.thulasiram,parimala.thulasiraman}@umanitoba.ca

Abstract. Portfolio diversification is a crucial strategy for mitigating risk and enhancing long-term returns. This paper introduces a unique approach to large-scale diversification using Ant Brood Sorting clustering, a nature-inspired algorithm, in conjunction with co-integration measure of time series. Traditional diversification strategies often struggle during uncertain market times. In contrast, the proposed method leverages Ant Brood Sorting to group similar stocks based on the co-integration of their closing prices. This approach allows for the creation of diversified portfolios from a wide range of stocks. The study presents promising results, with clusters of stocks showing both high correlation and cosine similarity, validating the effectiveness of the approach. Silhouette score, a measure of cluster quality, and inter-cluster analysis demonstrate support in validating the results of the study by displaying similarities between the stocks being clustered and distinctiveness with stocks in other clusters. The research contributes to the application of nature-inspired algorithms in large-scale portfolio diversification, offering potential benefits for investors seeking resilient and balanced portfolios.

Keywords: Ant Brood Sorting · Portfolio Diversification · Cointegration · Clustering

1 Introduction

Portfolio Diversification (PD) involves spreading investments across a variety and different types of assets to reduce the overall risk of the portfolio and enhance the potential for long-term returns. The idea behind PD is to avoid putting all the eggs into one basket, i.e. to avoid putting the bulk of the total portfolio budget in similar types of assets so that the risk exposure to any one kind of asset is limited. Different asset types perform differently under diverse market conditions. By diversifying the investments, investors mitigate the impact of individual assets or similar kinds of assets on the whole portfolio and safeguard the capital. While PD may not accurately predict the highest return on assets, it spreads the risk effectively. Diversification not only helps in mitigating risk but also provides the opportunity to reap benefits in different segments of assets, achieving a balanced and resilient portfolio that stands the test of time.

S. Smith et al. (Eds.): EvoApplications 2024, LNCS 14634, pp. 115–130, 2024.
https://doi.org/10.1007/978-3-031-56852-7_8

The process of building a portfolio involves the selection of assets and finding suitable weights to allocate to each asset. PD entails the selection of assets in order to spread the unsystematic risk of the portfolio across all the assets in the portfolio, whereas Portfolio Optimization entails deciding the appropriate weights of each asset in the portfolio to maximize the overall return of the portfolio while minimizing overall risk. With an increase in the number of assets under a portfolio, it becomes challenging when there is a vast number of assets to choose from [11]. In recent years, nature-inspired algorithms have been considered on a large scale in computational finance literature [4]. The benefits of using nature-inspired algorithms come from their ability to quickly explore the possible solutions to a problem and efficiently exploit the solutions to improve upon them.

The conventional way to attain diversification in a portfolio is to select stocks from different asset classes, different industry sectors, or different geographical regions [21]. Some of the common diversification strategies are based on concepts such as the law of large numbers, correlation, capital asset pricing model, and risk parity. These diversification strategies, however, have failed to work when diversification was needed the most for risk aversion [12]. In this study, we present a PD strategy to group similar types of stocks by applying a nature-inspired heuristics called Ant Brood Sorting clustering technique on the statistical property of stocks' closing prices.

The remainder of this paper is structured as follows: Sect. 2 presents the related works of using nature-inspired computing in financial time series and discusses the motivation behind this study. Section 3 presents the definitions of the methods used in this study. Section 4 presents the experiment setup in detail along with the implementation of the experiment. Section 5 shows the results obtained and Sect. 6 concludes this study.

2   Related Work and Motivation

The legacy portfolio creation used classical time series models in creating optimal portfolios. Many professionals still use these time series models in the stock selection process before forming a portfolio [19] despite the fact that time series models have been shown to be inferior to computational algorithmic models [8]. Stock selection has long been recognized as a difficult and crucial task. Choosing stocks for successful portfolio development is heavily reliant on trustworthy stock ranking. Recent breakthroughs in machine learning and data mining have created substantial opportunity to handle these difficulties more effectively [10]. Huang [10] used Support Vector Regression (SVR) and Genetic Algorithms (GAs) to create a stock selection model. Their model used SVR to forecast each stock's future return, while GA optimized model parameters and input data.

Portfolio selection using nature-inspired algorithms has shown advantage over traditional methods because of superior searching ability through the heuristics [1]. Oduntan et al. [21] used Ant Brood Sorting clustering method to gain intelligence from time series data and use that intelligence to form clusters of similar stocks to create diversified portfolios. Liu et al. [16] used a variation of Ant Brood Clustering (ABC) to cluster financial time series data and received a high-quality clustering result as depicted by the intra-cluster distance. ABC Sorting has also been found to have promising results when hybridized with other algorithms [20].

Meta-heuristic algorithms have been proven to find the best answers for a wide range of complex and unique portfolio models [11]. Durán et al. [5] explored using memetic algorithm for multiobjective investment portfolio optimization with cardinality restrictions in the context of the Markowitz model. Hasan et al. [9] used whale optimization algorithm, a nature-inspired approach that mimics the haunting process of the sea whale, for portfolio optimization on the data-set of DAX-100, the German stock exchange index consisting of 100 stocks. Oduntan et al. [20] tested using and brood sorting clustering algorithm based on grouping of broods amongst ants, to gather financial intelligence from time-series of 30 stocks for portfolio diversification. Meng et al. [15] used grey wolf optimizer, a meta-heuristic optimizing algorithm inspired by the hunting behavior of grey wolves, for stock selection out of 200 stocks. Mazumdar et al. [18] used swarm intelligence for portfolio optimization and construction from a pool of 100 stocks. Shahid et al. [25] presented a novel portfolio selection strategy using gradient-Based Optimizer on a data-set of 30 stocks and compared its performance with a particle swarm optimization approach. There are about 65,000 stocks listed in stock exchanges worldwide. To achieve an effective diversification, it is essential to consider a vast number of stocks. A broad selection of stocks across different sectors, industries, and market segments can help mitigate the impact of poor-performing stocks on the overall portfolio.

To best of our knowledge, there hasn't been any prior research that utilizes a nature-inspired algorithm for selecting stocks across a wide spectrum of stocks for portfolio diversification. Moreover, Ant Brood Clustering, one of the interesting nature-inspired algorithms, has not been used in prior studies for portfolio diversification with a large number of stocks, the focus of this study. Therefore, our study represents a distinctive contribution to the application of a nature-inspired algorithm for a large scale portfolio diversification.

3   Ant Brood Clustering

Deneubour et al. [6] proposed a computing model inspired by behavior of ant colonies that clean their nest by collecting and organizing corpses into piles. The core idea is that ants wander in the nest to pick corpses from isolated areas and drop corpses where more related and similar items are, as shown in Figs. 1 and 2, thus growing the clusters in the colony. The likelihood of ants picking and dropping corpses are calculated mathematically. One uniqueness of this clustering algorithm is that we do not predefine the number of clusters to be formed. The algorithm is agnostic of number of clusters, it creates the clusters as it deems necessary as per the underlying mathematical formulas.

3.1 Measuring Object Similarity for Clustering

Lumer and Faieta [17] proposed a variation of the work by Deneubour et al. [6] by introducing a way to measure similarity between objects in the swarm when clustering. Given a 2-d grid (m x m) of spacial terrain where elements/objects are laid out randomly, ants, also referred as agents, perform a random walk on the grid. When an unladen ant gets to a point in the grid which has an element present in that grid, the probability of an ant to pick that element is given by:

$$P_p = \left(\frac{k_1}{k_1 + f} \right)^2 \tag{1}$$

whereas when a laden ant reaches to an empty point in the grid, the probability of that ant to drop the element is given by:

$$P_d = \left(\frac{f}{k_2 + f} \right)^2 \tag{2}$$

where k_1 and k_2 are constants. f is the similarity density measure and is calculated as:

$$f(o_i) = \frac{1}{s^2} \sum_{o_i \in Neigh(s*s)(r)} \left[1 - \frac{d(o_i, o_j)}{\alpha} \right] if \ f > 0 \tag{3}$$

$$Otherwise f(o_i) = 0$$

where $d(o_i, o_j)$ is a measure of the similarity distance between the object o_i and another object o_j within its neighborhood, $s * s$ is the number of grids in the neighborhood of object o_i, and α is a parameter used to define the scale for dissimilarity, i.e. how close two items should be to be considered close. The similarity measure that we use in this study is the level of co-integration between the stocks and is described in detail in the following subsection.

Clustering algorithms are techniques used to group similar items together. We apply ant brood sorting clustering to identify group of similarly behaving assets for portfolio diversification, which will help in constructing diversified portfolios.

3.2 Co-integration

Co-integration refers to a long-term statistical relationship between two or more time series that move together in a stable way, though the time series individually may have short-term fluctuations or trends. Co-integration shows the equilibrium connection between different individual time series and helps explain

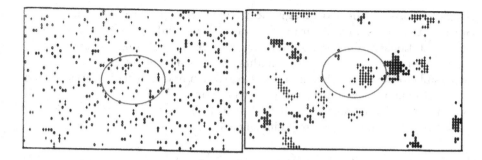

Fig. 1. Ant Brood Clustering Process (Adopted from [6])

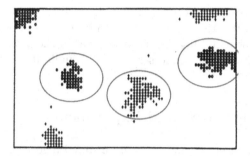

Fig. 2. Clusters of similar kind of items formed (Adopted from [6])

their behavior over time. The concept of co-integration was introduced by Engle and Granger [7] and is commonly known as Engle-Granger co-integration theory. This concept is heavily used in the finance industry, predominantly in a trading strategy called pair-trading [13,24,26], and [14].

Engle-Granger cointegration test performs the following two-step process that determines if there is co-integration between two time series [3]:

Step-1 Augmented Dickey-Fuller(ADF) Test: Conduct unit root test of both time series to determine if both time series have the same order of integration. The ADF test is applied using the model:

$$\Delta Y_t = \alpha + \beta t + \omega . Y_{t-1} + \sum_{i=1}^{k} \delta_i \Delta Y_{t-1} + \epsilon_t \tag{4}$$

The null hypothesis of the ADF test is that $\omega = 0$, which implies the presence of a unit root (non-stationarity), and evidence that $\omega < 0$ implies stationarity. Perform the ADF test for each time series and record the ADF test statistics using Eq. (5) and check if the null hypothesis can be rejected.

$$DF_\tau = \frac{\hat{\omega}}{SE(\hat{\omega})} \tag{5}$$

Step-2 Estimate the Co-integration Relationship Between Both Time Series: Use the standard Ordinary Least Square (OLS) regression and test for the stationarity in the residuals obtained by Eq. (6). OLS regression is a widely used statistical method for finding the best-fitting straight line as close as possible to the data-points in a linear regression model. It finds the estimated values of α and β by minimizing the error term ε_t. If the statistics value are lower than a critical value (usually 0.01 or 0.05) in the stationarity test (Eq. 7), we say that two time series are co-integrated.

$$Y_t = \alpha + \beta X_t + \varepsilon_t \rightarrow \varepsilon_t = Y_t - \alpha - \beta X_t \tag{6}$$

$$Check : \varepsilon_t \sim I(0) \tag{7}$$

For this study, we use an open-source python library [23] that runs the co-integration between two time series and provides t-statistic of unit-root test on residuals and P-value as results. We use the P-value, the measure of probability of cointegration between two time series as the similarity measure $d(o_i, o_j)$ in Eq. (3).

4 Dataset, Algorithm and Experiments

This study focuses on employing a nature-inspired algorithm to choose stocks from a broad range of options for the purpose of portfolio diversification.

4.1 Dataset

We use the individual stocks from the S&P 500 index for this study. The S&P 500 index is widely regarded as a benchmark for the overall performance of the United States of America (U.S) stock market. It consists of 500 large, established companies from various sectors, representing a good portion of the total market capitalization in the United States. It comprises companies from different sectors, including technology, finance, healthcare, consumer goods, etc., therefore, allowing us to examine the clustering on a broad scale of stocks. We use yfinance python library to download the daily adjusted closing prices of individual stocks of S&P 500 index for the past 8 years, from July 2015 to June 2023. We compute the P-value of all the pairs of stocks from these 500 stocks and save them in the cache memory to use in our experiment.

4.2 Implementation

We first implement the Ant Brood Sorting coupled with the co-integration test's P-value to test if this experiment can create cluster stocks of similar types of stocks. Algorithm 1 illustrates the steps that we did in this experiment to cluster similar types of stocks. We begin with initializing a 2-D grid $(m \times m)$ and place stocks and ants randomly on the grid. The value of m can be user-defined, we use Boryczka's [2] recommendation of $m = \sqrt{10 * n}$, where n is the number of stocks to be clustered.

Next, until the iteration termination condition is met, we keep looping through all the ants in each iteration. We check for each Ant if it's unladen and if there's a stock present at Ant's current location. If both conditions are true, we calculate the probability of Ant picking up that stock by comparing the similarity of the stock at Ant's location with stocks in its neighborhood. Similarly, if Ant is laden and its current location is free of any stock, we calculate the probability of Ant's dropping the laden stock by comparing the similarity of the stock with stocks in the neighborhood. If the probability is greater than a pre-determined user value, the ants pick or drop the stock at their position, respectively.

After picking/dropping the stock, the ants randomly move to a new spot in the grid within a predefined neighborhood. If the ant is laden, the priority is given to an empty site, and if the ant is unladen, we give priority to a site occupied by a stock. If no desirable sites are available, ant moves to any random site in the neighborhood.

To handle a situation of overlapping of a site that already has a stock with a laden ant moving to this site, we keep the stock laden by the ant hidden so that no other ant can pick this already laden stock and we also restrict the laden ant from dropping the stock at that site so that the site doesn't have two or more stocks at a single site. The neighborhood that the ants explore for their next step is bigger than the neighborhood used for calculating the probability of pick-up/drop-off actions. This improves the ants' ability to navigate through the spatial terrain for better and more efficient exploration of sites to pick/drop stocks. Ants move along the grid and perform pick-up and drop-off of stocks during each iteration based on the availability of stocks, empty sites, probability, and similarity/dissimilarity of stock within the neighborhood. The iteration terminates if, for a user-defined number of consecutive times, there is no pick or drop performed by the ants and all the ants are unladen. This termination condition makes sure that the iterations terminate when global optima is obtained. At the end of the iterations, we observe the resulting clusters and check for the validity of clusters if they have similar kinds of stocks.

4.3   Parameter Tuning

k_1 and k_2 are the threshold constants for picking and dropping, respectively, in Ant Brood Clustering. The value of these constants will have to be set in a way that when f (similarity measure) is $<< k_1$ then probability of picking up an item is close to 1, whereas when $f << k_2$ then probability of dropping an item is close to 0. R_p and R_d are comparator probability of pick and drop actions of the ants. These values are user-defined between 0 and 1 and are used to accelerate or brake the pick/drop speed of ants in the grid. We set these parameters in a way that the picking and dropping are set in a controlled yet loose fashion. We didn't intend to keep the movement too tight or too loose as it may cause a bottleneck or wandering explosion. The main objective of this algorithm is to see if it can cluster similar kinds of stocks, so we need a good flow of picking and dropping for forming clusters. For the process termination, we check for 1000 iterations for no pick or drop performed by the ants.

Algorithm 1. Ant Brood Clustering

INITIALIZE: Stocks and Ants randomly on the 2-d grid.

while Iterations termination condition is False **do**

    for each ant i **do**

        if Ant is unladen and the site at current location of ant has a stock **then**

            Compute $f(O_i)$ in the neighborhood using equation (3).

            Compute probability of picking up the item (P_p) using equation (1).

            Predetermine a pick-up probability comparator R_p between 0 and 1.

            if $R_p < P_p$ **then**

                Ant picks up the stock.

            end if

        else if ant is laden and the site at current location of ant is empty **then**

            Compute $f(O_i)$ in the neighborhood using equation (3).

            Compute probability of dropping the item (P_d) using equation (2).

            Predetermine a pick-up probability comparator R_d between 0 and 1.

            if $R_d < P_d$ **then**

                Drop the item at ants current site.

            end if

        end if

        Move the ant to next random site in the exploration neighborhood as per (4.2)

    end for

    Check for iteration termination condition mentioned in 4.2.

end while

Plot the clusters formed by final locations of stocks.

---

5   Results and Discussions

After parameter tuning and the successful termination of iterations, we capture the results of three random scenarios as shown in Fig. 3. The left sub-figures show the initial distribution of stocks and ants on the grid and the right sub-figures show the final results of the experiment. The blue scatters in the grid are stocks and the reds are ants. In Fig. 3, we provide results with 3 different random scenarios and it can be observed from the sub-figures that our experiments are successfully clustering a large number of stocks.

5.1   Heatmap and Cosine Similarity Results

The next step is to analyze the clusters to validate the clustering results. We use the correlation of each pair of stocks within individual clusters for validation. Figure 4 shows the heatmap of clusters from the results. It can be observed that at least 85% of the pairs of stocks within each cluster have a positive correlation.

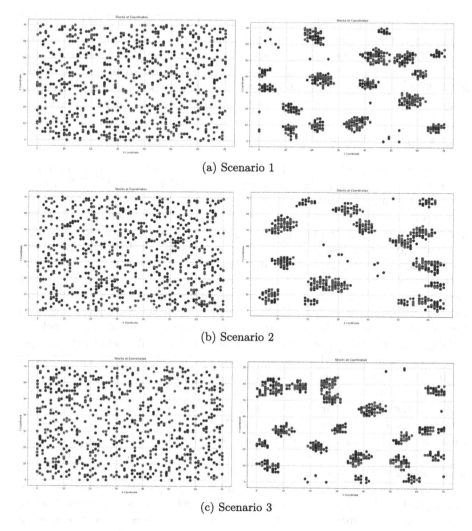

(a) Scenario 1

(b) Scenario 2

(c) Scenario 3

Fig. 3. Initial grid of ants and stocks (left) vs Final Clsuters obtained (right) in 3 random scenarios. Blue scatters represent stocks and red scatters represent ants. (Color figure online)

This explains that more than 85% of the stocks within each cluster have a similar magnitude. Since the clusters are made of time series, we use another validation measure known as cosine similarity. Cosine similarity is a measure of cosine of the angle between two vectors and is a measure of similarity in the directions of vectors [27]. Unlike Euclidian distance, cosine similarity is not highly sensitive to slight deformations such as seasonality in time series. The range of cosine similarity values typically falls between −1 and 1, where −1 suggests that the two vectors are diametrically opposed whereas 1 indicates that the two vectors being compared are identical in direction.

The average cosine similarity of stock pairs within each cluster is more than 0.9. For the three scenarios in Figure 3(a)-(c) the actual values are: 0.9221, 0.9319 and 0.9319 respectively. This validates our experimental results that not only the magnitude but also the directions of stocks within each cluster are similar.

5.2   Silhouette Score

To conduct a comprehensive validation of the end results, we also computed the Silhouette score of clusters formed as an alternative method to validate our results from multiple vantage points. Silhouette score [22] is a measure of the quality of clusters that is based on the tightness and separation of clusters. Silhouette score is calculated for each data point (stock in our case) by calculating the similarity of the data point with other data points in the same cluster and the dissimilarity with data points in other clusters. One importance of using silhouette scores for cluster validation is that they rely solely on the actual arrangement of items in clusters and are not influenced by the clustering algorithm used. Equation (8) presents the calculation for the Silhouette score $s(i)$ for a stock i where $a(i)$ is the mean distance between i and all other stocks in the same cluster and $b(i)$ is the mean distance of i from all stocks in the nearest neighbor cluster. We used Euclidean distance between stocks' closing prices to calculate the distance metric. The score ranges from -1 to $+1$, where a score near to $+1$ signifies a strong match of an item with its own cluster and a poor match to the neighbor cluster whereas a score near -1 means that the item is a better match for the neighbor cluster. We calculated the silhouette score for each individual cluster by taking the average silhouette score of all the stocks in that cluster. Table 1 shows the silhouette score obtained for the clusters of all 3 random scenarios presented in this study. The average silhouette scores for all 3 scenarios obtained are 0.43, 0.30, and 0.36. In general, a silhouette score of 0.30 and above is considered relatively moderate-high and indicates that the stocks within the clusters are relatively similar and there is a decent separation between clusters. This shows a positive validation for clusters formed, in that the clustering is effective and that the clusters are distinct.

$$s(i) = \frac{b(i) - a(i)}{max(a(i), b(i))} \tag{8}$$

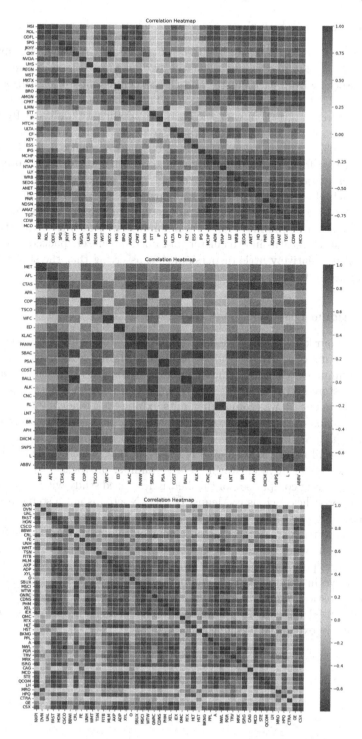

Fig. 4. Heatmap of Clusters

5.3   Inter-cluster Analysis

Table 2 presents an inter-cluster analysis of the results obtained in three (3) random scenarios. For all three (3) scenarios, we calculated the mean annualized returns of stocks within each cluster and the percentage of stock returns falling with one standard deviation of that mean. The analysis shows that the mean returns of different clusters in each scenario are significantly different from each other. For example, returns for clusters in scenario 1 range from 12% to 30%, from 10% to 30% in scenario 2, and from 9% to 23% in scenario 3. This validates that the clusters formed are different from each other in terms of annualized returns. It can also be observed that about 70%, on average, of the stocks' returns in each cluster fall within one standard deviation of the mean return of that cluster. In a normal distribution, about 68% of the data falls within one standard deviation of the mean, our result of 70% indicates that the standard deviation for stock returns in the clusters obtained are closely packed around the mean and show relatively little dispersion, thus asserting that the stocks within a cluster are close to each other.

5.4   Additional Discussion

To the best of our knowledge, the existing number of studies on the application of Ant brood sorting clustering for portfolio diversification for a direct comparison of results is limited, with no common metrics to compare this study. Oduntan et al. [21] clustered 30 stocks by running the experiment for 100,000 iterations. In their experiment, it was found that ants exhibited a tendency to allocate a significant portion of their time to random walks rather than effectively moving objects [16]. The number of actual iterations taken by our experiment prior to termination in all three (3) scenarios is less than 800, which explains that our experiment was efficient in clustering a large number of stocks.

Drawing a direct comparison of this heuristics-based study with deterministic clustering approaches such as K-Means, DBSCAN, etc., is not a sound approach due to their inherent differences. The data used by deterministic approaches generally contain multiple observations along with multiple features for each particular object, whereas we used just the time series of daily closing prices of stocks for this experiment so using the same data for deterministic methods will not be effective in generating and comparing results with deterministic methods. Deterministic methods aim to find accurate solutions, whereas Heuristics are typically used for complex problems where finding an optimal solution is computationally expensive or infeasible.

Table 1. Silhouette score for all 3 scenarios

Cluster ID	Scenario 1			Scenario 2			Scenario 3		
	(a)	(b)	**Score**	(a)	(b)	**Score**	(a)	(b)	**Score**
1	8470.66	5708.33	**−0.33**	1999.85	3376.69	**0.41**	2435.90	4079.50	**0.40**
2	3472.41	5708.33	**0.39**	9809.93	6521.73	**−0.34**	2300.42	3992.17	**0.42**
3	1908.44	5140.44	**0.63**	2399.20	3516.97	**0.32**	2127.45	3943.08	**0.46**
4	11195.34	7274.37	**−0.35**	1084.84	3074.51	**0.65**	2336.09	4082.30	**0.43**
5	2119.12	5164.97	**0.59**	1843.80	3348.83	**0.45**	2394.16	4092.61	**0.42**
6	1932.78	5103.87	**0.62**	2326.67	3472.49	**0.33**	1704.42	3814.09	**0.55**
7	2480.60	5374.99	**0.54**	2210.69	3449.42	**0.36**	5858.34	5329.69	**−0.09**
8	3184.92	5615.96	**0.43**	1593.67	3193.63	**0.50**	1681.57	3808.00	**0.56**
9	1872.99	5067.09	**0.63**	4726.10	3904.11	**−0.17**	2220.68	4020.99	**0.45**
10	1562.36	4960.98	**0.69**	2409.21	3511.90	**0.31**	5073.02	5329.69	**0.05**
11	3297.26	5597.41	**0.41**	2399.16	3510.56	**0.32**	2311.66	4022.25	**0.43**
12	1586.83	4955.55	**0.68**	3146.70	3904.11	**0.19**	3838.80	4700.80	**0.18**
13	2496.12	5374.92	**0.54**	1874.68	3293.18	**0.43**	974.01	3666.49	**0.73**
14	2374.82	5256.36	**0.55**	2292.58	3461.15	**0.34**	1438.64	3689.59	**0.61**
15	2807.78	5588.30	**0.50**	1727.58	3270.06	**0.47**	7620.32	6239.00	**−0.18**
Avg. Score	**0.43**			**0.30**			**0.36**		

Table 2. Inter-cluster analysis

Cluster ID	Scenario 1		Scenario 2		Scenario 3	
	Returns	% within 1 std	Returns	% within 1 std	Returns	% within 1 std
1	21.48	71.43	18.67	60.00	16.86	76.92
2	30.30	57.14	18.27	72.73	23.38	57.14
3	17.38	76.47	18.18	70.59	15.49	73.33
4	18.57	70.83	19.31	76.92	19.13	67.35
5	15.37	69.23	14.31	68.57	15.54	72.41
6	17.40	73.91	10.51	69.44	21.45	77.59
7	17.93	66.67	19.14	76.47	13.81	66.67
8	22.30	76.32	15.01	83.33	11.11	64.00
9	18.13	72.92	17.89	72.22	21.82	76.32
10	12.08	70.37	27.75	74.29	8.89	69.23
11	12.88	71.43	15.48	67.74	19.23	65.85
12	16.36	72.73	14.39	71.11	13.99	75.00
13	27.31	73.68	30.87	81.25	12.60	63.33
14	13.22	64.71	12.53	80.00	21.08	69.23
15	15.75	79.59	18.90	68.57	9.42	66.67

6   Conclusion

This study offers a unique contribution to the field of nature-inspired computation for large-scale portfolio diversification using Ant Brood Sorting clustering in conjunction with the co-integration of time series. The results demonstrate the algorithm's ability to effectively cluster stocks based on their similarity and the feasibility of using this method to create diversified portfolios from a large pool of stocks. This study represents a unique and valuable contribution to the field of portfolio diversification, offering a scalable approach to enhance risk management and potentially improve portfolio performance.

The correlation analysis demonstrates that over 85% of stock pairs within individual clusters exhibit positive correlations and an average cosine similarity of more than 0.9 further reinforces the consistency of stock behavior within clusters, thus validating the quality of the clusters formed and indicating that stocks within each cluster exhibited both similar magnitudes and directions. The Silhouette score analysis adds an additional layer of validation, affirming that the clusters are tightly packed and well-separated, with average scores exceeding 0.30. The inter-cluster analysis supports the validation of distinctiveness between each cluster by showcasing significant differences in mean annualized returns between clusters and more than 70% of stocks' returns in each cluster falling within one standard deviation of the cluster mean. This further confirms that the stocks within each cluster share similar financial performance characteristics and that the clusters are distinct from each other.

This approach holds promise for investors seeking to bring more robust and resilient diversification within their portfolios in diverse market conditions. For future studies, we intend to use the stocks from clusters formed in this study to create diversified portfolios and optimize the weights of the stocks using another nature-inspired algorithm called Particle Swarm Optimization. Further research in this study may also explore incorporating the seasonality factor of time series to update the clusters accordingly.

Acknowledgement. The first author acknowledges financial support from Professor Thulasiram and Graduate Enhancement of Tri-agency Stipends (GETS), University of Manitoba. The last two authors acknowledge the Discovery Grants from the Natural Sciences and Engineering Research Council (NSERC) Canada.

References

1. Arslan, H., Uğurlu, O., Eliiyi, D.T.: An overview of new generation bio-inspired algorithms for portfolio optimization, pp. 207–224. Springer Nature Singapore, Singapore (2022). https://doi.org/10.1007/978-981-16-8997-0_12
2. Boryczka, U.: Ant clustering algorithm. Intelligent Information Systems 1998 (01 2008)
3. Bui, Q., Ślepaczuk, R.: Applying hurst exponent in pair trading strategies on nasdaq 100 index. Phys. A: Stat. Mech. Appl. **592**, 126784 (2022). https://doi.org/10.1016/j.physa.2021.126784

4. Chen, Y., Zhao, X., Yuan, J.: Swarm intelligence algorithms for portfolio optimization problems: Overview and recent advances. Mobile Information Systems 2022 (07 2022). https://doi.org/10.1155/2022/4241049
5. Colomine Durán, F., Cotta, C., Fernández-Leiva, A.J.: Epoch-based application of problem-aware operators in a multiobjective memetic algorithm for portfolio optimization. In: Correia, J., Smith, S., Qaddoura, R. (eds.) Applications of Evolutionary Computation, pp. 210–222. Springer Nature Switzerland, Cham (2023)
6. Deneubourg, J.L., Goss, S., Franks, N., Sendova-Franks, A., Detrain, C., Chrétien, L.: The dynamics of collective sorting robot-like ants and ant-like robots. In: From Animals to Animats: Proceedings of the First International Conference on Simulation of Adaptive Behavior, pp. 356–365 (1991)
7. Engle, R.F., Granger, C.W.J.: Co-integration and error correction: representation, estimation, and testing. Econometrica 55(2), 251–276 (1987). http://www.jstor.org/stable/1913236
8. Freitas, F.D., De Souza, A.F., de Almeida, A.R.: Prediction-based portfolio optimization model using neural networks. Neurocomputing 72(10), 2155–2170 (2009). https://doi.org/10.1016/j.neucom.2008.08.019 lattice Computing and Natural Computing (JCIS 2007) / Neural Networks in Intelligent Systems Designn (ISDA 2007)
9. Hasan, F., Ahmad, F., Shahid, M., Khan, A., Ahmad, G.: Solving portfolio selection problem using whale optimization algorithm. In: 2022 3rd International Conference on Computation, Automation and Knowledge Management (ICCAKM), pp. 1–5 (2022). https://doi.org/10.1109/ICCAKM54721.2022.9990079
10. Huang, C.F.: A hybrid stock selection model using genetic algorithms and support vector regression. Appl. Soft Comput. 12(2), 807–818 (2012). https://doi.org/10.1016/j.asoc.2011.10.009
11. Kalayci, C.B., Ertenlice, O., Akbay, M.A.: A comprehensive review of deterministic models and applications for mean-variance portfolio optimization. Expert Syst. Appl. 125, 345–368 (2019). https://doi.org/10.1016/j.eswa.2019.02.011
12. Koumou, G.B.: Diversification and portfolio theory: a review. Fin. Markets. Portfolio Mgmt. 34(3), 267–312 (2020)
13. Krauss, C.: Statistical arbitrage pairs trading strategies: review and outlook. J. Econ. Surv. 31(2), 513–545 (2017). https://doi.org/10.1111/joes.12153
14. Liang, S., Lu, S., Lin, J., Wang, Z.: Hardware accelerator for engle-granger cointegration in pairs trading. In: 2020 IEEE International Symposium on Circuits and Systems (ISCAS), pp. 1–5 (2020). https://doi.org/10.1109/ISCAS45731.2020.9180586
15. Liu, M., Luo, K., Zhang, J., Chen, S.: A stock selection algorithm hybridizing grey wolf optimizer and support vector regression. Expert Syst. Appl. 179, 115078 (2021). https://doi.org/10.1016/j.eswa.2021.115078
16. Liu, Y.Y., Thulasiraman, P., Thulasiram, R.K.: Parallelizing active memory ants with mapreduce for clustering financial time series data. In: 2016 IEEE International Conferences on Big Data and Cloud Computing (BDCloud), Social Computing and Networking (SocialCom), Sustainable Computing and Communications (SustainCom) (BDCloud-SocialCom-SustainCom), pp. 137–144 (2016). https://doi.org/10.1109/BDCloud-SocialCom-SustainCom.2016.31
17. Lumer, E.D., Faieta, B.: Diversity and adaptation in populations of clustering ants. In: Proceedings of the third international conference on Simulation of adaptive behavior: from animals to animats 3: from animals to animats 3, pp. 501–508 (1994)

18. Mazumdar, K., Zhang, D., Guo, Y.: Portfolio selection and unsystematic risk optimisation using swarm intelligence. J. Bank. Financial Technol. 4 (01 2020). https://doi.org/10.1007/s42786-019-00013-x

19. Montgomery, D.C., Jennings, C.L., Kulahci, M.: Introduction to time series analysis and forecasting. John Wiley & Sons (2015)

20. Oduntan, O.I., Thulasiraman, P.: Hybrid metaheuristic algorithm for clustering. In: 2018 IEEE Symposium Series on Computational Intelligence (SSCI), pp. 1–9 (2018). https://doi.org/10.1109/SSCI.2018.8628863

21. Oduntan, O.I., Thulasiraman, P., Thulasiram, R.: Portfolio diversification using ant brood sorting clustering, pp. 256–261 (2014). https://doi.org/10.1109/NaBIC.2014.6921888

22. Rousseeuw, P.: Rousseeuw, p.j.: Silhouettes: A graphical aid to the interpretation and validation of cluster analysis. J. Comput. Appl. Math. **20**, 53–65 (11 1987). https://doi.org/10.1016/0377-0427(87)90125-7

23. Seabold, S., Perktold, J.: statsmodels: Econometric and statistical modeling with python. In: 9th Python in Science Conference (2010)

24. Sen, J.: Designing efficient pair-trading strategies using cointegration for the indian stock market. In: 2022 2nd Asian Conference on Innovation in Technology (ASIAN-CON), pp. 1–9 (2022). https://doi.org/10.1109/ASIANCON55314.2022.9909455

25. Shahid, M., Ashraf, Z., Shamim, M., Ansari, M.S.: A novel portfolio selection strategy using gradient-based optimizer. In: Saraswat, M., Roy, S., Chowdhury, C., Gandomi, A.H. (eds.) Proceedings of International Conference on Data Science and Applications: ICDSA 2021, Volume 2, pp. 287–297. Springer Singapore, Singapore (2022). https://doi.org/10.1007/978-981-16-5348-3_23

26. Tingjin Yan, M.C.C., Wong, H.Y.: Pairs trading under delayed cointegration. Quant. Finance **22**(9), 1627–1648 (2022). https://doi.org/10.1080/14697688.2022.2064760

27. Xia, P., Zhang, L., Li, F.: Learning similarity with cosine similarity ensemble. Inform. Sci. **307**, 39–52 (2015). https://doi.org/10.1016/j.ins.2015.02.024

Cellular Genetic Algorithms for Identifying Variables in Hybrid Gene Regulatory Networks

Romain Michelucci$^{(\boxtimes)}$ ⓘ, Vincent Callegari, Jean-Paul Comet ⓘ, and Denis Pallez ⓘ

Université Côte d'Azur, CNRS, I3S, Sophia Antipolis, France
{romain.michelucci,vincent.callegari,jean-paul.comet,
denis.pallez}@univ-cotedazur.fr

Abstract. The hybrid modelling framework of gene regulatory networks (hGRNs) is a functional framework for studying biological systems, taking into account both the structural relationship between genes and the continuous time evolution of gene concentrations. The goal is to identify the variables of such a model, controlling the aggregated experimental observations. A recent study considered this task as a free optimisation problem and concluded that metaheuristics are well suited. The main drawback of this previous approach is that panmictic heuristics converge towards one basin of attraction in the search space, while biologists are interested in finding multiple satisfactory solutions. This paper investigates the problem of multimodality and assesses the effectiveness of cellular genetic algorithms (cGAs) in dealing with the increasing dimensionality and complexity of hGRN models. A comparison with the second variant of covariance matrix self-adaptation strategy with repelling subpopulations (RS-CMSA-ESII), the winner of the CEC'2020 competition for multimodal optimisation (MMO), is made. Results show evidence that cGAs better maintain a diverse set of solutions while giving better quality solutions, making them better suited for this MMO task.

Keywords: cellular genetic algorithm · epistatic and multimodal optimisation problem · RS-CMSA-ESII · hybrid GRN · chronotherapy · real-world application

1 Introduction

Studying the dynamics of gene regulatory networks (GRNs) aims to understand the various cellular processes and pathways that empower a living organism to carry out essential functions, such as metabolic processes and the ability to adapt to environmental disturbances. Modelling such GRNs allows novel and better cognisance of disease initiation and progression, opening new perspectives in pharmacological fields such as chronotherapy, which can be viewed as the practice of administering medication at specific times during the day, taking into account the body's natural rhythms and the varying effects of the treatment. By logically following the activation or inhibition of genes and proteins under different conditions, biologist modellers can create models of these complex systems based on actual knowledge. That led to numerous modelling GRN frameworks such as *differential, stochastic* or *discrete* ones [22], each of them presenting its advantages and drawbacks. Whereas it is not too difficult to enumerate the different genes playing a role in a particular context as well as the known regulations between them, the common impediment remains the identification of the variables that govern the GRN dynamics.

In the present work, we consider *hybrid* frameworks [7] called hGRNs. They add to the discrete ones the time spent in each discrete state, allowing experimental observations to be represented as irregularly spaced time series of observable events. It has been shown that the hybrid model can exhibit these events in the same order and at the right time only if the dynamic variables that control the model behaviour satisfy a set of constraints. The design of these minimal constraints on the hGRN variables has been automated. An attempt has been made to use a continuous Constraint Satisfaction Problem (CSP) solver to extract solutions but faced difficulties when the number of variables increased [8]. Recently, [17] showed that the CSP, exhaustively characterising the set of solutions, can be expressed as a free optimisation problem (FOP) by indirectly handling constraints thanks to metaheuristics. The CSP was transformed into a non-separable, non-trivial, continuous, and single objective problem in which the search space increases exponentially with the number of genes in the hGRN. One limitation of this approach is that such algorithms are panmictic and can only identify one basin in the search space. From a modelling perspective, exhibiting a diverse sampling of biologically satisfactory solutions allows biologists to reason not only on one possible identification but also on a set of sensible ones. Therefore, this work focuses both on validating the previous approach on hGRNs involving more genes and complex dynamics and on the multimodal aspect of the identification problem. RS-CMSA-ESII is a new niching method for MMO that emerged as the most successful available method when robustness and efficiency are considered at the same time and does not make any assumptions such as distribution, shape, and size of the basins [2]. This CEC'2020 top niching-based algorithm is the logical choice to be tested as a baseline to gain more insights on its ability to find a set of solutions without having any assumptions on the modes. In the meantime, cGAs are well-known heuristics to tackle epistatic and multimodal tasks [4,5] since the diversity maintenance is guaranteed thanks to the structure and ratio of the population, unlike RS-CMSA-ESII which employs mechanisms with different sub-populations running in parallel. So, this research aims to address the problem of the hGRN variables identification to obtain a diverse set of quality solutions for increasingly complex models while seeking to identify the most suitable method for achieving these goals.

To meet these objectives and based on the research hypotheses set out above, the article is organised as follows: Sect. 2 describes the hGRN continuous optimisation problem by detailing: (i) the definition of the hybrid model along with its dynamics, (ii) the experimental observations that serve as input, and (iii) how this problem has been treated as an FOP. Section 3 encompasses an overview of RS-CMSA-ESII and cGAs from a multimodal perspective. Section 4 proposes experiments comparing CMA-ES, GA, multiple cGAs with varying ratios and structures, and RS-CMSA-ESII on three different hGRNs of increasing complexity. Experimental results and statistical tests are presented and discussed. Finally, conclusions are drawn in Sect. 5.

2   hGRN Variables Optimisation

2.1   Hybrid Gene Regulatory Networks

Hybrid modelling of gene regulatory networks (GRNs) aims to describe the effect of regulations between genes in a biological system by taking into account the continuous time component. Traditionally, a GRN is a directed graph in which vertices express abstractions of one or multiple biological genes (v_1, v_2), and edges that act as either activation ($\rightarrow$) or inhibition ($\dashv$) represent regulations (Fig. 1a). This static representation seems of limited interest since it does not integrate any dynamics. However, from Fig. 1a, the corresponding discrete dynamics (Fig. 1b) can be built. First, grey boxes are obtained from the previous GRN by enumerating all possible states $\mathbb{S}$: each grey square box identifies a *discrete state* $\eta \in \mathbb{S}$ defined by the level of the GRN genes. If we suppose that the maximum level of each gene v_i is 1, then the top right box is the state where each gene is expressed at its maximum level and is denoted by $\eta = (\eta_{v_1}, ..., \eta_{v_n})$. In Fig. 1, this state is $\eta = (\eta_{v_1}, \eta_{v_2}) = (1, 1)$. From this first step, transitions between discrete states can be drawn (black arrows) and symbolise the discrete evolution of the concentration of the gene products. Although the obtained discrete state graph of Fig. 1b is deeply interesting for logical reasoning about regulatory changes, it disregards temporal information, which is nevertheless crucial, for example, for optimising medical treatments by taking account of biological rhythms.

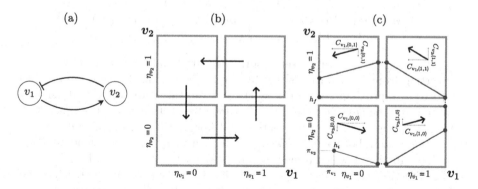

Fig. 1. Example of a GRN depicted as a directed graph (a), its discrete state graph (b), and a possible dynamic of its hybrid state graph (c) (taken from [17]).

The hybrid modelling framework adds the notion of temporal continuous evolution to the previous dynamics by adding linear continuous trajectories (red straight lines) to the discrete transitions of a GRN (pictured with dotted red lines in Fig. 1c). On a trajectory, a point is called a *hybrid state* and given by its position π within a discrete state η. As an example, the initial hybrid state h_i in Fig. 1c has the coordinates $\left((\eta_{v_1}, \eta_{v_2})^t, (\pi_{v_1}, \pi_{v_2})^t\right) = \left((0,0)^t, (0.25, 0.25)^t\right)$. To determine a complete trajectory through a set of discrete states, hGRN models require an initial hybrid state h_i and a vector of the evolution of concentrations in each discrete state, called *celerity vector*. This vector gives the direction and celerity of each gene $v \in V$ in a discrete state $\eta \in \mathbb{S}$, e.g. the celerity of v_1 in $\eta = (0,0)$ is denoted $C_{v_1,(0,0)}$. In the general case, the celerity of v in η is a floated value defined as $C_{v,\eta}$.

The aim is to identify celerity vectors to generate valid hGRN models of the biological system under study. Such a determination could help biologists make new interpretations about the possible dynamics of the system.

2.2 Biological Knowledge

The identification process requires some input data, which allows the modeller to validate or not a possible valuation of continuous variables. While much work [10,18,20,21] is based on gene expression data, our approach takes into consideration already-formalised information analysed by biologists derived from both biological data and expertise.

The formalism abstracts the knowledge extracted from biological experiments under the form of constraints on the global trajectory: it must (i) start from an initial hybrid state $h_i = (\eta_i, \pi_i)$, (ii) verify a triplet of properties in each successive discrete state $(\Delta t, b, e)$ where Δt expresses the time spent; b delineates the observed behaviours during the continuous trajectory; e specifies the next discrete state transition, and (iii) reach the final hybrid state $h_f = (\eta_f, \pi_f)$. Let us detail the biological knowledge (BK) used for the example of Fig. 1c:

$$\{h_i\} \begin{pmatrix} 5.0 \\ noslide\,(v_2) \\ v_1+ \end{pmatrix}; \begin{pmatrix} 7.0 \\ slide^+\,(v_1) \\ v_2+ \end{pmatrix}; \begin{pmatrix} 8.0 \\ noslide\,(v_2) \\ v_1- \end{pmatrix}; \begin{pmatrix} 4.0 \\ slide^-\,(v_1) \\ v_2- \end{pmatrix} \{h_f\}$$

$h_i = \left((0,0)^t, (\pi_{v_1}, \pi_{v_2})^t\right)$ represents both the initial and final state ($h_i = h_f$).

Starting from h_i, the time spent by the trajectory inside the discrete state $\eta = (0,0)$ is approximately 5 h ($\Delta t = 5.0$). Within this state, the celerity should move towards the next discrete state of v_1 (v_1+) so as to increase the concentration level of gene v_1 until it reaches the right border without touching either the top or the bottom border ($noslide(v_2)$) and then jump into the neighbour state $\eta = (1,0)$. In this new discrete state, the trajectory evolves for 7 h ($\Delta t = 7.0$) in the direction of $\eta_{v_2} = 1$ (v_2+) but, this time, the trajectory reaches the right border, which corresponds to the maximum admissible concentration of v_1 ($slide^+(v_1)$). This process continues until the trajectory reaches h_f. Any valuation of dynamic variables, i.e. celerity vectors, leading to a trajectory satisfying this BK is considered admissible.

2.3   Single Objective and Multimodal Optimisation Problem

Searching for celerity values that satisfy BK initially led to characterising the problem as a CSP and solving it by constraint-based programming [8]. On the one hand, this constraint-based programming method was able to exhaustively find the over-approximated sets of solutions, but as the number of dimensions increased, such a method was unable to extract even one particular solution.

A recent attempt [17] has recently formulated the problem as being single-objective by proposing an adequate fitness function consisting of three criteria and testing this approach on the hGRN model of Fig. 1c (only two genes). In this preliminary study, the decision vector to be optimised consisted of finding the initial hybrid state h_i and all celerity values of all discrete states:

$$h_i, \{C_{v,\eta} | v \in V, \eta \in \mathbb{S}\} \tag{1}$$

Thus, for example, finding an admissible valuation of Fig. 1c satisfying BK was equivalent to finding the optimal parameter set of:

$$x = (h_i; C_{v_1,(0,0)}; C_{v_2,(0,0)}; C_{v_1,(1,0)}; C_{v_2,(1,0)}; C_{v_1,(1,1)}; C_{v_2,(1,1)}; C_{v_1,(0,1)}; C_{v_2,(0,1)}).$$

In this previous work, the fitness function is defined as the sum of three distances, each corresponding to one of the criteria associated with BK:

$$f(x) = \sum_{\eta} d_{\Delta t}(tr, BK) + d_b(tr, BK) + d_e(tr, BK) \tag{2}$$

where $d_{\Delta t}(tr, BK)$ is the distance between the expected time given by BK (Δt) and the time spent in the current state by the considered trajectory; $d_b(tr, BK)$ represents the distance between the trajectory behaviour inside the discrete state and the property of BK; and $d_e(tr, BK)$ compares the expected next discrete state according to BK with the discrete state into which the considered trajectory enters. The function domain is $\left(\prod_{v \in V}[0, b_v]\right) \times [0,1]^n \times \mathbb{R}^{|C|}$ where n is the number of genes and $|C|$ is the total number of celerities to identify, i.e. the length of the decision vector. The codomain is $\mathbb{R}^+$.

Minimising these three criteria led to the identification of admissible celerity values. However, the optimisation problem becomes increasingly complex when considering hGRN models with many genes. It implies more celerity values to identify and more complex interactions, leading to harder implicit constraints. The continuous CSP solver was unable to extract even one particular solution when considering a model with five genes, leading to 240 variables in the decision vector. Furthermore, the task is multimodal: it is interesting to find diverse solutions to provide biologists with evidence for different interpretations of hGRN dynamics. The approach proposed by [17] did not address this issue. The peculiarities of this optimisation problem are: (i) there is an infinite number of solutions that satisfy the BK constraints, and (ii) the optima solutions lie on a neutral landscape, i.e. a plateau. Indeed, solutions form a measure zero set due to the equality constraints on the time criterion in the fitness function. Therefore, the optimisation procedure requires the ability to sample, in a continuous landscape, global and local optima plateaus of measure zero. These considerations specific to this optimisation problem cannot be addressed only by panmictic schemes. Therefore, the limits of the mentioned approach are tested by introducing experiments with well-known multimodal heuristic algorithms on higher dimensional hGRNs.

3   RS-CMSA-ESII and cGAs for MMO

RS-CMSA-ES [1] was designated the most successful niching method for the CEC'2013 MMO test suite. In this initial version, several parallel subpopulations, each following the evolution scheme of CMSA-ES [9], aim at finding distinct global minima. CMSA-ES is an adapted version of CMA-ES [14], diminishing the complexity of the adaptation process and implying fewer hyperparameters tuning. RS-CMSA-ES gathers several techniques and encompasses them as a new algorithm for MMO without making any assumption about the fitness landscape: taboo points (points from which the offspring of a subpopulation must maintain a sufficient distance, i.e. the centre of the fitter subpopulations and the previously identified basins), the normalised Mahalanobis distance, and the Ursem's hill-valley function [23]. The new variant RS-CMSA-ESII [2] introduces an update of the adaptation schemes for the normalised taboo distances, new termination criteria for subpopulation evolution, and an improvement of the time complexity thanks to (i) a new initialisation strategy of subpopulations, and (ii) a more accurate metric for the determination of critical taboo regions thanks to the properties of Mahalanobis distance. The RS-CMSA-ESII superiority over successful niching methods in static MMOs made it an ideal candidate for this study.

cGAs are well-known methods for addressing multimodal and epistatic problems [4,5]. They are a subclass of GAs in which the population is structured in a specified topology, allowing individuals to interact only with their neighbours. The topological structure defines a connected graph where a vertex represents an individual, and an edge represents the possibility of interaction between two individuals: each individual, in this graph, can only mate with its neighbours. Therefore, in a cGA, the choice of the population topology and the neighbourhood are two parameters that guide the search and control the solutions' diffusion speed along the graph. The *radius* introduced in [5] directs the dispersion strength based on the chosen neighbourhood: the higher the radius, the more spread out a neighbourhood's pattern is, and so the easier a good solution will reach other individuals of the population because there will be less intermediate individuals to the most distant individual. Furthermore, [19] introduced the *ratio* measure controlling the balance between exploration and exploitation. It is defined as a trade-off between the radii of the neighbourhood and the population structure: reducing the ratio leads to the promotion of exploration. Overlapping neighbourhoods also help to explore the search space because the slow diffusion of solutions through the population allows exploration by preserving diversity [3,4]. On the one hand, this leads cGAs to find several optima compared to GAs and to be well suited for complex problems. On the other hand, this is often at the expense of slower convergence towards global optima.

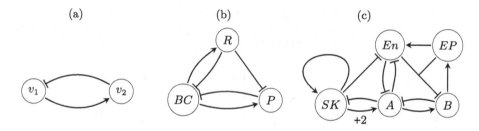

Fig. 2. Interaction graphs of the 2G (a), 3G (b), and 5G (c) hGRN.

Table 1. Description of hGRN models.

Name	Nb. genes	Decision vector len.	BK
example cycle (2G)	2	8	Given in Sect. 2.2
circadian cycle (3G)	3	20	[7]
cell cycle (5G)	5	240	[6]

In the following section, tests have been set up to compare the RS-CMSA-ESII performance along with cGAs to demonstrate which method is best suited to our multimodal task. Different structure and ratio values for cGA are experimented with to evaluate their performance. We compared all the results with standard panmictic metaheuristics on three hGRN models of increasing complexity to assess the suitability of their diversity mechanism for such MMO problems.

4   Experimental Study

The three hybrid models of GRN are depicted in Figure 2 and described in Table 1 in terms of (i) the number of genes, (ii) the length of the decision vector to optimise, and (iii) constraints from BK utilised for evaluating candidate solutions.

4.1   Optimisation Methods and Parameters Search

The comparison is carried out between $(\mu+\lambda)$ GA, CMA-ES, six synchronous cGAs with different ratios and neighbourhood structures, and RS-CMSA-ESII.

The two continuous metaheuristic implementations come from PyMoo [11], and each of the hyperparameters chosen is identical to those detailed in [17]. Their population size is also 500. Since we were interested in observing the influence of the cGAs parameters to find those most suitable for solving the different hGRN problems, multiple sets of parameters were tested (listed in Table 2). The names of the neighbourhoods follow the classical notation: the label Ln (linear) for the neighbourhoods composed by the n nearest neighbours in a given axial

Table 2. Description of tested cGAs parameters.

Name	Population	Neighbourhood	Ratio
cGAL5	5×10	L5	0.279
cGAL9	10×10	L9	0.367
cGAL29	15×15	L29	0.719
cGAL41	21×21	L41	0.851
cGAL13	$7 \times 7 \times 7$	L13	0.607
cGAC9	$7 \times 7 \times 7$	C9	0.408

direction (north, south, west and east) while the label Cn (compact) designates the neighbourhoods containing the $n - 1$ nearer individuals to the considered one (in horizontal, vertical, and diagonal directions). The population size and the neighbourhood structure vary so that we can test (i) low ratio cGAs with a small population size and, conversely, (ii) high ratio cGAs with a larger population, both in a toroidal 2G square grid, and (iii) 3G neighbourhood structure. To ensure fair results, their implementation is also based on the standard GA implementation provided in PyMoo. RS-CMSA-ESII implementation is taken from [2] with the control parameters set to their default values.

Each experiment is run 50 times to obtain statistically significant results. The termination criteria chosen is the number of function evaluations (NFE): $100,000$ for 2G and 3G and $200,000$ for 5G. These values were chosen based on the relative complexity and the decision vector length.

4.2   Results

For each algorithm, problem dimension and at each generation, we compute the best candidate solution so far, repeat executions 50 times, and compute the Mean Best Fitness (MBF). The monotonic evolution of all algorithms is shown in the left column of Fig. 3. It can be observed that (i), as expected, panmictic metaheuristics perform worse than cGAs in all cases since they reach a plateau faster and attain a higher fitness score after convergence; (ii) cGAL13, cGAL29, and cGAL41 stand out among the algorithms tested since, on the one hand, they have a slower convergence, and on the other hand, even when the maximum budget is attained, their curves show that the search process could have pursued its convergence; and (iii) RS-CMSA-ESII performs worse than CMA-ES.

In addition, Cumulative Distribution Function (CDF) curves are constructed on the right side of Fig. 3 for each hGRN considered. Each CDF curve describes the probability of finding a solution at, or below, a given fitness score. For instance, in 3G, there is almost an 80% probability that a user will obtain a solution with a fitness score less than or equal to 10^{-4} with cGAL9 given 100,000 NFE. From these plots, (i) cGAs don't often find the overall best solution (the one with the lowest fitness score) but results are rarely unsatisfactory (>1), (ii) in all cases, CMA-ES can deliver top results (satisfactory and precise solutions) as it is of poor performance (not solving the problem), (iii) RS-CMSA-ESII similarly to CMA-ES has mixed performance and does not find any single satisfactory valuation in 5G.

In MMO, the chi-square-like performance statistic and maximum peak ratio are common measures to identify a maximum number of optima (local and global). However, both of these measures assume the number and locations of the global optima are known *a priori*. This assumption does not hold in our case, so the scoring function used is introduced in [16] and defined as:

$$sc(P, \theta_l, \theta_u) = \sum_{B_j \in Bin_k(clust_\sigma(P), \theta_l, \theta_u)} w_j |B_j| \tag{3}$$

This alternative performance measure suggests the selection of a threshold interval $[\theta_l, \theta_u]$ covering all fitness score values considered interesting by an expert. θ_l is the ideal point while θ_u is an upper bound below which fitness values are judged satisfactory. In our case, $\theta_l = 0$ and $\theta_u = 10^{-2}$. 10^{-2} is a precision error coherent with biological expertise. For instance, a trajectory which would slide in a state during a fraction of seconds ($<\theta_u$) before going to the

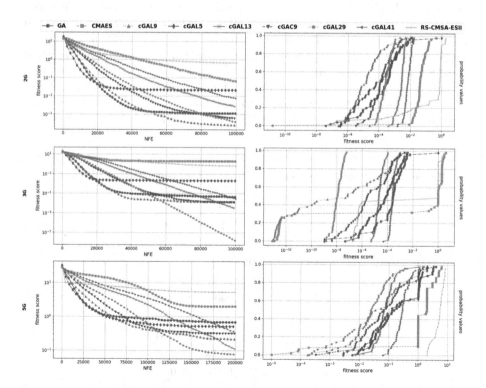

Fig. 3. Monotonic evolution of MBF values (left) and CDF curves of overall best results (right) for the three hGRNs.

next discrete state is a satisfying trajectory despite BK stating $noslide(v)$. The score measurement uses density-based clustering with parameter σ to remove redundancy between candidate solutions clustered closely around the same local optimum. In this study, DBSCAN [13] is parametrised with $\sigma = 10^{-1}$ which is the maximum Euclidean distance between two samples for one to be considered as in the neighbourhood of the other. Equidistant binning is then used to adapt the distribution weights: more emphasis is put on higher quality optima than lower ones. The number of bins is kept at 16. This score assesses the combined quality of the found candidate solutions while it is not prone to be misled by redundancy. Table 3 shows numerical values for the mean scores where bold results highlight the best performance for each model dimension. The comparison indicates that small ratio cGAs (cGAL9 and cGAL5) are to be preferred for 2G, whereas cGAs with a higher ratio perform better in the 3 and 5G cases, as shown by cGAL41 and cGAL29. It should also be noted that, in 5G, the extrema ratio values (cGAL5 and cGAL41) are penalised for being too exploratory or exploitative. cGAC9 has interesting results in all three cases but never stands out.

Table 4 summarises statistics of the last population clustered: it contains only the fitness values of the best candidate solutions $(<\theta_u)$ gathered around each distinct optima found by clustering. The best results (column by column) are shown in bold. The average of the mean and standard deviation of the clustered results is reported, as well as the overall minimum fitness scores (the reader can refer to the leftmost point of each corresponding CDF curve). When considering one particular run, it may appear that an algorithm did not find any solution below θ_u. In such cases, the maximum value θ_u is considered: this results in a normalised average with the ideal value being θ_l, and θ_u the nadir one. It can be observed that cGAL9 finds, on average, higher quality optima than other algorithms in 2 and 5G. In 3G, cGAL29 identifies satisfying solutions with a lower fitness score on average.

Table 3. Overview of the average performance measurement over 50 runs.

Algorithms	2G	3G	5G
GA	12.13	148.09	8.29
CMA-ES	3e-3	0.275	29.32
cGAL9	**19.99**	63.47	21.11
cGAL5	16.82	30.07	2.04
cGAL13	7.33	290.59	0.03
cGAC9	13.78	162.28	34.11
cGAL29	13.90	182.18	**49.81**
cGAL41	1.56	**311.07**	0.0
RSCMSAII	1e-2	0.275	0.0

Table 4. Summary of clustered results.

Algorithms	2G		3G		5G	
	mean ± std	min	mean ± std	min	mean ± std	min
GA	1e-3 ± 2e-4	4e-8	4e-4 ± 2e-6	5e-8	99e-4 ± 94e-4	6e-3
CMA-ES	98e-4 ± 96e-4	**2e−11**	7e-3 ± 2e-13	1e-13	9e-3 ± 9e-3	**1e−5**
cGAL9	**8e−4±8e−4**	3e-7	1e-4 ± 7e-6	**8e−14**	**85e−4±7e−3**	2e-4
cGAL5	1e-3 ± 9e-4	3e-7	9e-4 ± 2e-4	1e-9	95e-4 ± 94e-4	3e-3
cGAL13	6e-3 ± 2e-3	4e-4	3e-4 ± 4e-4	5e-6	1e-2 ± 98e-4	7e-3
cGAC9	1e-3 ± 1e-3	3e-6	1e-4 ± 8e-6	4e-9	9e-3 ± 8e-3	3e-4
cGAL29	3e-3 ± 2e-3	3e-5	**9e−7±1e−5**	1e-9	79e-4 ± 7e-3	2e-4
cGAL41	8e-3 ± 3e-3	2e-3	2e-3 ± 16e-4	5e-5	1e-2 ± 0	1e-2
RSCMSAII	8e-3 ± 1e-20	8e-8	7e-3 ± 2e-13	1e-13	1e-2 ± 0	1e-2

4.3   Statistical Analysis

A statistical validation campaign was conducted to evaluate the observed differences in the reported performance values of all algorithm pairs for each different hGRN. We consider two null hypotheses H_0^1 which states that the observed performance scores are equal, and H_0^2 which states that the average fitness scores obtained by clustering are similar. These null hypotheses are duplicated for each of the hGRN dimensions considered. To test them, we first employed the Friedman rank-sum test to assess whether at least two methods exhibit significant differences. The p-values for the null hypotheses show, at a $\alpha = 5\%$ confidence level, that the differences are significant. The choice between parametric and non-parametric tests is made according to the independence of the samples (seeds are different), whether or not the data samples are normally distributed, and the homoscedasticity of the variances [12]. As neither normality nor homoscedasticity conditions required for the parametric tests application hold, the non-parametric Wilcoxon signed-rank test was performed. In a complementary way, to reduce the issue of Type I errors in multiple comparisons, the Bonferroni correction method was applied. [15] gives the score +1 (resp. -1) for the superior (resp. inferior) algorithm whenever the considered null hypothesis could be significantly rejected. A score of 0 is assigned when neither algorithm is significantly better than the other. Since we have three different case studies (2G, 3G, 5G), for each pair of algorithms and each null hypothesis, we sum the three obtained scores to estimate which one is globally better considering the three hGRNs. Table 5 (resp. Table 6) show these sums according to the pairwise Wilcoxon tests (resp. Bonferroni correction): a positive number for algorithm in line l shows that it was significantly better than the algorithm in column c (considering the three hGRNs). For example, according to the Bonferroni correction applied on H_0^1, we can state that cGAL29 is significantly better than RS-CMSA-ESII for the three study cases but compared to cGAL41, we can only say that it is globally better:

Table 5. Pairwise Wilcoxon statistical tests of H_0^1 (left) and H_0^2 (right).

	CMA-ES		cGAL9		cGAL5		cGAL13		cGAC9		cGAL29		cGAL41		RSCMSAII	
GA	+2	+2	0	−2	0	+1	0	+1	−2	−1	−2	−1	0	+2	+2	+2
CMA-ES			−2	−2	−2	−2	−1	−1	−2	−2	−2	−2	−1	−1	0	+1
cGAL9					+3	+2	+1	+3	0	+1	0	0	+1	+3	+3	+3
cGAL5							0	0	−1	0	−1	−1	0	+2	+2	+2
cGAL13									−1	−3	−1	−3	0	+2	+2	+2
cGAC9											−1	−1	+1	+3	+3	+3
cGAL29													+1	+3	+3	+3
cGAL41															+2	+1

Table 6. Bonferroni post-hoc analysis of H_0^1 (left) and H_0^2 (right) with bolded differences compared to Table 5.

	CMA-ES		cGAL9		cGAL5		cGAL13		cGAC9		cGAL29		cGAL41		RSCMSAII	
GA	+2	+2	0	**0**	+1	0	0	+1	**0**	**0**	−1	0	0	+2	+2	**+1**
CMA-ES			−2	−2	−2	−2	**−2**	**−2**	−2	−2	−2	−2	**−2**	**−2**	0	**0**
cGAL9					**+1**	**+1**	0	**+2**	0	**0**	0	+1	**0**	**+2**	+2	+2
cGAL5							0	**+1**	−1	**−1**	−1	0	0	+2	+2	**+1**
cGAL13									**0**	−2	−1	−3	0	+2	+2	**0**
cGAC9											**0**	**0**	0	+2	+2	+2
cGAL29													+1	+3	+3	+3
cGAL41															+2	**0**

cGAL29 may have scored +2 and cGAL41 +1 or cGAL29 may have scored +1 and cGAL41 0.

If we analyse the conclusions supported by the tests, based on the acceptance or rejection of the above hypotheses, we arrive at the following findings: on the different tasks, cGAL9 and cGAL29 are more competitive in finding more optima than other algorithms with better fitness values on average. RS-CMSA-ESII lags as the panmictic algorithms maintain greater diversity in their population across different hGRN landscapes.

4.4   Visualisation

Figure 4 shows the diversity of solutions of cGAL9 tested on hGRNs with 2, 3 and 5 genes. Please note that three different graph types are modelled to emphasize the same phenomenon: the evolution of gene products concentration. In 2G (Fig. 4a) and 3G (Fig. 4b), the discrete states can be represented as squares and cubes. However, in 5G (Fig. 4c), the choice has been made to represent the evolution of concentration (in y-axis) as a function of the time spent (in x-axis) for the different genes. This visually confirms that the application of evolutionary computation allows us to exhibit very different solutions, each consistent with BK.

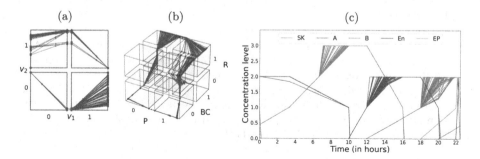

Fig. 4. Admissible trajectories obtained with cGAL9 on the 2G (a), 3G (b), and 5G (c) hGRN.

5   Conclusion

hGRN variable identification is framed as an ideal tool to help biologists develop hypotheses and facilitate the design of their experiments. This study proposes an improvement to [17] since (i) it shows that evolutionary computation can outperform constraint-based approach by dealing with higher dimensional models, the 5G cell cycle in this study, and (ii) it is now able to find a diverse set of optima solutions instead of a unique one. CGAs have shown superiority over the best available niching-based algorithm (RS-CMSA-ESII) by maintaining diversity within the population structure. Surprisingly, RS-CMSA-ESII does not ensure diversity in the results: only one solution is found. In our case, optima are located on a neutral landscape: there is an infinite number of solutions forming a null set. Therefore, for sampling a continuous landscape with global and local optima plateaus of measure zero, the mechanisms employed by RS-CMSA-ESII are not suitable. Because the Ursem's hill-valley test fails, it ensures that only one subpopulation at a time evolves, leading to a single solution. That entails the degenerate use of the metaheuristic, explaining the disappointing results of RS-CMSA-ESII. In the case of cGAs, maintaining diversity through population structure helps to preserve diversity in the parameter space and thus enables us to obtain a diversity in the phenotype space. Future works will consider the development of specific diversity mechanisms to better leverage the multimodality issue on a neutral landscape: the design of an appropriate *self-adaptive* cGA to obtain quality results while maximising the number of optima. At the same time, introducing larger biological systems will lead to applying *large-scale* optimisation.

Acknowledgments. This work has been supported by the French government, through the France 2030 investment plan managed by the Agence Nationale de la Recherche, as part of the "UCA DS4H" project, reference ANR-17-EURE-0004.

References

1. Ahrari, A., Deb, K., Preuss, M.: Multimodal optimization by covariance matrix self-adaptation evolution strategy with repelling subpopulations. Evol. Comput. (2017). https://doi.org/10.1162/evco_a_00182
2. Ahrari, A., Elsayed, S., Sarker, R., Essam, D., Coello, C.A.C.: Static and dynamic multimodal optimization by improved covariance matrix self-adaptation evolution strategy with repelling subpopulations. IEEE Trans. Evol. Comput. (2021). https://doi.org/10.1109/TEVC.2021.3117116
3. Alba, E., Dorronsoro, B.: Solving the vehicle routing problem by using cellular genetic algorithms. In: European Conference on Evolutionary Computation in Combinatorial Optimization (2004). https://doi.org/10.1007/978-3-540-24652-7_2
4. Alba, E., Dorronsoro, B.: Introduction to cellular genetic algorithms. In: Cellular Genetic Algorithms (2008). https://doi.org/10.1007/978-0-387-77610-1_1
5. Alba, E., Troya, J.M.: Cellular evolutionary algorithms: evaluating the influence of ratio. In: International Conference on PPSN (2000). https://doi.org/10.1007/3-540-45356-3_3
6. Behaegel, J., Comet, J.P., Bernot, G., Cornillon, E., Delaunay, F.: A hybrid model of cell cycle in mammals. In: 6th International Conference on Computational Systems-Biology and Bioinformatics (2015). https://doi.org/10.1142/S0219720016400011
7. Behaegel, J., Comet, J.P., Folschette, F.: Constraint identification using modified Hoare logic on hybrid models of gene networks. In: Proceedings of the 24th Int. Symposium TIME (2017). https://doi.org/10.4230/LIPIcs.TIME.2017.5
8. Behaegel, J., Comet, J.P., Pelleau, M.: Identification of dynamic parameters for gene networks. In: Proceedings of the 30th IEEE International Conference ICTAI (2018). https://doi.org/10.1109/ICTAI.2018.00028
9. Beyer, H.G., Sendhoff, B.: Covariance matrix adaptation revisited - the cmsa evolution strategy. In: International Conference on PPSN (2008). https://doi.org/10.1007/978-3-540-87700-4_13
10. Biswas, S., Acharyya, S.: Neural model of gene regulatory network: a survey on supportive meta-heuristics. Theory Biosci. (2016). https://doi.org/10.1007/s12064-016-0224-z
11. Blank, J., Deb, K.: pymoo: Multi-objective optimization in python. IEEE Access (2020)
12. Eftimov, T., Korošec, P.: Statistical analyses for meta-heuristic stochastic optimization algorithms: GECCO Tutorial (2020). https://doi.org/10.1145/3377929.3389881
13. Ester, M., Kriegel, H.P., Sander, J., Xu, X., et al.: A density-based algorithm for discovering clusters in large spatial databases with noise. In: KDD (1996)
14. Hansen, N., Auger, A.: Cma-es: evolution strategies and covariance matrix adaptation. In: Proceedings of the 13th Annual Conference Companion on Genetic And Evolutionary Computation (2011). https://doi.org/10.1145/2001858.2002123

15. Kronfeld, M., Dräger, A., Aschoff, M., Zell, A.: On the benefits of multimodal optimization for metabolic network modeling. In: German Conference On Bioinformatics (2009)

16. Kronfeld, M., Zell, A.: Towards scalability in niching methods. In: IEEE CEC (2010). https://doi.org/10.1109/CEC.2010.5585916

17. Michelucci, R., Comet, J.P., Pallez, D.: Evolutionary continuous optimization of hybrid gene regulatory networks. In: EA 2022. https://doi.org/10.1007/978-3-031-42616-2_12

18. Mitra, S., Biswas, S., Acharyya, S.: Application of meta-heuristics on reconstructing gene regulatory network: a bayesian model approach. IETE J. Res. (2021). https://doi.org/10.1080/03772063.2021.1946433

19. Sarma, J., De Jong, K.A., et al.: An analysis of local selection algorithms in a spatially structured evolutionary algorithm. In: ICGA, pp. 181–187. Citeseer (1997)

20. da Silva, J.E.H., Betnardino, H.S., Helio J.C., B., Vieira, A.B., Luciana C.D., C., de Oliveira, I.L.: Inferring gene regulatory network models from time-series data using metaheuristics. In: IEEE CEC (2020). https://doi.org/10.1109/CEC48606.2020.9185572

21. Sun, J., Garibaldi, J., Hodgman, C.: Parameter estimation using meta-heuristics in systems biology: a comprehensive review. IEEE/ACM Trans. Comput. Biology Bioinform. (2012). https://doi.org/10.1109/TCBB.2011.63

22. Tenazinha, N., Vinga, S.: A survey on methods for modeling and analyzing integrated biological networks. IEEE/ACM Trans. Comput. Biol. Bioinform. (2011). https://doi.org/10.1109/TCBB.2010.117

23. Ursem, R.K.: Multinational evolutionary algorithms. In: Proceedings of CEC (1999). https://doi.org/10.1109/CEC.1999.785470

Evolving Artificial Neural Networks for Simulating Fish Social Interactions

Lea Musiolek[1,5(✉)] [iD], David Bierbach[2,5] [iD], Nils Weimar[3], Myriam Hamon[4],
Jens Krause[2,5] [iD], and Verena V. Hafner[1,5] [iD]

[1] Adaptive Systems Group, Humboldt-Universität zu Berlin, Berlin, Germany
lea.musiolek@hu-berlin.de
[2] Department of the Biology and Ecology of Fish, Humboldt-Universität zu Berlin,
Berlin, Germany
[3] Zoology Department, Universität Bonn, Bonn, Germany
[4] Bernstein Center for Computational Neuroscience, Berlin, Germany
[5] Science of Intelligence, Research Cluster of Excellence, Marchstr. 23, 10587 Berlin,
Germany
https://www.scienceofintelligence.de

Abstract. Can we use computational modeling to infer whether fish can remember or anticipate each other's movements? What minimum of temporal input and internal complexity is sufficient to model a specific fish, or to produce generally "fish-like" behavior? Agent-based modeling to emulate biological behavior has been used to great effect, both in real-world and simulated experiments. We present feedforward neural network architectures for simulating fish social interactions, evolved using evolution strategies in two different experiments. Evolution of the temporal input of the partner fish's position when testing models on labeled data uncovers anticipation or memory capacities used by a focal fish. When testing via a general discriminator for fish-like trajectories, the right neural network architecture and temporal input are shown to be a necessary, but insufficient condition for highly lifelike simulations. Lifelike simulations for some datasets are possible as simple functions of the input, showing variability in the complexity of individual fish's social behaviors.

Keywords: Evolution Strategy · Fish · Social Interactions ·
Agent-Based Modeling · Artificial Agents

1 Introduction

Can a small freshwater fish anticipate a partner's movements in social interaction? Does it need to, in order to behave "like a fish"? Finding the answers is

Funded by the Deutsche Forschungsgemeinschaft (DFG, German Research Foundation) under Germany's Excellence Strategy - EXC 2002/1 "Science of Intelligence" - project number 390523135.

Supplementary Information The online version contains supplementary material available at https://doi.org/10.1007/978-3-031-56852-7_10.

not as easy as it sounds. It has been shown through tailored biological experiments that *Poecilia reticulata* (guppies) are capable of leading and following each other, socially learning food locations, anti-predator behavior and other useful skills [7], choosing interaction and mating partners based on past observations of conspecifics [3], and learning another's movement patterns in order to precede it to a goal destination [2]. However, inferring from such behaviors that the fish are socially anticipating requires making many assumptions about their motivations and information processing. Pezzulo [17] and others define anticipation as using "predictive capabilities to optimize behavior and learning to the best of [one's] knowledge". In this study we show how computational modeling and evolution strategies can be leveraged to find out how guppies use temporal information about a partner to inform their actions *a posteriori* from recordings of freely moving fish pairs, without experimental manipulation or additional assumptions. Evolving model architectures by using certain fitness functions enables us to draw conclusions similar to those of evolutionary biologists based on real animals' anatomies. To our knowledge, a similar approach has only been tried by Olivares et al. [16] in recent years.

We take inspiration from the long tradition of using simple circuits and rules to model animal behavior in response to certain stimuli, both in real-life and simulated experiments. Grey Walter's work on the Machina speculatrix and the Machina docilis [24] showed that very simple electronic circuits can be sufficient to produce some behaviors reminiscent of innate animal behaviors and even classical conditioning. Note that the "learning circuits" were the result of an analysis of the operations "involved in establishing a connection between different stimuli to achieve a conditioned response". This means that they tell us something about the behaviors that can be achieved with simple circuitry, but not much about the way animal bodies and brains achieve similar behavior. Another example are Braitenberg vehicles. Initially conceived as a thought experiment, they are minimal robots whose wheel actuators are directly coupled to simple sensors. By varying the sensor-actuator connections, they can be made to show behaviors associated with living organisms, such as an enduring attraction or aversion to a light source, obstacle avoidance and chemotaxis [5,22].

This naturally leads to the question of how architecturally complex agents even need to be in order to show successful behaviors in their environment, including the social environment. Simulations of simple recurrent neural networks with two or three hidden nodes which control Braitenberg-like agents emitting acoustic signals have shown that social interaction itself can increase a network's complexity (as measured by the entropy between the internal nodes) and lead to interesting (though not necessarily lifelike) behavior patterns when two such agents interact [9,18,19].

To our knowledge, most of the work done in this area uses present-time sensory input and, if at all, introduces a temporal aspect through the use of recurrencies in the artificial neural network architecture. In Couzin et al.'s [11] zone model of fish social behavior, individual fish react almost instantaneously to others' movements. However, possible memory or prediction capabilities of individual fish are not examined. On the other hand, Murakami, Niizato and Gunji [15] as well as Strömbom and Antia [23] model swarm dynamics using anticipation.

Given our interest in memory and anticipation of a social partner's actions, we are looking for a way in which temporal dynamics (such as delayed or anticipatory reactions to a partner's actions) may be detectable from an outside perspective in pre-recorded behavior data, without needing to experimentally manipulate the behavior. In addition, we want to find out what role such dynamics play in producing lifelike simulations of fish movements.

Based on the work reviewed so far, we decided to simulate fish interactions using minimal (at least at first) feedforward artificial neural networks and treating memory and prediction capacities as external "modules" rather than as properties of the neural architecture or internal model (as do, for example, Blum, Winfield and Hafner [4]. This is in line with Reynolds' [20] approach of giving his artificial agents "approximately the same information that is available to a real animal as the end result of its perceptual and cognitive processes" (more details in the Methods section). This comes with the drawback of not being able to simulate two partners freely interacting: In order to have "predictions" of a partner, the focal agent model must interact with a pre-recorded one. However, this allowed us to directly compare the individual performance of models with the exact same partner and based on the exact memory/prediction timesteps they are receiving as input, while keeping their neural architecture simple.

The presence or absence of the temporal input modules were determined by an optimization process. In this process, the complexity of the neural network architecture (as measured by the number of hidden layers and their respective nodes) would be adapted in order to better approximate the "decision making function" mapping input states to motor actions. As target functions to optimize, we used two different measures in two separate experiments: In experiment 1, we used the framewise deviation of a model's predictions from the original fish track it was trained on. In experiment 2, we used the "fish-likeness" of a model's simulation, as rated by a long short-term memory (LSTM) discriminator trained for this purpose. Both were traded off against the number of trainable parameters in the models in order to encourage simplicity. By thus tweaking our modeling choices and examining the best fitting models, we hoped to answer the following questions: **Experiment 1:** Can we show in how far a given fish predicts or remembers another's movements and changes its behavior accordingly? **Experiment 2:** What minimum of temporal input and internal complexity (taking into account Occam's razor [6]) is sufficient to model a specific fish, or to produce generally "fish-like" behavior? In keeping with the origin of the biological agents we were modeling, we chose to use a custom Evolution Strategy [1] in order to optimize our model architectures while keeping them simple.

2   Methods

Please find a diagram of our workflow in Fig. 1c.

2.1   Ground Truth Data

Our efforts in both experiments were aimed at successfully modeling existing data from five live pairs of Trinidad guppies, filmed and tracked while moving freely in an experimental tank. Their movements were filmed in a white square tank of 88 cm width and 7.5 cm water depth (see Fig. 1a) and tracked at 30 frames per second (fps) using the BioTracker movement tracking software [14]. In every evolution run, one of the two fish per pair (the "focal fish") was used as ground truth data to be modeled, while the other live fish's trajectory functioned as the "social partner".

2.2   Neural Network Models

We configured all models as multi-layer perceptrons using Keras [10]. Independently of the number of layers and the number of nodes in each layer, all hidden layers were densely connected and used leaky ReLU with an alpha of 0.01 as an activation function. The case of 0 hidden layers implemented a linear mapping of input to output (please see supplementary material for an illustration), with one output being clipped to minimum 0 (ReLU). All models were trained using the Keras adaptive moment estimation ("Adam") optimizer with a clipping norm of 2.0, a maximum of 100 epochs but early stopping in the case of stagnant validation loss (with a patience parameter of 5). The two output nodes represented 1) the length/magnitude of the predicted movement, and 2) a change in heading direction in 2-argument arctangent (atan2), respectively. The movement length node was a ReLU to prevent "negative" (backwards) movements. The direction change output node was not regularized. We used polar coordinates as this was computationally easier when using a field of view. Please see Fig. 1b for an example. If used for simulation, the predictions of the network were transformed into x and y displacement coordinates in the global cartesian coordinates. This was done by a custom function which also clipped the maximum length of the movement to 4cm to regularize the simulations and prevent "jumps" across the tank.

2.3   Input Information

We implemented the inputs to the network as higher-level "spatial awareness" modules: a wall detection module and a partner detection module. We decided to "outsource" the fish's awareness of walls and partners in this way and give the model precise information in the same spatial format as the required output instead of, say, using simulated raycasting to detect walls and partners and letting the model combine the ray information into a spatial representation to act upon. This meant that a model only had to make the movement prediction, saving it some computation and eliminating one potential source of error. Input information was precise within a field of view of 172° on each side (which is realistic for this species of fish). For information on wall vision, please see the supplementary material. Figure 1b illustrates our input modules.

Partner Detection Module (*Partner*). The partner detection module computed the position of the partner fish from the simulated fish's point of view and passed the polar coordinates as input to the neural network. It implemented the capability to "remember" or "predict" the partner fish's position at other points in time. Given an integer $-a$, the module would compute the position of the partner fish a timesteps in the past relative to the *current* position of the focal fish, essentially allowing the model to remember past positions of the fish from its current perspective. Given an integer a, the module would compute the position of the partner fish a timesteps in the future (this information was available as our tracks were pre-recorded), allowing the model to predict the partner's position from the perspective of the present. One model could use this module with more than one timestep at once depending on its hyperparameters, adding 2 input nodes (for the two coordinates) to the model per timestep used. The timesteps used by a given model were encoded as a list of integers in the hyperparameter "genome" which was subject to our evolutionary strategy.

2.4 Evolution Strategy for Neural Network Architecture Search

Evolution strategies are a class of optimization methods first developed in the 1960s by Rechenberg and others [1]. Like other evolutionary computation methods, they use ways of "reproducing" and "selecting" artificial entities based on certain traits across multiple loops or "generations", inspired by natural evolution. Unlike other methods, they usually evolve the entities' mutation rates along with the other traits. The model hyperparameters subject to our evolution strategy were the following:

- The entirety of hidden layers and the number of nodes in each layer
- The partner perception timesteps used by the agent
- The presence or absence of wall input
- The magnitude of the noise added to all model inputs (more precisely, the factor by which the standard deviation of the entire input for a given feature was multiplied to form the standard deviation of a Gaussian with mean 0 to draw noise samples from)
- The noise level added to model outputs when simulating (more precisely, the factor by which the standard deviation of the original framewise fish movements was multiplied to form the standard deviation of a Gaussian with mean 0 to draw noise samples from)

These hyperparameters indirectly encoded the model architecture, and each model was trained using gradient descent with randomly initialized weights before testing it for selection. The outcomes we aimed to minimize simultaneously were a) the testing loss as measured in two different ways and b) the number of trainable parameters in the model (its complexity).

Reproduction. As is recommended for a combinatorial task [1], we used a $(\mu + \lambda)$ evolution strategy, in which μ parents are used to create λ offspring in

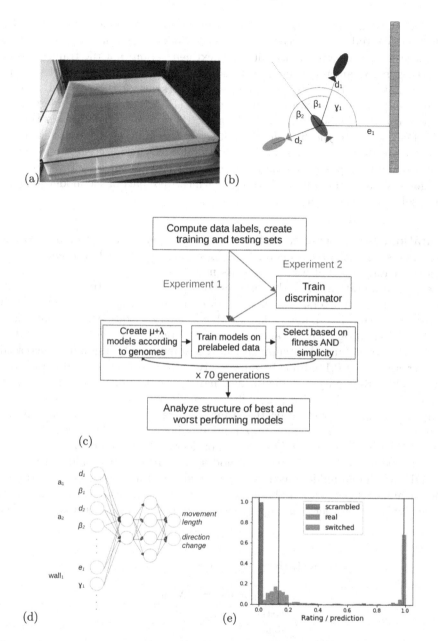

Fig. 1. a) Test tank for filming fish movements. b) Focal fish (red) and example input from partner fish (blue) d_1, d_2 and β_1, β_2: Partner distances and angles for two example timesteps. e_1 and γ_1: Wall distance and angle (only one wall shown). c) Workflow for both experiments. d) Example feedforward neural network (FNN) with several partner input timesteps, wall vision and hidden layers [2, 4]. e) Performance of lifelikeness discriminator: Relative frequency histogram of the ratings of the discriminator on unused real, scrambled and switched data. (Color figure online)

each generation, and the parents and offspring are then pooled and tested for selection. We did not use recombination to produce the offspring, but rather each selected parent was copied into the next generation (not the trained model but the hyperparameter "genome"), in addition to two mutated versions. Thus, $\mu = 8$ and $\lambda = 16$. All copies and mutant versions of the selected parents were then built as feedforward neural networks (FNNs) and trained via gradient descent using pre-labeled data. Afterwards, they were tested for selection as described below. In the first generation, $\mu = 8$ default models were initiated according to the settings in Table 1, and two mutant versions created for each to form the initial population. The mutation itself was governed by two "strategy parameters" (mutation rate and strength) which were part of the model genome, and evolved along with it.

Mutation. Mutation was carried out as follows. In addition to the model hyperparameters (or "object parameters" in ES speak), we evolved two "endogenous strategy parameters" of each model: its mutation strength s and its mutation rate p. The mutation strength was a factor applied to mutation steps in real-valued model parameters, and the mutation rate used as p for sampling from a Bernoulli distribution to determine whether a binary mutation would take place or not. For creating each mutant version of a parent, we randomly increased or decreased s by 1 (and then clipped it at a minimum of 1), randomly increased or decreased p by 0.1 (clipped to stay between 0.1 and 1), flipped wall input on/off with probability p, added a new partner perception timestep randomly chosen from $-150,150$ with probability p, randomly subtracted or added s to a randomly chosen existing partner time step, removed one randomly chosen partner timestep with probability p, randomly increased or decreased the input noise factor by $s/20$, randomly increased or decreased the motor noise factor by $s/20$, added a hidden layer with $s+1$ nodes with probability p, added a node to a randomly chosen hidden layer with probability p, and removed one randomly chosen hidden layer with probability p (provided there was one). In order to avoid duplicate partner timesteps, the set of partner timesteps was used after mutation.

Table 1. Hyperparameters of default model.

Hyperparameter	Value
Partner timesteps	[0]
Wall vision	False
Input noise factor	0.0
Motor noise factor	0.0
Hidden layers	[0]
Mutation rate	0.3
Mutation strength	1

Selection. We performed two neuroevolution experiments, using two different forms of testing loss according to the different study aims. In both methods, the testing loss was traded off against the number of model parameters to select models for reproduction. All our datasets were used in both experiments. In **experiment 1**, we used the loss from testing each model on prelabeled, unseen data from the same dataset. This served to gauge in how far the models were able to approximate the framewise movement "decisions" of the original fish, as the model received input based on the original fish's real positions. The model architectures evolved by this method served to illuminate the individual behavior of the respective fish, and its unique dynamics with the partner fish. For each dataset, we did one evolution run using the first fish as the focal fish, and another using the second fish. In **experiment 2**, we used the negative ratings of the discriminator model described in Sect. 2.8 as the testing loss. Only the first fish of each dataset was used here. Each model was still built and trained on prelabeled data in the same way as in experiment 1, but then made to freely simulate 400 frames of fish track given the original partner input, to be rated by the discriminator. This served to select for models what were able to produce the most fish-like trajectories and interactions with the partner fish, thus giving us more general insights about the behavior of these guppies.

In both experiments, at each generation of an evolution run we computed the Pareto optimal model genomes according to both minimal testing loss and minimal number of trainable parameters in the model. A (strongly) Pareto optimal data point in a two-dimensional feature space is one where, if there exists another data point with a better value on one feature, that data point must rate worse on the other feature [13]. This allowed us, in the first round, to select models according to both our desired outcomes (low testing loss *and* low complexity) without having to weigh them against each other. However, as the number of Pareto optimal models (also called the Pareto front) is not necessarily $= \mu$, we used the average of the min-max scaled test loss and complexity for each model as a secondary criterion for ranking the models, and then used the μ best models for reproduction, making this an elitist selection technique. The procedure was repeated for 70 generations per run.

2.5 Data Labels

We trained all models in both experiments using labels based on the ground truth data. The labels represent the direction change of the original fish in radians and its movement length in cm at each frame, with the mean squared error of both output nodes as the loss function for each one. As the two outputs were roughly on the same scale (movement length in cm and direction change in radians were both expected to be in $[0, 1]$), we weighted the output nodes equally for the final model loss. Training loss was defined as the deviation of the predicted move from the ground truth of the recorded focal fish. While this is obviously not biologically realistic as no animal has a deterministic trajectory to follow, our whole study rests on the assumption that a fish's framewise movements can be at least partly predicted based on its perception of the partner's position and the tank walls.

2.6 Simulation

Each trained model could be used to simulate a fish trajectory given the partner and wall input suitable to its input configuration. Together with the partner trajectory, this could then be passed to a pre-trained discriminator model to be rated for "lifelikeness". The framewise outputs of the simulator model would be transformed into movement coordinates. A movement was only performed if the resulting fish position was inside the tank walls of the dataset currently used; if not, the simulated focal agent simply remained in the same place but the simulation continued, feeding it new input for new potential movements.

2.7 Analysis of Evolution Results

As it was difficult in our application to verify the correct running of the Evolution Strategy objectively, we relied on the testing loss development across generations. Runs in which testing loss did not sink towards the final generations, did not reach levels below 0.5 or oscillated greatly throughout were regarded as "not converged". However, by analyzing the best and worst performing models across all generations, we can still glean some cautious insights into the solutions found by these runs. Comparing the best models for the two fish in each dataset allowed us to draw inferences about the dynamic between the two fish. What all the evolution runs seemed to have in common was a marked bimodal distribution of testing loss across generations and model hyperparameters (see plots in the supplementary material). Given this pattern in the testing loss, we decided to compare the ensembles of the 30 best performing and the 30 worst performing models for each fish with two-sample t tests to gain information on the hyperparameter choices for modeling that fish. Comparing the partner timestep histograms between the best and worst models for each fish provides information on the partner timestep input likely used by that fish. While experiment 2 was more exploratory, we formulated specific predictions for experiment 1: If both fish in a pair mostly ignored each other, we would expect the timestep histograms to be inconclusive for both fish respectively, as the evolution strategy would not find an adequate model of either fish's movements based on the other one. If one fish in a pair mostly remembered or anticipated the other's movements but was itself largely ignored, then the timestep histograms for the first fish should show a predominance of the best models within a certain time range. The other, "careless" fish in this scenario should either have inconclusive timestep histograms, or a predominance of timesteps exactly the opposite of the first fish. If both fish coordinated their movements closely, with one mostly reacting and the other mostly anticipating, we would also expect a "mirrored" pattern in their timesteps. Such a mirrored pattern would therefore be difficult to interpret causally, while the other patterns would present a clear picture.

2.8   Simulation Discriminator

In order to be able to automatically evaluate the lifelikeness of a simulated fish pair trajectory, we built a discriminator LSTM model trained to distinguish real pre-recorded fish pairs from a) datasets with scrambled frames (i.e. completely un-fish-like trajectories) and b) real focal and partner fish trajectories switched between different datasets (i.e. fish-like but not interacting). It was loosely inspired by the discriminator part of Gupta et al.'s [12] Social GAN for producing socially acceptable walking trajectories. The discriminator contained one densely connected layer of 64 LSTM cells and one output node regularized with a sigmoid function. We trained using binary cross entropy loss and the "Adam" optimizer. The discriminator was trained on 35 datasets of recorded fish pairs not used in the later study. When testing on unused cuts of the training data sets, the discriminator did a good job separating the real fish pair data from scrambled and switched tracks (see Fig. 1e). For additional information, please see the supplementary material.

Code Availability.   All our code is available at gitlab.com/leamusi/ fish_simulation (main project code) and gitlab.com/leamusi/fish_movements (additional tools for processing fish movements).

3   Results

3.1   Experiment 1: Selection Through Testing on Prelabeled Data

For each dataset of two fish interacting, we did one evolution run using the first fish as the focal fish, and another using the second fish. Please find detailed results and plots in the supplementary material. For all runs, the majority (at least 60%, usually more) of the best performing models had no wall vision. Results on motor noise are not reported when testing on labeled data, as they only come to bear when testing in simulation. For dataset N07P3, convergence was unclear while evolving models for either fish. Partner timesteps 50 frames in the future were advantageous for modeling the first fish, while input 50 frames from the past helped the best models of the second fish. For N12P4, the evolution processes for both fish achieved solutions with comparatively low testing loss. The best models of the first fish were dominated by inputs 50–100 frames or 1.6–3.3 s in the past, while the temporal inputs for the second fish were inconclusive. For dataset N13P3, it appears that only the evolution process for modeling the first fish converged on good solutions. The timestep results for both fish were inconclusive. In dataset N13P4, only the evolution strategy for the first fish achieved solutions with low loss. The first fish's temporal input was dominated by information 0 to 100 frames in the future, while the second fish's was dominated by past input (100 to 0 frames back). For dataset N16P4, convergence of the evolution process was unclear for the first fish, but achieved good solutions for the second one. The results regarding temporal input were inconclusive for both fish.

T tests on the other hyperparameters between the best and worst performing models were generally nonsignificant except for input noise factor (which was always lower in the best models) and the number of hidden nodes (which was significantly higher in the best models for several fish).

3.2 Experiment 2: Selection Via Discriminator Ratings

Fixed Strategy Parameters. The performance of the evolution strategy using discriminator ratings fell short of our expectations as we had reached ratings of up to 0.9 in a pilot study with 40 generations and *fixed* strategy parameters: the mutation strength was set at 1 and the mutation rate at 0.3. We therefore report a selection of these results here (Fig. 2a–d), including the resulting simulations (see Fig. 2g–h). For dataset N07P3, there was no significant difference on any hyperparameter except input noise factor ($\delta = -0.09, t(58) = 2.4, p < 0.05$), while both the best and the worst models had an average of 31 hidden nodes and 20 partner timesteps. Timestep distributions of the best and worst models were identical. For dataset N12P4, there was no significant difference on any variable, no clear picture in the timestep distribution, and some of the best rated models had 0 hidden nodes and a minimum of 7 partner timesteps. The best models for the other datasets only achieved top ratings of 0.4. For the results obtained using evolving strategy parameters, please see the supplementary material.

4 Discussion

4.1 Experiment 1

The bimodal distribution of the testing loss seems to indicate that for each dataset, there are two attractors for a model's testing loss to gravitate towards, with the lower one being a natural baseline loss. This baseline may be due to the fact that a) there may be systematic input factors or modeling choices which our study does not account for or b) no focal fish's behavior is entirely deterministic, and therefore no model can be expected to recreate it perfectly. The absence of wall vision in most best performing models may be easily explained by the fact that partner fish (like focal fish) usually stayed close to the walls, meaning that information about the walls was contained in partner input, and wall vision input would thus have been redundant. The partner timestep results for both fish in dataset N07P3 clearly mirror each other. This makes it difficult to state whether one fish was anticipating the other, the second fish was following the first reactively, or both. However, it is clear that the two fish's movements have a strong temporal connection. The results for N12P4 suggest that the first fish appears to have used its memory to follow its partner, while the partner itself was not minding the first fish at all. In N13P4, the temporal results again mirror each other, making causal inference difficult but showing a clear temporal link

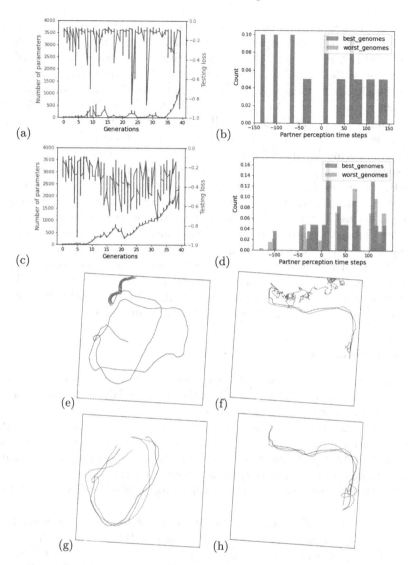

Fig. 2. a)–d): Results of neuroevolution using selection via discriminator with fixed strategy parameters. Left column: Change of testing loss (blue) and model complexity (trainable parameters, red) across evolution generations. Right column: Histograms of partner input timesteps for the 30 best (blue) and the 30 worst (orange) models. a)–b) Dataset N07P3. c)–d) N12P4. **e)–h):** Simulations by different models: original focal fish in green, partner in blue, simulation in red. e) Simulation by one of the best 30 models for the first fish of dataset N07P3 evolved using prelabeled data. f) Simulation by the best model for the dataset N12P4 evolved using the discriminator, with evolving strategy parameters. Discriminator rating 0.5. g) Simulation by the best model for dataset N07P3 evolved using the discriminator, with fixed strategy parameters. Discriminator rating 0.91. h) Simulation by the best model for dataset N12P4 evolved using the discriminator, with fixed strategy parameters. Discriminator rating 0.92. (Color figure online)

in the pair's behavior. From the lack of convergence and inconclusive histograms for the two fish pairs in datasets N13P3 and N16P4, it appears that the two partners did not mind each other much in either case, which is reflected in a visualization of the raw data (see supplementary material, Fig. 1).

The results also indicate that models seem to benefit from a low input noise factor when tested on labeled data, although the average input noise factor in the best models was still nonzero for all datasets. This suggests a sweet spot for input noise, at which the noise increases model robustness but does not distort predictions too much. Apart from this, the results on the general hyperparameters suggest an advantage of having more hidden nodes for modeling some datasets. Taken together, these results show how evolving the architecture and input configuration of an FNN can be used to infer the interaction dynamics of fish pairs. It must be remarked that even the best models evolved with this selection method did not produce very realistic-looking simulations (see Fig. 2e). This is likely due to the fact that training and testing were done on input computed frame-wise from the perspective of the pre-recorded focal fish. When simulating freely, the agent may move into positions relative to the partner fish that would be unusual for the real focal fish, meaning that the model did not have "experience" with such input, and was not selected or trained for it.

4.2   Experiment 2

Using the simulation discriminator for selection in the evolution strategy means that the results cannot tell us much about the individual datasets being modeled. The discriminator rates the partner trajectory together with the simulated focal fish on how much they resemble a real fish pair, but we cannot say for certain what this rating captures, or what influence the specific partner trajectory has. Therefore, the best performing genomes can teach us something about creating lifelike fish simulations in general, but not about imitating a specific focal fish. We therefore interpret the results jointly, without focusing on each specific dataset.

With Fixed Strategy Parameters. The similarity of the best and worst model architectures is an astonishing result. It indicates that the only thing which made a difference here was the random weight initiation when compiling the models, which led some to learn very high quality simulation skills, while others possibly got caught in a local minimum. This leads us to the conclusion that a suitable model architecture is a necessary but not a sufficient condition for good model performance, and that the contribution of the model training should not be underestimated. While in experiment 1 specific timesteps and low input noise were advantageous for modeling specific fish, it appears that something else is necessary (but not sufficient) when trying to fool a discriminator: the number and variety of partner timesteps. Given this, hidden nodes may not even be necessary, and a very simple function of the partner input may produce highly lifelike simulations.

5   Summary

We show that by evolving model architectures and input configurations in experiment 1, we can capture how some guppies' movements can be predicted through their memory and others' through their anticipation of a partner's movements, with the input-output relationship flexibly determined by an FNN as a function approximator. In general, more hidden nodes in an FNN did not necessarily seem to bring an advantage in all pair interactions: the behavior rules leading from partner perception to the fish's own behavior can be very simple, confirming the findings of previous models of fish behavior. For future studies, being able to infer the temporal dynamics between two fish without experimental manipulation means that we have a new, more efficient way of gauging aspects of a fish's social personality (does it tend to react or anticipate?). When attempting to build socially competent fish robots, we can then adapt their behavior to what we already know about the specific social partner. Just as in experiment 1, the results of experiment 2 show that very simple fish behavior models can produce lifelike simulations. We also learn from this experiment that the architecture itself does not determine model performance: rather, the right architecture seems to be necessary but training is key.

6   Limitations and Outlook

One obvious limitation of our study is the fact that our models do no guarantee causality: finding out that a given FNN approximates the function connecting hypothetical visual input received by a fish to its movements does not mean that the real fish was actually using such inputs and performing such computations. In the case of partner input from the future, for example, it is theoretically possible that rather than the focal fish predicting the partner's future position, the partner is systematically positioning itself a certain way in relation to where it saw the focal fish looking seconds ago. However, the benefits of such behavior would be unclear. Another limitation of our approach should be addressed, as it arises naturally from the work we reviewed above: Our models are clearly not embodied (not even within the simulation), and we do not account for the physical dynamics of fish movements at all. This means that our study is only a first step towards building accurate models of fish movement, which also happens to answer some more questions about the original data. In future, our models could undergo further evolution when combined with a model of fish's bodies and movement dynamics. An example for this could be the models of burst-coast swimming dynamics presented by Calovi et al. [8] and Sbraraglia et al. [21]. It is to be expected that such a joint evolution of social and physical movement dynamics may produce different solutions than this study, and produce new insights.

References

1. Beyer, H.G., Schwefel, H.P.: Evolution strategies - a comprehensive introduction. Nat. Comput. **1**(1), 3–52 (2002). https://doi.org/10.1023/A:1015059928466
2. Bierbach, D., Gómez-Nava, L., Francisco, F.A., Lukas, J., Musiolek, L., Hafner, V.V., Landgraf, T., Romanczuk, P., Krause, J.: Live fish learn to anticipate the movement of a fish-like robot. Bioinspiration Biomimetics **17**(6), 065007 (2022). https://doi.org/10.1088/1748-3190/ac8e3e
3. Bierbach, D., Sassmannshausen, V., Streit, B., Arias-Rodriguez, L., Plath, M.: Females prefer males with superior fighting abilities but avoid sexually harassing winners when eavesdropping on male fights. Behav. Ecol. Sociobiol. **67**(4), 675–683 (2013). https://doi.org/10.1007/s00265-013-1487-8
4. Blum, C., Winfield, A.F.T., Hafner, V.V.: Simulation-Based Internal Models for Safer Robots. Frontiers in Robotics and AI 4 (2018). https://doi.org/10.3389/frobt.2017.00074
5. Braitenberg, V.: Vehikel: Experimente mit künstlichen Wesen. LIT Verlag Münster (2004)
6. Britannica, E.: Occam's razor | Origin, Examples, & Facts | Britannica, August 2023. https://www.britannica.com/topic/Occams-razor
7. Brown, C., Laland, K.N.: Social learning in fishes: a review. Fish Fish. **4**(3), 280–288 (2003). https://doi.org/10.1046/j.1467-2979.2003.00122.x
8. Calovi, D.S., Litchinko, A., Lecheval, V., Lopez, U., Escudero, A.P., Chaté, H., Sire, C., Theraulaz, G.: Disentangling and modeling interactions in fish with burst-and-coast swimming reveal distinct alignment and attraction behaviors. PLoS Comput. Biol. **14**(1), e1005933 (2018). https://doi.org/10.1371/journal.pcbi.1005933
9. Candadai, M., Setzler, M., Izquierdo, E.J., Froese, T.: Embodied dyadic interaction increases complexity of neural dynamics: a minimal agent-based simulation model. Front. Psychol. **10**, 540 (2019). https://doi.org/10.3389/fpsyg.2019.00540
10. Chollet, F.: Keras (2015). https://keras.io/
11. Couzin, I.D., Krause, J., James, R., Ruxton, G., Franks, N.R.: Collective memory and spatial sorting in animal groups. J. Theor. Biol. **218**(1), 1–11 (2002). https://doi.org/10.1006/jtbi.2002.3065
12. Gupta, A., Johnson, J., Fei-Fei, L., Savarese, S., Alahi, A.: Social GAN: Socially Acceptable Trajectories with Generative Adversarial Networks, March 2018. https://doi.org/10.48550/arXiv.1803.10892
13. Mock, W.B.T.: Pareto optimality. In: Chatterjee, D.K. (ed.) Encyclopedia of Global Justice, pp. 808–809. Springer, Netherlands, Dordrecht (2011). https://doi.org/10.1007/978-1-4020-9160-5_341
14. Mönck, H.J., et al.: BioTracker: an open-source computer vision framework for visual animal tracking, March 2018. https://doi.org/10.48550/arXiv.1803.07985
15. Murakami, H., Niizato, T., Gunji, Y.P.: Emergence of a coherent and cohesive swarm based on mutual anticipation. Sci. Rep. **7**(1), 46447 (2017). https://doi.org/10.1038/srep46447
16. Olivares, E., Izquierdo, E.J., Beer, R.D.: A Neuromechanical Model of Multiple Network Rhythmic Pattern Generators for Forward Locomotion in C. elegans. Frontiers in Computational Neuroscience 15 (2021)
17. Pezzulo, G., Butz, M.V., Castelfranchi, C.: The anticipatory approach: definitions and taxonomies. In: Pezzulo, G., Butz, M.V., Castelfranchi, C., Falcone, R. (eds.) The Challenge of Anticipation. LNCS (LNAI), vol. 5225, pp. 23–43. Springer, Heidelberg (2008). https://doi.org/10.1007/978-3-540-87702-8_2

18. Reséndiz-Benhumea, G.M., Sangati, E., Froese, T.: Levels of coupling in dyadic interaction: an analysis of neural and behavioral complexity. In: 2020 IEEE Symposium Series on Computational Intelligence (SSCI), pp. 2250–2256, December 2020. https://doi.org/10.1109/SSCI47803.2020.9308429

19. Reséndiz-Benhumea, G.M., Sangati, E., Sangati, F., Keshmiri, S., Froese, T.: Shrunken social brains? a minimal model of the role of social interaction in neural complexity. Frontiers in Neurorobotics 15 (2021)

20. Reynolds, C.W.: Flocks, herds and schools: a distributed behavioral model. ACM SIGGRAPH Comput. Graph. 21(4), 25–34 (1987). https://doi.org/10.1145/37402.37406

21. Sbragaglia, V., Klamser, P.P., Romanczuk, P., Arlinghaus, R.: Evolutionary Impact of Size-Selective Harvesting on Shoaling Behavior: Individual-Level Mechanisms and Possible Consequences for Natural and Fishing Mortality. Am. Nat. 199(4), 480–495 (2022). https://doi.org/10.1086/718591

22. Shaikh, D., Rañó, I.: Braitenberg Vehicles as Computational Tools for Research in Neuroscience. Frontiers in Bioengineering and Biotechnology 8 (2020)

23. Strömbom, D., Antia, A.: Anticipation induces polarized collective motion in attraction based models. Northeast J. Complex Syst. 3(1), March 2021. https://doi.org/10.22191/nejcs/vol3/iss1/2

24. Walter, W.G.: A machine that learns. Sci. Am. 185(2), 60–64 (1951)

Heuristics for Evolutionary Optimization for the Centered Bin Packing Problem

Luke de Jeu and Anil Yaman[✉][iD]

Vrije Universiteit Amsterdam, De Boelelaan 1111,
1081 HV Amsterdam, The Netherlands
a.yaman@vu.nl

Abstract. The Bin Packing Problem (BPP) is an optimization problem where a number of objects are placed within a finite space. This problem has a wide range of applications, from improving the efficiency of transportation to reducing waste in manufacturing. In this paper, we are considering a variant of the BPP where irregular shaped polygons are required to be placed as close to the center as possible. This variant is motivated by its application in 3D printing, where central placement of the objects improves the printing reliability. To find (near) optimum solutions to this problem, we employ Evolutionary Algorithms, and propose several heuristics. We show how these heuristics interact with each other, and their most effective configurations in providing the best solutions.

Keywords: Bin Packing Problem · Heuristics · Evolutionary Algorithms

1 Introduction

The Bin Packing Problem (BPP) is an optimization problem in which a number of objects are placed within a space to maximize a certain objective function. The BPP has a wide range of applications in many fields. Some examples include: reducing material cost in both additive and subtractive manufacturing (for arranging object placement in 3D printing or cutting sheet material) [17,18], reducing costs in transportation (for arranging packages in shipping containers) [14], and warehouse optimization (for optimal placement and storage of goods) [21].

The dimensionality of the BBP problem, in terms of space and objects, can differ depending on the nature of the application domain. In this paper, we consider the field of Fused Filament Fabrication (FFF) 3D printing for additive manufacturing where three-dimensional irregular shaped objects are placed on a printing build plate. The objective in this case is to place the objects as close to the center of the placement area as possible. Due to mechanical limitations of non-industrial and non-professional FFF 3D printers, the heated build plate is

Supplementary Information The online version contains supplementary material available at https://doi.org/10.1007/978-3-031-56852-7_11.

sensitive to convection and thermal expansion. This causes heat dissipation and warping of the printing build plate. This effect is observed at the edge of the build plate the most. This is shown to be one of the major causes of printing failure [23,27]. As the popularity of FFF 3D printing has been increasing in recent years, there is a need for more convenient and effective methods for more reliable printing [20]. In this paper, we focus on a possible solution to this problem, where a variant of the BPP aims to place the objects as close to the center as possible. Other possible applications of this problem could also be found in areas such as electronic manufacturing, design of circular circuit boards and urban planning.

Our use case is concerned with the placement of irregular polygons at the center of a printing build plate. Therefore, we formalize this problem as the 2D Irregular Centered BPP (2DICBBP). An example illustration of results from two commonly used software, in comparison with the results of our approach, are shown in Fig. 1. A clear distinction between these results is that, the algorithm in Fig. 1a can handle only convex shapes, whereas, the algorithm in Fig. 1b can also handle concave shapes. In Fig. 1c, we show the results of our proposed approach that can yield placement that appears visually denser than the others.

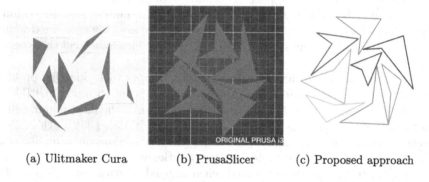

(a) Ulitmaker Cura   (b) PrusaSlicer   (c) Proposed approach

Fig. 1. Visual comparison of the auto-arrange functionality between two currently available slicer software, and a solution found by the methods proposed in this research. Each example is over the same concave shapes.

The BBP is difficult to solve computationally. Finding an optimal arrangement is NP-hard [10]. Since computing the optimal solution is challenging, heuristic search algorithms, such as Evolutionary Algorithms (EA) [4], become feasible alternatives for reducing computational time and finding approximate solutions [26].

Many 2DBPP solving methods only focus on packing rectangles of varying dimensions [15]. However, some solving methods are designed for irregular shapes, including concave, convex and non-symmetric polygons of varying number of vertices [24]. The most common application for the 2DBPP is found in industrial environments such as production (which requires the efficient cutting of materials), warehouse management, and transport [8,22]. Due to the nature of these applications, the vast majority of the 2DBPP solving methods attempt to leave as much unused space as possible on one of the sides of the "bin", while packing as densely as possible from the opposite side. For these reasons, the heuristics such as the "bottom-left" heuristic where the placement of objects starts from the bottom-left and proceeds to top-right, have been used widely [5,6,16].

Guo et al. (2022) presents an extensive review of the decades of research and literature on the 2D Irregular Bin Packing Problem (2DIBPP) [13]. The wide range of environments, methods, and parameters across the literature demonstrate the variability within the 2DIBPP, as well as the wide range of applications. The designed application of the BPP variation is in essence a 3D packing problem simplified to a 2D environment. Because of this, it shares similarities with both 2D and 3D related work. The applications of the 2DIBPP generally lay in removing/cutting material (such as CNC machining or cutting wood, paper, sheet metal, etc.), with the intended purpose being to minimize waste material [19]. The applications of the 3DBPP and 3DIBPP tend to arrange objects (such as filling containers), with the intended purpose being to maximize the transport efficiency [11,12]. While the methods in this paper present a solution in a 2D environment, the underlying purpose is similar to that of a 3D BPP, which is to place items. Additionally, the goal is to place finite number objects within a space, as close to the center as possible. However, we are only concerned with a single bin, thus, placement of the objects in multiple bins is beyond the scope of this paper.

Currently, slicer software for 3D printers attempts to tackle this problem through auto-arrange features. However, this process is not optimized. UltiMaker Cura and PrusaSlicer (Fig. 1) are considered to be two of the most popular slicer software. Cura has millions of users, according to UltiMaker [3]. Both of these slicers use the same fundamental process for this auto-arrange feature, libnest2d [1,2]. While relatively simple and low computational cost, this implementation is non-optimal and inherently flawed when applied to irregular polygons. For instance, this method is unable to utilize the space within a concave shape, and will always use a convex bounding box of each shape [19]. During the course of this research, a new version of PrusaSlicer was released which tackles this exact problem (version 2.6.1). Through the use of numerous optimizers, the newly improved auto-arrange feature is much more effective and can utilize concave shapes, but still shows room for improvement. The recency of this feature demonstrates how improvements on the BPP around a center point are currently in development, in demand, and being utilized.

The 2DBBP with the object of center placement remains largely unexplored. One of the variations of the BPP that shares similar characteristics is referred to as the Circle Packing Problem (CPP). The CPP packs circles within a bin, and in some cases, this bin is also circular. In these cases, the objective is to minimize the size of the circular bin, similar to our objective. Evolutionary methods such as genetic algorithms and differential evolution have shown to be effective for this problem [9]. Additionally, the load-balanced BPP attempts to pack items in such a way that the center of mass of all the items in the bin is as close to the center of the bin as possible, for which a hill-climbing search method is used [25]. This is one of the few BPP variations for which the item's relation to the center of the bin plays a role in the objective. Thus, extending this literature, in this work, we investigate several heuristics combined with the Evolutionary Algorithms to find effective solutions for the 2-Dimensional Bin Packing Problem for irregular shapes that are required to be placed as close to the center as possible. We focus on four-sided irregular polygons to strike a balance between complexity and feasibility.

2   Method

The problem we tackle in this paper is a version of the BPP where we aim to place polygons within a 2D space such that their distances to the center is minimized. This 2D space will be referred to as the bin. In addition, the polygons should not overlap. Thus, we can state this problem formally as:

Let n be the number of polygons. Each polygon is represented as P_i for i in $[1, n]$. Each polygon P_i has vertices represented as V_{ij} for j in $[1, k]$, where k is the number of vertices in polygon P_i. Let (x_{ij}, y_{ij}) represent the coordinates of the j-th vertex of i-th polygon. Let (X_c, Y_c) represent the coordinates of the center of the bin. The objective function can be formulated as:

Minimize $D = \max(\sqrt{(x_{ij} - X_c)^2 + (y_{ij} - Y_c)^2})$ for all i in $[1, n]$ and all j in $[1, k]$.

Subject to:

- The polygons do not overlap.
- The polygons are within the bin.

To find a (near-)optimum solution to this problem, we use EAs in combination with a set of heuristics we proposed. In the context of the EAs, we refer to a candidate solution as PolyGroup, which encodes an arrangement of the polygons. We can generate new PolyGroups from existing ones using evolutionary operators that are informed by our proposed heuristics.

To calculate the fitness of a PolyGroup, an important evaluation metric is the distance from the center to the furthest coordinate of the polygons within the PolyGroup. This distance can also be described as the radius of the circle that circumscribes the PolyGroup (see Fig. 3 for an example illustration, where circumscribed circles of PolyGroups are shown in red). As a fitness value, it is more meaningful to have an indication of the density of the polygons, as opposed to an absolute value such as this distance/radius. This can help with comparison of performance of two different evolutionary processes that involve different shape initializations. Therefore, to reduce differences in fitness across the use of different polygons, and to gain more insight on the packing density instead of absolute distance, another method is used. The fitness is calculated

as the ratio between the surface area of the polygons, and the surface area of
the circumscribed circle.

$$fitness = \frac{\text{surface area of polygons}}{\text{surface area of circumscribed circle}} \qquad (1)$$

Within a run, the surface area of the polygons will remain constant. The
surface area of the circumscribed circle should decrease, resulting in a higher fitness value. Compared to using the absolute value of the distance to the furthest
away coordinate, this ratio method performs the crucial part of improving when a
desired change is made (a change which brings the furthest coordinate(s) closer),
while additionally providing insight into the density and allowing for more accurate comparisons between runs with different shapes. For comparing the results
of different heuristics, while testing over the same shapes, simply using an absolute value would still show the same performance difference. However, this fitness
method has as added value that it could somewhat be compared with tests using
completely different scales or sizes.

The genotype representation of a PolyGroup encodes the four coordinates of
each polygon on 2-dimensions. We also add a rotation parameter to specify the
rotation angle of polygons. We fix the number of polygons to 5 and thus the
representation of a PolyGroup consists of 45 real-valued numbers.

2.1 Baseline EA

Baseline EA provides a bare minimum algorithm for building our heuristics. This
also provides a point of comparison for the success of the proposed heuristics.
The algorithm consists of the following steps that are standard in EAs:

Initialization. An initial population of PolyGroups is generated by loading
in a pre-generated random set of polygons for each PolyGroup. Not only does this
random polygon selection contain the data of the polygon shapes, but also the
positional and orientation data which is used as the initial placement, in order
to make the experiment repeatable. For a different run, a different randomly
generated set of shapes (and starting positions) is initialized, in order to examine
the EA behavior over altered starting conditions.

Evolutionary Operators. The Baseline EA involves only the mutation
operator to keep it as simple as possible. Other operators, such as recombination,
are implemented and discussed in the Heuristics section. Through the mutation
operator, a new PolyGroup is generated by duplicating a random PolyGroup
from the initialization, or from the survivors of the previous generation. This
newly created PolyGroup will then attempt to undergo mutations in the form
of changing the position and orientation of the polygons within this duplicate.
These changes in position and orientation are made randomly within pre-defined
bounds. Only the mutations which do not cause any overlap between the polygons are accepted. A limited number of mutation attempts are made per polygon
(i.e., 10 attempts), in order to prevent exceeding long runtimes when one or
more polygons have a small chance of mutating to an unoccupied space, as a
result of being surrounded by other polygons. In the instance that none of the

mutations are accepted, the duplicate PolyGroup will have identical polygon placement and orientation as the PolyGroup it was originally duplicated from.

Selection. We use a $(\mu+\lambda)$ selection strategy where λ number of offspring are generated from μ number of parents, the new generation is formed by selecting top μ individuals (ranked based on fitness) from the concatenated set of these two populations [7].

2.2   Heuristics

We propose four heuristics. Each of them is expected to have a positive effect on the overall performance. However, the exact effect of the heuristics may change when applied in combination with one or multiple other heuristics.

Recombination. The Recombination heuristic creates a new PolyGroup from two randomly selected parents. The resulting child PolyGroup is a mix between its two parents. Each polygon in the child receives a center coordinate and a rotation value in between the center and rotational values of the corresponding polygon of the parents. The shape of the corresponding polygon is then built at the assigned center point with the assigned rotation. This allows the Recombination heuristic to place the child polygon at a location and orientation depending on the location and orientation of the corresponding polygon of the parents, while preserving the shapes encoded into the genotypes.

Formally, the Recombination heuristic can be stated as follows: let G be a PolyGroup, containing an arranged set of polygons. Let n be the number of polygons in the PolyGroup. Each polygon P_i for $i \in [1, n]$ consists of a set of four tuples and a rotation value. Each tuple contains the x- and y-coordinate of a vertex of the polygon. These four vertices can be used to calculate a center coordinate, which combined with the rotation value forms (x_i, y_i, t_i). A new PolyGroup G', consisting of polygons P_i' each with a center coordinate and a rotation in degrees (x_i', y_i', t_i'), is generated from two parents $G^{(1)}$ and $G^{(2)}$ as follows:

$$x_i' = x_i^{(2)} + (x_i^{(1)} - x_i^{(2)})/2 + r(x_i^{(1)} - x_i^{(2)})/2,$$

$$y_i' = y_i^{(2)} + (y_i^{(1)} - y_i^{(2)})/2 + r(y_i^{(1)} - y_i^{(2)})/2$$

where r is a random value sampled from a triangular distribution between -1 and 1. For t_i' in P_i', a similar calculation is used, using the same value for r, but utilizing additional steps to ensure the polygon rotates the direction that is shortest (clockwise or counter-clockwise), depending on the shortest rotation between $t_i^{(1)}$ and $t_i^{(2)}$. After the generation of the center coordinate and the rotation value for each P_i' in G', the shape of the corresponding P_i of either parent $G^{(1)}$ or $G^{(2)}$ is imposed on the center coordinate, to regenerate the coordinates of the four vertices of P_i'.

As the creation of polygons P_i' does not check if the newly created polygons are overlapping with any other polygons in G', additional steps are required to correct any overlapping polygons, and to ensure none of the polygons in G' are overlapping. These additional steps are similar to the random mutation operator, except for a distinct difference. The random mutation would only attempt a number of times to move and rotate a polygon randomly within the pre-defined range, the additional steps to separate any overlapping polygons will continue until no polygons are overlapping, and can move polygons beyond the pre-defined range. Instead of reverting the polygon back to its original position after a failed mutation attempts, such as done by the random mutation, these additional steps continue to move the polygon from its new still overlapping position, after every failed mutation. Through this method, an overlapping polygon will randomly "wander" its vicinity until it no longer overlaps with any other polygons.

Mutation Direction. The mutation operator implemented in Baseline EA moves and rotates each polygon randomly within given parameters. However, this mutation method can be altered to improve the fitness. The Mutation Direction heuristic performs a local search on a polygon based on its local fitness, and accepts solutions only if the local fitness is improved. The local fitness is the fitness of an individual polygon, which is defined as the maximum distance from a vertex of the polygon to the center of the bin. Identical to the mutation operator, the Mutation Direction heuristic has a fixed number of attempts. However, with the Mutation Direction heuristic, only the mutations which improve the local fitness of the polygon are accepted.

The local fitness of a polygon is calculated as: each polygon P has vertices $V_i = (x_k, y_k)$ for k in $[1, m]$ where m is the number of vertices of each polygon. Let (X_c, Y_c) represent the coordinates of the center of the bin. For all i in $[i, n]$

$$fitness_{local} = \max(\sqrt{(x_k - X_c)^2 + (y_k - Y_c)^2}), \text{ for all } k \in [1, m].$$

Mutation Order. For most operations, the order of the polygons and Poly-Groups is changed to prevent any unintentional biasing. However, the Mutation Order heuristic offers a different approach. The heuristic sorts polygons by their local fitness (the distance from the center to the furthest coordinate) in ascending order before mutation. This is thought to enhance performance as central polygons can move inward first, freeing up space for outer polygons. For instance, if polygon A is blocked by polygon B from moving closer to the center, polygon A cannot move inward. But if polygon B had moved inward first, polygon A could have followed suit. This heuristic aims to improve such situations where a movable polygon blocks another polygon's mutation. By mutating polygons in ascending local fitness order, polygons closer to the center mutate first, potentially creating space for outer polygons to move inward.

Variable Step Size. The translation and rotation amount for each mutation is random, within bounds. This random mutation strength is then multiplied by the step size. The step size can be adjusted to strengthen or weaken the maximum amount by which a polygon can be mutated. For the Baseline EA, this step size remains constant throughout all generations. The Variable Step Size heuristic introduced a step size which changes depending on the current generation number. The variable step size is determined according to the following formula:

$$\text{variable step size} = \frac{\text{absolute initial step size}}{\sqrt{\text{current generation number} + 1}} \qquad (2)$$

The result of this variable step size formula is that the step size will decrease over the course of the generations. The step size will be the largest for the first generation, and will be most drastically reduced over the first few generations, after which the reduction in step size will become less drastic.

3   Experimental Setup

Each of the mentioned heuristics can be active or inactive, regardless of the activation state of the other heuristics. Therefore, to examine the effect of each heuristic on the fitness of the EA, as well as examining the effect in combination with other heuristics, a binary decision table can be constructed with 16 configurations for the four heuristics. This can be seen in Table 1. We run 30 independent evolutionary processes for each configuration for 500 generations with a population size of 50. Using our implementation, a single evolutionary process, of 500 generations, over one configuration, resulted in a runtime between 30 and 120 s depending on the number of heuristics used. Our implementation is on Python 3.10, and ran on an AMD Ryzen 5 3600X processor, without parallelization. The algorithm could further be optimized to increase the speed. In comparison, other available software packages (demonstrated in Fig. 1a and 1b) were implemented using C++ and can provide results in seconds. However, they do not employ an evolutionary optimization approach to provide better placement.

During the initialization process of each run, all configurations are initialized with the same set of randomly generated shapes and starting positions. Therefore, all configurations within a run share the exact same starting parameters. For each different run, a different set of random shapes and starting positions is used. All starting conditions are equal within a run, but are different across runs.

The final fitness value of the 30 runs of a configuration, can be compared to the 30 final fitness values of another configuration using the Wilcoxon signed-rank test. This test can be used to compare each configuration with every other configuration, and examine its statistical significance.

4   Results

4.1   Configuration Performance Results

After running each configuration, fitness progression plots have been created for each individual configuration. The fitness plot of each individual configuration, along with their average final fitness (AFF - average of final fitness value at the end of the evolutionary processes), and the standard deviation range of the 30 runs, can be found in Supplementary Material. Additionally, the complete

Table 1. The Configurations Setup table shows the properties of each configuration. The name of each configuration is shortened to *config* and a number. When a heuristic is stated as *True*, it is applied to that configuration. When a heuristic is set to *False*, it is not.

Configurations Setup				
Configura-tions	Recombi-nation	Mutation Direction	Mutation Order	Variable Step Size
config 1	False	False	False	False
config 2	False	False	False	True
config 3	False	False	True	False
config 4	False	False	True	True
config 5	False	True	False	False
config 6	False	True	False	True
config 7	False	True	True	False
config 8	False	True	True	True
config 9	True	False	False	False
config 10	True	False	False	True
config 11	True	False	True	False
config 12	True	False	True	True
config 13	True	True	False	False
config 14	True	True	False	True
config 15	True	True	True	False
config 16	True	True	True	True

evolutionary process of a single run over all configurations can be seen in this video: https://rb.gy/7i9diu.

The numerical results can be found in Table 2, sorted by (AFF) in descending order. The results show emerging groups of four configurations, which each have similar average final fitness values, and share heuristic properties. Configurations 10, 12, 2, and 4 form Group A, and have the highest AFF. All four configurations of Group A have Mutation Direction set to False, and Variable Step Size set to True. Configurations 14, 16, 15 and 13 form Group B. All configurations in Group B have Recombination and Mutation Direction set to True. Configurations 5, 6, 7 and 8 form Group C. All configurations in Group C have Recombination set to False and Mutation Direction set to True. Finally, configurations 11, 1, 9 and 3 form Group D. All configurations in Group D have Mutation Direction and Variable Step Size set to False.

Figure 2 shows the average fitness trends during the evolutionary processes. It is clear that each configuration within a group shows similar behavior/progression as the other configurations in its group. Additionally, each group shows different behavior/progression compared to the other groups. Group A and D show similar behavior, as they both show relatively slow progression over

Table 2. The numerical results of the runs from Fig. 2, sorted by AFF in descending order. The column labeled as "Statistical Test" shows the statistical significance of the results relative to the first row ("=" and "+" representing statistical similarity or difference respectively) based on the Wilcoxon signed-rank test, and a p-value of 0,05. The groups are defined based on the ranking on the AFF of the configurations.

Configurations Results by AFF								
Configura-tions	Heuristics				AFF	Std Range	Statis-tical Test	Group
	Recom-bination	Muta-tion Direc-tion	Muta-tion Order	Variable Step Size				
config 10	True	False	False	True	**0,633**	0,108		
config 12	True	False	True	True	0,624	0,084	=	A
config 2	False	False	False	True	0,611	0,1	=	
config 4	False	False	True	True	0,609	0,077	+	
config 14	True	True	False	True	0,562	0,121	+	
config 16	True	True	True	True	0,546	0,117	+	B
config 15	True	True	True	False	0,529	0,144	+	
config 13	True	True	False	False	0,518	0,122	+	
config 5	False	True	False	False	0,483	0,123	+	
config 6	False	True	False	True	0,481	0,135	+	C
config 7	False	True	True	False	0,48	0,117	+	
config 8	False	True	True	True	0,476	0,128	+	
config 11	True	False	True	False	0,453	0,089	+	
config 1	False	False	False	False	0,441	0,076	+	D
config 9	True	False	False	False	0,439	0,074	+	
config 3	False	False	True	False	0,427	0,065	+	

the evaluations, but continue to increase throughout all 500 generations. The main performance difference between Group A and D is that Group A has a much stronger fitness increase in the first few generations. The only difference in heuristics between these two groups, is that Group A has Variable Step Size set to True, while Group D has Variable Step Size set to False. Both overlap in sharing Mutation Direction to be set as False. Group B and C also show comparable differences in their behavior. Both show a very strong increase in fitness in the first few generations. The configurations in these groups increase their performance to such an extent that their fitness temporarily surpasses the fitness of the configurations in Group A (which will in later generations overtake Group B and C in fitness). The only difference in heuristics between Group B and C is that Group B has Recombination set to True, while Group C has Recombination set to False. Both overlap in sharing Mutation Direction as True.

To have a visual understanding of the difference in performance between each group, the polygon positioning of the first and last generation can be observed for a configuration of each group in Fig. 3. The example configurations are config

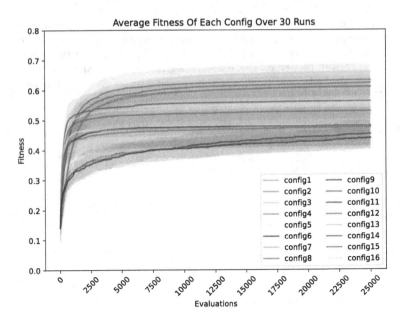

Fig. 2. The average fitness of each configuration, using 30 runs per configuration. Each run lasted 500 generations (25000 evaluations in total for population size of 50). Using evaluations on the x-axis allows for more accurate plot comparison when other parent/children sizes are used.

12 for Group A, config 16 for Group B, config 6 for Group C and config 1 for Group D. These configurations are the second best in each group, to represent the average performance of the group. The visualizations of the polygon positioning of all configurations, in addition to the group examples of two other runs, can be found in Supplementary Material.

The Std range of each configuration, as seen in Table 2, gives an indication as to the consistency between runs. This shows how sensitive the configurations are to changes in the shapes/initial positions, and how much a configuration relies on randomness and its vulnerability towards local maxima. The Std range between the configurations within each group tends to be similar. The average Std range for groups A, B, C and D are 0.092, 0.126, 0.126 and 0.076 respectively.

4.2   Individual Heuristic Effects

While the results in Fig. 2 and Table 2 show which combinations of heuristics result in the highest fitness value, the effect of the individual heuristics must also be examined. For example, when examining the effect of Recombination, one of the pairs to examine is the pair of config 1 and config 9. As seen in Table 1, these configurations share the same heuristics, but config 9 is with the Recombination heuristics whilst config 1 is without. Pairs such as these can be examined to gain better insight into what the effect is of each individual heuristic on different

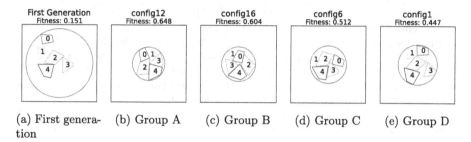

(a) First genera-   (b) Group A   (c) Group B   (d) Group C   (e) Group D
tion

Fig. 3. The polygon positioning of the first and last generation of one configuration per group. Each of the visualizations is over the same run (run 1), therefore having the same shapes and starting position.

heuristic configurations. These pairs, and the effect of each individual heuristic, can be seen in Table 3.

Table 3a examines the effect of the Recombination heuristic. For each use of the Recombination heuristic, it shows a positive or insignificant effect. This indicates that the Recombination heuristic as implemented in this research did not have a measurable negative effect, and only had a positive or insignificant effect. Only one of the configuration pairs between 1–4 has a significant positive effect, while all configuration pairs between 5–8 have a significant positive effect. This would seem to indicate that the effect of the Recombination heuristic is influenced by the presence of the Mutation Direction heuristic. Table 3b examines the effect of the Mutation Direction heuristic. The effect of this heuristic is significant for each configuration pair, indicating it has a strong effect on the performance, regardless of the other heuristics. However, whether the Mutation Direction heuristic has a significant positive or negative effect seems to be reliant on the presence of other heuristics. For configuration pairs 1, 3, 5 and 7, the heuristic has a significant positive effect, while for pairs 2, 4, 6, and 8, the heuristic has a significant negative effect. This clearly shows that the Mutation Direction heuristic as implemented in this research is heavily reliant on the Variable Step Size heuristic. When the Variable Step Size heuristic is False, then the presence of the Mutation Direction heuristic shows a significant positive effect on the final fitness. However, when the Variable Step Size heuristic is True, then the presence of the Mutation Direction heuristic shows a significant negative effect on the final fitness. Table 3c examines the effect of the Mutation Order heuristic. The Mutation Order Heuristic has no significant effect for configuration pairs 1–4, while showing more significant effect in configurations 5–8. The only two configuration pairs with a significant positive effect, pair 5 and 7, seem to indicate that the Mutation Order heuristic only has a somewhat significant effect when there is Recombination and no Variable Step Size present.

Finally, Table 3d examines the effect of the Variable Step Size heuristic. The results clearly demonstrate that the Variable Step Size heuristic only has a positive effect on the final fitness when the Mutation Direction heuristic is False. If

Table 3. Each sub-table shows the numbers of configuration pairs which are identical, besides being with or without a single focused heuristic. The presence of this focused heuristic either has a positive (P), negative (N), or statistically insignificant (I) effect on the AFF. The non-focused heuristics are kept constant between the two configurations of each pair. The heuristics are shortened to a letter for compactness. Recombination (R), Mutation Direction (D), Mutation Order (O), Variable Step Size (V). The use of these heuristics is denoted as either True (T) or False (F).

Pair nr.	Without	With	D	O	V	Effect on AFF
1	config 1	config 9	F	F	F	I
2	config 2	config 10	F	F	T	I
3	config 3	config 11	F	T	F	P
4	config 4	config 12	F	T	T	I
5	config 5	config 13	T	F	F	P
6	config 6	config 14	T	F	T	P
7	config 7	config 15	T	T	F	P
8	config 8	config 16	T	T	T	P

(a) Focused heuristic: Recombination

Pair nr.	Without	With	R	O	V	Effect on AFF
1	config 1	config 5	F	F	F	P
2	config 2	config 6	F	F	T	N
3	config 3	config 7	F	T	F	P
4	config 4	config 8	F	T	T	N
5	config 9	config 13	T	F	F	P
6	config 10	config 14	T	F	T	N
7	config 11	config 15	T	T	F	P
8	config 12	config 16	T	T	T	N

(b) Focused heuristic: Mutation Direction

Pair nr.	Without	With	R	D	V	Effect on AFF
1	config 1	config 3	F	F	F	I
2	config 2	config 4	F	F	T	I
3	config 5	config 7	F	T	F	I
4	config 6	config 8	F	T	T	I
5	config 9	config 11	T	F	F	P
6	config 10	config 12	T	F	T	I
7	config 13	config 15	T	T	F	P
8	config 14	config 16	T	T	T	N

(c) Focused heuristic: Mutation Order

Pair nr.	Without	With	R	D	O	Effect on AFF
1	config 1	config 2	F	F	F	P
2	config 3	config 4	F	F	T	P
3	config 5	config 6	F	T	F	I
4	config 7	config 8	F	T	T	I
5	config 9	config 10	T	F	F	P
6	config 11	config 12	T	F	T	P
7	config 13	config 14	T	T	F	I
8	config 15	config 16	T	T	T	I

(d) Focused heuristic: Variable Step Size

the Mutation Direction heuristic is True, then the Variable Step Size heuristic has no significant effect.

5   Discussion

In conclusion, the 2DICBPP, as implemented in this paper, demonstrates consistent positive and negative changes when certain heuristic combinations are applied. However, these effects are not always the same, and can vary greatly depending on the presence of other heuristics. Each heuristic impacts the performance and behavior of the EA differently. Additionally, the effect of the heuristics themselves are impacted by the behavior of the EA.

The Recombination heuristic has only shown positive or insignificant effects. Its positive effects are most notable when the EA without the Recombination heuristic is prone to getting stuck at a local optimum. This indicates that the main positive effect of the Recombination heuristic comes from introducing randomness and new variations. The effect of the Mutation Direction and Variable Step Size heuristics showed high dependence on the presence of one another. The presence of the Mutation Direction will make the effect of the Variable Step Size insignificant, which otherwise would be positive. The presence of the Variable Step Size will make the Mutation Direction have a negative effect, which would otherwise have a positive effect. The Mutation Order heuristic generally had a very insignificant and mild effect, but seems more influential in EAs with more variety among the population.

These findings demonstrate how the effect of individual heuristics on the performance of the EA can be highly dependent on the presence of other heuristics, within the confines of the 2DICBPP.

6   Conclusion

The objective of this paper is to formalize a variant of a 2-dimensional bin packing problem that can be applied to 3D printing for object placement to improve efficiency and reliability. We propose several heuristics within the EAs to improve the placement of polygon shaped objects as close to the center as possible, in order to reduce the effect of mechanical limitations.

We demonstrate the effects of the proposed heuristics, when they are applied individually or in combination. Each heuristic contains some variables or design decisions, which may or may not form an accurate representation of similar heuristics. Additional research that can examine varieties of other types of heuristics may lead to new insights or allow for heuristic combinations that can have a positive impact. Although our visual inspection of the results of the existing software packages indicated that they could be improved, quantitative assessment of their performance and comparing to our results can provide further insights.

The main comparison between the configurations was using the fitness values after 500 generations. While it is likely that many applications which require some solution for the 2DICBPP have the time to run an extensive EA with hundreds of generations, some applications may be more time sensitive and require a much shorter run. Researching the result of the heuristics on multiple generation intervals may give additional insights into effective heuristics for different use cases.

Finally, an optimization problem such as the 2DICBPP may benefit from stepping away from traditional EA implementations and applying more versatile methods. For example, the Mutation Direction heuristic may be used mutualistically with the Variable Step Size heuristic, or a type of shuffling/randomization heuristic, when being alternated at crucial moments. EAs which apply different heuristics throughout the same run, depending on generation intervals or if certain triggers are met, may result in more efficient solving methods, but will need further research.

References

1. GitHub - tamasmeszaros/libnest2d: 2D irregular bin packaging and nesting library written in modern C++ – github.com. https://github.com/tamasmeszaros/libnest2d. Accessed 13 Oct 2023
2. GitHub - Ultimaker/pynest2d: Python bindings for libnest2d – github.com. https://github.com/Ultimaker/pynest2d. Accessed 12 Nov 2023
3. UltiMaker Cura – ultimaker.com. https://ultimaker.com/software/ultimaker-cura/. Accessed 30 Oct 2023
4. Bäck, T., Schwefel, H.P.: An overview of evolutionary algorithms for parameter optimization. Evol. Comput.$\mathbf{1}$(1), 1–23 (1993). https://doi.org/10.1162/evco.1993.1.1.1. https://doi.org/10.1162/evco.1993.1.1.1
5. Berkey, J.O., Wang, P.Y.: Two-dimensional finite bin-packing algorithms. J. Operational Res. Soc. $\mathbf{38}$(5), 423–429 (1987). https://doi.org/10.1057/jors.1987.70
6. Burke, E., Hellier, R., Kendall, G., Whitwell, G.: A new bottom-left-fill heuristic algorithm for the two-dimensional irregular packing problem. Oper. Res. $\mathbf{54}$(3), 587–601 (2006). https://doi.org/10.1287/opre.1060.0293
7. Costa, L., Oliveira, P.: An evolution strategy for multiobjective optimization. In: Proceedings of the 2002 Congress on Evolutionary Computation. CEC'02 (Cat. No.02TH8600). IEEE (2002). https://doi.org/10.1109/cec.2002.1006216
8. Côté, J.F., Haouari, M., Iori, M.: Combinatorial benders decomposition for the two-dimensional bin packing problem. INFORMS J. Comput.$\mathbf{33}$(3), 963–978 (2021). https://doi.org/10.1287/ijoc.2020.1014
9. Flores, J.J., Martínez, J., Calderón, F.: Evolutionary computation solutions to the circle packing problem. Soft. Comput. $\mathbf{20}$, 1521–1535 (2016)
10. Garey, M.R., Johnson, D.S.: Approximation algorithms for bin packing problems: a survey. Analysis and Design of Algorithms in Combinatorial Optimization, pp. 147–172 (1981). https://doi.org/10.1007/978-3-7091-2748-3_8
11. Gonçalves, J.F., Resende, M.G.: A biased random key genetic algorithm for 2d and 3d bin packing problems. Int. J. Prod. Econ. $\mathbf{145}$(2), 500–510 (2013). https://doi.org/10.1016/j.ijpe.2013.04.019
12. Griffiths, V., Scanlan, J.P., Eres, M.H., Martinez-Sykora, A., Chinchapatnam, P.: Cost-driven build orientation and bin packing of parts in selective laser melting (SLM). Eur. J. Oper. Res. $\mathbf{273}$(1), 334–352 (2019). https://doi.org/10.1016/j.ejor.2018.07.053
13. Guo, B., Zhang, Y., Hu, J., Li, J., Wu, F., Peng, Q., Zhang, Q.: Two-dimensional irregular packing problems: a review. Front. Mech. Eng. 8, August 2022. https://doi.org/10.3389/fmech.2022.966691
14. Kang, K., Moon, I., Wang, H.: A hybrid genetic algorithm with a new packing strategy for the three-dimensional bin packing problem. Appl. Math. Comput.$\mathbf{219}$(3), 1287–1299 (2012). https://doi.org/10.1016/j.amc.2012.07.036
15. Kao, C.Y., Horng, J.T.: On solving rectangle bin packing problems using genetic algorithms. In: Proceedings of IEEE International Conference on Systems, Man and Cybernetics (1994). https://doi.org/10.1109/icsmc.1994.400073
16. Laabadi, S., Naimi, M., Amri, H.E., Achchab, B.: A binary crow search algorithm for solving two-dimensional bin packing problem with fixed orientation. Procedia Comput. Sci. $\mathbf{167}$, 809–818 (2020). https://doi.org/10.1016/j.procs.2020.03.420
17. Lamas-Fernandez, C., Bennell, J.A., Martinez-Sykora, A.: Voxel-based solution approaches to the three-dimensional irregular packing problem. Oper. Res. $\mathbf{71}$(4), 1298–1317 (2023). https://doi.org/10.1287/opre.2022.2260

18. Lodi, A., Martello, S., Vigo, D.: Approximation algorithms for the oriented two-dimensional bin packing problem. Eur. J. Oper. Res. **112**(1), 158–166 (1999). https://doi.org/10.1016/s0377-2217(97)00388-3 ¡error l="308" c="Invalid ¡error l="306" c="Invalid ¡error l="307" c="Invalid
command: paragraph not started." /¿ command: paragraph not started." /¿ command: paragraph not started." /¿ ¡error l="308" c="Invalid ¡error l="306" c="Invalid
command: paragraph not started." /¿ command: paragraph not started." /¿

19. Lopez, E., Ochoa, G., Terashima-Marín, H., Burke, E.: An effective heuristic for the two-dimensional irregular bin packing problem. Ann. Oper. Res. **206**, 241–264 (2013). https://doi.org/10.1007/s10479-013-1341-4

20. Mpofu, T.P., Mawere, C., Mukosera, M.: The impact and application of 3d printing technology. Int. J. Sci. Res., June 2014

21. Munien, C., Ezugwu, A.E.: Metaheuristic algorithms for one-dimensional bin-packing problems: a survey of recent advances and applications. J. Intell. Syst. **30**(1), 636–663 (2021). https://doi.org/10.1515/jisys-2020-0117

22. Puchinger, J., Raidl, G.R., Koller, G.: Solving a real-world glass cutting problem. In: Evolutionary Computation in Combinatorial Optimization, pp. 165–176. Springer, Heidelberg (2004). https://doi.org/10.1007/978-3-540-24652-7_17

23. Tamir, T.S., Xiong, G., Fang, Q., Dong, X., Shen, Z., Wang, F.Y.: A feedback-based print quality improving strategy for FDM 3d printing: an optimal design approach. The International J. Adv. Manuf. Technol. **120**(3-4), 2777–2791 (2022). https://doi.org/10.1007/s00170-021-08332-4

24. Terashima-Marín, H., Ross, P., Farías-Zárate, C., López-Camacho, E., Valenzuela-Rendón, M.: Generalized hyper-heuristics for solving 2d regular and irregular packing problems. Ann. Oper. Res. **179**, 369–392 (2010)

25. Trivella, A., Pisinger, D.: The load-balanced multi-dimensional bin-packing problem. Comput. Oper. Res. **74**, 152–164 (2016). https://doi.org/10.1016/j.cor.2016.04.020

26. Volna, E.: Genetic algorithms for two dimensional bin packing problem. In: AIP Conference Proceedings, vol. 1648, p. 550002 (2015). https://pubs.aip.org/aip/acp/article/1648/1/550002/802815/Genetic-algorithms-for-two-dimensional-bin-packing

27. Zhang, J., Wang, X.Z., Yu, W.W., Deng, Y.H.: Numerical investigation of the influence of process conditions on the temperature variation in fused deposition modeling. Mater. Des. **130**, 59–68 (2017). https://doi.org/10.1016/j.matdes.2017.05.040

A Hierarchical Approach to Evolving Behaviour-Trees for Swarm Control

Kirsty Montague[(✉)], Emma Hart[iD], and Ben Paechter[iD]

Edinburgh Napier University, Edinburgh, Scotland
{k.montague,e.hart,b.paechter}@napier.ac.uk

Abstract. Behaviour trees (BTs) are commonly used as controllers in robotic swarms due their modular composition and to the fact that they can be easily interpreted by humans. From an algorithmic perspective, an additional advantage is that extra modules can easily be introduced and incorporated into new trees. Genetic Programming (GP) has already been shown to be capable of evolving BTs to achieve a variety of sub-tasks (primitives) of a higher-level goal. In this work we show that a hierarchical controller can be evolved that first uses GP to evolve a repertoire of primitives expressed as BTs, and then to evolve a high-level BT controller that leverages the evolved repertoire for a foraging task. We show that the hierarchical approach that uses BTs at two levels outperforms a baseline in which the BTs are evolved using only low-level nodes. In addition, we propose a method to improve the quality of the primitive repertoire, which in turn results in improved high-level BTs.

Keywords: Swarm-robotics · Quality-Diversity · Genetic-Programming

1 Introduction

Collective intelligence arises in a swarm via the interaction of multiple agents that act individually according to their current perception of the environment. Many approaches to designing a control mechanism that results in a desired behaviour at the swarm level exist in the literature. A common means of control is to use a *hierarchical controller* in which a set of low-level control modules referred to as *primitives* are combined into a complex controller, referred to as an *arbitrator* [6,11,14,21]. The low-level modules (primitives) can either be hand-designed [11] or auto-designed, e.g. using evolution to evolve a neural-network controller [6,14] or a behaviour-tree (BT) [24]. Quality-diversity approaches such as MAP-Elites and Novelty search are increasingly being used to generate repertoires of primitives: Montague *et. al.* [24] use genetic programming combined with MAP-Elites to generate BT primitives for a foraging task, while in [8,15] a repertoire of neural controllers is generated using novelty-search and neuro-evolution. Arbitrators can also take many different forms: the AutoMoDe family of controllers mainly uses probabilistic finite-state machines (PFSMs) [15], while

BTs are used in [17–19, 21]. Neural networks can also be evolved as arbitrators, e.g. [13].

It is clear from the above that in developing a hierarchical controller, there are many choices in terms of the representation of both primitives and arbitrators. We suggest that BTs are an obvious choice for describing both primitives and arbitrators. A BT is itself a hierarchical model which consists of actions, conditions, and operators connected by directed edges [4]. They are modular in the sense that the set of actions available to the tree can be easily modified, any part of the tree can be extracted and reused, and the modules themselves can be generated by multiple means. In addition, although neural controllers are more common, they are not well-suited to crossing the reality-gap (i.e. obtaining consistent performance between a simulated experiment and a physical experiment) due to the fine-tuned precision obtained in simulation. Conversely, Francesca *et. al.* [11] propose that increasing bias by constraining the evolutionary search to pre-defined behaviour modules reduces variance and therefore sensitivity to the reality gap. Furthermore, a neural network can also be difficult to analyse or modify whereas BTs are human-readable and therefore go some way towards being explainable [16].

In this paper, we build on an existing line of work in hierarchical controller development by using evolutionary methods to develop a hierarchical control system for a swarm which has an *interpretable* controller. Specifically, we use two evolutionary methods to learn a hierarchical controller whose primitives and arbitrator *are both* represented as BTs: to the best of our knowledge, there is no existing hierarchical controller of this form. We leverage a set of primitives which are evolved to fulfil several manually defined objectives for this purpose, using (1) a multi-task version of GP (MTGP) and (2) MAP-Elites, both described by [24]. We first extend the work of [24] in improving the quality of the primitive set learned by MTGP by introducing the notion of *compatible objectives* (see Sect. 3.2). We then use GP to evolve a BT arbitrator for a foraging task, comparing the use of the different primitives' repertoires evolved in the previous step as input. We find that the arbitrator using primitives as nodes significantly outperforms a learned controller that uses low-level behaviours directly. Secondly, our results demonstrate that using a repertoire that contains multiple, diverse versions of each primitive leads to higher performing arbitrators than using a repertoire containing only a single high-performing primitive for each of the desired sub-behaviours.

2   Background

Hierarchical forms of control in which a desired task is decomposed into a set of simpler sub-tasks are common in many areas of robotics. Typically, an *arbitrator* is designed that selects *primitives* that execute individual behaviours, where the arbitrator can be informed by inputs from the environment or the robot's perceptions.

A long line of work in developing hierarchical controllers was spawned with AutoMoDe [11] and its sequence of successors [3, 10, 19–22]. This family of methods almost all generate probabilistic finite-state machine (PFSM) arbitrators, optimised using iterated F-race [23], with some exceptions e.g. IcePop [20] which uses Simulated Annealing as the optimiser.

In another example Cully *et. al.* [6] evolve a large set of walking gaits using a quality-diversity algorithm (MAP-Elites) for a legged robot, and an arbitrator (using Bayesian optimisation) selects the most appropriate primitive given the state of the robot and environment. Duarte *et. al.* [7] propose EvoRBC which also uses a QD algorithm (Novelty-Search) to first evolve a repertoire of low-level locomotion patterns (represented as vectors of parameters supplied to the robot's actuation system), and then to evolve a neural-network which acts as an arbitrator. EvoRBC-II extended the EvoRBC approach to include the use of closed-loop primitives. In the previously mentioned work, the *goal* of each primitive is hand-designed: that is, the decomposition of the desired high-level behaviour into primitives is performed by a human with knowledge of the desired task, e.g. for a foraging task, specifying primitives such as 'go-to-food' or 'go-to-nest'. In [15], a new approach is proposed which is mission agnostic, i.e. does not rely on the definition of task specific primitives. Their framework 'Nata' automatically generates probabilistic finite-state machines (arbitrators) in which states are selected from a repertoire of neural networks, and transition conditions are selected from a set of rules based on the sensory capabilities of the robotic platform considered [15]. A QD method is again used to generate a repertoire, after which Iterated F-Race [23] is used to assemble them into PFSMs.

Many of the methods just mentioned use neural network controllers either to create the repertoire of primitives or as the arbitrator. However, some concerns have been raised that such finely tuned precision is not suited to crossing the reality gap [11] while an additional concern is that a neural-network is a black-box, i.e. it is difficult to analyse or modify. PFSMs go someway towards addressing this criticism but require a compromise between reactivity and modularity: they cannot easily be broken down into their constituent parts because of dependencies between components and they do not scale well as the number of states grows [5]. On the other hand, Behaviour Trees (BTs) have an inherent capacity to reproduce the same functionality as PFSMs [4], but they maintain independence between components which removes these trade-offs and constraints. As noted by [16], they are also human-readable and therefore can be useful in explaining behaviours. Montague *et. al.* [24] proposed a method of evolving BT primitives using GP, exploring a multi-task GP method as well as MAP-Elites, but did not extend this to evolving an arbitrator. MAPLE [18] and Cedrata [19], both from the AutoMoDe family, use iterated F-Race to evolve BT arbitrators but not primitives. In Kuckling *et. al.* [21], two new variants of Cedrata are proposed, Cedrata-GP and Cedrata-GE which are based on genetic programming and grammatical evolution, respectively. The performance of the evolved BTs is compared against the performance of solutions created by a human designer, showing that Cedrata finds solutions that are also reliably found by human designers. However, the automatić design methods fail to discover the same communication strategies as the human designers.

In this paper we propose an approach in which for the first time both the primitives and arbitrator are represented as BTs, leading to increased transparency in interpreting them. We first extend the work described in [24] that uses GP and QD methods to evolve a repertoire of primitives to improve the quality of the repertoire. Then we evolve a BT arbitrator that leverages this repertoire using GP, evaluating it on a foraging task that is common in swarm robotics.

3   Methodology

The goal of this work is to evolve a hierarchical controller for a foraging task which is composed of BTs at both the primitive and arbitrator level, evolved by GP in both cases. We build directly on previous work by Montague *et. al.* [24] which demonstrated that GP could be used to generate BT primitives for a foraging task but stopped short of generating the high-level arbitrator. We make the following contributions:

- Evolve an extended set of primitives to enlarge the repertoire available to the arbitrator using (1) MAP-Elites in conjunction with GP (a Quality Diversity algorithm denoted QD) and (2) a multi-task GP method (denoted MTGP[1]) that simultaneously evolves for multiple task fitnesses using an implicit diversity mechanism. Specifically, we extend the set of primitives from *increase-neighbourhood-density*, *go-to-nest* and *go-to-food* to include *reduce-neighbourhood-density*, *go-away-from-nest* and *go-away-from-food*.
- Propose an approach to improve the quality of the repertoires generated by the multi-task method MTGP that only considers *compatible* objectives in its task set, i.e. does not include for example *go-to-food* and *go-away-from-food* which cannot be satisfied by the same controller.
- Use GP to evolve a high-level BT controller for a foraging task leveraging the primitive repertoires as input, comparing repertoires created by the different methods outlined above.

3.1   Setup

We consider a foraging task in which the objective is for each robot in the swarm to visit the food region and then the nest region as many times as possible over the course of each simulated trial. A more detailed description can be found in Sect. 4.2.

As per [24], a swarm is composed of nine footbot robots (Fig. 1, [1]) deployed in the arena shown in Fig. 2. Controllers are evaluated using the ARGoS simulator [25]. We use the same set up as described in [24], where the robots navigate

---

[1] Note that the authors in [24] referred to this method as e.g. $GP_{O1,O2,O3}$ however we believe it is more correctly described as a multi-task algorithm, e.g. [26].

by estimating the distance and direction of points of interest using information from their neighbours, while a blackboard provides an interface between the BT controllers and the footbots' sensor data. The reader is referred to the publication of [24] for full details.

In each set of experiments (to evolve primitives or arbitrators) each controller is evaluated over ten trials with randomised starting positions divided between two predefined arena configurations. The only difference in the way that arbitrators are evaluated compared with the primitives is that the length of each trial is increased from 20 s to 100 s.

Fig. 1. A screenshot of a footbot robot in the arena taken in ARGoS.

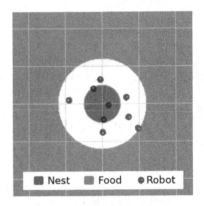

Fig. 2. The arena layout, with nine robots initialised in random starting positions.

In all of our experiments we evolve BT controllers with GP implemented using DEAP [9]. For the quality-diversity approach we combine GP with MAP-Elites using QDpy[2]. We use the same evolutionary parameters and BT implementation as described in [24], except that we add a new condition - *ifGotFood* - which indicates whether the robot has visited the food region since its last visit to the nest region. In doing so, we introduce a new internal state. Tables 1 and 2 list the nodes for evolving primitives for ease of reference. The reader is referred to [24] for detailed descriptions of the algorithms and GP implementation.

Each algorithm is run with ten different random seeds for each objective (or combination of objectives) for 1000 generations. The GP population size and the MAP-Elites batch size are both set to 25, while MTGP is assigned a population of 75 to reflect that it generates controllers for three objectives at once (therefore does not have to be run 3 times as with the other methods). All parameter settings are taken from [24]. Performance for each primitive is defined according to the median over 10 runs of the metrics described in [24] for the three primitive behaviours defined in [24] and for the three new primitive behaviours as defined in Sect. 3.2 which we introduce in this paper. The fitness of a BT arbitrator is defined in Sect. 4.2.

---

[2] https://pypi.org/project/qdpy/.

3.2   Evolving New Primitives

As noted above, we first use Map-Elites (denoted QD from herein) and MTGP to evolve three new primitives which provide the opposite behaviour to the original primitives, *increase density, go to nest* and *go to food*. The motivation behind this is to increase the number of options available to the arbitrator which in turn might find better behaviours. We opted for 'obvious' objectives at this stage, but there is clearly room for considering either further hand-crafted ones or auto-generating them in future work. These primitives are described in detail below:

– **Decrease neighbourhood density** maximises the difference between the density of neighbouring robots at the beginning and end of each trial, which we calculate by subtracting the final density from the initial density.

– **Move away from the nest region** maximises the difference in distance estimated by each robot at the start and end of each trial based on the shortest route by hops via neighbouring robots. The difference is calculated by subtracting the final distance from the initial distance.

– **Move away from the food region** maximises the difference in the robots' absolute distance to the food region at the beginning and end of each trial, calculated by subtracting the initial distance from the final distance.

These primitives are evolved using the same nodes as in [24], as shown in Tables 1 and 2.

Table 1. Condition Nodes

If on food	Returns success if the robot is within the food region
If food to left	Returns success if the shortest route to the food region is to the robot's left
If food to right	Returns success if the the shortest route to the food region is to the robot's right
If in nest	Returns success if the robot is within the nest region
If nest to left	Returns success if the shortest route to the nest region is to the robot's left
If nest to right	Returns success if the shortest route to the nest region is to the robot's right
If robot to left	Returns success if the nearest robot is to this robot's left
If robot to right	Returns success if the nearest robot is to this robot's right

The choice of primitives to be evolved can cause issues for MTGP: at each iteration, this algorithm randomly selects one of the objective functions and assigns a fitness based on the chosen function. This encourages generalisation and

Table 2. Action Nodes

Stop	No movement for one tick
Forwards	Move forwards for one tick
Forwards left	Right wheel forwards for one tick, rotating the robot anti-clockwise
Forwards right	Left wheel forwards for one tick, rotating the robot clockwise
Reverse	Move backwards for one tick
Reverse left	Right wheel in reverse for one tick, rotating the robot clockwise
Reverse right	Left wheel in reverse for one tick, rotating the robot anti-clockwise

was shown by [24] to improve performance for some objectives compared to a GP algorithm that evolved for each objective individually. However, it should be clear that some objectives are incompatible as previously mentioned. We therefore evaluate two versions of MTGP: one in which only compatible objectives are used, and another which includes objectives which are mutually exclusive. Hence, the following algorithms for evolving primitives are compared:

– **GP**: A baseline GP algorithm that evolves controllers for one objective at a time and is repeated for each primitive.

– **MTGP**: An algorithm that evolves controllers for multiple objectives at once, selecting one objective at random as the fitness function for each tournament used to select parents. We compare its performance using both *incompatible* (dubbed MTI) and *compatible* (dubbed MTC) combinations of objectives.

– **QD**: A MAP-Elites algorithm that evolves a collection of solutions for one objective at a time whose behaviours are diverse with respect to a set of user-defined characteristics. The characteristics which distinguish them are taken from [24] and are: difference in the ratios of forwards and backwards movement; ratios of clockwise and anti-clockwise rotations; the ratio of condition nodes and action nodes which are executed during simulation.

We define (*increase density, go to nest* and *go away from food*) as one set of compatible objectives, and (*reduce density, go away from nest* and *go to food*) as another[3]. We also define two sets of incompatible objectives: (*increase density, go to nest* and *go to food*) and (*reduce density, go away from nest* and *go away from food*).

3.3 Evolving an Arbitrator

To evolve a high-level foraging behaviour that leverages a repertoire of primitives evolved above, we use the single objective GP algorithm denoted GP above.

---

[3] Obviously objectives such as *increase density* and *decrease density* are mutually exclusive and therefore are never considered together.

This is exactly the same algorithm proposed by [24] except that the low-level action nodes are replaced by the primitives in the chosen repertoire and the set of condition nodes is restricted. We use the same evolutionary parameters as used to evolve the primitive repertoire, maintaining the population size of 25 individuals and running the algorithm for 1000 generations. Foraging also requires longer simulations so we increase the length of each trial in the arena from 20 s for primitives to 100 s for the arbitrator.

The objective function for the arbitrator rewards robots for each visit to the nest region which follows a visit to the food region. Upon visiting the food region, a robot is considered to be carrying food. If it then enters the nest region it reverts back to its default state and its score is incremented by one.

The fitness score S is defined as the number of times any robot carrying food arrives in the nest f divided by the number of robots in the arena r (which is nine in these experiments), i.e. $S = f/r$

We compare the following repertoires as input to this method:

- R1: the highest performing behaviour for each primitive found by QD.
- R2: the highest performing behaviour for each primitive returned by MTGP using compatible behaviours[4].
- R3: a repertoire containing multiple diverse BTs for each primitive. This is obtained by dividing the whole container returned by MAP-Elites into two equally sized bins along each axis, selecting the best controller from each of the eight resulting bins. This results in 8 behaviours for each of the 6 objectives, i.e. a total of 48 action nodes in the repertoire.
- R4: a repertoire containing eight diverse BTs for each primitive obtained by casting all individuals found by MTGP to the same MAP-Elites grid and dividing the axes in the same way, retrieving the best BT from each of the resulting eight bins.
- R5: a baseline experiment which uses the low-level action nodes used to evolve the primitives.

Note that using the repertoires of evolved primitives makes several of the condition nodes used by [24] obsolete. For example, condition nodes concerned only with navigation such as 'ifRobotToLeft/Right' are irrelevant at the arbitrator's level of abstraction and are therefore removed from the condition lists. This leaves just three condition nodes: (1) Is this robot in the food region; (2) Is this robot in the nest region; (3) Is this robot carrying food. The mutation operators insert condition or action nodes with equal probability.

4   Results

We first evaluate the proposed approaches for improving the primitive repertoires, i.e. by adding additional objectives, and using two versions of MTGP with compatible or incompatible subsets of objectives.

---

[4] Experiments in Sect. 4.1 showed that the performance of MTGP using compatible objectives was often better than using incompatible objectives.

4.1   Extending and Improving the Primitive Repertoire

Figure 3 shows boxplots of fitness over 10 repetitions of the performance of each algorithm listed in Sect. 3.2 for each of the six primitive behaviours evolved. To compare pairs of algorithms, a Shapiro-Wilk test was performed to check for normality, after which either a Student t-test if the data was judged to be normal or a Mann-Whitney test otherwise. The results of these tests are shown in Table 3: a confidence level of 0.05 is used to test for significance.

Table 3. Statistical testing results showing pairwise comparisons for different combinations of objectives. Statistically significant results within a confidence interval of 0.05 are shown in bold. The type of test applied is shown in italics: italicised = Mann-Whitney, non-italics=T-test

	GP vs MTI	GP vs MTC	GP vs QD	MTI vs MTC	MTI vs QD	MTC vs QD
Increase density	*0.7913*	0.0580	*0.5575*	*0.0640*	*0.3847*	0.0760
Go to nest	*0.4727*	**0.0452**	*0.5708*	0.2204	*0.0757*	**0.0257**
Go to food	*0.3217*	0.3240	**0.0270**	*0.0526*	***0.0168***	0.2199
Reduce density	0.8067	0.3838	**0.0352**	0.4535	**0.0037**	**0.0018**
Go away from nest	***0.0172***	***0.0211***	***0.0017***	0.6980	0.9482	0.6372
Go away from food	**0.0009**	**0.0259**	0.2002	**<0.0001**	**<0.0001**	0.1414

Table 4. Median values for each algorithm with the highest median in bold

	GP	MTI	MTC	QD
Increase density	0.585309	0.582788	**0.594218**	0.585442
Go to nest	0.826672	0.839357	**0.855242**	0.826128
Go to food	0.847747	0.842908	0.850078	**0.856015**
Reduce density	0.535612	0.533996	0.533903	**0.537092**
Go away from nest	0.792973	0.811753	0.803499	**0.813578**
Go away from food	0.624641	0.609507	**0.638132**	0.629626

Figure 3 shows that the highest median performance is obtained by *MTC* for three objectives (*increase density, go to nest, go away from food*) and by QD for the remaining three (*reduce density, go away from nest, go to food*). However, as shown in Table 3, the result is not always significant[5]. For four objectives, the

---

[5] Further work should increase the number of runs from the 10 performed to ascertain whether we should be confident in this result.

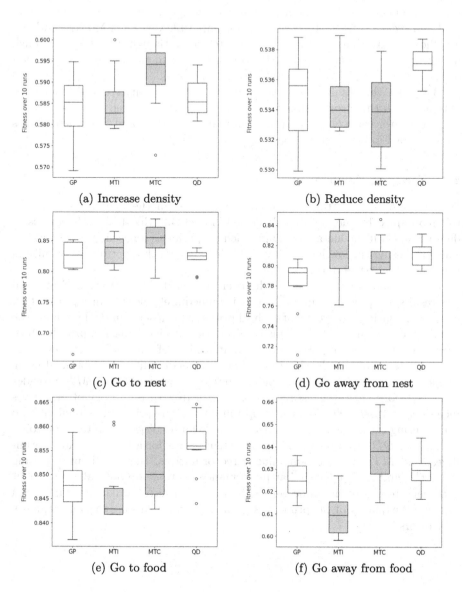

Fig. 3. Box-plots of the best fitness obtained for all three algorithms including two variations of the multi-task algorithm (MTI/MTC). Where its three objectives are compatible the performance is shown in blue; subsets of incompatible objectives are shown in red. (Color figure online)

median performance obtained from the compatible version of MTGP is higher than that of the incompatible version, although again the difference is not significant except in the case of *go away from food*. Surprisingly, for the *go away from nest* and *reduce density* objectives, the incompatible version of MTGP produces a higher median than its compatible counterpart although its variance is much higher. For *go away from food* and *go away from nest*, the incompatible version of MTGP performs significantly worse than the baseline.

Based on the results just described, we decide to discard the repertoires obtained with the incompatible version of MTGP and proceed to evolve a high-level arbitrator using only repertoires obtained from MTC and QD. These results are described in the next section.

4.2   Foraging Experiments

We compare BT arbitrators evolved using GP from the four different repertoires obtained by the methods described in Sect. 4.2. Recall that the goal is to determine: (1) if the hierarchical approach outperforms a baseline that evolves a single BT using the low-level nodes in Table 2; (2) which repertoire of primitives results in the best performing arbitrator.

Figure 4 shows boxplots of results over 10 repeated experiments. Statistical test results are presented in Table 5. It is immediately clear from Fig. 4 that all experiments using repertoires of evolved primitives outperform the baseline[6] (the first four entries in Table 5). This confirms that a hierarchical approach which leverages a repertoire of pre-evolved primitives is preferable to directly evolving an arbitrator using low-level actions. The best median fitness is obtained using a repertoire containing 8 diverse behaviours per objective (QD_8). MT_8 provides similar performance, suggesting that having a diverse repository of primitives including multiple behaviours that optimise the same primitive is preferable to simply using the single best primitive available for an objective in the repertoire. Recall that QD produces the highest median fitness for a behaviour for 3 primitives, and MTC for the remaining three primitives. Hence it is unsurprising that QD_8 and MT_8 have similar performance, as they are both able to exploit good repertoires. In the same vein, QD_1 and MT_1 have similar performance in terms of the quality of the primitives in the repertoire, leading to similar quality arbitrators.

---

[6] All experiments were run for the same amount of computational time taking into account the time taken to evolve the primitives: thus the baseline experiments are run for more generations than the arbitrator.

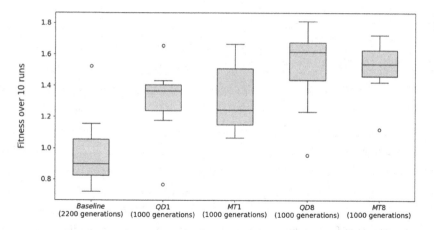

Fig. 4. A baseline algorithm which evolves a foraging behaviour from primitive actions nodes (go forwards, etc.) compared with ones that use the best of each of the sub-behaviours from the QD and MTGP repertoires instead, and ones which use eight versions of each sub-behaviour from each of those repertoires.

Table 5. Statistical testing results showing pairwise comparisons for foraging. Statistically significant results within a confidence interval of 0.05 are shown in bold.

Comparison	p-value	Type of test
Baseline vs QD repertoires of one	**0.0049**	**T-test**
Baseline vs MT repertoires of one	**0.0032**	**T-test**
Baseline vs QD repertoires of eight	**0.0001**	**T-test**
Baseline vs MT repertoires of eight	**<0.0001**	**T-test**
QD repertoires of one vs MT repertoires of one	0.9302	T-test
QD repertoires of eight vs MT repertoires of eight	0.9822	T-test
QD repertoires of one vs QD repertoires of eight	0.0665	T-test
MT repertoires of one vs MT repertoires of eight	**0.0316**	**T-test**
QD repertoires of one vs MT repertoires of eight	**0.0312**	**T-test**
MT repertoires of one vs QD repertoires of eight	0.0701	T-test

4.3   Readability

One of the main advantages of using BTs as opposed to NNs is that they are amenable to being understood by humans. Figure 5b shows an example of one of the high fitness BTs evolved, first in the full form return by the GP algorithm and then with its redundant nodes pruned by hand. The latter can be interpreted as follows: 'If you have food *go away from food*, and then if you are not in the nest, or you did not have food, check again if you have food. If you do then *go to food*, otherwise *reduce density*'.

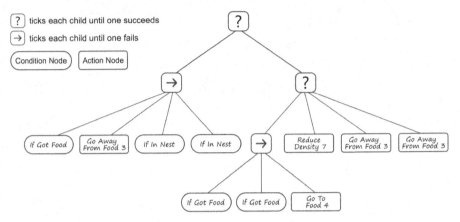

(a) A verbose tree for foraging. Some of the condition nodes are duplicated and all action nodes return success, so any subsequent children of a select node (denoted "?") will never be reached.

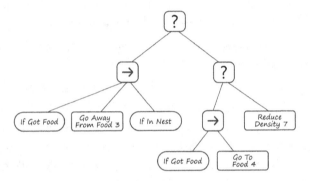

(b) The same tree with redundant nodes removed.

Fig. 5. Generated with repertoires of eight BTs per sub-behaviour from the QD repertoire.

A cursory examination will reveal that the use of *go to food* seems nonsensical. However, there is no requirement for the arbitrator to use the sub-behaviours for the purpose imagined by the designer or rewarded by the fitness function: for example we could speculate that *go to food* in Fig. 5b is being used simply to propel the robot forwards or backwards, since this is often what 'going to food' amounts to. This in itself is a useful behaviour.

5   Conclusions and Further Work

This paper builds on a line of work that uses a hierarchical method of developing a control system for a swarm of robots. At the primitive level, a set of controllers are created that optimise sub-tasks of the desired goal. A higher-level controller known as an arbitrator then combines the previously generated primitives into a controller that executes the defined goal. Although the use of hierarchical methods is well-known (particularly regarding the AutoMoDe [12] series of control software), previous methods have tended to use neural-networks or PFSMs as arbitrators, with a small number of recent papers proposing BTs [21]. In this paper, we propose a method that uses BTs at both levels of the hierarchy, i.e. to evolve the primitives and then the arbitrator. As noted by [16], BTs offer considerably more explanatory power than neural-networks.

Building on previous work by Montague *et. al.* [24] that proposed using BTs to evolve primitives, we extend this work in several ways. First, we extended the set of sub-tasks described in [24] to provide new primitives that could be useful in a foraging task. Secondly we proposed an amendment to the multi-task GP approach proposed in [24] that only considers compatible behaviours when generating multiple primitives simultaneously. Finally, we evolved an arbitrator as a BT using GP that exploits the new evolved repertoires, showing that repertoires that contain multiple BTs per primitive that achieve the same objective in different ways produce the highest performing controllers. We provide an example of a BT to illustrate that it can be easily read and analysed to understand the evolved behaviour.

There is much potential for future work. Rather than evolving primitives then an arbitrator sequentially, a meta-evolutionary algorithm could be used to search for the set of primitives that produce the best arbitrator, following a similar process to [2]. While BTs are inherently readable, it would be interesting to investigate the trade-off between readability and performance: replacing the BT arbitrator with a neural-network or PFSM and repeating the experiment would illustrate any such trade-off. The function of each of the primitives at the lower level of the hierarchy is human-designed, as are the action and condition nodes used by the GP algorithm to evolve primitives. A first step in removing the need for human expertise has recently been described by Hasslemann *et. al.* [15] which tries to automatically define primitives. This type of approach could also be integrated with our proposed methodology. Finally, we proposed a first naïve approach to selecting primitives for a repertoire from the much larger container of solutions generated by both the QD and MTGP algorithms. An approach that tried to maximise diversity might yield better results, or could itself be subject to a search process, given that large containers are generated. Finally, repeating the experiments in other collective tasks would provide further insights into the generality of the approach.

References

1. Bonani, M., et al.: The marxbot, a miniature mobile robot opening new perspectives for the collective-robotic research. In: 2010 IEEE/RSJ International Conference on Intelligent Robots and Systems, pp. 4187–4193 (2010). https://doi.org/10.1109/IROS.2010.5649153
2. Bossens, D.M., Mouret, J.B., Tarapore, D.: Learning behaviour-performance maps with meta-evolution. In: Proceedings of the 2020 Genetic and Evolutionary Compu-

tation Conference, pp. 49–57. GECCO '20, Association for Computing Machinery, New York, NY, USA (2020). https://doi.org/10.1145/3377930.3390181

3. Cambier, N., Ferrante, E.: AutoMoDe-pomodoro: an evolutionary class of modular designs, pp. 100–103 (2022). https://doi.org/10.1145/3520304.3529031

4. Colledanchise, M., Ögren, P.: How behavior trees modularize hybrid control systems and generalize sequential behavior compositions, the subsumption architecture, and decision trees. IEEE Trans. Rob. **33**(2), 372–389 (2017). https://doi.org/10.1109/TRO.2016.2633567

5. Colledanchise, M., Ögren, P.: Behavior trees in robotics and AI: an introduction. CoRR abs/1709.00084 (2017). http://arxiv.org/abs/1709.00084

6. Cully, A., Clune, J., Tarapore, D., Mouret, J.B.: Robots that can adapt like animals. Nature **521**(7553), 503–507 (2015)

7. Duarte, M., Gomes, J., Oliveira, S., Christensen, A.: EvoRBC: evolutionary repertoire-based control for robots with arbitrary locomotion complexity (2016). https://doi.org/10.1145/2908812.2908855

8. Duarte, M., Gomes, J., Oliveira, S.M., Christensen, A.L.: Evolution of repertoire-based control for robots with complex locomotor systems. IEEE Trans. Evol. Comput. **22**(2), 314–328 (2018). https://doi.org/10.1109/TEVC.2017.2722101

9. Fortin, F.A., De Rainville, F.M., Gardner, M., Parizeau, M., Gagné, C.: DEAP: evolutionary algorithms made easy. J. Mach. Learn. Res. Mach. Learn. Open Source Softw. **13**, 2171–2175 (2012)

10. Francesca, G., et al.: AutoMoDe-chocolate: automatic design of control software for robot swarms. Swarm Intell. 9 (2015)

11. Francesca, G., Brambilla, M., Brutschy, A., Trianni, V., Birattari, M.: AutoMoDe: a novel approach to the automatic design of control software for robot swarms. Swarm Intell. **8**, 1–24 (2014). https://doi.org/10.1007/s11721-014-0092-4

12. Francesca, G., Brambilla, M., Brutschy, A., Trianni, V., Birattari, M.: Automode: a novel approach to the automatic design of control software for robot swarms. Swarm Intell. **8**, 89–112 (2014). https://doi.org/10.1007/s11721-014-0092-4

13. Gomes, J., Christensen, A.L.: Task-agnostic evolution of diverse repertoires of swarm behaviours. In: Dorigo, M., Birattari, M., Blum, C., Christensen, A.L., Reina, A., Trianni, V. (eds.) Swarm Intelligence, pp. 225–238. Springer International Publishing, Cham (2018). https://doi.org/10.1007/978-3-030-00533-7_18

14. Gomes, J., Oliveira, S.M., Christensen, A.L.: An approach to evolve and exploit repertoires of general robot behaviours. Swarm Evol. Comput. **43**, 265–283 (2018)

15. Hasselmann, K., Ligot, A., Birattari, M.: Automatic modular design of robot swarms based on repertoires of behaviors generated via novelty search. Swarm Evol. Comput. **83**, 101395 (2023). https://doi.org/10.1016/j.swevo.2023.101395

16. Hogg, E., Hauert, S., Harvey, D., Richards, A.: Evolving behaviour trees for supervisory control of robot swarms. Artif. Life Robot. **25**, 569–577 (2020)

17. Kuckling, J., Ligot, A., Bozhinoski, D., Birattari, M.: Behavior trees as a control architecture in the automatic modular design of robot swarms. In: Dorigo, M., Birattari, M., Blum, C., Christensen, A.L., Reina, A., Trianni, V. (eds.) Swarm Intelligence, pp. 30–43. Springer International Publishing, Cham (2018). https://doi.org/10.1007/978-3-030-00533-7_3

18. Kuckling, J., Ligot, A., Bozhinoski, D., Birattari, M.: Behavior trees as a control architecture in the automatic modular design of robot swarms. In: Dorigo, M., Birattari, M., Blum, C., Christensen, A.L., Reina, A., Trianni, V. (eds.) ANTS 2018. LNCS, vol. 11172, pp. 30–43. Springer, Cham (2018). https://doi.org/10.1007/978-3-030-00533-7_3

19. Kuckling, J., van Pelt, V., Birattari, M.: Automatic modular design of behavior trees for robot swarms with communication capabilites. In: Castillo, P.A., Jiménez Laredo, J.L. (eds.) Applications of Evolutionary Computation, pp. 130–145. Springer International Publishing, Cham (2021). https://doi.org/10.1007/978-3-030-72699-7_9

20. Kuckling, J., Ubeda Arriaza, K., Birattari, M.: AutoMoDe-icepop: automatic modular design of control software for robot swarms using simulated annealing. In: Bogaerts, B., et al. (eds.) Artificial Intelligence and Machine Learning, pp. 3–17. Springer International Publishing, Cham (2020). https://doi.org/10.1007/978-3-030-65154-1_1

21. Kuckling, J., Van Pelt, V., Birattari, M.: AutoMoDe-cedrata: automatic design of behavior trees for controlling a swarm of robots with communication capabilities. SN Comput. Sci. 3(2), 136 (2022). https://doi.org/10.1007/s42979-021-00988-9

22. Ligot, A., Hasselmann, K., Birattari, M.: AutoMoDe-arlequin: neural networks as behavioral modules for the automatic design of probabilistic finite-state machines. In: Dorigo, M., et al. (eds.) Swarm Intelligence, pp. 271–281. Springer International Publishing, Cham (2020)

23. López-Ibáñez, M., Dubois-Lacoste, J., Pérez Cáceres, L., Birattari, M., Stützle, T.: The irace package: iterated racing for automatic algorithm configuration. Oper. Res. Perspect. 3, 43–58 (2016). https://doi.org/10.1016/j.orp.2016.09.002, https://www.sciencedirect.com/science/article/pii/S2214716015300270

24. Montague, K., Hart, E., Nitschke, G., Paechter, B.: A quality-diversity approach to evolving a repertoire of diverse behaviour-trees in robot swarms. In: Correia, J., Smith, S., Qaddoura, R. (eds.) Applications of Evolutionary Computation, pp. 145–160. Springer Nature Switzerland, Cham (2023). https://doi.org/10.1007/978-3-031-30229-9_10

25. Pinciroli, C., et al.: ARGoS: a modular, parallel, multi-engine simulator for multi-robot systems. Swarm Intell. 6, 271–295 (2012). https://doi.org/10.1007/s11721-012-0072-5

26. Wei, T., Wang, S., Zhong, J., Liu, D., Zhang, J.: A review on evolutionary multi-task optimization: trends and challenges. IEEE Trans. Evol. Comput. (2021)

Evolutionary Algorithms for Optimizing Emergency Exit Placement in Indoor Environments

Carlos Cotta[1,2](✉)(iD) and José E. Gallardo[1,2](iD)

[1] Departamento Lenguajes y Ciencias de la Computación, ETSI Informática, Campus de Teatinos, Universidad de Málaga, 29071 Málaga, Spain
{ccottap,pepeg}@lcc.uma.es
[2] ITIS Software, Universidad de Málaga, Málaga, Spain

Abstract. The problem of finding the optimal placement of emergency exits in an indoor environment to facilitate the rapid and orderly evacuation of crowds is addressed in this work. A cellular-automaton model is used to simulate the behavior of pedestrians in such scenarios, taking into account factors such as the environment, the pedestrians themselves, and the interactions among them. A metric is proposed to determine how successful or satisfactory an evacuation was. Subsequently, two metaheuristic algorithms, namely an iterated greedy heuristic and an evolutionary algorithm (EA) are proposed to solve the optimization problem. A comparative analysis shows that the proposed EA is able to find effective solutions for different scenarios, and that an island-based version of it outperforms the other two algorithms in terms of solution quality.

Keywords: Pedestrian Evacuation · Cellular Automata · Greedy Heuristics · Evolutionary Algorithms

1 Introduction

In the event of an emergency, the rapid and orderly evacuation of crowds from enclosed spaces is essential to minimize casualties and ensure public safety. Needless to say, it can also become a critical challenge requiring meticulous planning at different levels, in order to avoid panic, bottlenecks, and potential harm to people in a potentially chaotic scenario [9]. There are different factors that need being taken into account depending on the specificities of each situation (e.g., what the particulars of the environment are, what the typical size and composition of the crowd is, and so on), and the level at which the planning is done (e.g., architectural decisions, signaling, etc.). In this work we are specifically concerned about the placement of emergency exits in the most convenient way to facilitate the efficient evacuation of the crowd.

This work is supported by Spanish Ministry of Science and Innovation under project Bio4Res (PID2021-125184NB-I00 - http://bio4res.lcc.uma.es) and by Universidad de Málaga, Campus de Excelencia Internacional Andalucía Tech.

In order to approach any evacuation optimization problem –such as the one considered here– and attain safe and efficient evacuation plans, understanding and predicting the behavior of pedestrians is of paramount importance. However, pedestrian evacuation is a complex and dynamic process, influenced by many factors, such as the environment, the pedestrians themselves, and the interactions among them. Therefore, modeling pedestrian evacuation is a challenging task that requires a balance between simplicity and realism. There are different tools that can be used for this purpose, depending on the scope of the simulation. Thus, whereas macroscopic approaches will often consider the crowd as a continuous medium whose flow is to be modeled, e.g., see [2,8], microscopic models will focus on the pedestrians –the individual components of the crowd– and model the crowd behavior as an emergent property of the collective behavior of those individual agents. The latter models can be further divided into two major categories, namely models based on social forces (in which pedestrians are particles in a continuous space, subject to different forces resulting from their interaction with the environment and other particles, e.g., [3,12]), and cellular-automaton (CA) models (in which the environment is modeled as a discrete grid, and pedestrians transition between these following some predefined rules, e.g., [16,18]). We refer to [4,13] for a more in-depth survey of all these approaches.

We have precisely considered the CA approach in this work, and devised a model for modeling the behavior of a crowd evacuating an indoor environment (see Sect. 3). Using this tool, we aim to find which would be the most appropriate location for emergency exits. This also entails defining appropriate metrics to assess to which extent an evacuation was successful/satisfactory or not. We do this in Sect. 2. Subsequently, we consider different algorithmic approaches to tackle this problem. To be precise, we devise an iterated greedy heuristic and an evolutionary algorithm (EA) for this purpose (see Sect. 4). We conduct an extensive experimentation to analyze the performance of these algorithms (as well as an island-based version of the EA) in Sect. 5. Our main aim in this work is to determine the effectiveness of these approaches for this particular optimization setting, as a stepping stone for devising more powerful approaches and tackling more complex evacuation scenarios. We close this work with a critical outlook of the results and an overview of the following steps in this research.

2   Problem Statement

In order to model the evacuation problem, we need to start by formalizing the indoor space from which the evacuation is attempted. To this end, let $\mathcal{A}$ be this space, which we will assume to be a rectangular area of width w and height h. This rectangular area represents the floor plan of an enclosed space and therefore all its boundaries are assumed to be blocked (i.e., to be non-traversable), except in specific locations which will be denoted as *accesses*. More precisely, we can define an access α as a pair (p_α, w_α), where p_α denotes a point along the perimeter (i.e., a value between 0 and $2(w + h)$, where 0 corresponds to a certain predefined reference point (e.g., the bottom-left corner of $\mathcal{A}$) of the area at which the access is anchored, and w_α denotes the width of the access along the perimeter, that is, the access extends from p_α to $p_\alpha + w_\alpha$[1]. Now, within $\mathcal{A}$ there may be a number of *obstacles*. Each obstacle $o \subseteq \mathcal{A}$ denotes a non-traversable region (representing real-world objects such as walls or furniture). Therefore, the whole environment can be represented as a tuple (w, h, A, O), where:

- w and h are the width and height of $\mathcal{A}$ respectively.
- $A = \{\alpha_1, \ldots, \alpha_k\}$ is a collection of accesses.
- $O = \{o_1, \ldots, o_m\}$ is a collection of obstacles.

This environment is crowded with n pedestrians (they represent the users of said environment, i.e., residents, workers, customers, etc. depending on what it is being modeled) distributed along traversable areas of $\mathcal{A}$. At time $t = 0$, an emergency is declared and the evacuation of the place begins. Let $\mathbb{M}$ be a model that can be used to predict the behavior of pedestrians in this context, and how the evacuation process would then be conducted (cf. Sect. 3). Let $\rho_i(t)$ represent the position coordinates of the i-th pedestrian at time t, and let T be the maximum time up to which the model is simulated. Then, we can split the collection of pedestrians into two sets:

- *evacuees* $\xi^+ = \{ i \mid 1 \leqslant i \leqslant n, \exists t_i \leqslant T : \exists \alpha \in A : \rho_i(t_i) \in \alpha \}$, i.e., all pedestrians i who manage to reach an access before T.
- *non-evacuees* $\xi^- = \{1, \ldots, n\} \backslash \xi^+$, i.e., the pedestrians who could not reach an access before T. Given a non-evacuee i, we can define $d_i = \min_{\alpha \in A} \|\alpha - \rho_i(T)\|$, i.e., their distance to the nearest exit at the end of the simulation.

In order to quantify the extent to which the evacuation is successful, different metrics could be used. We consider the following hierarchy of objectives:

1. The first goal is to minimize the number of non-evacuees $|\xi^-|$. This has the highest priority.

The next levels of the hierarchy depend on whether the first goal could be accomplished or not. In the first case ($\xi^- = \emptyset$), we consider:

2a. Minimize the time at which the last pedestrian left the area, i.e., minimize $t^* = \max_{1 \leqslant i \leqslant n} t_i$.
3a. Minimize the average time at which pedestrians left the area, i.e., minimize $\bar{t} = \frac{1}{n} \sum_{1 \leqslant i \leqslant n} t_i$.

If the evacuation was however not complete, then:

2b. Minimize the minimum distance between a non-evacuee and an access, i.e., minimize $d^* = \min_{i \in \xi^-} d_i$.
3b. Minimize the average distance between non-evacuees and accesses, i.e., minimize $\bar{d} = \frac{1}{n} \sum_{i \in \xi^-} d_i$.

---

[1] Note that since the perimeter is closed, the sum is to be understood as cycling back to 0 when reaching $2(w + h)$.

This hierarchy of goals can be combined into a single numerical value by using appropriate weights that ensure that any comparison respects said hierarchy. To be precise, let $\sigma(\mathcal{A}, S)$ be a tuple containing the evacuation status of each pedestrian and the corresponding value of d_i or t_i at the end of the simulation, given that $S = [\rho_1(0), \ldots, \rho_n(0)]$ are the initial positions in $\mathcal{A}$ of the pedestrians at time $t = 0$. Then, we define:

$$f(\sigma(\mathcal{A}, S)) = |\xi^-| + [\xi^- = \emptyset]\left(\frac{1}{T}\max_{1 \leqslant i \leqslant n} t_i + \frac{1}{nT^2}\sum_{1 \leqslant i \leqslant n} t_i\right) + \\ + [\xi^- \neq \emptyset]\left(\frac{1}{D}\min_{i \in \xi^-} d_i + \frac{1}{nD^2}\sum_{i \in \xi^-} d_i\right) \tag{1}$$

where $[\cdot]$ are Iverson brackets, and $D = \sqrt{w^2 + h^2}$ is the diagonal of the area. Now, we can formally define the OPTIMAL EVACUATION PROBLEM (OEP) as:

Instance: a tuple $(\mathcal{A}, \mathbb{S}, k, \omega)$, where
- $\mathcal{A} = (w, h, A, O)$ is the environment.
- $\mathbb{S} = \{S_1, \ldots, S_l\}$ is a collection of initial configurations of n pedestrians, i.e., for all $1 \leqslant i \leqslant l$, $|S_i| = n$.
- $k \in \mathbb{N}$ is a non-zero value that indicates the number of emergency exits whose location is sought.
- $\omega > 0$ is the width of emergency exits.

Solution: a collection $E = \{e_1, \ldots, e_k\} \subset [0, 2(w+h)]$, where each e_i represents the location of an emergency exit and such that

$$\psi(E) = \frac{1}{l}\sum_{1 \leqslant i \leqslant l} f(\sigma(\mathcal{A}', S_i)) \tag{2}$$

is minimal, where $\mathcal{A}'$ is obtained from $\mathcal{A}$ by adding $\{(e_1, \omega), \ldots, (e_k, \omega)\}$ to the existing accesses.

Having defined the problem, let us turn our attention to how pedestrian behavior is modeled in next section.

3   A CA for Modeling Pedestrian Evacuation

Cellular automata are simple and powerful tools to simulate complex systems, as they can capture the emergence of global patterns from local interactions. In this section, we describe the details of our CA model for pedestrian evacuation.

3.1   State of the CA

The state of the CA is the state of each cell in the environment (represented by a regular lattice of square cells). Each cell can be in one of three states:

- empty: The cell is empty and can be occupied by a pedestrian.
- occupied: The cell is occupied by a pedestrian.

– `obstacle`: The cell is occupied by an obstacle and cannot be occupied by a pedestrian.

Some cells in the environment are marked as *exit* cells. These are the cells that the pedestrians want to reach to leave the environment. We assume that pedestrians are rational and will try to find the shortest path to the nearest exit. However, the presence of obstacles and other pedestrians can affect their movement and make them choose alternative paths. To capture this behavior, we define two concepts for each cell: the *static field* and the *crowd repulsion*. The former is a measure of how close a cell is to an exit. The latter is a measure of how crowded the neighborhood of a cell is, taking into account obstacles and other pedestrians. We use these two concepts to calculate the *desirability* of a cell, which is the probability that a pedestrian will move to that cell.

The static field of a cell is computed using Dijkstra's algorithm, which is a well-known algorithm for finding the shortest path between two nodes in a weighted graph [6]. We consider the environment as a graph, where nodes are cells and edges are connections between neighboring cells. The weight of an edge is the geometric distance between the cell centers, if the target cell is not an obstacle and infinity otherwise. Formally, we define the graph as $G = (V, E)$, where V is the set of cells in the environment and E is the set of edges between neighboring cells. The weight function is $w : E \to \mathbb{R}^+$, such that $w(v_i, v_j)$ is the geometric distance between cells v_i and v_j, as defined before. Let $\mathcal{SP}_{i,j}$ be the length of the shortest path from cell (i, j) to its nearest exit as computed by Dijkstra's algorithm. The static field of a cell (i, j) is then defined as:

$$\mathcal{SF}_{i,j} = 1 - \frac{\mathcal{SP}_{i,j}}{\mathcal{SP}_{\max}} \tag{3}$$

where $\mathcal{SP}_{\max}$ is the larger shortest path from any cell in the environment to its nearest exit. This definition makes the static field be in [0,1] and only depend on the relative distance of a cell to its nearest exit. The higher the static field, the closer the cell is to an exit. Notice that, as this field is static, it does not change over time and is only computed once before the simulation.

The crowd repulsion of a cell is computed using the number of reachable cells in its neighborhood. A cell is reachable if it is currently empty and not blocked by an obstacle. For each occupied cell (i, j), let $\mathcal{N}_{i,j}$ be the set of reachable cells in its neighborhood. The repulsion of a cell (i, j) is defined as the inverse of one plus the number of reachable cells in this neighborhood:

$$\mathcal{R}_{i,j} = (1 + |\mathcal{N}_{i,j}|)^{-1} \tag{4}$$

where $|\cdot|$ denotes the cardinality of a set. This definition makes the repulsion be in (0,1] and depend on how crowded the neighborhood of a cell is. The higher the repulsion, the more crowded the neighborhood is.

The desirability of a cell is computed using a combination of the static field and the crowd repulsion. We introduce two parameters to weight the importance of these two factors: the *field attraction bias* ϕ and the *crowd repulsion bias* ζ.

The field attraction bias reflects how strongly the pedestrians are attracted to the exit cells, while the crowd repulsion reflects how strongly the pedestrians are repelled by the crowded cells. We firstly define the attraction of a cell (i, j) as:

$$\mathcal{A}_{i,j} = \exp(\phi \cdot \mathcal{SF}_{i,j} - \zeta \cdot \mathcal{R}_{i,j}) \tag{5}$$

In this way, the attraction of a cell is a positive number that increases with the static field and decreases with the crowd repulsion. The higher the attraction, the more desirable the cell is. However, we can make the pedestrian behavior more realistic and adaptive by reducing the reliance on the global knowledge of the environment and by making use of the information available in the local neighborhood. As the attraction of a cell is not enough to capture this behavior, we need to consider instead the *desirability* of a cell, which is defined as the gradient of its attraction. The desirability of a cell reflects how the attraction changes locally by comparing the attraction of the cell with the minimum attraction in its reachable neighborhood. Let $\mathcal{A}_{\min_{i,j}}$ denote the minimum attraction in neighborhood of cell (i, j), which is defined as:

$$\mathcal{A}_{\min_{i,j}} = \min_{(k,l) \in \mathcal{N}_{i,j}} \mathcal{A}_{k,l} \tag{6}$$

Then, the desirability of cell (i, j) is defined as:

$$\mathcal{D}_{i,j} = \epsilon + \mathcal{A}_{i,j} - \mathcal{A}_{\min_{i,j}} \tag{7}$$

where ϵ is a small number which is added to avoid the desirability being zero ($\epsilon = 10^{-5}$ in our implementation). In this way, the desirability of a cell is a positive number that increases with the gradient of the attraction. The higher the desirability, the more likely a pedestrian will move to that cell. The desirability of a cell is the main input of the local rule that updates the state of each cell on each time step. The local rule is based on a probabilistic transition function that determines the probability of a pedestrian moving from one cell to another.

3.2 Update Procedure

The update procedure is the procedure that is used to update the state of the CA on each time step. The procedure is as follows:

1. We start by marking as empty in the next state the cells that are currently occupied by pedestrians, as they may change depending on their movement.
2. We then mark the cells that are occupied by obstacles in the current state as obstacle in the next state. These cells will not change, as they cannot be occupied by pedestrians.
3. We also mark the exit cells that are occupied by pedestrians in the current state as empty in the next state. This models the evacuation of the pedestrians through the exits. We assume that once a pedestrian reaches an exit, they leave the environment and do not come back.

4. For any other cell that is occupied by a pedestrian in the current state, we compute the desirabilities of reachable neighboring cells. We use the desirability as the probability of a pedestrian moving to that cell and randomly select one neighboring cell according to these probabilities. If the selected cell is not occupied by another pedestrian in the next state, we mark it as occupied by the pedestrian in the next state. This means that the pedestrian moves to that cell. Otherwise, we mark the current cell as occupied by the pedestrian in the next state, i.e., the pedestrian stays in the same cell. This way, we avoid collisions between pedestrians and ensure that each cell can have at most one pedestrian. To ensure fairness among pedestrians, we shuffle the order in which we process occupied cells on each time step.

We consider that each cell in the environment is a square and we denote by cl its side length. We denote the time elapsed for each time step as Δt. The speed of a pedestrian that moves to a neighboring cell on each time step is then $cl/\Delta t$. We call this the *reference speed* of a pedestrian, and denote it by v. However, not all pedestrians may move at the same speed (for instance, some pedestrians may move slower than the reference speed, due to physical or psychological factors). To model this, we introduce for each pedestrian a parameter called *velocity percent* (v_p), which is a percentage of the reference speed. For example, if $v_p = 0.5$ for a pedestrian, their speed would be $0.5v$. We model this by letting v_p be the probability of a pedestrian moving to a neighboring cell on each time step so that, on average, their speed would be $v_p \cdot v$.

3.3   Transition Function

The transition function is the function that determines the probability of a pedestrian moving from one cell to another. The function is based on the desirability of the neighboring cells. The function is defined as follows:

$$T(c_i, c_j) = \begin{cases} P_{i,j} \cdot v_p, & \text{if } c_j \text{ is empty or an exit in the current state} \\ 0, & \text{otherwise} \end{cases} \qquad (8)$$

where c_i and c_j are two neighboring cells, $P_{i,j}$ is the probability of agent in cell c_i to move to cell c_j based on its desirability:

$$P_{i,j} = \frac{\mathcal{D}_{c_j}}{\sum_{c \in \mathcal{N}_{c_i}} \mathcal{D}_c} \qquad (9)$$

and v_p is the velocity percent of the pedestrian in cell c_i. The transition function returns the probability of the pedestrian in cell c_i moving to cell c_j on the next time step. The function is zero if cell c_j is blocked or already occupied by another pedestrian in the current state, or if the pedestrian in cell c_i does not move in this time step, which happens with probability $1 - v_p$. The transition function is applied to each occupied cell in the current state, after shuffling the order of the cells. The result of the function and the procedure to avoid collisions (step

---

Algorithm 1: Greedy constructive heuristic

Data: an instance OEP($\mathcal{A}, \mathbb{S}, k, \omega$)

$E \leftarrow \emptyset$;

$\eta \leftarrow \lceil 2(w+h)/\omega \rceil$;

for $i \leftarrow 1$ **to** k **do**

    $p \leftarrow \text{rand}(0, 2(w+h))$;

    $best \leftarrow \infty$;

    for $j \leftarrow 1$ **to** η **do**

        $cur \leftarrow \psi(E \cup \{p\})$;

        if $cur < best$ **then** $best \leftarrow cur$; $e \leftarrow p$;

        $p \leftarrow p + \omega$;

        if $p > 2(w+h)$ **then** $p \leftarrow p - 2(w+h)$

    end

    $E \leftarrow E \cup \{e\}$;

end

return E

---

4. in Sect. 3.2) is used to update the state of the CA on the next time step. The update procedure is repeated until all the pedestrians have evacuated or a maximum number of time steps (corresponding to time T) is reached.

4   Algorithms for Emergency Exit Optimization

As indicated in Sect. 2, a solution to problem instance OEP($\mathcal{A}, \mathbb{S}, k, \omega$) is a set $E = \{e_1, \ldots, e_k\} \subset [0, 2(w+h)]$. The mapping between solutions and their associated objective functions values is not just non-linear, but also not available in closed form, and only computable via a stochastic simulation. Thus, it is complex to design low-level heuristics to construct such solutions. We can however engineer a constructive approach on top of the simulations, based on greedy principles. The core of this approach is shown in Algorithm 1.

This procedure starts by picking a random initial point p along the perimeter. Then all points $p, p+\omega, p+2\omega, \ldots, p+\eta\omega$ are potential candidates to place an exit, where the addition is assumed to wrap around the length of the perimeter, and η is picked so as to ensure that we cover the whole perimeter. For each candidate, we simulate the system with an emergency exit in the corresponding location (in addition to any other exits that might have been considered in previous steps), and keep the one that returns the best value of the objective function. This is repeated as many times as needed (i.e., k times) to construct the solution. Notice that this procedure involves computing the value of the objective function $\eta \cdot k$ times. Also, this is a randomized procedure and therefore can be iterated as many times to desired to obtain different greedy solutions. We will denote this latter iterated procedure as **greedy**.

As an alternative to this greedy heuristic, we consider an EA approach. This is a real-coded EA in which individuals are vectors of k values in the range $[0, 2(w+h)]$. We can initially generate such vectors by sampling uniformly at random the

---

Algorithm 2: Set-based recombination

Data: two sets $E = \{e_1, \ldots, e_k\}$ and $E' = \{e'_1, \ldots, e'_k\}$
$C \leftarrow E \cup E'$; $S \leftarrow \emptyset$;
for $i \leftarrow 1$ **to** k **do**
   | $e \leftarrow$ PICK (C); // makes random selection
   | $S \leftarrow S \cup \{e\}$; $C \leftarrow C \setminus \{e\}$;
end
return S

---

search space. Notice that we do not introduce any constraint regarding the non-overlap of exits. Having two overlapping exits is equivalent within the simulation to having a single exit of width $2\omega - \text{overlap}$. We pose that this is less convenient than having two exits back to back without overlapping, or those two exits strategically placed somewhere else. For this reason, we expect evolution will get rid of those suboptimal solutions without the need of introducing an explicit constraint. As to mutation, we have opted for a Gaussian perturbation of a single exit, whose amplitude is a certain percentage γ of its current value, i.e.,

$$e' \leftarrow e \cdot (1 + \gamma \mathcal{N}(0,1)) \tag{10}$$

where $\mathcal{N}(0,1)$ is a normally distributed random value of mean 0 and variance 1. As usual, the value of the variable will wrap around $[0, 2(w+h)]$. As for recombination, we have opted for a discrete set-based approach, since standard operators for continuous variables require a meaningful matching between homologous variables in the parental solutions which is not possible (or at least non-trivial) in this problem. Our recombination algorithm is depicted in Algorithm 2. It creates a set of candidate locations from the union of the individuals being recombined, and makes a sequence of random picks without replacement from this candidate set. The resulting operator is therefore transmitting and assorting, but not necessarily respectful [15]. Besides these operators, our EA uses binary tournament solution, and elitist generational replacement. We have also considered an island version of this EA [1], which divides the population into a number of separate demes (arranged following a certain topology – a bidirectional ring in our case) which evolve in partial isolation, and periodically migrate the best solution to neighboring demes, who accept these in substitution of their current worst solutions. We will denote our EA and our island-based EA as EA and iEA respectively. All algorithms are available in our GitHub repository[2].

5   Experimental Results

The different algorithms described in the previous section have been put to test on a collection of problem instances with different features. These instances and the remaining experimental parameters are described in Sect. 5.1. Subsequently, the numerical results will be reported and analyzed in Sect. 5.2.

---

[2] https://github.com/Bio4Res/pedestrian-evacuation-optimization.

5.1   Experimental Setup

To evaluate the performance of different algorithms, we have generated several environments that simulate evacuation scenarios. Our instance generator discretizes the evacuation area in the same fashion our CA does (see Sect. 3), and places obstacles randomly in the domain, avoiding overlaps and ensuring a minimum distance between them. The obstacles are rectangular and their dimensions are randomly generated as follows: the width of the obstacle can be either one or two cells, if the obstacle is vertical, or between one and 25 cells, if the obstacle is horizontal. The height of the obstacle is inversely proportional to the width, and it can be between one and half of the rows of the domain. The orientation of the obstacle is also randomly chosen, with a 50% probability of being vertical or horizontal. The position of the obstacle is randomly selected, with the condition that the obstacle does not exceed the boundaries of the domain, and that there is a minimum distance of two cells between the obstacle and any other obstacle, so that the agents can always move around them. The purpose of the obstacles is to create a realistic, diverse, and challenging environment for the agents, by obstructing their movement and forcing them to find alternative paths.

We have generated three sets of instances, each containing five environments with different characteristics depending on the number $|O|$ of obstacles:

- *low-density*: $|O| \in \{20, \ldots, 30\}$. A low density of obstacles implies that the agents have more space to move and less chances of colliding with them.
- *mid-density*: $|O| \in \{50, \ldots, 75\}$. A medium density of obstacles means that the agents have less space to move and more chances of colliding with them, but still have some room for maneuvering and finding alternative paths.
- *high-density*: $|O| \in \{100, \ldots, 150\}$. A high density of obstacles results in the agents having very little space to move and very high chances of colliding with them, facing a lot of congestion and bottlenecks in their movement.

In all cases, the width and height are picked from $[40, 50]$ and $[20, 30]$, the side of the square cells is 0.5 m and no exits are initially placed. Hence, evacuation will only proceed through the emergency exits placed by the optimization algorithms. We consider three setting in this regard, namely $k \in \{3, 4, 5\}$ exits. The width of the emergency exits is set to $\omega = 2$ m. All the instances are publicly available in our data repository [5]. For each instance, we have randomly generated 1000 initial pedestrian configurations. 20 are used as training set for the optimization algorithms, and the remaining ones are used as test set. In every case we have considered 100 pedestrians. Each of them has a reference velocity $v = 1.3$ m/s, a velocity percent $v_p \in [0.5, 1]$, field attraction bias $\phi \in [1.5, 2]$, and crowd repulsion bias $\zeta \in [0.25, 0.5]$. The simulation is run up to $T = 60$ s.

Regarding the algorithms, in all cases we consider *maxevals* = 20000. The EA has a population size $\mu = 100$, recombination probability $p_X = 0.9$, mutation probability equivalent to a mutation rate $1/\ell$ per variable, where ℓ is the number of variables, and gaussian mutation amplitude $\gamma = 0.05$. As to the iEA, it considers 4 islands of size $\mu = 25$, and migration frequency of 10 generations. No fine tuning of these parameters has been attempted. For each algorithm, floor plan and number of exits sought, we perform 20 runs.

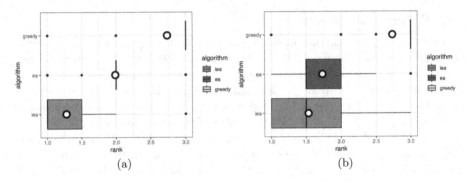

Fig. 1. (a) Rank distribution of the different algorithms on the training set. (b) Rank distribution of the best solution of each algorithm on the test set.

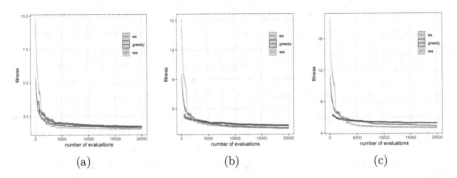

Fig. 2. Evolution of fitness in three of the instances. (a) low-density (b mid-density (c) high-density

5.2   Results

Table 1 shows the summary of results over the 20 runs of the algorithms. As it can be seen there is a general superiority of iEA over all types of instances and number of exits, and even more clearly for $k \geqslant 4$ exits. This superiority is not just clear on a head-to-head basis with respect to EA and greedy on specific instances, but it is also globally significant. To show this, we rank each algorithm on each problem instance, and determine the distribution of ranks – see Fig. 1a. These ranks show statistically significant differences according to Quade test [14] (Quade $F = 37.351$, p-value $= 1.803e{-}12$). Subsequently, we conduct Holm test with Bonferroni correction [7,10] using iEA as control algorithm. The test is passed against both EA and greedy with p-value $= 7.433e{-}4$. These results indicate that the evolutionary search, and in particular the island-based EA, is capable of effectively navigating the search space and finding solutions that perform satisfactorily on the training set. Fig. 2 shows an example of the evolution of fitness as a function of the number of solution evaluations for the three algorithms. As it can be seen, greedy often starts with good quality solutions, typically better than those of EA and iEA for a similar computational effort.

Table 1. Results of the algorithms (out of 20 runs) on the training set. Each column depicts the best (x^*), median ($\tilde{x}$), mean ($\bar{x}$) and standard error of the mean ($\sigma_{\bar{x}}$). For each instance, the algorithm with the best mean is marked with a star ($\star$), and the remaining algorithms are marked with a symbol that denotes whether the differences are statistically significant at $\alpha = 0.01$ (■), $\alpha = 0.05$ (•), and $\alpha = 0.1$ (○) according to a Wilcoxon rank sum test [17].

instance	greedy x^*	$\tilde{x}$	$\bar{x} \pm \sigma_{\bar{x}}$		EA x^*	$\tilde{x}$	$\bar{x} \pm \sigma_{\bar{x}}$		iEA x^*	$\tilde{x}$	$\bar{x} \pm \sigma_{\bar{x}}$	
low-density-1-3	8.109	8.410	8.493 ± 0.045	■	7.111	7.111	7.171 ± 0.023	★	7.111	7.111	7.184 ± 0.058	
low-density-2-3	7.511	8.436	8.662 ± 0.216	■	6.417	6.467	6.520 ± 0.020	★	6.417	6.488	6.534 ± 0.024	
low-density-3-3	9.462	9.584	9.606 ± 0.020		9.462	9.660	9.687 ± 0.035	•	9.409	9.660	9.598 ± 0.030	★
low-density-4-3	11.309	11.309	11.759 ± 0.115	■	9.259	9.556	9.657 ± 0.069	■	9.259	9.309	9.438 ± 0.072	★
low-density-5-3	8.510	8.862	8.867 ± 0.035	■	4.066	4.318	4.411 ± 0.052	○	4.066	4.315	4.279 ± 0.030	★
mid-density-1-3	13.104	13.482	13.435 ± 0.039	■	10.709	10.709	10.894 ± 0.118	★	10.709	11.509	11.613 ± 0.210	•
mid-density-2-3	15.306	15.432	15.477 ± 0.046	■	14.507	14.507	14.507 ± 0.000	★	14.507	14.507	14.507 ± 0.000	
mid-density-3-3	24.656	25.007	24.919 ± 0.055	■	15.555	15.555	15.688 ± 0.061	★	15.555	15.658	15.696 ± 0.057	
mid-density-4-3	14.758	15.006	15.016 ± 0.019	■	11.704	12.058	12.198 ± 0.100		11.704	11.757	12.089 ± 0.128	★
mid-density-5-3	13.656	14.383	14.195 ± 0.107	■	12.557	12.557	12.557 ± 0.000	★	12.557	12.557	12.642 ± 0.075	
high-density-1-3	16.909	17.804	17.736 ± 0.083	■	15.005	15.005	15.206 ± 0.126	★	15.005	15.005	15.436 ± 0.144	
high-density-2-3	17.556	17.607	17.813 ± 0.110	■	17.556	17.556	18.034 ± 0.477		17.556	17.556	17.556 ± 0.000	★
high-density-3-3	25.757	25.982	26.055 ± 0.072	■	18.757	19.307	19.341 ± 0.097		18.757	19.005	19.218 ± 0.111	★
high-density-4-3	17.205	17.831	17.629 ± 0.084	■	15.105	15.356	15.959 ± 0.216	★	15.105	15.306	16.069 ± 0.236	
high-density-5-3	13.406	13.508	13.613 ± 0.056	■	13.006	13.006	13.325 ± 0.095		13.006	13.006	13.129 ± 0.038	★
low-density-1-4	1.316	1.510	1.492 ± 0.025	★	1.333	1.668	1.700 ± 0.060	•	1.263	1.738	1.611 ± 0.051	
low-density-2-4	3.015	3.369	3.359 ± 0.040	■	2.472	2.690	2.802 ± 0.084	•	2.378	2.577	2.588 ± 0.031	★
low-density-3-4	2.216	2.888	2.793 ± 0.076	■	1.827	2.110	2.152 ± 0.039	■	1.808	1.967	2.078 ± 0.131	★
low-density-4-4	2.774	3.246	3.171 ± 0.041	■	2.006	2.145	2.144 ± 0.024		1.869	2.122	2.123 ± 0.034	★
low-density-5-4	1.100	1.170	1.166 ± 0.009	■	1.100	1.189	1.205 ± 0.023		1.053	1.089	1.106 ± 0.013	★
mid-density-1-4	3.515	3.963	3.926 ± 0.041	■	2.867	3.342	3.367 ± 0.048	■	2.961	3.190	3.213 ± 0.045	★
mid-density-2-4	5.261	5.984	5.808 ± 0.078	○	5.261	5.637	5.769 ± 0.118		5.261	5.470	5.618 ± 0.106	★
mid-density-3-4	6.161	6.513	6.523 ± 0.046	■	5.610	6.111	6.100 ± 0.041		5.610	5.860	5.870 ± 0.044	★
mid-density-4-4	5.011	5.188	5.221 ± 0.046	■	2.929	3.120	3.156 ± 0.026	■	2.666	3.062	3.005 ± 0.041	★
mid-density-5-4	5.114	5.345	5.351 ± 0.030	•	4.967	5.188	5.266 ± 0.044		4.915	5.263	5.247 ± 0.036	★
high-density-1-4	4.611	5.090	5.151 ± 0.082	■	4.110	4.487	4.564 ± 0.055		4.110	4.440	4.484 ± 0.068	★
high-density-2-4	7.507	7.663	7.742 ± 0.057	■	7.112	7.509	7.492 ± 0.047	★	7.112	7.360	8.889 ± 0.523	
high-density-3-4	7.009	7.259	7.229 ± 0.041	■	6.859	7.209	7.189 ± 0.038	■	6.712	6.985	6.980 ± 0.031	★
high-density-4-4	5.206	5.487	5.614 ± 0.073	■	4.711	4.944	5.098 ± 0.090	★	4.519	4.786	5.424 ± 0.193	
high-density-5-4	5.513	5.863	5.898 ± 0.032	■	5.014	5.816	5.787 ± 0.070		4.662	5.640	5.523 ± 0.083	★
low-density-1-5	0.985	1.053	1.055 ± 0.010	■	0.968	1.025	1.035 ± 0.012	•	0.948	0.979	1.005 ± 0.012	★
low-density-2-5	1.241	1.455	1.423 ± 0.017	★	1.295	1.561	1.608 ± 0.050	■	1.151	1.611	1.513 ± 0.052	
low-density-3-5	1.417	1.691	1.707 ± 0.035	■	1.419	1.608	1.609 ± 0.020	•	1.352	1.531	1.520 ± 0.024	★
low-density-4-5	1.130	1.263	1.253 ± 0.019	•	1.192	1.250	1.254 ± 0.012	■	1.062	1.168	1.198 ± 0.028	★
low-density-5-5	0.942	0.964	0.965 ± 0.002	■	0.924	0.945	0.950 ± 0.005	○	0.908	0.935	0.941 ± 0.006	★
mid-density-1-5	1.552	1.812	1.787 ± 0.025	■	1.265	1.534	1.536 ± 0.029	○	1.323	1.416	1.489 ± 0.040	★
mid-density-2-5	3.520	3.971	3.980 ± 0.047	■	3.318	3.820	4.067 ± 0.120		3.272	3.744	3.838 ± 0.081	★
mid-density-3-5	3.515	4.309	4.255 ± 0.057	■	3.515	4.036	4.016 ± 0.058		3.513	3.912	3.990 ± 0.078	★
mid-density-4-5	2.064	2.380	2.355 ± 0.050	■	1.298	1.505	1.522 ± 0.035	○	1.201	1.366	1.433 ± 0.035	★
mid-density-5-5	2.215	2.393	2.517 ± 0.063	■	1.467	1.771	1.774 ± 0.032	■	1.460	1.666	1.657 ± 0.022	★
high-density-1-5	2.112	2.411	2.419 ± 0.051	•	2.012	2.385	2.504 ± 0.084	•	1.760	2.090	2.201 ± 0.078	★
high-density-2-5	4.967	5.410	5.479 ± 0.062		4.811	5.511	5.507 ± 0.093		4.465	5.514	5.384 ± 0.101	★
high-density-3-5	4.911	5.237	5.246 ± 0.039	■	4.513	4.914	4.895 ± 0.045	■	4.113	4.611	4.655 ± 0.068	★
high-density-4-5	3.218	3.573	3.599 ± 0.044	■	2.010	2.511	2.468 ± 0.069	■	1.905	2.126	2.234 ± 0.072	★
high-density-5-5	2.060	2.410	2.363 ± 0.047	■	1.520	1.864	1.881 ± 0.042		1.512	1.812	1.829 ± 0.034	★

Table 2. Test results of the best solution found by each algorithm during training. The meaning of symbols is the same as in Table 1.

instance	greedy x^*	$\tilde{x}$	$\bar{x} \pm \sigma_{\bar{x}}$		EA x^*	$\tilde{x}$	$\bar{x} \pm \sigma_{\bar{x}}$		iEA x^*	$\tilde{x}$	$\bar{x} \pm \sigma_{\bar{x}}$	
low-density-1-3	2.000	9.001	8.816 ± 0.093	■	0.972	8.001	7.866 ± 0.087	★	0.972	8.001	7.866 ± 0.087	
low-density-2-3	2.015	8.011	8.278 ± 0.084	■	1.010	7.521	7.680 ± 0.083	★	1.010	7.521	7.680 ± 0.083	
low-density-3-3	3.013	11.001	10.839 ± 0.098		3.013	11.001	10.839 ± 0.098		2.009	11.001	10.673 ± 0.097	★
low-density-4-3	4.067	12.026	12.687 ± 0.110	■	2.000	10.010	10.399 ± 0.099	★	2.000	10.010	10.399 ± 0.099	
low-density-5-3	1.005	9.011	8.962 ± 0.091	■	0.918	6.001	5.958 ± 0.074	★	0.918	6.001	5.958 ± 0.074	
mid-density-1-3	5.011	15.001	14.707 ± 0.112	■	3.010	11.010	11.340 ± 0.103	★	3.010	11.010	11.34 ± 0.103	
mid-density-2-3	6.010	16.001	15.970 ± 0.110	■	5.010	15.011	15.391 ± 0.112	★	5.010	15.011	15.391 ± 0.112	
mid-density-3-3	13.010	25.002	25.242 ± 0.137	■	8.001	17.009	17.296 ± 0.119	★	8.001	17.009	17.296 ± 0.119	
mid-density-4-3	4.028	16.001	15.648 ± 0.117	■	4.019	13.001	12.831 ± 0.107	★	4.019	13.001	12.831 ± 0.107	
mid-density-5-3	4.011	15.001	15.022 ± 0.112	■	3.000	13.001	12.954 ± 0.101	★	3.000	13.001	12.954 ± 0.101	
high-density-1-3	7.010	17.001	16.852 ± 0.117	■	6.011	15.010	15.336 ± 0.108	★	6.011	15.010	15.336 ± 0.108	
high-density-2-3	7.011	18.001	18.038 ± 0.118	★	7.011	18.001	18.038 ± 0.118		7.011	18.001	18.038 ± 0.118	
high-density-3-3	16.001	27.002	27.097 ± 0.137	■	8.009	20.018	20.685 ± 0.127	★	8.009	20.018	20.685 ± 0.127	
high-density-4-3	7.010	18.510	18.585 ± 0.120	■	7.013	17.010	17.227 ± 0.115	★	7.013	17.010	17.227 ± 0.115	
high-density-5-3	5.011	14.011	14.365 ± 0.107		5.011	14.011	14.35 ± 0.109	★	5.011	14.011	14.350 ± 0.109	
low-density-1-4	0.819	2.000	1.915 ± 0.035	★	0.863	2.010	2.098 ± 0.039	■	0.809	2.010	2.056 ± 0.037	■
low-density-2-4	0.896	4.010	3.982 ± 0.059	■	0.852	3.011	3.039 ± 0.050	★	0.874	3.011	3.144 ± 0.054	
low-density-3-4	0.853	3.009	2.973 ± 0.049	■	0.863	2.029	2.669 ± 0.049	★	0.875	3.001	2.885 ± 0.049	■
low-density-4-4	0.917	4.001	3.875 ± 0.060	■	0.809	2.042	2.714 ± 0.049	★	0.906	3.010	3.164 ± 0.053	■
low-density-5-4	0.754	1.062	1.664 ± 0.030	■	0.754	1.021	1.422 ± 0.025		0.787	1.024	1.417 ± 0.024	★
mid-density-1-4	0.917	4.020	4.429 ± 0.063	○	0.885	4.020	4.446 ± 0.064	○	0.917	4.014	4.300 ± 0.064	★
mid-density-2-4	0.983	6.010	6.329 ± 0.077	★	0.983	6.010	6.329 ± 0.077		0.983	6.010	6.329 ± 0.077	
mid-density-3-4	2.000	7.013	7.408 ± 0.082	■	1.009	6.009	6.216 ± 0.075	★	1.009	6.009	6.216 ± 0.075	
mid-density-4-4	0.994	6.010	6.166 ± 0.074	■	0.907	4.010	4.201 ± 0.062		0.939	4.010	4.177 ± 0.062	★
mid-density-5-4	1.010	7.000	6.691 ± 0.078	■	0.994	7.001	6.904 ± 0.081	■	0.885	6.001	5.617 ± 0.070	★
high-density-1-4	0.929	5.014	5.449 ± 0.070	■	0.907	5.010	4.978 ± 0.068	★	0.907	5.010	4.978 ± 0.068	
high-density-2-4	1.029	8.010	8.432 ± 0.088		0.972	8.010	8.251 ± 0.089	★	0.972	8.010	8.251 ± 0.089	
high-density-3-4	0.962	8.009	8.281 ± 0.088	■	0.972	7.020	7.627 ± 0.083	★	1.090	8.001	8.035 ± 0.087	■
high-density-4-4	1.000	6.014	6.496 ± 0.078	■	0.961	6.009	6.021 ± 0.077	★	1.005	6.010	6.063 ± 0.074	
high-density-5-4	1.037	7.011	7.151 ± 0.079	■	0.972	6.011	6.102 ± 0.073	★	1.000	6.001	5.851 ± 0.071	★
low-density-1-5	0.775	1.022	1.482 ± 0.026	■	0.754	1.010	1.242 ± 0.019	★	0.743	1.003	1.174 ± 0.017	★
low-density-2-5	0.743	1.041	1.570 ± 0.027	★	0.732	1.042	1.638 ± 0.030		0.786	1.042	1.630 ± 0.030	
low-density-3-5	0.852	2.012	2.217 ± 0.043	■	0.863	2.009	2.087 ± 0.039	•	0.819	2.001	1.988 ± 0.038	★
low-density-4-5	0.797	1.027	1.486 ± 0.026		0.775	1.021	1.475 ± 0.026	★	0.743	1.027	1.540 ± 0.028	
low-density-5-5	0.742	1.011	1.23 0± 0.019	■	0.655	0.970	1.046 ± 0.012	★	0.689	0.971	1.06 0± 0.013	
mid-density-1-5	0.808	2.010	2.119 ± 0.039	■	0.831	2.000	1.922 ± 0.036		0.809	2.000	1.906 ± 0.036	★
mid-density-2-5	0.984	5.000	4.713 ± 0.069	■	0.896	4.010	4.231 ± 0.064	★	0.917	4.017	4.452 ± 0.066	•
mid-density-3-5	1.000	5.009	5.243 ± 0.069	■	1.000	5.009	5.243 ± 0.069	■	0.972	5.000	4.785 ± 0.067	★
mid-density-4-5	0.820	2.029	2.577 ± 0.046	■	0.797	1.044	1.703 ± 0.032	★	0.765	1.069	1.844 ± 0.035	■
mid-density-5-5	0.884	3.011	3.119 ± 0.052	■	0.830	2.011	2.182 ± 0.041		0.830	2.011	2.142 ± 0.039	★
high-density-1-5	0.863	2.022	2.514 ± 0.045	■	0.842	2.028	2.577 ± 0.045		0.863	2.014	2.218 ± 0.042	★
high-density-2-5	1.010	7.000	6.700 ± 0.081	■	0.928	6.000	5.803 ± 0.074	★	0.950	6.001	5.943 ± 0.074	
high-density-3-5	0.994	6.009	6.148 ± 0.075	■	0.950	6.000	5.755 ± 0.073	■	0.929	5.009	5.201 ± 0.070	★
high-density-4-5	0.885	3.036	3.581 ± 0.059	■	0.896	3.000	2.802 ± 0.048	★	0.852	3.000	2.819 ± 0.049	
high-density-5-5	0.830	2.021	2.436 ± 0.044		0.830	2.024	2.511 ± 0.045	○	0.831	2.021	2.379 ± 0.042	★

However, in the long run the evolutionary approaches are capable of outperforming the greedy heuristic.

Subsequently, we move to the test phase. To this end, we select the solution that has the best fitness on each instance for each algorithm, and evaluate it on all the test cases. Table 2 shows the resulting results. Note that iEA remains superior in general, and both EAs outperform greedy. However, the differences are less marked. This is better seen in Fig. 1b, where the rank distribution of the different algorithms according to the performance of their solution on the test set is shown. Again, these ranks show statistically significant differences according to Quade test (Quade $F = 50.376$, p-value $= 2.618e-15$), and iEA remains the algorithm with the best mean rank, so it is chosen as control algorithm for Holm test. Now, the test is passed against greedy (p-value ≈ 0), but not against EA (p-value $= 3.428e-1$). We believe this may be an indication that the training set is not large enough and therefore iEA may be overfitting its solutions.

6   Conclusions

Optimizing the placement of emergency exits in indoor environments is not just a problem of importance for public safety, but also poses a challenging optimization task. We have conducted a comparative analysis of two different optimization approaches, namely an iterated greedy heuristic and an evolutionary algorithm (in two variants, both panmictic and island-based). This analysis indicates the superiority of the evolutionary approaches, both on the training and test phases, underpinning the need for powerful global optimization techniques in this context. It also hints at the need of using larger training sets, which of course will have a toll on computational cost. This makes a strong case for directing effort into solutions of computational nature (such as parallel computing) and solutions of algorithmic nature (e.g., lightweight simulations or surrogate models [11]).

In addition to the research directions sketched above, it is clear that the evacuation scenario can be enriched with additional layers of complexity. While we have here assumed situations of orderly evacuation as an initial base case, we can go on to consider situations in which the cause of the emergency does pose a visible threat (e.g., a rampant fire, or ongoing explosions) that might disrupt the evacuation process or the flow of people. Such scenarios may be in need of more sophisticated approaches, and this work has paved the way for hybrid approaches that combine greedy components within an evolutionary search engine. Work is in progress in this area.

Acknowledgments. The authors thank the Supercomputing and Bioinnovation Center (SCBI) of the University of Malaga for their provision of computational resources (the Picasso supercomputer http://www.scbi.uma.es).

References

1. Alba, E., Tomassini, M.: Parallelism and evolutionary algorithms. IEEE Trans. Evol. Comput. **6**(5), 443–462 (2002)
2. Bellomo, N., Bellouquid, A., Knopoff, D.: From the microscale to collective crowd dynamics. Multiscale Model. Simul. **11**(3), 943–963 (2013)

3. Cao, R.F., et al.: Development of an agent-based indoor evacuation model for local fire risks analysis. J. Safety Sci. Resilience 4(1), 75–92 (2023)

4. Chen, J., Shi, T., Li, N.: Pedestrian evacuation simulation in indoor emergency situations: approaches, models and tools. Saf. Sci. **142**, 105378 (2021)

5. Cotta, C., Gallardo, J.E.: Instance dataset for the pedestrian evacuation problem (2023). https://osf.io/cnh7u/

6. Dijkstra, E.W.: A note on two problems in connexion with graphs. Numer. Math. **1**(1), 269–271 (1959)

7. Dunn, O.J.: Multiple comparisons among means. J. Am. Stat. Assoc. **56**(293), 52–64 (1961)

8. Golas, A., Narain, R., Lin, M.C.: Continuum modeling of crowd turbulence. Phys. Rev. E **90**(4), 042816 (2014)

9. Haghani, M.: Optimising crowd evacuations: mathematical, architectural and behavioural approaches. Saf. Sci. **128**, 104745 (2020)

10. Holm, S.: A simple sequentially rejective multiple test procedure. Scand. J. Stat. **6**(2), 66–70 (1979)

11. Jin, Y.: Surrogate-assisted evolutionary computation: recent advances and future challenges. Swarm Evol. Comput. **1**(2), 61–70 (2011)

12. Li, Z., Xu, C., Bian, Z.: A force-driven model for passenger evacuation in bus fires. Phys. A **589**, 126591 (2022)

13. Martinez-Gil, F., Lozano, M., García-Fernández, I., Fernández, F.: Modeling, evaluation, and scale on artificial pedestrians: a literature review. ACM Comput. Surv. **50**(5), 72:1–72:35 (2017)

14. Quade, D.: Using weighted rankings in the analysis of complete blocks with additive block effects. J. Am. Stat. Assoc. **74**(367), 680–683 (1979)

15. Radcliffe, N.J.: The algebra of genetic algorithms. Ann. Math. Artif. Intell. **10**(4), 339–384 (1994)

16. Shi, M., Lee, E.W.M., Ma, Y.: A dynamic impatience-determined cellular automata model for evacuation dynamics. Simul. Model. Pract. Theory **94**, 367–378 (2019)

17. Wilcoxon, F.: Individual comparisons by ranking methods. Biometrics Bull. **1**(6), 80 (1945)

18. Zheng, Y., Li, X.G., Jia, B., Jiang, R.: Simulation of pedestrians' evacuation dynamics with underground flood spreading based on cellular automaton. Simul. Model. Pract. Theory **94**, 149–161 (2019)

Finding Sets of Solutions for Temporal Uncertain Problems

Jens Weise[✉][ID] and Sanaz Mostaghim[ID]

Faculty of Computer Science, Otto von Guericke University, Magdeburg, Germany
{jens.weise,sanaz.mostaghim}@ovgu.de
https://ci.ovgu.de

Abstract. The multi-objective pathfinding problem is a complex and NP-hard problem with numerous industrial applications. However, the number of non-dominated solutions can often exceed human comprehension capacity. This paper introduces a novel methodology that leverages the concept of a Pareto graph to address this challenge. Unlike previous approaches, our method constructs a graph that relates paths where there is potential for change between them and applies a graph community algorithm to identify solution subsets based on specific aspects defined by a decision-maker. We describe the construction of a Route Change Graph (RCG) to represent possible route changes. A matrix is constructed to save the number of possible change opportunities between two routes, which is then used to construct the RCG. We propose using a threshold value for edge weights in the graph construction, balancing between minimising the number of edges and maintaining connectivity. Following the construction of the RCG, we apply a community detection algorithm to identify closely related solutions, using Leiden algorithm due to its efficiency and refinement phase. We propose calculating various metrics on these communities, including Density, Average Cluster Coefficient, Group Betweenness Centrality, and Graph Degree Centrality, to provide insights into the network structure and interconnectivity. This methodology offers a more manageable set of solutions for decision-makers, enhancing their ability to make informed decisions in complex multi-objective pathfinding problems.

Keywords: pathfinding · decision-support · Pareto graph

1 Introduction

Finding a path from one point to another while optimising multiple objectives is known as the multi-objective pathfinding problem and is considered NP-hard [20]. In several industries this technique can be applied, e.g., route planning, aviation, networking or medical applications [20]. For instance, there are multiple

---

This work is funded by the German Federal Ministry of Education and Research through the 6G-ANNA project (grant no. 16KISK092).

objectives to consider, when planning a logistic trip for a truck, e.g., curvature of the road, ascent and length of a route. All these objectives should be considered simultaneously. In medical applications, inserting a needle to perform a minimal invasive tumour therapy can include objectives such as distance to the vessel system or damaged tissue. Often, these objectives are in conflict. Applying multi-objective optimisation techniques to such problems, can give a decision-maker (DM) a better insight into the problem. The result of such an optimisation is a set of non-dominated solutions, where no solution is better than the other.

However, the cardinality of the obtained set of non-dominated solutions can exceed the number of solutions, a DM can comprehend [20]. According to Miller, humans can comprehend 7 ± 2 information chunks, although more recent research indicates that this number is lower (approx. 3 to 4 chunks) [11,16]. Various reduction techniques have been proposed that identify important and interesting solutions.

In this paper, we propose a new methodology that utilises the concept of a Pareto graph [13]. In contrast to the original approach, we construct a graph that sets paths into relation when there is the possibility to change between them. Furthermore, we apply various graph community algorithms to identify subsets of solution that comply with various aspects which can be given by a DM. In contrast to other approaches, our proposed methodology does not reduce the whole set of non-dominated solutions, but finds subsets from which DMs can choose.

The paper is structured as follows. In Sect. 2, we describe the necessary background, while Sect. 3 is dedicated to the related work. Section 4 presents our proposed methodology, divided into graph construction and community detection and analysis. In Sect. 5, we evaluate the results and give a conclusion and outlook in Sect. 6.

2 Background

In this section, we present the relevant background, i.e., the multi-objective pathfinding problem, various aspects of graph theory, including community detection.

2.1 Graph Theory

Graphs are used to represent the relations between entities. A graph G consists of a set of vertices that are the representations of such entities and a set of edges that denote the relations. An edge usually consists of an unordered or ordered set of two vertices. Formally, a directed graph G is a pair $G = (V, E)$, where V denotes the set of vertices and E is the set of edges, where $E \subseteq \{(n, n') \mid (n, n') \in V^2, n \neq n', n, n' \in V\}$. Note that E consists of two-element ordered subsets of V^2, which renders a graph *directed*. In undirected graphs, by contrast, E consists of two-element unordered subsets of V^2 [23].

Connected Components. For a graph $G = (V, E)$, a connected component c of G is a subgraph $c = (V', E')$, where $V' \subseteq V$ and $E' \subseteq E$. For any two vertices $u, v \in V'$, there exists a sequence of vertices $(v_1, v_2, ..., v_n)$ and a sequence of edges $(e_1, e_2, ..., e_{(n-1)})$ such that: $v_1 = u$ and $v_n = v$ (i.e., the first vertex is u, and the last vertex is v). For each $i, 1 \leq i \leq n - 1, e_i$ is an edge in E' that connects v_i to $v_{(i+1)}$. Furthermore, we define C as the set of all connected components. Therefore $C = \{c_i\}$, with $i = 1, \cdots, k_{comp}$, where k_{comp} is the number of connected components [23].

Communities. Communities in graphs refer to subsets of nodes within a larger network that exhibit higher intra-connectivity compared to interconnectivity. The detection and analysis of communities play a crucial role in understanding the structure and function of complex systems, including social networks, biological networks, and information networks. Various algorithms and methods have been developed to uncover communities in graphs, with the common objective of identifying densely connected subgraphs. A fundamental concept used in community detection is modularity, which measures the quality of a partition of nodes into communities. The modularity of a graph partition is defined as:

$$Q = \frac{1}{2|E|} \sum_{i,j} \left(A_{ij} - \frac{deg(i)\, deg\, deg(j)}{2|E|} \right) \delta(\mathfrak{C}_i, \mathfrak{C}_j) \tag{1}$$

where A_{ij} represents elements of the adjacency matrix, $deg(i)$ and $deg(j)$ are the degrees of node i and j, m is the total number of edges, $\mathfrak{C}_i$ and $\mathfrak{C}_j$ are the communities of nodes i and j, and $\delta(\mathfrak{C}_i, \mathfrak{C}_j)$ is the Kronecker delta function that equals 1 if $\mathfrak{C}_i = \mathfrak{C}_j$ and 0 otherwise.

The modularity optimization problem aims to find the partition that maximizes Q, indicating strong community structure. Beyond modularity, other methods like spectral clustering, hierarchical clustering, and random-walk-based approaches have been developed to uncover communities. The study of communities in graphs has provided valuable insights into the organization of networks and has practical applications in recommendation systems, information diffusion modelling, and network analysis [12].

2.2 Multi-objective Pathfinding

The multi-objective route planning problem, hereafter called the *pathfinding problem*, can be defined as a network flow problem [14, 15]. The goal is to find a set of optimal paths (routes) $P* = \{p_1, \cdots, p_L\}$ in a graph

$$G = \left(V, E, \phi, \vec{f}, \iota_V(\mathfrak{P}), \iota_E(\mathfrak{P}), n_s, n_e \right) \tag{2}$$

where V is the set of vertices or nodes, E represents the set of edges and ϕ represents a function mapping every edge to an ordered pair of nodes n and n'; hence $\phi : E \rightarrow \{(n, n') \mid (n, n') \in V^2\}$. A path p_i is the sequence of nodes from a starting node $n_S \in V$ to a predefined end node $n_{End} \in V$, i.e., $p_i = (n_i, n_{i+1} \cdots, n_k)$, where $n_S = n_i$ and $n_{End} = n_k$ and $n_i \in V$ for $i = 1, 2, \cdots, k$ and $\exists \phi(e_{i,i+1}) = (n_i, n_{i+1}) \in E$ for $i = 1, 2, \cdots, k - 1$. Such a path p is called a path of length $k - 1$ from n_1 to n_k. A path p_i is here represented as a list of nodes in a graph. Another representation is a list of edges to traverse; hence $p_i = (e_1, \cdots, e_{k-1})$ where $n_S = \phi(e_1)(1)$ and $n_{End} = \phi(e_k)(2)$ and $e_i \in E$ for $i = 1, 2, \cdots, k$. Following the definition of a multi-objective optimisation problem, the decision variable x is a path p in search space Ω [20].

3   Related Work

In this section, we present the related work about decision support systems (DSSs) that is used to decrease the number of solutions a DM has to choose from. Furthermore, we give a short overview on methodologies related to the concept of a Pareto graph, in which non-dominated solutions are put into relation using a graph structure.

3.1   Pareto Set Reduction as a DSS

In real-world applications, the Pareto set can be vast, making it challenging for decision-makers to analyse and select a preferred solution. To address this challenge, Pareto set reduction techniques have been developed as a decision support tool, aiming to reduce the size of the Pareto set while preserving its essential characteristics [9].

Pareto set reduction methods, utilized to provide decision-makers with a more manageable set of solutions, are divided into clustering-based and representative selection approaches. Clustering-based methods amalgamate similar solutions within the Pareto set into clusters, selecting a representative solution from each cluster, and have been further explored through various subsequent works focusing on the clustering of non-dominated solutions and the application of graph-based representations in Multi-Objective Optimization (MOO) [2,20]. For instance, a graph-theoretical clustering approach has been proposed to identify a reduced set encapsulating extreme solutions of Pareto optimal solutions for MOO problems [8]. Another technique employs clustering in both the objective and decision spaces to find intersection sets, aiding a DM in electing the optimal solution [20]. Conversely, representative selection strategies try to directly select a subset of solutions embodying the diversity and distribution of the entire Pareto set [8]. Through these methodologies, both approaches facilitate simplified analysis and decision-making by rendering a condensed yet diverse set of solutions for evaluation.

3.2   Pareto Graphs

In multi-objective optimization (MOO), obtaining a well-distributed set of non-dominated solutions is a crucial goal. Paquete and Stützle extended the concept of Pareto graphs [5] (also known as *efficient graphs*) to represent relationships among solutions in the objective space. Each node in the Pareto graph corresponds to a solution, and each directed edge represents whether one solution can be reached from another within a certain distance. They conducted an experimental analysis on the properties of the Pareto graph induced by the set of efficient solutions for multi-objective combinatorial optimization problems, observing that the Pareto graph contains clusters of non-dominated solutions which are tightly connected subsets of solutions [13]. Furthermore, Liefooghe et al. proposed to use a graph in which edges represent the potential ability of a search algorithm to jump from one solution to another [10].

4   Finding Related Paths

In this section, we describe how pairs of paths can be identified that share common sub paths and how a respective graph from this information can be constructed. Furthermore, we propose to use community detection algorithms to find interesting subsets of paths. These communities can help a DM to make a more informed decision.

4.1   Constructing the Route-Change-Graph

To represent possible changes of routes, we can construct a Route-Change-Graph (RCG), that is a graph $G = (V, E)$ where each $v \in V$ represents a single path from the designated start to the goal node and each $e \in E$ represents a change opportunity between two routes (two nodes).

   Such a graph is constructed by analysing a set of possible routes and identifying their pairwise common contiguous nodes (excluding start and end). For each pair of routes (r_i, r_j), where a route $r = (n_s, \cdots, n_e)$, we construct the intersection of their subsets of contiguous nodes, excluding n_s and n_e. Let r_i and r_j be the ordered sets of their respective points. Therefore, $I_{ij} = r_i \setminus \{n_s, n_e\} \cap r_j \setminus \{n_s, n_e\}$ is the intersection of the two sets without the start and end nodes. We create a such an intersection for each route pair. Each set I_{ij} contains nodes and, therefore, subroutes, that are present and shared in two routes. However, instead of obtaining the cardinality of the intersection set I_{ij}, i.e. $|I_{ij}|$, we save the number of common contiguous subroutes $|\mathfrak{S}_{ij}|$ (between r_i and r_j, in a matrix M, where each column and each row represents a route. Therefore, the $n \times n$ matrix M is symmetric.

   To obtain $|\mathfrak{S}_{ij}|$, we consider two ordered sets r_i and r_j, where each r_i consists of a sequence $(n_1, \cdots, n_k)$ with k being variable. The task is to find the number of common contiguous subsequences between r_i and r_j. A contiguous subsequence of r_i is any sequence $(n_{i_a}, \cdots, n_{i_b})$ where $1 \leq a < b \leq k$, and the indices a and b form a contiguous range.

   Let's denote by $S(r_i)$ the set of all contiguous subsequences of r_i, i.e., $S(r_i) = \{s | s$ is a contiguous subsequence of $r_i\}$. We are interested in finding the cardinality of the intersection of $S(r_i)$ and $S(r_j)$, denoted $|\mathfrak{S}_{ij}| = |S(r_i) \cap S(r_j)|$.

In Algorithm 1, we show pseudocode to compute the common contiguous subsequences.

Algorithm 1. Common Contiguous Subsequences of Ordered Sets r_1 and r_2

```
 1: function COMMONCONTIGUOUSSUBSEQUENCES(r₁, r₂)
 2:     S(r₁) ← GetContiguousSubsequences(r₁)
 3:     S(r₂) ← GetContiguousSubsequences(r₂)
 4:     common_count ← 0
 5:     for each s₁ in S(r₁) do
 6:         for each s₂ in S(r₂) do
 7:             if s₁ = s₂ then
 8:                 common_count ← common_count + 1
 9:             end if
10:         end for
11:     end for
12:     return common_count
13: end function

14: function GETCONTIGUOUSSUBSEQUENCES(r)
15:     subsequences ← ∅
16:     for i = 1 to length(r) do
17:         for j = i + 1 to length(r) + 1 do
18:             subsequences ← subsequences ∪ {(rᵢ, ..., r_{j-1})}
19:         end for
20:     end for
21:     return subsequences
22: end function
```

Each element in M contains then the number of possible change opportunities between two routes. In the following, we only consider one half of the matrix, as it is symmetric. The intersection of a route to itself is the route itself and is not considered in the following analysis (the respective matrix cells are set to 0). The matrix M looks as follows.

$$
M = \begin{array}{c|ccccc}
 & r_1 & r_2 & \cdots & r_{n-1} & r_n \\
\hline
r_1 & 0 & |\mathfrak{S}_{12}| & |\mathfrak{S}_{13}| & |\mathfrak{S}_{14}| & |\mathfrak{S}_{15}| \\
r_2 & - & 0 & |\mathfrak{S}_{23}| & |\mathfrak{S}_{24}| & |\mathfrak{S}_{25}| \\
\vdots & \vdots & - & 0 & |\mathfrak{S}_{34}| & |\mathfrak{S}_{35}| \\
r_{n-1} & \vdots & \vdots & - & 0 & |\mathfrak{S}_{45}| \\
r_n & - & \cdots & \cdots & - & 0
\end{array}
$$

With the obtained route change matrix M, we can now construct the RCG, i.e. $G_{RCG} = (V_{RCG}, E_{RCG})$. We assume a bidirectional possibility to change between two routes. Each route r is represented by a node $v_{r_i} \in V_{RCG}$. Each element in the matrix M represents an edge in E_{RCG} between two routes (column

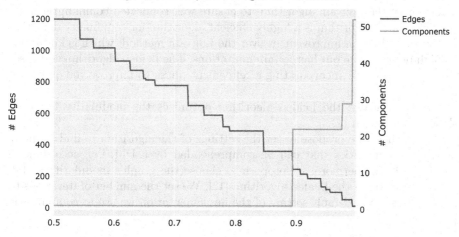

Fig. 1. $|E|$ and $|C|$ in relation to τ for $G_{RCG,\tau}$

and row), and the value represents the edge's weight. As the matrix has as many rows and columns as there are routes, the resulting graph can be substantially large. Therefore, we propose to use a threshold value τ for the edges' weights that are constructed in the graph. The threshold value τ is determined using quantiles on the matrix' values. An edge is constructed if the respective value is over the specified threshold. However, constructing fewer edges can result in a disconnected graph and, therefore, in having multiple connected components. Nevertheless, our proposed RCG should have the least possible number of connected components while also having the least possible number of edges, keeping it less dense. The graph can be constructed for various thresholds, and the value that maintains both properties low can then be identified. In Fig. 1, such an analysis is shown. With an increasing threshold, the number of edges decrease while there are more connected components. In the given example, we can decide on a threshold quantile of 0.891, which results in one connected component and 363 edges in E_{RCG}. However, for a different set of routes, it can happen that the possibility of having only one connected component is not given. Then, a τ should be chosen, that minimizes $|C|$.

4.2 Community Detection and Analysis

After constructing the RCG, which nodes represent paths, which edges represent change possibilities and which edges' weights represent how often a route can be changed, we can apply community detection algorithms to identify closely related solutions. Furthermore, we propose to use various metrics of these communities to identify subsets of solutions that are presented to a DM. In addition, we propose three strategies a DM can utilise to identify a feasible and fitting community.

Community Detection. We propose to use the Leiden algorithms, which is an extension of the Louvain algorithm, to ensure well-connected communities [18].

The Leiden algorithm is a highly efficient algorithm for community detection in networks. It is an improvement over the Louvain method, which is known for its high performance but has certain limitations. The Leiden algorithm addresses these limitations by incorporating a refinement phase to improve the quality of partitions.

Mathematically, the Leiden algorithm optimizes the modularity function, shown in Equation (1).

We furthermore propose to set the settings of the algorithm to find as many different communities that can be comprehended by a DM, i.e., according to [16], 3 to 4. Furthermore, we propose to choose the graph's modularity as the quality function of the Leiden algorithm [12]. We set the number of iterations to 2, since this is the default setting of the implementation we use to compute the partition [19].

Community Analysis. After finding a good number of communities, we propose to compute various metrics on these. These metrics should reflect how intra-connected the communities are, but also if they are interconnected to other communities. We have decided to compute four metrics for each community:

1. **Density** [3]. The density of a graph structure, denoted as $\rho(G)$, is a measure that provides insight into how many edges are present in the graph relative to the maximum possible number of edges. For an undirected simple graph with $|V|$ vertices, the maximum number of edges is $\frac{|V|(|V|-1)}{2}$. Thus, the density is defined as:

$$\rho(G) = \frac{2|E|}{|V|(|V|-1)} \tag{3}$$

   where $|E|$ represents the number of edges in the graph. For directed graphs, the maximum number of edges is $|V|(|V|-1)$, and the density is calculated as:

$$\rho(G) = \frac{|E|}{|V|(|V|-1)} \tag{4}$$

   Consequently, a graph's density ranges from 0 (for an empty graph) to 1 (for a complete graph).

2. **Average Cluster coefficient** [17]. The average clustering coefficient, $\langle \mathscr{C} \rangle$, quantifies the degree of clustering in a network. It's calculated as:

$$\langle \mathscr{C} \rangle = \frac{1}{|V|} \sum_{v_i \in G} \mathscr{C}(v_i) \tag{5}$$

   Where:
   - $|V|$ is the total number of nodes

- v_i represents each node in the graph G
- $\mathscr{C}(v_i)$ is the clustering coefficient of node v_i

For a given node v_i, $\mathscr{C}(v_i)$ is the node's clustering coefficient, i.e., a proportion of existing links between its neighbours over the total possible links.

Given a node v with k_v neighbours, the cluster coefficient $\mathscr{C}(v)$ for that node can be calculated using the following equation:

$$\mathscr{C}(v) = \frac{2\Upsilon_v}{k_v(k_v - 1)}$$

where Υ_v represents the number of edges between the neighbours of v. This equation calculates the ratio between the number of actual edges Υ_v and the maximum possible number of edges between k_v nodes. In other words, it is the ratio of actual triangles that node involved in and the number of possible triangles.

If a node has less than two neighbours, its clustering coefficient is 0. This measure provides an overall sense of the network's *cliquishness*.

3. **Group Betweenness Centrality** [6]. Everett and Borgatti proposed the concept of Group Betweenness Centrality as a measure to identify the most central group within a network. It extends the idea of individual node centrality to encompass groups of nodes. The Group Betweenness Centrality of a group of nodes is defined as the sum of the fraction of shortest paths between all pairs of nodes in the network that pass through at least one node in the group. This measure reflects the extent to which a group collectively acts as a bridge or gatekeeper between other nodes in the network. Given a group of nodes, $\mathcal{V}$, the betweenness centrality of this group, denoted as $bc(\mathcal{V})$, is given by:

$$bc(\mathcal{V}) = \sum_{n_{\text{from}} \neq v \neq n_{\text{to}}} \left(\frac{\sigma(n_{\text{from}}, n_{\text{to}} | \mathcal{V})}{\sigma(n_{\text{from}}, n_{\text{to}})} \right) \tag{6}$$

Where:

- $\sigma(n_{\text{from}}, n_{\text{to}})$ is the total number of shortest paths from node n_{from} to node n_{to}
- $\sigma(n_{\text{from}}, n_{\text{to}} | \mathcal{V})$ is the number of those paths that pass through some node in group $\mathcal{V}$

Notice that $n_{\text{from}} \neq v \neq n_{\text{to}}$ means that we take all pairs of nodes except those pairs where either node is in the group $\mathcal{V}$. In contrast to the other three metrics, we compute the Group Betweenness Centrality for a community in the scope of the whole graph, while the other metrics are calculated using solely the nodes and edges of the respective community.

4. **Graph Degree Centrality** [7]. The degree centrality of a graph is a measure of the overall connectivity of the graph. It is an average of the degree centralities of all nodes in the graph.

It is defined as:

$$dc(G) = \frac{\sum_{i=1}^{|V|}[dc(v^*) - dc(v_i)]}{|V|^2 - 3|V| + 2} \tag{7}$$

Where:

- $dc(G)$ represents the degree centrality of the graph G
- $dc(v^*)$ and $dc(v_i)$ denote the degree centralities of the node with the highest degree (v^*) and each other node (v_i) respectively
- $|V|$ is the total number of nodes in the network

This formula calculates the sum of differences between the degree centrality of the node with the highest degree and that of every other node. This sum is then normalized by dividing it by $|V|^2 - 3|V| + 2$, which is derived from the maximum possible sum of differences.

In this case, a higher degree centrality indicates that one node (the one with the highest degree) is significantly more connected than others, while a lower degree centrality suggests a more evenly distributed network where no single node dominates in terms of connections.

Community Selection. After computing the four different metrics for each community, we can use one or more of these measurements to select a community that is being presented to a DM. With our approach, we shift the DM's task from the objective space (and where possible interesting areas are) to a space where they have to decide on specific properties of subsets of solutions. Especially for problems similar to the multi-objective pathfinding problem, that can be highly uncertain from a temporal perspective, it can be beneficial to choose a subset of solutions rather than a single solution to have alternatives ready when the solution is executed but not feasible any more. For instance, when traversing a path, a DM may get the information that a chosen segment on a later stage of the path is not traversable any more. With a pre-computed set of alternative solutions, the DM can still choose from various non-dominated solutions. As follows, we present three strategies to use the proposed community metrics. We propose to apply non-dominated sorting of the set of communities using their respective metric values, and then to use a combination of the four metrics.

Always alternatives. If a DM aims for solutions that have always alternatives when being traversed, we propose to choose a community with a high density and low graph degree centrality. An example of such a community is presented in *Community 1* in Fig. 2.

Main route, but possible dead ends in alternatives. A star shape community represents a set of alternatives with one main route and adjacent solutions. Choosing such a set may result from a high priority on a specific route. However, depending on the number of rays of the star, i.e., the alternatives, a different route might be available with the sacrifice of having no more alternatives afterwards. Nevertheless, a return to the main route can be possible. An example of such a community is presented in *Community 2* in Fig. 2.

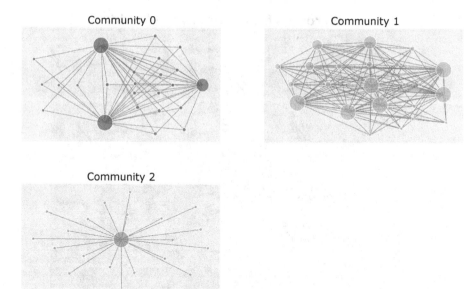

Fig. 2. The three obtained communities. Force-directed layout for visualisation.

Few central solutions, few alternatives. A community can exist with multiple central solutions, where each solution has a substantially high number of alternatives, and that community has also a few additional solutions with a lower number of available alternatives. An example of such a community is presented in *Community 0* in Fig. 2.

5   Evaluation and Discussion

In this section, we apply the proposed methodology to an instance of the pathfinding problem that has been proposed in [20] and which has been published in [22].

The instance of the problem represented the task of finding the set of Pareto optimal routes within the European road network from Warsaw to Madrid. The final network consisted of 1.14×10^8 nodes and 1.46×10^8 edges and a variation of the NSGA-II algorithm [4,20] was applied to optimise (minimization) four objectives, i.e., length of the route, time to traverse it, positive ascent and the curvature. For a detailed description, the interested reader is referred to [20]. The authors have obtained 69 different and non-dominated routes that are shown in Fig. 3. Although the routes are very similar from a visual perspective, there are small differences in various locations.

We can now construct all intersection sets $\mathfrak{S}_i$, using our proposed methodology, and build the matrix M from it. The result is a 69×69 matrix, which elements represent possible changes between routes and the value related to the number of possible changes. From this adjacency matrix, we construct the

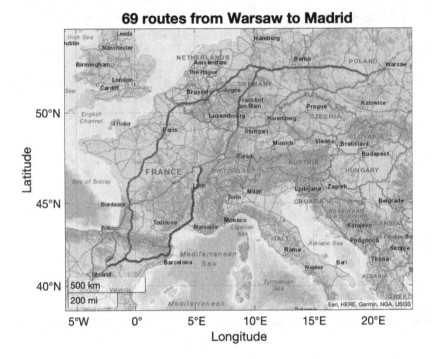

Fig. 3. All obtained Pareto-optimal routes for four objectives [20]

respective RCG, shown in Fig. 4. The graphical representation was created using a force directed layout [1].

As described in Sect. 4.1, we use the 0.891-quantile as a threshold so that our graph has exactly one connected component. In Fig. 4, we have also coloured the communities, that have been found when applying the Leiden algorithm. In Fig. 2, we show each community separately, also arranged using a force-directed layout. From a visual approach, the structural differences of the communities are already visible.

To compare the communities, we can now compute the proposed metrics, i.e., density, average cluster coefficient, group betweenness centrality and graph degree centrality. In Fig. 5, we show these metrics for each community. It should be noted, that, in terms of these metrics, all communities are non-dominated. We propose to only use non-dominated communities. From a visual perspective, community 2 is structurally different compared to the other two. It has a rather high graph degree centrality and group betweenness centrality, but also a low density and an average cluster coefficient of 0, as the community does not contain any triangles.

We assume that a DM should decide on one specific community. However, also the linking between communities can be of interest. The DM can utilise the Group Betweenness Centrality to estimate how well a change between commu-

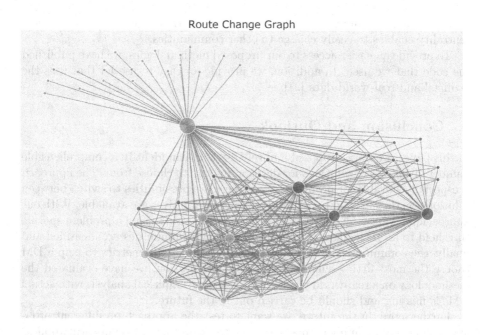

Fig. 4. The obtained RCG from the real-world example. Layout obtained by applying a force directed algorithm

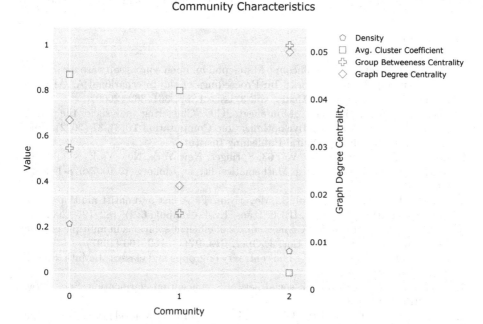

Fig. 5. Various characteristics of each community. The right y-axis shows Graph Degree Centrality, the left axis shows the other metrics.

nities can be done. In other words, a community with a high group betweenness centrality enables to easily change to other communities.

To provide an easier access to our proposed methodologies, we have published the code that we used. In addition, we provide an easy-to-use UI that uses the artificial and real-world data [21].

6   Conclusion and Outlook

In this paper, we have proposed a novel DSS that can identify a comprehensible number of subsets of solutions for decision-makers to choose from. The approach is especially suitable for problems where there are possibilities to switch between solutions, as they are temporal uncertain and alternatives are available. With our approach, an Route-Change-Graph (RCG) is generated using a problem specific threshold to keep the number of edges low, then communities are identified and finally, the communities are analysed using various graph metrics to help a DM choose the most fitting subset of solutions. In addition, we have evaluated the methodology on a real-world problem. However, an empirical analysis with actual DMs is missing and should be carried out in the future.

Furthermore, in the future, we want to test the approach on different problems than route planning on maps, e.g., network routing or also medical applications. Moreover, other graph related metrics than the four that we have utilised, should be evaluated in the future. We see our proposed methodology as a starting point to more problem-centric DSS instead of general applicable approaches.

References

1. Bastian, M., Heymann, S., Jacomy, M.: Gephi: an open source software for exploring and manipulating networks. In: Proceedings of the International AAAI Conference on Web and Social Media, vol. 3, issue 1, pp. 361–362 (2009)
2. Bejarano, L.A., Espitia, H.E., Montenegro, C.E.: Clustering analysis for the pareto optimal front in multi-objective optimization. Computation **10**(3), 37 (2022). Publisher: Multidisciplinary Digital Publishing Institute
3. Bollobás, B.: Graph Theory, vol. 63. Springer, New York, New York, NY (1979), series Title: Graduate Texts in Mathematics. https://doi.org/10.1007/978-1-4612-9967-7
4. Deb, K., Pratap, A., Agarwal, S., Meyarivan, T.: A fast and elitist multiobjective genetic algorithm: NSGA-II. IEEE Trans. Evol. Comput. **6**(2), 182–197 (2002)
5. Ehrgott, M., Klamroth, K.: Connectedness of efficient solutions in multiple criteria combinatorial optimization. Eur. J. Oper. Res. **97**(1), 159–166 (1997)
6. Everett, M.G., Borgatti, S.P.: The centrality of groups and classes. J. Math. Sociol. **23**(3), 181–201 (1999)
7. Freeman, L.C.: Centrality in social networks conceptual clarification. Soc. Netw. **1**(3), 215–239 (1978)
8. Kahagalage, S., Turan, H.H., Jalalvand, F., El Sawah, S.: A novel graph-theoretical clustering approach to find a reduced set with extreme solutions of Pareto optimal solutions for multi-objective optimization problems. J. Global Optim. **86**(2), 467–494 (2023)

9. Korhonen, P., Wallenius, J.: A pareto race. Naval Res. Logist. (NRL) **35**(6), 615–623 (1988)
10. Liefooghe, A., Derbel, B., Verel, S., López-Ibáñez, M., Aguirre, H., Tanaka, K.: On pareto local optimal solutions networks. In: Auger, A., Fonseca, C.M., Lourenço, N., Machado, P., Paquete, L., Whitley, D. (eds.) Parallel Problem Solving from Nature - PPSN XV. LNCS, pp. 232–244. Springer International Publishing, Cham (2018)
11. Miller, G.A.: The magical number seven, plus or minus two: some limits on our capacity for processing information. Psychol. Rev. **63**(2), 81–97 (1956)
12. Newman, M.E.J., Girvan, M.: Finding and evaluating community structure in networks. Phys. Rev. E **69**(2), 026113 (2004), publisher: American Physical Society
13. Paquete, L., Stützle, T.: Clusters of non-dominated solutions in multiobjective combinatorial optimization: an experimental analysis. In: Barichard, V., Ehrgott, M., Gandibleux, X., T'Kindt, V. (eds.) Multiobjective Programming and Goal Programming. LNEMS, pp. 69–77. Springer, Berlin, Heidelberg (2009)
14. Pulido, F.J.J., Mandow, L., Pérez-De-La-Cruz, J.L.L.: Dimensionality reduction in multi objective shortest path search. Comput. Oper. Res. **64**, 60–70 (2015)
15. Raith, A., Ehrgott, M.: A comparison of solution strategies for biobjective shortest path problems. Comput. Oper. Res. **36**(4), 1299–1331 (2009)
16. Rouder, J.N., Morey, R.D., Cowan, N., Zwilling, C.E., Morey, C.C., Pratte, M.S.: An assessment of fixed-capacity models of visual working memory. Proc. Natl. Acad. Sci. **105**(16), 5975–5979 (2008)
17. Schank, T., Wagner, D.: Approximating clustering coefficient and transitivity. J. Graph Algorithms Appl. **9**(2), 265–275 (2005)
18. Traag, V.A., Waltman, L., van Eck, N.J.: From Louvain to Leiden: guaranteeing well-connected communities. Sci. Rep. **9**(1), 5233 (2019). Publisher: Nature Publishing Group
19. Traag, V., et al.: vtraag/leidenalg: 0.10.0 (2023)
20. Weise, J.: Evolutionary many-objective optimisation for pathfinding problems. Doctoral Thesis, Otto von Guericke University Magdeburg, Magdeburg (2023). Accepted: 2023-03-14T10:25:49Z. ISBN: 9781839107955
21. Weise, J.: Pareto graph analysis (2023). https://doi.org/10.5281/zenodo.10044244
22. Weise, J., Mostaghim, S.: Dataset (Pareto fronts and sets) for the Multi-objective pathfinding problem (2023). https://doi.org/10.5281/ZENODO.10008219
23. Wilson, R.J.: Introduction to Graph Theory. Longman (2010)

Interpretable Solutions for Breast Cancer Diagnosis with Grammatical Evolution and Data Augmentation

Yumnah Hasan[(✉)] [iD], Allan de Lima [iD], Fatemeh Amerehi [iD],
Darian Reyes Fernández de Bulnes [iD], Patrick Healy [iD], and Conor Ryan [iD]

University of Limerick, Limerick, Ireland
{Yumnah.Hasan,Allan.Delima,Fatemeh.Amerehi,Darian.Reyesfernandezdebulnes,
Patrick.Healy,Conor.Ryan}@ul.ie

Abstract. Medical imaging diagnosis increasingly relies on Machine Learning (ML) models. This is a task that is often hampered by severely imbalanced datasets, where positive cases can be quite rare. Their use is further compromised by their limited interpretability, which is becoming increasingly important. While *post-hoc* interpretability techniques such as SHAP and LIME have been used with some success on so-called black box models, the use of inherently understandable models makes such endeavours more fruitful. This paper addresses these issues by demonstrating how a relatively new synthetic data generation technique, STEM, can be used to produce data to train models produced by Grammatical Evolution (GE) that are inherently understandable. STEM is a recently introduced combination of the Synthetic Minority Over-sampling Technique (SMOTE), Edited Nearest Neighbour (ENN), and Mixup; it has previously been successfully used to tackle both between-class and within-class imbalance issues. We test our technique on the Digital Database for Screening Mammography (DDSM) and the Wisconsin Breast Cancer (WBC) datasets and compare Area Under the Curve (AUC) results with an ensemble of the top three performing classifiers from a set of eight standard ML classifiers with varying degrees of interpretability. We demonstrate that the GE-derived models present the best AUC while still maintaining interpretable solutions.

Keywords: Augmentation · Breast Cancer · Ensemble · Grammatical Evolution · STEM

1 Introduction

In medical imaging diagnoses, where decisions can have significant implications for individual's health, it is essential to gain a thorough understanding of the factors influencing these decisions. While Machine Learning (ML) models have proven effective in diagnosing a variety of medical conditions in medical imaging [29], their limited interpretability poses a challenge to their broader adoption. Moreover, the recently introduced European Union (EU) Communication

S. Smith et al. (Eds.): EvoApplications 2024, LNCS 14634, pp. 224–239, 2024.
https://doi.org/10.1007/978-3-031-56852-7_15

on Fostering a European approach to AI [1] specifically targets explainability as a key concern for the deployment of ML and Artificial Intelligence (AI) models.

Another prevalent challenge in the medical imaging domain is the issue of class imbalance within the dataset. Methods such as Synthetic Minority Over-sampling Technique (SMOTE), Edited Nearest Neighbour (ENN), and Mixup combined together as STEM [16], which leverages the full distribution of minority classes, can effectively address both inter-class and intra-class imbalances. In [16], STEM was applied in-conjunction with an ensemble of ML classifiers, producing promising outcomes. However, understanding the reasoning behind ML model predictions remains a complex task. Furthermore, as the volume of instances and the specificity of problems grow, the complexity of the derived solutions also increases.

Building trust in ML classifiers and understanding the behaviour of the solutions is pivotal to their broader acceptance. Employing inherently explainable models is a useful strategy when generating Explainable AI models. Grammatical Evolution (GE) [26], an Evolutionary Computation (EC) technique, has been used to leverage grammars to define and constrain the syntax of potential solutions, producing inherently explainable models [22].

To address these challenges, we developed a classification system based on GE. Our study includes a comprehensive comparison with an ensemble of other ML classifiers. Notably, GE models show enhanced interpretability compared to other traditional ML models. GE provide solutions in the form of symbolic expressions, offering a more intuitive understanding of the decision-making process. This emphasis on interpretability is crucial, especially in healthcare, where understanding the rationale behind decisions is of paramount importance.

Our research hypothesises that the use of the STEM augmentation technique combined with an approach rooted in GE produces more interpretable solutions as compared to the other ensemble ML classifiers.

The contributions of this paper are as follows. Firstly, we develop a method that combines a GE classifier with STEM, outperforming an ensemble of ML classifiers, as indicated by the superior AUC. Secondly, our approach distinguishes itself by offering more interpretable solutions compared to the ensemble method. Finally, the paper presents rigorous statistical analyses to comprehensively evaluate the performance of implemented data augmentation techniques on each data setup.

The rest of the paper is structured as follows: Sect. 2 reviews the existing literature. Section 3 outlines the proposed methodology, and Sect. 4 addresses experimental details performed in this work. Results and discussion are described in Sect. 5, and Sect. 6 presents the conclusion and future guidelines.

2   Literature Review

In the realm of medical applications, particularly in the context of breast cancer diagnosis, the issue of imbalanced datasets is a critical concern. Imbalances, where one class significantly outweighs the other, can introduce biases and compromise the reliability of ML models. Implementing effective strategies for class balancing, such as oversampling, undersampling, and their combination, results in a more balanced and representative training dataset [9]. Previous studies [14,17] have recognized the impact of class imbalance in medical datasets for ML tasks.

Moreover, ML algorithms have demonstrated notable efficiency in the classification of medical data. A compelling study showcases the effectiveness of ensembles, where Bayesian networks and Radial Basis Function (RBF) classifiers with majority voting resulted in an accuracy of 97% [20] when applied to the Wisconsin Breast Cancer (WBC) dataset. Furthermore, an approach that combined linear and non-linear classifiers using Micro Ribonucleic Acid (miRNA) profiling achieved an impressive accuracy of 98.5% [28].

While these findings are promising, ML algorithms may struggle to contextualize information and are susceptible to unexpected or undetected biases originating from input data. Additionally, they often lack transparent justifications for their predictions or decisions [25]. In response to this, employing GE can yield interpretable solutions. As a variant of Genetic Programming, GE evolves human-readable solutions, offering explanations for the rationale behind its classification decisions, which is a significant advantage over current paradigms in unsupervised and semi-supervised learning [10].

Previous studies have already demonstrated the effectiveness of GE across a range of ML tasks. It has proven valuable for feature generation and feature selection [11], as well as for hyperparameter optimization [24]. The GenClass system [3], built upon GE, demonstrates promising outcomes and outperforms RBF networks in certain classification problems. They utilized thirty benchmark datasets from the UCI and KEEL repositories, including Haberman, which consists of breast cancer instances. While it has excelled in these areas, there are still avenues for further exploration.

In this paper, we aim to investigate the efficiency of utilizing GE as a medical imaging classifier combined with STEM to handle imbalance distributions of data samples, particularly in breast cancer diagnosis. Leveraging the interpretive and adaptable features of GE, our objective is to achieve accurate and reliable outcomes that can be easily explained.

3   Methodology

For analysis, we utilize two primary breast cancer datasets. One consists of images, the Digital Database for Screening Mammography (DDSM) [18], while the other consists of tabular data, the WBC [31] dataset. *DDSM* is a comprehensive collection of mammograms, encompassing both normal and abnormal images. For this study, we focused on *DDSM's* Cancer 02 volume and three volumes of normal samples (Volume 01-03). By selecting one volume of cancer images compared to three volumes of normal images, we maintain a realistic class imbalance ratio. These images come from the Craniocaudal (CC) and Mediolateral Oblique (MLO) views of both the left and right breasts. We work with

152 cancerous images and 876 healthy ones from volumes 1-3. Each image was divided into four segments: the entire breast (**I**), the top segment (**It**), the middle segment (**Im**), and the bottom segment (**Ib**).

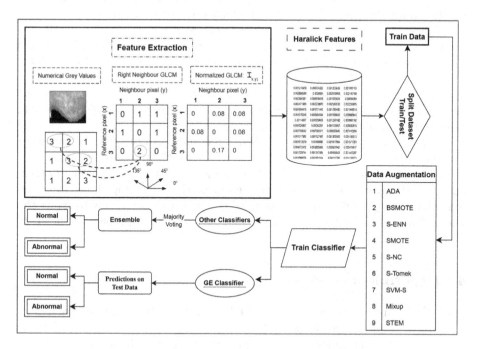

Fig. 1. Outline of the proposed approach for breast cancer classification using GE and other classifiers.

The *WBC* dataset consists of 30 features derived from Fine Needle Aspiration (FNA) samples of breast masses, categorising patients into benign (non-cancerous) and malignant (cancerous) cases. It comprises 212 malignant samples and 357 benign samples.

To create a dataset containing breast cancer images from the *DDSM* image for evaluating the proposed methodology, we first need to extract features that will be used for training. This involves isolating the breast region, eliminating irrelevant background data, segmenting the breast region, and extracting pertinent features to generate a comprehensive training dataset of breast segments. Initially, a median filter is applied to reduce noise within the images. Subsequently, non-essential background data, often containing machine-generated labels such as 'CC' or 'MLO', is removed. For this step, we employed a precise Otsu thresholding technique. Following this, the segmenting process proposed in [27] effectively partitioned images into three overlapping segments.

Feature extraction is the next critical phase. In our study, we extracted a set of Haralick's Texture Features [15] from both whole and segmented images. The selection of these features is based on the hypothesis that there are discernible textural differences between normal and abnormal images. Specifically, we compute thirteen distinct Haralick features from the Gray-Level Co-Occurrence (GLCM) matrix, employing four orientations corresponding to two diagonal (grey-level numeric values of the images) and two adjacent neighbours. This process results in generating a total of 52 features per segment or image.

High class imbalance present in the utilized datasets poses a significant challenge in developing robust and accurate predictive models. Therefore, explicit data augmentation has been implemented in the training set to effectively address this class imbalance challenge. Using nine distinct augmentation approaches outlined in Sect. 4.3, synthetic samples are generated to enrich the dataset with more discriminative information, ultimately improving the learning capabilities of the model.

In the last step, the GE classifier and an ensemble of other ML classifiers are trained separately to make predictions on the test set. Augmented training data is used, while the original imbalanced test set is used for testing. For ensembling, eight ML classifiers are used as mentioned in Sect. 4.5. The top three classifiers, based on AUC, are selected and combined through majority voting to create the final predictions. The complete pipeline of the proposed approach is shown in Fig. 1.

4   Experimental Details

The DDSM and WBC datasets are used to evaluate the proposed technique. The study employs five different data setups to train the classifiers. For the WBC dataset, a single setup is utilized, consisting of 30 breast mass features per sample acquired through FNA.

In contrast, the DDSM dataset includes images from two views, CC and MLO. To conduct experiments, the dataset is categorized into four distinct configurations based on these views. In the initial setup, denoted as "S_{CC}", data is exclusively extracted from segments of the CC view. Conversely, the second category, "S_{MLO}", comprises segmented images exclusively from the MLO view. The third configuration, "S_{CC+MLO}", combines segments from both views. Lastly, the fourth setup, "F_{CC+MLO}", considers the full image (non-segmented) features from both the CC and MLO views for comprehensive analysis. The number of features for each segment or image is 52, used in all these setups

We divided the datasets into training and testing sets at an 80:20 ratio, respectively. Notably, all $DDSM$ configurations exhibit significant class imbalances, with class ratios ranging from 6:94 S_{CC}, S_{MLO} and S_{CC+MLO} setups. For F_{CC+MLO} the ratio between the positive versus negative class is 15:85. Likewise, the WBC dataset has a class distribution of 37% positive and 63% negative classes as illustrated in Fig. 2.

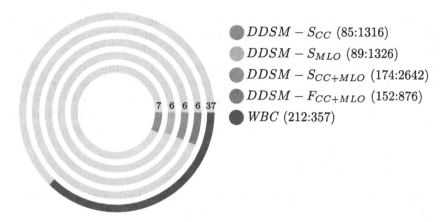

Fig. 2. Concentric ring chart for setup description. Rings are setups, and the coloured areas indicate training positive percent. Legend includes the training positive and negative total samples.

4.1   System Settings

All the ML experiments were conducted using the PyCaret library [2]. The GRAPE [8] framework was used to perform GE experiments. For statistical analysis, we employed the AutoRank Python library [19] to evaluate the performance of the implemented augmentation approaches. Our code, along with our dataset configurations, is available in our GitHub repository[1].

4.2   Performance Metric

To evaluate the performance of the designed approach, AUC has been selected as the assessment metric which uses Trapezoidal rule for its computation. AUC has become a widely accepted performance measure in classification problems due to its reliability, particularly in the context of imbalanced datasets [13,21].AUC serves as a comprehensive metric, encompassing both sensitivity (Eq. 1) and specificity (Eq. 2), considering various threshold values. T_{Pos} denotes true positives, T_{Neg} true negatives, F_{Pos} false positives, and F_{Neg} denotes false negatives.

$$Sensitivity = \frac{T_{Pos}}{T_{Pos} + F_{Neg}} \tag{1}$$

$$Specificity = \frac{T_{Neg}}{T_{Neg} + F_{Pos}} \tag{2}$$

4.3   Class Balancing

The methods utilized for generating synthetic data with the aim of equalizing the class distribution ratio include the Synthetic Minority Oversampling Technique

---

[1] https://github.com/yumnah3/Interpretable-Breast-Cancer-Diagnosis.git.

(SMOTE) [7], Borderline SMOTE (BSMOTE) [14], SMOTENC (S-NC) [7], Support Vector Machine SMOTE (SVM-S) [23], Mixup [32], and ADASYN (ADA) [17]. Additionally, three hybrid methods, SMOTE Edited Nearest Neighbour (S-ENN) [30] SMOTE-Tomek (S-Tomek) [5] and combination of SMOTE, ENN, and Mixup (STEM) are also implemented to compare against each other.

Notably, STEM generates a balanced number of samples for each class. Compared to other methods, it demonstrates the ability to increase the number of data samples more extensively, resulting in improved model performance.

4.4   Grammatical Evolution

GE's grammars are typically defined in Backus-Naur Form (BNF), a notation represented by the tuple N, T, P, S, where N is the set of $non-terminals$, transitional structures usually with semantic meaning, T is the set of $terminals$, items in the phenotype, P is a set of production rules, and S is a start $non-terminal$. The following simple grammar was created to evolve solutions for the first four data setups with 52 numerical features, whereas, for the last setup, 30 numerical features were used:

$$\langle expression \rangle ::= \langle operator \rangle (\langle expression \rangle, \langle expression \rangle \mid \langle operand \rangle$$
$$\langle operator \rangle ::= \texttt{add} \mid \texttt{sub} \mid \texttt{mul} \mid \texttt{pdiv}$$
$$\langle operand \rangle ::= \langle x \rangle \mid \langle digit \rangle \langle digit \rangle . \langle digit \rangle \langle digit \rangle$$
$$\langle x \rangle ::= \texttt{x[0]} \ldots \texttt{x[51]}$$
$$\langle digit \rangle ::= 0 \mid 1 \mid 2 \mid 3 \mid 4 \mid 5 \mid 6 \mid 7 \mid 8 \mid 9$$

This grammar permits the use of basic arithmetic operations (addition, subtraction, multiplication, and division –protected in case the divisor is equal to 0) and the inclusion of real numbers constants. These constants are helpful because GE can explore beyond the parameter space given to minimize the error between expected and predicted outputs, something that does not happen with other ML classifiers. The $non-terminal$ X encompasses the fifty-two numerical features for the first four setups of the DDSM dataset and the thirty numerical features for the WBC dataset.

The output domain of the evaluations is $o \in [-\infty, \infty]$. Subsequently, a sigmoid function is applied to constrain the values to $\sigma(o) \in [0, 1]$. For binary classification, the typical interpretation of the sigmoid function is the probability of belonging to class 1, and therefore we use $\sigma(o)$ to calculate AUC. Table 1 presents the experimental parameters used in this work:

Table 1. List of parameters used to run GE

Parameter type	Parameter value
Number of runs	30
Number of generations	100
Population size	200
Mutation probability	0.01
Crossover probability	0.8
Elitism size	1
Codon size	255
Initialisation	Sensible
Maximum initial depth	10
Maximum depth	35
Wrapping	0

4.5 Other Classifiers

We also used the augmented training data to train a diverse ensemble of eight ML classifiers. This ensemble includes Random Forest (RF), Linear Discriminant Analysis (LDA), Quadratic Discriminant Analysis, LightGBM, XGBoost, AdaBoost, KNN, and Extra Trees models. Initially, a comprehensive model is trained using all eight classifiers. Subsequently, based on the AUC metric, the three best-performing models are selected. These selected models are then combined through a majority voting approach. The final predictions are made on the test dataset, which consists of imbalanced and unseen samples.

5 Results and Discussion

To evaluate the performance of the proposed method, five distinct data setups are employed. Four configurations are derived from the DDSM dataset, considering variations in views, segments, and full images. The fifth setup is from the WBC dataset. To enhance the robustness of the training setups, nine augmentation approaches are applied and compared. The assessment is conducted using an ensemble of other ML classifiers, alongside GE.

The performance of the classifiers is compared based on AUC for each dataset. The ensemble classifiers are denoted by their respective initials: L_d for Linear Discriminant Analysis, Q for Quadratic Discriminant Analysis, E for ExtraTree, R for Random Forest, L_i for Lightgbm, K for KNN, A for Adaboost, and X for Xgboost. It is important to note that the AUC values of the other ensemble classifiers are presented for a single run, and they are then compared against the median AUC derived from 30 runs conducted with GE.

Table 2 provides an overview of the results. In the first setup, S_{CC}, an AUC of 0.91 was achieved, outperforming the ensemble of L_dQE, which obtained an

AUC of 0.90. Similarly, in the second setup, S_{MLO}, an AUC of 0.90 was attained, while the ensemble of L_dQE achieved a slightly lower AUC of 0.84.

For the third setup S_{CC+MLO}, an AUC of 0.92 was observed using the GE classifier, outperforming other classifiers that yielded the highest AUC of 0.87 using L_dQE. When the classifiers were trained on full image features in setup F_{CC+MLO}, the highest AUC values were 0.94 and 0.85, obtained by the GE classifier and the ensemble of L_dQE, respectively.

When comparing the AUC using the WBC dataset, both GE and the ensemble of AKL_r achieved an AUC of 0.99.

Table 2. A comparison of the AUC for GE and the ensemble approaches using the nine different augmentation techniques for each data setup.

Setups	Classifiers	ADA	BSMOTE	S-ENN	SMOTE	S-NC	S-Tomek	SVM-S	Mixup	STEM
S_{CC}	GE	0.91	0.90	0.89	0.91	0.90	0.90	0.90	0.91	0.90
	Others	0.76	0.73	0.93	0.77	0.82	0.77	0.73	0.90	0.90
		L_dQE	L_dQE	L_dQE	L_dQE	L_dQE	L_dQE	L_dQE	L_dQE	L_dQE
S_{MLO}	GE	0.90	0.90	0.90	0.90	0.87	0.90	0.89	0.90	0.89
	Others	0.80	0.80	0.80	0.82	0.78	0.82	0.81	0.81	0.84
		EL_iR	EL_iR	L_dQE	EL_iR	EL_iX	EL_iX	EL_iR	L_dQE	L_dQE
S_{CC+MLO}	GE	0.91	0.91	0.92	0.91	0.92	0.91	0.91	0.90	0.91
	Others	0.75	0.68	0.77	0.75	0.70	0.76	0.62	0.76	0.87
		EL_iR	EL_iR	EL_iR	EL_iR	EL_iX	EL_iR	EL_iR	EL_iR	L_dQE
F_{CC+MLO}	GE	0.93	0.91	0.90	0.92	0.93	0.94	0.93	0.93	0.93
	Others	0.78	0.84	0.72	0.81	0.82	0.82	0.82	0.81	0.85
		EQR	EL_iR	ERX	EQR	EL_iQ	EL_iR	EQR	L_iQL_d	L_dQE
WBC	GE	0.98	0.98	0.99	0.98	0.99	0.98	0.99	0.98	0.99
	Others	0.94	0.94	0.94	0.94	0.95	0.94	0.94	0.94	0.99
		L_dQE	L_dQE	EKL_i	L_dQE	L_dQE	L_dQE	L_dQE	L_dEL_i	AKL_r

The augmentation approaches are compared using the boxplot presented in Fig. 3. The plot indicates the AUC obtained from all nine augmentation approaches for each setup across all 30 runs. The horizontal line in red indicates the median value of the respective group.

GE provides valuable insights into the most informative features used in the solutions, as demonstrated in Table 3, which present the most frequently used features for each setup. The features extracted and presented in these tables are sorted by their impact on the solutions. Common features consistently found in Table 3 for the $DDSM$ dataset include "Inverse Difference Moment (IDM)" (feature 17), "Contrast" (feature 5), and "Difference Entropy" (feature 41). Both contrast and IDM represent the difference in grey levels between pixels, while entropy indicates the level of randomness in the grey levels.

For the WBC dataset, as shown in Table 3, the top three features that consistently appear in the solutions are 21, 20, and 27, corresponding to "Concave Point Worst", "Fractal Dimension", and "Radius Worst" respectively. The concave point worst feature indicates the severity of the concave portion of the shape,

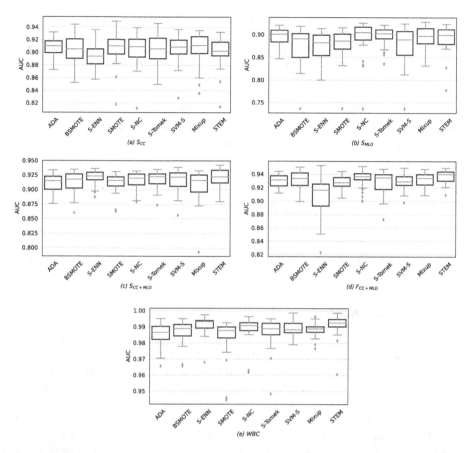

Fig. 3. Boxplot analysis comparing opponent approaches and their AUC distributions across multiple runs

with "worst" denoting the highest mean value. The "fractal dimension" is a crucial characteristic that provides information related to the geometric shape of the fractals. The third feature, radius worst, represents the largest mean value for the distances from the centre to points on the perimeter.

While other ML models may share the feature of interpretability, they often present challenges that GE does not encounter. Decision trees and RF, though interpretable, lose clarity with complex structures and aggregation [4]. LDA relies on the Gaussian distribution of the data and assumes that the covariance of two classes is the same [12], limiting its applicability. In contrast, GE does not depend on these factors and maintains transparency throughout its evolution, even when addressing complex and non-linear problems.

Table 3. This analysis unveils prevalent features used by GE in all five setups. For S_{CC} and S_{MLO}, percentages are computed from 8684 and 7945 occurrences. Likewise, contributions to S_{CC+MLO} and F_{CC+MLO} are based on 8138 and 8522 occurrences, respectively. The features of *WBC* are also examined, with percentages drawn from 9076 appearances.

S_{CC}		S_{MLO}		S_{CC+MLO}		F_{CC+MLO}		*WBC*	
Feature	Usage	Feature	Usage	Feature	Usage	Feature	Usage	Feature	Usage
17	6.22%	5	4.93%	17	5.35%	41	3.78%	21	7.83%
41	4.87%	4	4.46%	5	4.36%	37	3.71%	20	7.64%
38	4.58%	41	3.95%	7	4.33%	4	3.63%	27	6.47%
5	4.19%	7	3.75%	41	4.02%	38	3.46%	24	5.56%
18	3.88%	17	3.65%	18	3.93%	11	3.38%	1	5.1%
7	3.50%	34	3.34%	38	3.55%	17	3.18%	13	4.87%
36	2.73%	45	3.15%	11	3.08%	5	3.11%	7	4.23%

5.1 Statistical Analysis

The statistical comparison of implemented data augmentation techniques involved a non-parametric Bayesian signed-rank test [6] applied to each dataset. In our analysis, conducted on nine augmentation techniques with 30 paired AUC samples each, the test distinguished between methods being pair-wise larger, smaller or inconclusive. The approaches listed in the rows are compared with the methods presented in the corresponding column. The subsequent Bayesian signed-rank test revealed significant distinctions among the techniques. In the cases where STEM has outperformed the other approaches are underlined in the Table 4.

In the S_{CC} setup, as illustrated in Table 4(a), STEM, Mixup, SMOTE, ADA, S-NC, SVM-S, S-Tomek and BSMOTE all exhibit larger medians than S-ENN.

The statistical comparison of medians depicted in Table 4(b) among various augmentation populations reveals notable differences for S_{MLO} setup. STEM, S-NC, S-Tomek, ADA, and Mixup exhibit larger medians compared to BSMOTE, SVM-S, SMOTE, and S-ENN.

Similarly, for setup S_{CC+MLO} in the Table 4(c) STEM again showcases its effectiveness by outperforming S-NC, BSMOTE, Mixup, SMOTE, and ADA in medians. Additionally, S-ENN demonstrates superiority by exhibiting larger medians than Mixup, SMOTE, and ADA. Additionally, S-Tomek outperforms SMOTE in median values. SVM-S, in particular, stands out with a larger median than ADA.

Moreover, STEM stands out by consistently surpassing S-Tomek, Mixup, BSMOTE, ADA, SVM-S, SMOTE, and S-ENN in median values presented in Table 4(d) for F_{CC+MLO} . Additionally, S-NC demonstrates superiority over SMOTE and S-ENN, while S-Tomek outperforms S-ENN in median values. Mixup, BSMOTE, ADA, SVM-S, and SMOTE all exhibit larger medians than S-ENN.

Finally, in the *WBC* setup, as depicted in Table 4(e), STEM emerged as the top-performing method, surpassing S-NC, BSMOTE, S-Tomek, Mixup, SVM-S, SMOTE, and ADA. S-NC exhibited a higher median than SMOTE and ADA, while Mixup outperformed SMOTE in median value. SVM-S demonstrated a larger median than SMOTE and ADA.

The Bayesian analysis results are summarized in Fig. 4. It reveals that STEM, a combination of S-ENN and Mixup, emerges as the top-ranking approach. This result underscores the effectiveness of this combined strategy in enhancing performance. Notably, S-ENN and Mixup individually secure the second and third positions, further affirming the significance of this ensemble approach.

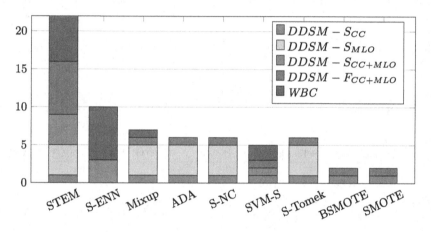

Fig. 4. The illustration of the overall results acquired from the Bayesian signed-rank test is shown here. The cumulative score is the total number of times one approach outperforms the other. STEM obtained a cumulative score of 23 where the maximum possible is 40 (comparing one versus another 8 approaches in 5 setups), outperforming the other approaches. Each color represents distinct test setups used for the evaluation.

Table 4. The results of the Bayesian signed-ranked test are summarized here for the nine augmentation approaches for each data setup. Arrows indicate the direction of differences: ⇑ for larger, ⇓ for smaller, - for inconclusive, and N/A for not applicable results. A family-wise significance level of $\alpha \equiv 0.05$ is employed.

(a) S_{CC}

	STEM	Mixup	SMOTE	ADA	S-NC	SVM-S	S-Tomek	BSMOTE	S-ENN
STEM	N/A	–	–	–	–	–	–	–	⇑
Mixup	–	N/A	–	–	–	–	–	–	⇑
SMOTE	–	–	N/A	–	-	–	–	–	⇑
ADA	–	–	–	N/A	–	–	–	–	⇑
S-NC	–	-	-	-	N/A	-	-	-	⇑
SVM-S	–	–	–	–	–	N/A	-	–	⇑
S-Tomek	–	–	-	-	–	–	N/A	–	⇑
BSMOTE	–	–	–	–	–	–	–	N/A	⇑
S-ENN	⇓	⇓	⇓	⇓	⇓	⇓	⇓	⇓	N/A

(b) S_{MLO}

	STEM	Mixup	S-NC	S-Tomek	ADA	BSMOTE	SVM-S	SMOTE	S-ENN
STEM	N/A	–	–	–	–	⇑	⇑	⇑	⇑
Mixup	–	N/A	–	–	–	⇑	⇑	⇑	⇑
S-NC	–	–	N/A	–	–	⇑	⇑	⇑	⇑
S-Tomek	–	–	-	N/A	–	⇑	⇑	⇑	⇑
ADA	–	–	–	–	N/A	⇑	⇑	⇑	⇑
BSMOTE	⇓	⇓	⇓	⇓	⇓	N/A	–	–	–
SVM-S	⇓	⇓	⇓	⇓	⇓	–	N/A	–	–
SMOTE	⇓	⇓	⇓	⇓	⇓	–	–	N/A	–
S-ENN	⇓	⇓	⇓	⇓	⇓	–	–	–	N/A

(c) S_{CC+MLO}

	STEM	S-ENN	S-Tomek	SVM-S	S-NC	BSMOTE	Mixup	SMOTE	ADA
STEM	N/A	–	–	–	–	⇑	⇑	⇑	⇑
S-ENN	–	N/A	–	–	–	–	⇑	⇑	⇑
S-Tomek	–	–	N/A	-	-	–	–	–	–
SVM-S	–	–	–	N/A	-	-	-	-	⇑
S-NC	–	–	–	–	N/A	–	–	–	–
BSMOTE	⇓	-	–	–	–	N/A	–	–	–
Mixup	⇓	⇓	–	–	–	–	N/A	–	–
SMOTE	⇓	⇓	—	–	–	–	–	N/A	–
ADA	⇓	⇓	–	⇓	–	–	–	–	N/A

(d) F_{CC+MLO}

	STEM	S-NC	Mixup	S-Tomek	BSMOTE	ADA	SVM-S	SMOTE	S-ENN
STEM	N/A	–	⇑	⇑	⇑	⇑	⇑	⇑	⇑
S-NC	–	N/A	–	–	–	–	–	–	⇑
Mixup	⇓	–	N/A	–	–	–	–	–	⇑
S-Tomek	⇓	-	-	N/A	-	-	-	-	⇑
BSMOTE	⇓	–	–	–	N/A	–	–	–	⇑
ADA	⇓	–	–	–	–	N/A	–	–	⇑
SVM-S	⇓	–	–	–	–	–	N/A	–	⇑
SMOTE	⇓	–	–	–	–	–	–	N/A	⇑
S-ENN	⇓	⇓	⇓	⇓	⇓	⇓	⇓	⇓	N/A

(*continued*)

Table 4. (*continued*)

(e) *WBC*

	STEM	S-ENN	S-NC	SVM-S	Mixup	ADA	BSMOTE	S-Tomek	SMOTE
STEM	N/A	–	⇑	⇑	⇑	⇑	⇑	⇑	⇑
S-ENN	–	N/A	⇑	⇑	⇑	⇑	–	⇑	⇑
S-NC	⇓	⇓	N/A	–	–	–	–	–	–
SVM-S	⇓	⇓	–	N/A	–	⇓	–	–	⇓
Mixup	⇓	⇓	–	–	N/A	–	–	–	⇑
ADA	⇓	⇓	–	⇓	–	N/A	–	–	–
BSMOTE	⇓	–	–	–	–	–	N/A	–	–
S-Tomek	⇓	⇓	–	–	–	–	–	N/A	–
SMOTE	⇓	⇓	–	⇓	⇓	–	–	–	N/A

6 Conclusion and Future Work

In this study, we addressed class imbalance and interpretability challenges in medical imaging diagnosis by using GE to produce classifier trained on data augmented by the recently-introduced STEM technique. Our approach not only delivers interpretable solutions but also outperforms an ensemble of other ML classifiers in terms of performance. The analysis conducted on the *DDSM* and *WBC* datasets emphasizes the effectiveness of GE, as evidenced by improvements in AUC and its ability to identify critical data features. Notably, our inclusion of Bayesian signed-rank test results confirms that STEM emerges as the best-performing approach for augmentation. The improved AUC and enhanced interpretability of our approach can help build trust and facilitate informed decisions. Thus, our study validates the proposed hypothesis, demonstrating the efficacy of the combined GE and STEM approach.

For future research, we suggest improving performance by incorporating additional image attributes, such as wavelet transformations and local binary patterns, to enhance the feature set and dataset diversity. Furthermore, exploring the mixture of different datasets to assess the robustness of our approach across various image data sources would be interesting.

Acknowledgements. The Science Foundation Ireland (SFI) Centre for Research Training in Artificial Intelligence (CRT-AI), Grant No. 18/CRT/6223 and the Irish Software Engineering Research Centre (Lero), Grant No. 16/IA/4605, both provided funding for this study.

References

1. Communication on Fostering a European approach to Artificial Intelligence — Shaping Europe's digital future (Apr 2021)
2. Ali, M.: Pycaret: an open source, low-code machine learning library in python version 2.3 (2020)

3. Anastasopoulos, N., Tsoulos, I.G., Tzallas, A.: Genclass: a parallel tool for data classification based on grammatical evolution. SoftwareX **16**, 100830 (2021)

4. Arrieta, A.B., et al.: Explainable artificial intelligence (xai): Concepts, taxonomies, opportunities and challenges toward responsible ai. Inform. Fusion **58**, 82–115 (2020)

5. Batista, G.E., Bazzan, A.L., Monard, M.C., et al.: Balancing training data for automated annotation of keywords: a case study. Wob **3**, 10–8 (2003)

6. Benavoli, A., Corani, G., Mangili, F., Zaffalon, M., Ruggeri, F.: A bayesian wilcoxon signed-rank test based on the dirichlet process. In: International Conference on Machine Learning, pp. 1026–1034. PMLR (2014)

7. Chawla, N.V., Bowyer, K.W., Hall, L.O., Kegelmeyer, W.P.: Smote: synthetic minority over-sampling technique. J. Artifi. Intell. Res. **16**, 321–357 (2002)

8. de Lima, A., Carvalho, S., Dias, D.M., Naredo, E., Sullivan, J.P., Ryan, C.: GRAPE: grammatical Algorithms in Python for Evolution. Signals **3**(3), 642–663 (2022). https://doi.org/10.3390/signals3030039

9. Fernández, A., López, V., Galar, M., Del Jesus, M.J., Herrera, F.: Analysing the classification of imbalanced data-sets with multiple classes: binarization techniques and ad-hoc approaches. Knowl.-Based Syst. **42**, 97–110 (2013)

10. Fitzgerald, J.M., Azad, R.M.A., Ryan, C.: GEML: Evolutionary unsupervised and semi-supervised learning of multi-class classification with Grammatical Evolution. In: 2015 7th International Joint Conference on Computational Intelligence (IJCCI), vol. 1, pp. 83–94 (Nov 2015)

11. Gavrilis, D., Tsoulos, I.G., Dermatas, E.: Selecting and constructing features using grammatical evolution. Pattern Recogn. Lett. **29**(9), 1358–1365 (2008). https://doi.org/10.1016/j.patrec.2008.02.007

12. Ghojogh, B., Crowley, M.: Linear and quadratic discriminant analysis: Tutorial. arXiv preprint arXiv:1906.02590 (2019)

13. Halimu, C., Kasem, A., Newaz, S.S.: Empirical comparison of area under roc curve (auc) and mathew correlation coefficient (mcc) for evaluating machine learning algorithms on imbalanced datasets for binary classification. In: Proceedings of the 3rd International Conference on Machine Learning and Soft Computing, pp. 1–6 (2019)

14. Han, H., Wang, W.-Y., Mao, B.-H.: Borderline-SMOTE: a new over-sampling method in imbalanced data sets learning. In: Huang, D.-S., Zhang, X.-P., Huang, G.-B. (eds.) ICIC 2005. LNCS, vol. 3644, pp. 878–887. Springer, Heidelberg (2005). https://doi.org/10.1007/11538059_91

15. Haralick, R.M., Shanmugam, K., Dinstein, I.H.: Textural features for image classification. IEEE Trans. Syst. Man Cybernet. 610–621 (1973)

16. Hasan, Y., Amerehi, F., Healy, P., Ryan, C.: Stem rebalance a novel approach for tackling imbalanced datasets using smote, edited nearest neighbour, and mixup (2023). https://arxiv.org/abs/2311.07504

17. He, H., Bai, Y., Garcia, E.A., Li, S.: Adasyn: adaptive synthetic sampling approach for imbalanced learning. In: 2008 IEEE International Joint Conference on Neural Networks (IEEE World Congress on Computational Intelligence), pp. 1322–1328. IEEE (2008)

18. Heath, M., et al.: Current status of the digital database for screening mammography. In: Digital Mammography: Nijmegen, pp. 457–460. Springer (1998). https://doi.org/10.1007/978-94-011-5318-8_75

19. Herbold, S.: Autorank: a Python package for automated ranking of classifiers. J. Open Source Softw. **5**(48), 2173 (2020). https://doi.org/10.21105/joss.02173

20. Jabbar, M.A.: Breast cancer data classification using ensemble machine learning. Eng. Appli. Sci. Res. **48**(1), 65–72 (2021)
21. Liang, X., Jiang, A., Li, T., Xue, Y., Wang, G.: Lr-smote-an improved unbalanced data set oversampling based on k-means and svm. Knowl.-Based Syst. **196**, 105845 (2020)
22. Murphy, A., Murphy, G., Amaral, J., MotaDias, D., Naredo, E., Ryan, C.: Towards incorporating human knowledge in fuzzy pattern tree evolution. In: Hu, T., Lourenço, N., Medvet, E. (eds.) EuroGP 2021. LNCS, vol. 12691, pp. 66–81. Springer, Cham (2021). https://doi.org/10.1007/978-3-030-72812-0_5
23. Nguyen, H.M., Cooper, E.W., Kamei, K.: Borderline oversampling for imbalanced data classification. Inter. J. Knowl. Eng. Soft Data Paradigms **3**(1), 4–21 (2011). https://doi.org/10.1504/IJKESDP.2011.039875
24. Noorian, F., de Silva, A.M., Leong, P.H.W.: gramEvol: grammatical evolution in R. J. Stat. Softw. **71**, 1–26 (2016). https://doi.org/10.18637/jss.v071.i01
25. Rashed, B.M., Popescu, N.: Machine learning techniques for medical image processing. In: 2021 International Conference on E-Health and Bioengineering (EHB), pp. 1–4 (Nov 2021). https://doi.org/10.1109/EHB52898.2021.9657673
26. Ryan, C., Collins, J.J., Neill, M.O.: Grammatical evolution: Evolving programs for an arbitrary language. In: Banzhaf, W., Poli, R., Schoenauer, M., Fogarty, T.C. (eds.) EuroGP 1998. LNCS, vol. 1391, pp. 83–96. Springer, Heidelberg (1998). https://doi.org/10.1007/BFb0055930
27. Ryan, C., Krawiec, K., O'Reilly, U.-M., Fitzgerald, J., Medernach, D.: Building a stage 1 computer aided detector for breast cancer using genetic programming. In: Nicolau, M., et al. (eds.) EuroGP 2014. LNCS, vol. 8599, pp. 162–173. Springer, Heidelberg (2014). https://doi.org/10.1007/978-3-662-44303-3_14
28. Sharma, S.K., Vijayakumar, K., Kadam, V.J., Williamson, S.: Breast cancer prediction from microRNA profiling using random subspace ensemble of LDA classifiers via Bayesian optimization. Multimedia Tools Appli. **81**(29), 41785–41805 (2022). https://doi.org/10.1007/s11042-021-11653-x
29. Varoquaux, G., Cheplygina, V.: Machine learning for medical imaging: methodological failures and recommendations for the future. npj Digital Med. **5**(1), 1–8 (2022). https://doi.org/10.1038/s41746-022-00592-y
30. Wilson, D.L.: Asymptotic properties of nearest neighbor rules using edited data. IEEE Trans. Syst. Man Cybernet., 408–421 (1972)
31. Wolberg, W.H., Street, W.N., Mangasarian, O.L.: Breast cancer wisconsin (diagnostic) data set [uci machine learning repository] (1992)
32. Zhang, H., Cisse, M., Dauphin, Y.N., Lopez-Paz, D.: Mixup: Beyond Empirical Risk Minimization (Apr 2018). https://doi.org/10.48550/arXiv.1710.09412

Applying Graph Partitioning-Based Seeding Strategies to Software Modularisation

Ashley Mann[✉], Stephen Swift, and Mahir Arzoky

Brunel University London, UB8 3PH Uxbridge, UK
{Ashley.Mann,Stephen.Swift,Mahir.Arzoky}@brunel.ac.uk

Abstract. Software modularisation is a pivotal facet within software engineering, seeking to optimise the arrangement of software components based on their interrelationships. Despite extensive investigations in this domain, particularly concerning evolutionary computation, the research emphasis has transitioned towards solution design and convergence analysis rather than pioneering methodologies. The primary objective is to attain efficient solutions within a pragmatic timeframe. Recent research posits that initial positions in the search space wield minimal influence, given the prevalent trend of methods converging upon akin local optima. This paper delves into this phenomenon comprehensively, employing graph partitioning techniques on dependency graphs to generate initial clustering arrangement seeds. Our empirical discoveries challenge conventional insight, underscoring the pivotal role of seed selection in software modularisation to enhance overall outcomes.

Keywords: Software Engineering · Heuristic Search · Software Modularisation · Graph Partitioning

1 Introduction

1.1 General Background

As software systems grow, maintenance becomes challenging for incoming engineers unfamiliar with the original code, often leading to the need for significant overhauls or discontinuation of extensive legacy systems. To ensure sustainable management, creating modular subsystems is crucial. Instead of portraying them as clusters in the source code, a more practical approach is representing dependencies in graph form. Mancoridis et al. define the software modularisation problem as arising from the exponential complexity of interconnected software module relationships within evolving systems. This is often approached as a heuristic-search-based clustering problem to identify optimal representations by clustering subsystems based on the strength of their relationships [26].

The escalating complexity, often addressed through evolutionary computation, is evident in various software implementations, including both single-objective [3,16] and multi-objective [18,27] approaches. Pioneering methodologies aim to enhance the structure of software systems. Optimisation of subsystems extends to diverse attributes, such as classes, methods, and variables.

S. Smith et al. (Eds.): EvoApplications 2024, LNCS 14634, pp. 240–258, 2024.
https://doi.org/10.1007/978-3-031-56852-7_16

Methodological advancements now consider type-based dependence analysis [22], multi-pattern clustering [8], and effort estimation [32]. These efforts explore pre-processing and post-processing improvements alongside optimisation strategies.

The modularisation of software, mainly through heuristic search and evolutionary computation methodologies, extensively incorporates graph theory and data clustering. Academic works commonly use graph representations of software systems, employing data clustering for nodes and implementing algorithms to assess cluster quality [2,19,37]. Despite graph creation not inherently enhancing software engineers' understanding of architecture structure, language-independent graphs can focus on specific relationships or entire systems [10,30,35]. Clustering arrangements can be portrayed through various methods, such as a one-dimensional vector, a two-dimensional cluster-based structure, or a one-dimensional constrained representation known as a restricted growth function, which, despite its constraints, exhibits distinctive properties [7]. Clustering arrangement measurement typically addresses cohesion and coupling, striving for optimal cohesion within clusters and minimal coupling between clusters, fostering the creation of clearly defined groups [2].

1.2   Motivation

A recent study conducted by our research group explores varied representations for clustering arrangements and different starting points, providing insights into the search space of software systems [25]. The study highlights the list-of-lists representation as the most robust, emphasising its significance in problem-solving. Notably, the paper suggests that the starting point choice is inconsequential, as various representations converge towards similar outcomes regarding final fitness, especially one and two-dimensional list-based ones.

This paper is motivated by exploring converging results based on starting points. Our primary objective is to determine whether alternative starting positions can replicate or potentially improve previous findings. If diverse starting positions tend to converge toward a similar region in the search space, we aim to uncover the reasons behind this convergence. Is there a basin of attraction leading to a potential global optimum solution, or do these methods unintentionally get stuck in closely adjacent local optima?

In recent years, a discernible research gap has emerged in clustering arrangement representation and software graph representation. Additionally, up to the present time, there is a notable absence of publications in the field of software modularisation specifically dedicated to addressing the concept of starting points. While we recognise that meta-heuristics, such as Iterated Local Search [21], can generate seeded starting points based on previous experimental iterations, our reference pertains to the primary initial search, distinct from subsequent iterations.

Building upon this motivation, we aim to explore innovative approaches for generating starting points that surpass the performance of previous experiments. If our findings suggest the existence of a basin of attraction, our goal is to devise more efficient methods to reach this point faster than conventional approaches.

However, even if the evidence points in a different direction, our overarching objective is to develop a more efficient method for navigating and exploring the search space.

This paper focuses on enhancing software system clustering by integrating graph partitioning techniques with seeded search methods applied to graph-based representations. Situated within Search-Based Software Engineering, our research particularly centres on software modularisation. To achieve our goal, we begin with a domain background, introduce innovative concepts, outline our experimental procedure, and present our results.

2   Related Work

2.1   Bunch and Munch

Exploring software modularisation can be achieved using tools such as Bunch [23] and Munch [4] [5]. Bunch, developed by Mancoridis et al., combines a Steepest Ascent Hill Climbing (SAHC) and Genetic Algorithms for improved clustering arrangements [23,24]. On the other hand, Arzoky et al., Munch employs Random Mutation Hill Climbing (RMHC) for enhanced performance and ease of implementation [4]. Both strategies use different fitness functions - Bunch utilises the *MQ* fitness function, while Munch employs *EVM* and *EVMD* [4,24,34]. Despite employing different measurement strategies, MQ, EVM and EVMD yield similar clustering results [17]. However, the exhaustive nature of Bunch may hinder performance when runtime is a critical consideration.

2.2   Starting Points and Search Space

In the context of a heuristic-search-based clustering problem, the quest for optimal solutions necessitates delving into the search space, which comprises all conceivable arrangements of a clustering configuration. This exploration entails generating an initial clustering arrangement known as the starting point. Subsequently, through mutation (searching), this arrangement is modified and compared to the graph representation of the software. The goal is to enhance the clustering of nodes that demonstrate robust relationships. Before embarking on a search, a crucial decision lies in determining the optimal starting point for seeking an improved clustering arrangement.

Several starting points are available when searching for local optima, which, in our context, represents the nearest approximation to the optimal clustering arrangement that maximises the cohesion of each cluster within the search space. We provide three illustrative examples: we can cluster all nodes individually for maximum coupling (Fig. 1), together for maximum cohesion (Fig. 2), or randomly (Fig. 3).

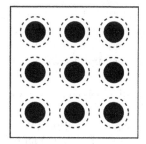

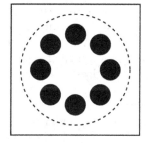

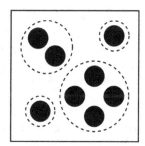

Fig. 1. Independent   **Fig. 2.** All In One   **Fig. 3.** Random

3   Research Questions

We aim to address research inquiries regarding our endeavour to discover improved starting points for software modularisation. We aim to uncover more effective strategies for achieving optimal outcomes. In this paper, we outline the following research questions that we intend to investigate:

1. What is the performance difference between graph-partitioned clustering arrangements and randomly generated ones when applied to large and small software systems?
    (a) When using hill climbing with various initial clustering arrangements on the same software system, do the solutions converge to similar outcomes or do disparities persist?
    (b) How do the runtimes of searches using graph partition and randomly generated clustering arrangements vary, and are there trade-offs between runtime and solution quality?
2. Is there a significant disparity between the Weighted Kappa[1] values of the final clustering arrangements and a gold standard[2], and what is the nature of this comparison?

Initially, we aim to evaluate whether the graph-based initial clustering arrangements result in enhanced outcomes compared to randomly generated configurations through Munch. We aim to contrast the clustering patterns derived from graph partitioning with those generated randomly across software systems of varying sizes. Alongside assessing our fitness function, we analyse the documented improvements at the final iteration. This entails identifying the convergence point and scrutinising the runtime of the search, which encompasses both the initialisation of the starting configuration and the subsequent search process. By possessing information about the ultimate fitness value and the corresponding

---

[1] Weighted Kappa is employed to assess the similarity of clustering arrangements and is applied in Sects. 5.3, 6, and 7.

[2] A gold standard represents the theoretical best solution for a given problem, a rarity in real-world datasets where it is seldom known.

iteration when it is achieved, we aim to discern the genuine impact of the initial clustering configurations on the search dynamics. We aim to determine whether specific clustering arrangements contribute to a faster convergence, enabling us to refine our search methodology for reaching the convergence point earlier and mitigating the risk of potential time loss.

In addition to assessing the effectiveness of our initial clustering configurations based on fitness, convergence, and runtime, we also evaluate the final clustering arrangements against gold standards using Weighted Kappa (WK) [1]. WK serves as a measure of agreement between two clustering arrangements, explicitly focusing on modularisation. As the WK values increase, the level of agreement between the two solutions also rises. A WK value 1 signifies identical clustering arrangements, while 0 indicates empirical dissimilarity. A WK value of 0.5 or higher indicates a robust structural similarity between the two clustering configurations. We opt for WK over other methods, such as Adjusted Rand [29], due to its ease of implementation, longstanding presence in the field, and well-established interpretability/quality scale. The authors also note that WK and Adjusted Rand are identical.

4   Methods

Our focus now shifts towards the methodologies aligned with our exploration of optimal starting points for software modularisation. We present our selected search method and detail our implementation of graph partitioning designed to yield appropriate starting points.

4.1   Munch

As previously indicated, software modularisation is characterised as a heuristic-search-based clustering problem. Therefore, our initial consideration lies in devising a strategy for heuristic search before delving into the discussion of our implementation of graph partitioning for generating starting points. We adopt a reverse-engineered adaptation of Arzoky et al.'s Munch to address this [5]. This adaptation has been enhanced to afford us the flexibility to determine the commencement of our exploration and the nature of our search strategy. We will now delve into an exploration of the various components that constitute Munch.

Foremost, Munch uses Module Dependency Graphs (MDG) as our graph-based representation of software systems. As defined by Mancoridis et al., $MDGs$ illustrate subsystem connections to gauge relationships between components. In the context of our current research, we designate the nodes of MDG as software classes, and the edges represent interconnected relationships. $MDGs$ prove versatile, capable of describing software structure over time or facilitating the segmentation of extensive software systems for enhanced comprehension. Let MDG M be an n by n symmetric binary matrix, where a 1 at row x and column y (M_{xy}) indicates a relationship between software components x and y, and 0 indicates that there is no relationship. To avoid confusion throughout this paper, MDG and graph are considered synonymous.

$$M_{xy} = \begin{cases} 1 & \text{if a relationship exists between } x \text{ and } y \\ 0 & \text{otherwise,} \end{cases}$$

For Munch, we adopt a list-of-list-based cluster representation based on its ease of implementation. A list-of-list clustering arrangement (C) is defined as a list $([C_1, ..., C_k])$, with each subset list/cluster (C_i) containing $1, 2, ..., n$ elements. These subsets must be non-empty $(C_i \neq \emptyset)$, and they should not share any common items $(C_i \cap C_j = \emptyset)$ for different subsets. Effective optimisation problem-solving requires consideration of the search space, exploration strategy, and fitness function. Equation 1 illustrates all possible ways to partition C_k clusters containing n elements. Note that $1 \leq k \leq n$. We justify opting for lists over sets in the implementation, emphasising the advantages of simpler implementation and reduced computational complexity, particularly in scenarios involving non-indexed sets. Since each cluster and cluster element requires indexing, the search space aligns with Eq. 1, deviating from the set nature characterised by $\text{Bell}(n)$.

$$\sum_{k=1}^{n} \left(\frac{n!}{k! \cdot (n-k)!} \cdot k! \right) \qquad (1)$$

Before delving into the search strategy of Munch, it is essential to define the fitness function. The primary goal of a search strategy is to uncover a clustering arrangement that most effectively aligns with the ideal modular structure of the software system. The assessment entails analysing the subsets $[C_1, ..., C_k]$, where the elements (C_i), representing $1, 2, ..., n$, illustrate their relationships within the MDG. To avoid confusion, we will refer to the subsets as clusters.

For our replication of Munch, it is unsurprising we introduce EVM as our selected fitness function. We opt for EVM over Bunch's MQ due to its demonstrated robustness against noise and suitability for real-world software systems, as substantiated by research [17]. When provided with an arrangement C and an MDG, EVM evaluates and scores each cluster by considering the number of intra-relationships in the MDG. To prevent any potential confusion, we establish the definition of EVM as the aggregate of individual cluster scores, denoted as SubEVM (refer to Eqs. 2 and 3). EVM aims to maximise the score of relationships within a specified clustering arrangement. However, a potential drawback exists, as EVM may mistakenly assign high scores to clustering arrangements with high cohesion. Even minor adjustments to a solution can significantly enhance its fitness.

$$EVM(C, MDG) = \sum_{i=1}^{k} SubEVM(C_i, MDG) \qquad (2)$$

$$SubEVM(C_i, MDG) = \sum_{a=1}^{|C_i|-1} \sum_{b=a+1}^{|C_i|} (2M(C_{ia}, C_{ib}) - 1) \qquad (3)$$

To enhance the efficiency of Munch, we incorporate Arzoky et al.'s *EVMD*. This method generates a score aligning with *EVM* by integrating past *EVM* outcomes and determining the new result based on the classes designated for exchange. It demonstrates computational efficiency by computing the new fitness before implementing any modifications. Throughout this paper, we choose to utilise *EVM* as a collective term, encompassing both *EVM* and *EVMD*, to prevent potential confusion in future discussions about *EVM*.

Concerning the mentioned modifications, the inclusion of *EVMD* enables the execution of "Try/Do Moves." This variant of Small-Change, involving the random mutation of clustering arrangements, reduces computational overhead by initially testing the result of a small change (Try Move) before actual implementation (Do Move). To effectively utilise *EVMD*, the small-change process is limited to two elements simultaneously.

Finally, our focus shifts to the heuristic search. As mentioned earlier, Arzoky et al.'s Munch primarily employ RMHC as its heuristic search method. Despite implementing the ability to alter the heuristic search in our Munch, we opt to persist with RMHC. This choice is motivated by its reliability, ease of implementation, and superior performance compared to stochastic heuristics, such as SAHC. Below, we present Algorithm 1, elucidating how Munch searches for enhanced clustering arrangements. For practical reasons, we choose to employ EVM in the pseudocode example, even though we leverage Arzoky et al.'s EVMD fitness function to enhance performance:

Algorithm 1. Munch

1: **function** MUNCH(*Iterations, MDG*)
2:     Let C be a clustering arrangement   ▷ Random or Seed Starting Point
3:     Let $F = \text{EVM}(C, MDG)$ ▷ Current Fitness
4:     **for** $i = 1$ to *Iterations* **do**
5:         Let $C' = C$ ▷ Copy of C
6:         Choose two random clusters X and Y $(X \neq Y)$ from C, ▷ Move Operator
7:         Move a random variable from cluster X to Y in C' ▷ Move Operator
8:         Let $F' = \text{EVM}(C', MDG)$ ▷ New Fitness
9:         **if** $F' \geq F$ **then**   ▷ Compare Fitness
10:             Let $C = C', F = F'$ ▷ Continue using best Solution
11:         **end if**
12:     **end for**
13:     **return** C ▷ Output is C
14: **end function**

4.2   Graph Partitioning

So far, we have established the importance of graphs and clustering arrangements regarding software modularisation. Now, we focus on using the structure graphs to discover new clustering arrangement starting points. Specifically, our focus shifts to the Fiedler Vector [12]. This vector is linked to the second smallest eigenvalue, the Fiedler Eigenvalue, of a Laplacian Matrix [13]. Denoted as $L_{n \times n}$, a Laplacian Matrix is defined as $L = D - A$ where D represents the degree matrix of A which represents the connections between nodes [9]. In this context, $A \equiv MDG$. The Fiedler Vector is distinctive in its capability to enable a nearly perfect binary split of any given matrix. With this characteristic in mind, we have developed a tool that generates starting points through the recursive decomposition of graphs until no more Fiedler Vectors can be produced.

We generate a tree structure to facilitate the recursive decomposition of input graphs. The root of the tree is our input software graph and clustering arrangement. The clustering arrangement must begin with all nodes placed in a single cluster. This initial cluster will be subsequently split alongside the graph, ultimately leading to our final clustering arrangement, representing a fully decomposed software graph.

Leveraging our understanding of the Fiedler Vector, we identify the Fiedler eigenvalue of its attributed graph at each tree node, deduce its associated eigenvectors, and establish a well-balanced, binary-split graph partition. Simultaneously, we split the associated cluster with each partition, ensuring a one-to-one relationship between the subgraph's nodes and the associated cluster concerning the root MDG. This approach allows us to maintain traceability as we proceed with the decomposition. The new branches that emerge from the root node are reintroduced into a recursive function that continues to iterate until it identifies all possible partitions.

4.3   Starting Points

After generating a tree, we have two starting point approaches. Algorithm 2 illustrates the initial method for creating a "Leaf" arrangement. We gathered all leaf nodes from the tree, identified by their lack of children. Subsequently, we arrange these leaf nodes in ascending order based on SubEVM (see Eq. 3) and then incorporate nodes with unique clusters into our clustering arrangement. We organise all leaf nodes in ascending order to prevent branches from becoming disconnected at different depths, possibly leading to duplicate values. In an ideal scenario, all leaf nodes, regardless of depth, should be unique, and therefore, we incorporate this logic for peace of mind.

Algorithm 3 exemplifies our alternative approach to constructing a clustering arrangement. In this method, we recursively traverse the tree, evaluating the cohesion of each node in comparison to its children. This ensures the creation of a clustering arrangement containing all unique values, emphasising the highest possible cohesion within the context of the MDG for a given tree. We refer to this starting point as our "Max" arrangement.

Apart from Leaf and Max, our modified version of Munch can generate clustering arrangements randomly distributed uniformly, denoted as "Random." Due to publication constraints, we abstain from delving into the intricacies of this method. In summary, a "uniformly distributed random" arrangement is defined by a clustering setup generated through the utilisation of Bell Numbers, Stirling

---

Algorithm 2. BuildLeaf

1: **function** BUILDLEAF(*root*)
2:     Let *leafNodes* be a stack of all leaves of root
3:     Sort *leafNodes* in descending order of SubEVM   ▷ see Eq. 2
4:     Let *C* be empty clustering arrangement
5:     **while** *leafNodes* is not empty **do**
6:         Let *node* be popped *leafNodes*   ▷ Pop top *node* from stack
7:         **if** *node* is not in *C* **then**
8:             Let *C* = [*C* [*node*]]   ▷ Add unique cluster (*node*) to arrangement (*C*)
9:         **end if**
10:    **end while**
11:    **return** *C*   ▷ Output is *C*
12: **end function**

Algorithm 3. BuildMax

1: **function** BUILDMAX(*root*)   ▷ Start Recursion
2:     Let *C* = empty clustering arrangement
3:     Let *C* = POPULATEARRANGEMENT(*root, C*)
4:     **return** *C*   ▷ Output is *C*
5: **end function**
6: **function** POPULATEARRANGEMENT(*node, C*)   ▷ Recursive Function
7:     **if** *node has children* **then**
8:         Let *Fp* be SubEVM of parent *node*   ▷ See Eq. 2
9:         Let *Fc* be sum SubEVM of *children*   ▷ See Eq. 2
10:        **if** *Fp* > *Fc* **then**   ▷ If the SubEVM of parent *node* is greater
11:            Let *C* = [*C* [*node*]]   ▷ Add node to arrangement *C*
12:        **else**   ▷ Continue Recursion
13:            Let POPULATEARRANGEMENT(*left child of node, C*)
14:            Let POPULATEARRANGEMENT(*right child of node, C*)
15:        **end if**
16:    **else**
17:        Let *C* = [*C* [*node*]]   ▷ Add leaf *node* to arrangement *C*
18:    **end if**
19:    **return** *C*   ▷ Output is *C*
20: **end function**

Numbers of the Second Kind [20,33,36], and their interconnected relationships [11].

5   Experimental Setup

Before presenting the Munch results of our graph partitioning tool, we need to establish an empirical framework.

5.1   Graph Collection and Pre-processing

First, we must collect software systems. Throughout our research, we developed a specialised tool that extracts open-source software systems using GitHub's RESTful API [14]. GitHub is our platform of choice for several compelling reasons. With a substantial user base exceeding 94 million developers, a continuously growing number of 52 million open-source repositories, and a cumulative total of 413 million contributions [15], we have access to a wide and diverse range of graphs.

Collecting and forming these graphs is often neglected in academic literature, creating a challenge in determining the authenticity of these systems - whether they are genuinely open-source, artificially generated, or specific to certain industries. Generating MDGs requires understanding the relationships between each class within a given software system. This can be achieved using software metric tools such as SciTools Understand [31], which provide pairwise relationships to build a symmetric graph. After our extractor downloads the desired software system, we manually process each system using SciTools Understand. Future efforts will explore using GitHub's TreeSitter parsing system [6] to automatically generate MDGs.

In this experiment series, we collect 50 "Small" open-source MDGs with class counts from 100 to 300, chosen based on relevance and high popularity ("stars") using the GitHub API. Due to storage constraints and the laborious manual MDG creation, we aim to develop an automated MDG generator, contemplating additional storage allocation pending study outcomes. Additionally, we have five "Big" MDGs (class counts: 1000 to 1500) sourced from prior research and industry collaboration, allowing exploration of size and characteristic-based result variations. Refer to Appendix A for a detailed breakdown. The terms "Small 50" and "Big 5" distinguish the two MDG groups in this paper.

5.2   Experiment Setup

Our experiments are described as follows. First, we collect Munch results for each MDG using all starting point combinations and iterations, as outlined below. Secondly, we collect Gold Standard results involving high-iteration/high-fitness outcomes to compare with our initial experiments. Finally, we analyse the results and present our findings concerning our outlined Research Questions.

1. For each experiment, for every graph (Small 50 and Big 5):
   (a) Select one of three starting points (Leaf, Max, Random)
   (b) Select one of three iterations (10k, 100k, 1m)
   (c) Run Munch
   (d) Document final iteration statistics and associated clustering arrangement
   (e) Repeat Steps a-c 250 times
2. Repeat Step 1 until all starting points and iterations are explored.

For our gold standards, we generate a Random starting point for each graph and run Munch for 100 million iterations, collecting the same information as in our initial experiments. We repeat the process 250 times to ensure that we compare our initial experiment clustering arrangements to an absolute-best gold standard. Although conducting more iterations would have been preferable, it was impractical due to the extended runtime, taking several days per graph. To streamline experimental runs with our chosen iteration increments, we implemented parallel thread management, allowing multiple instances of Munch to run concurrently while optimising CPU and memory usage.

5.3   Data Collection and Analysis

We gather data on the fitness scores of the ultimate clustering configurations, pinpoint the convergence point (the last iteration demonstrating improved fitness), and gauge the runtime. Furthermore, we document the final clustering configurations into text files. Employing these files, we have crafted a bespoke tool to methodically evaluate the WK between the ultimate configurations derived from our initial points in contrast to our gold standards.

We have compiled a dataset of 275,000 files, combining the initial experiment results and gold standards. To enhance the manageability of these results for analysis, we employ MS Access, MS Excel, and Python for data processing. Due to the extensive volume of results and constraints in page space, our principal methodology involves computing averages across all data. Additionally, we streamline our findings by identifying and formatting the optimal results, providing a count of these instances per starting point type, thereby highlighting the suitability of each.

6   Results

Initially, we present RMHC results for each starting point category across selected iterations. Our research evaluates the performance of diverse starting points in searches across graphs of varying sizes, considering fitness, convergence, and runtime. The goal is to identify similarities or disparities in these aspects based on our predefined research questions. When implemented on large and small software systems, the performance differences among Leaf, Max, and Random are apparent in Tables 1[3] and 2[4]. Max consistently demonstrates superior fitness across iterations, as evidenced by the average final fitness values obtained from our three starting points over the specified iterations.

Table 3[5] details the average final convergence statistics across iterations, indicating the iteration where improvement was observed. A strong resemblance between the average fitness and convergence strongly implies a correlation, potentially indicating a basin of attraction where all solutions converge. Compared to Random in Tables 1 and 2, Leaf and Max achieve final fitness levels more rapidly across all iterations, notably enhancing results. For smaller graphs,

---

[3] Values formatted bold in Table 1 signify the highest average final fitness.

[4] Values formatted bold in Table 2 signify the highest average final fitness.

[5] Values formatted bold in Table 3 signify the shortest average convergence point.

Table 1. Average Final Fitness using Starting Point by Iterations

Size	10k			100k			1m		
	Leaf	Max	Random	Leaf	Max	Random	Leaf	Max	Random
102	60.664	**60.972**	60.516	72.660	**73.032**	72.544	73.248	73.236	**72.984**
105	**56.060**	56.016	56.052	**68.024**	67.696	67.844	**69.356**	69.192	69.344
112	88.616	**90.116**	87.532	99.512	**100.000**	99.920	101.040	**101.268**	101.108
119	66.968	**68.096**	66.204	80.100	**80.224**	80.168	80.960	**81.048**	80.900
123	99.516	**102.200**	98.068	**115.932**	115.688	114.896	117.808	**117.908**	117.660
127	82.340	**89.820**	79.452	105.680	**106.072**	105.272	107.216	**107.556**	107.040
129	81.796	**86.764**	80.500	98.544	**99.196**	98.848	99.800	99.856	**100.072**
130	37.612	**37.828**	35.592	53.016	53.064	**53.204**	53.832	53.660	**53.968**
135	97.748	**105.836**	96.864	120.684	**120.796**	120.636	122.200	122.052	**122.228**
135	101.480	**114.520**	100.760	133.832	**134.052**	133.908	**135.168**	134.920	135.000
141	74.428	**77.876**	71.392	88.224	88.552	**88.832**	89.832	**89.952**	89.928
145	**51.880**	51.856	47.848	**65.812**	65.624	65.672	**66.468**	66.380	66.372
151	105.948	**115.840**	102.016	**137.616**	137.008	137.224	139.276	139.132	**139.324**
158	84.080	**90.924**	76.424	112.664	112.532	**112.828**	115.260	115.292	**115.368**
161	85.580	**90.444**	82.204	**115.968**	114.972	115.748	118.912	118.848	**119.232**
163	83.980	**92.048**	73.400	110.248	**110.640**	109.728	**112.864**	112.780	112.192
164	79.768	**80.568**	75.164	105.464	**105.940**	105.408	107.884	107.928	**108.020**
169	42.464	**45.196**	35.432	73.156	**73.528**	73.212	75.316	**75.424**	75.372
172	86.924	**94.396**	82.732	123.580	122.596	**124.684**	128.760	128.228	**129.076**
172	100.636	**116.444**	91.548	137.196	**137.936**	137.168	**141.620**	141.372	140.928
172	15.804	**16.088**	11.796	**41.640**	41.488	41.372	44.280	44.388	**44.464**
175	80.768	**84.536**	71.776	**125.328**	125.196	125.108	133.864	133.884	**133.948**
175	108.704	**125.236**	101.228	164.620	**166.920**	164.512	171.248	**171.840**	171.032
176	82.360	**96.328**	72.364	112.776	112.844	**113.164**	115.760	**115.764**	115.688
179	101.700	**123.148**	93.196	148.364	148.160	**148.752**	151.900	151.336	**151.940**
198	98.008	**114.484**	84.572	139.196	**140.164**	139.192	144.092	144.276	**144.552**
199	90.328	**99.060**	83.392	153.600	150.504	**153.812**	166.760	**167.100**	166.664
200	91.704	**104.004**	70.864	128.124	**128.336**	127.912	131.672	131.268	**131.832**
206	113.880	**137.076**	103.772	182.220	**184.900**	181.952	191.008	**191.616**	190.964
208	62.528	**69.752**	52.000	127.376	**127.772**	127.512	130.656	**130.760**	130.724
209	82.896	**90.680**	59.672	131.196	**131.596**	129.776	137.768	**138.024**	137.244
210	65.216	**77.376**	48.576	102.340	**103.320**	101.980	106.336	**106.576**	106.404
210	76.324	**88.716**	54.560	116.984	**117.444**	116.688	118.224	**118.272**	118.060
211	66.624	**73.748**	44.256	113.068	113.320	**113.408**	120.872	120.880	**122.092**
214	118.440	**132.280**	100.060	187.688	**188.472**	187.412	198.188	**198.980**	198.084
216	43.364	**52.080**	47.336	120.720	**121.368**	120.732	126.764	126.764	**126.780**
224	110.880	**140.856**	89.708	184.596	**187.608**	184.308	197.920	**198.212**	198.144
233	92.680	**120.196**	75.740	173.608	**175.156**	172.776	**184.992**	184.808	184.600
234	89.564	**100.176**	57.732	143.928	**144.152**	142.660	**150.776**	150.756	150.652
234	111.308	**127.900**	84.244	184.216	**186.116**	183.916	**196.420**	196.260	195.996
235	93.872	**108.148**	65.808	158.052	**160.152**	158.588	172.260	**173.060**	172.596
240	82.872	**95.224**	54.924	148.992	**151.108**	147.784	159.104	159.128	**159.236**
240	126.100	**143.636**	98.224	**214.848**	213.208	213.488	234.844	**235.044**	234.472
242	108.404	**139.252**	76.388	177.708	**179.356**	177.092	**186.904**	187.304	187.280
246	72.876	**80.716**	41.404	139.548	**139.996**	139.340	148.820	148.524	**148.876**
252	99.484	**110.880**	61.124	159.868	**161.400**	159.000	174.656	**175.284**	174.616
252	134.832	**171.012**	108.420	243.296	**254.296**	242.448	279.192	**281.980**	278.472
258	123.152	**157.012**	94.904	208.428	**208.740**	206.764	**227.840**	225.928	227.144
264	114.188	**168.672**	72.400	227.132	**229.244**	226.552	239.488	238.788	**239.720**
294	131.708	**182.384**	75.388	251.476	**255.352**	249.916	276.912	276.692	**276.972**
Count	2	**48**	0	8	**34**	8	10	**20**	**20**

Table 2. Average Final Fitness using Starting Point by Iterations

Size	10k			100k			1m		
	Leaf	Max	Random	Leaf	Max	Random	Leaf	Max	Random
1037	170.924	**329.764**	−1360.344	345.940	**471.344**	123.476	731.892	**755.984**	737.256
1164	163.252	**276.048**	−1853.428	286.364	**376.868**	−188.728	588.528	**606.000**	589.680
1311	87.880	**225.908**	−2152.008	284.280	**400.460**	−297.580	681.848	**719.780**	675.820
1440	124.664	**283.652**	−2553.976	351.456	**479.476**	−520.688	741.544	**789.288**	708.468
1441	54.932	**146.112**	−2648.656	230.028	**304.932**	−718.304	543.324	**571.064**	493.336
Count	0	**5**	0	0	**5**	0	0	**5**	0

Table 3. Convergence Statistics

Iterations	Starting Point	Small 50			Big 5			Count
		Min	Max	Avg	Min	Max	Avg	
10k	Leaf	8970.444	9889.524	9560.283	9670.992	9887.296	9816.322	0
	Max	**8892.732**	**9873.904**	**9430.464**	**9634.588**	9876.404	9799.476	6
	Random	9129.948	9917.304	9710.809	9971.996	9977.348	9974.834	0
100k	Leaf	**53996.848**	96390.908	80512.020	98831.280	99120.060	98980.680	1
	Max	54158.648	**95036.248**	**79139.873**	98552.824	99091.176	98786.134	5
	Random	54598.832	96333.324	80817.637	99769.868	99858.020	99828.650	0
1m	Leaf	84590.520	**691299.112**	375566.554	989895.556	993509.896	991965.960	1
	Max	**77164.980**	706030.088	**368672.253**	**987032.192**	**991801.380**	**989805.248**	5
	Random	85098.216	692030.048	377192.918	990564.152	994979.076	993397.874	0

convergence is reached well before the considered iterations. Although final iterations align for smaller graphs, more iterations could enhance the likelihood of reaching local optima in larger datasets.

In contrast to average final fitness and convergence, Table 4[6] highlights cumulative average runtimes presented for each start and subsequent search at various iterations, measured in milliseconds. Notably, these reported runtimes represent summed average runtimes, excluding additional computational overhead related to data I/O. While Random clustering allows faster processing, the overall statistical significance of runtimes is debatable. This prompts consideration of potential trade-offs between runtime efficiency and solution quality.

Table 5[7] displays WK results, juxtaposing clustering configurations resulting from our initial starting points against gold standards. In multiple statistics and iterations, Leaf and Max consistently surpass Random. The notable closeness between Max results and their corresponding gold standards in smaller graphs contrasts the generally low agreement observed for larger graphs.

---

[6] Values formatted bold in Table 4 signify the shortest runtime in milliseconds.
[7] Values formatted bold in Table 5 signify the highest Weighted Kappa agreement.

Table 4. Average sum of runtime in milliseconds

Iterations	Small 50			Big 5		
	Leaf	Max	Random	Leaf	Max	Random
10k	298.797	332.881	**185.662**	24.231	**24.180**	199.044
100k	792.818	833.359	**408.770**	**119.465**	120.744	259.044
1m	5389.751	5607.426	**2574.516**	996.988	1012.166	**673.654**
Count	0	0	**3**	**1**	**1**	**1**

Table 5. *WK* against Gold Standard Statistics

Iterations	Starting Point	Small 50				Big 5				Count
		Min	Max	Avg	StDev	Min	Max	Avg	StDev	
10k	Leaf	0.215	0.565	0.386	**0.092**	0.002	0.002	0.002	**0.000**	2
	Max	**0.241**	**0.609**	**0.440**	0.094	**0.040**	**0.156**	**0.093**	0.043	6
	Random	0.198	0.580	0.358	0.102	0.010	0.027	0.015	0.007	0
100k	Leaf	0.398	**0.847**	0.637	**0.097**	0.012	0.021	0.016	**0.005**	3
	Max	0.376	0.835	**0.643**	0.099	**0.059**	**0.176**	**0.117**	0.044	4
	Random	**0.402**	0.843	0.635	0.097	0.058	0.107	0.077	0.022	1
1m	Leaf	0.490	0.897	0.715	0.084	0.128	0.317	0.198	0.079	0
	Max	0.480	**0.904**	**0.715**	0.086	**0.211**	0.359	0.264	**0.060**	4
	Random	**0.497**	0.897	0.715	**0.083**	0.197	**0.386**	**0.267**	0.084	4

7   Summary of Main Findings

In summary, we aimed to show that graph-partitioning can generate starting points capable of improving the results of software modularisation. We encapsulate the findings to address the research inquiries in the following manner:

- Max starting point:
  - Attains the highest average fitness over 10k, 100k, and 1m iterations, with a pronounced emphasis on lower iteration counts.
  - Attains the highest count of average convergence across all iterations while sustaining the optimal average final fitness.
  - Attains the maximum average agreement (WK) with gold standards across 10k and 100k for both Small 50 and Big 5 graphs, highlighting noteworthy performance, especially in lower iterations.
- Leaf starting point:
  - Demonstrate fitness levels equal to or surpassing Random across all iterations, especially in the early stages
  - Surpasses Random with higher average fitness levels on large datasets at 10k and 100k iterations
  - Consistently exhibits faster convergence compared to Random.
- Random starting point:

- Shows a quicker average total runtime in milliseconds compared to Leaf and Max.
- Better suited for smaller datasets; however, an improvement over Max and Leaf necessitates higher iterations.
- Demonstrates greater resemblance to the gold standard than Max in larger systems at 1m iterations.

Distinct fitness variations emerge among Leaf, Max, and Random, with Max consistently outperforming over 10k, 100k, and 1m iterations, notably in Small 50 vs. Big 5 comparisons. Random outperforms Max and Leaf at 1 million iterations for large datasets. However, Max proves to be more suitable for average fitness and faster convergence across iterations and graph sizes. Since there are currently no guidelines for determining the number of iterations based on the size or properties of an MDG, the most prudent approach would be to initiate seeding with Max before executing Munch. WK comparisons show Max starting points yield higher average agreements, with potential improvements around 70% and significant opportunities at 90% agreement in 1m iterations. Thorough exploration is vital for understanding software system graph intricacies. Our commitment to accelerating software modularisation drives deeper exploration, with partition-based clustering performing significantly, especially at smaller iterations, making it compelling for future software optimisation.

8   Generalisability

This publication focuses on utilising graph partitioning for software modularisation. However, the application of graph partitioning for optimising initial positions can extend to other graph-based optimisation problems, contingent on the chosen fitness function. Although we prioritise EVM for its simplicity, other alternatives like MQ are viable. Our aim is to inspire exploration of graph partitioning for seeded optimisation.

9   Future Work

We plan to integrate our graph-based initial clustering with metaheuristics, specifically incorporating seeded starting points into the history of Iterated Local Search, as part of our ongoing investigation [28]. This initiative seeks to evaluate the potential improvement in the exploration of the search space and overall efficiency. Furthermore, our goals include delving deeper into software systems' search space, exploring graph structure, convergence prediction, and other avenues for enhancing software modularisation.

10   Appendix A

This Appendix showcases details about the software system MDG used in our experiments. Below, we showcase the following statistics for each software system:

1. ID
   - Each software system is assigned a unique identifier. We choose not to use the actual names of our software systems because our collection is sourced randomly from GitHub. These software system names can exhibit variation, and we intend to maintain professionalism and steer clear of potentially inappropriate names and software tools.
2. Nodes
   - Also known as vertices, these signify the number of software components (classes) within our Module Dependency Graphs (MDGs).
3. Edges
   - Denotes the number of relationships between software components.
4. Clustering Coefficient:
   - The extent to which nodes tend to cluster. A high score indicates a strong cohesion, while a low score indicates a higher coupling level. We present this statistic as these software systems exhibit remarkably low coefficients, indicating a high coupling level and a deficiency in the initial modular structure. There is potential here to investigate the nature of software structure over time, especially concerning the analysis of open-source software systems (Tables 6 and 7).

Table 6. "Big 5" Software MDG Statistics

Identification	Nodes	Edges	Avg Degree	Clustering Coefficient
1	1037	5470	5.274	0.000
2	1164	2072	1.780	0.000
3	1311	5630	4.294	0.000
4	1440	4889	3.395	0.000
5	1441	3058	2.122	0.000

Table 7. "Small 50" Open-Source Software MDG Statistics

Identification	Nodes	Edges	Avg Degree	Clustering Coefficient
01	102	312	0.061	0.007
02	105	257	0.047	0.002
03	112	436	0.070	0.007
04	119	343	0.049	0.002
05	123	440	0.059	0.004
06	127	409	0.051	0.004
07	129	357	0.043	0.002
08	130	225	0.027	0.001
09	135	387	0.043	0.002
10	135	510	0.056	0.004
11	141	333	0.034	0.001
12	145	274	0.026	0.001
13	151	595	0.053	0.002
14	158	422	0.034	0.001
15	161	413	0.032	0.001
16	163	414	0.031	0.001
17	164	686	0.051	0.002
18	169	387	0.027	0.001
19	172	609	0.041	0.002
20	172	591	0.040	0.002
21	172	357	0.024	0.000
22	175	737	0.048	0.004
23	175	749	0.049	0.003
24	176	371	0.024	0.000
25	179	467	0.029	0.001
26	198	552	0.028	0.001
27	199	1002	0.051	0.003
28	200	450	0.023	0.000
29	206	964	0.046	0.003
30	208	643	0.030	0.001
31	209	732	0.034	0.001
32	210	518	0.024	0.000
33	210	323	0.015	0.000
34	211	599	0.027	0.001
35	214	834	0.037	0.002
36	216	740	0.032	0.001
37	224	937	0.038	0.002
38	233	818	0.030	0.001
39	234	521	0.019	0.000
40	234	930	0.034	0.002
41	235	823	0.030	0.001
42	240	898	0.031	0.001
43	240	1115	0.039	0.002
44	242	836	0.029	0.001
45	246	672	0.022	0.001
46	252	1033	0.033	0.001
47	252	1591	0.050	0.004
48	258	1477	0.045	0.002
49	264	730	0.021	0.001
50	294	1275	0.030	0.001

References

1. Altman, D.: Skewed distributions. Practical statistics for medical research, pp. 60–63. Chapman & Hall, London (1997)
2. Arasteh, B.: Clustered design-model generation from a program source code using chaos-based metaheuristic algorithms. Neural Comput. Appl. **35**(4), 3283–3305 (2023)
3. Arasteh, B., Seyyedabbasi, A., Rasheed, J.M., Abu-Mahfouz, A.: Program source-code re-modularization using a discretized and modified sand cat swarm optimization algorithm. Symmetry **15**(2), 401 (2023)
4. Arzoky, M., Swift, S., Tucker, A., Cain, J.: Munch: an efficient modularisation strategy to assess the degree of refactoring on sequential source code checkings. In: 2011 IEEE Fourth International Conference on Software Testing, Verification and Validation Workshops, pp. 422–429. IEEE (2011)
5. Arzoky, M., Swift, S., Tucker, A., Cain, J.: A seeded search for the modularisation of sequential software versions. J. Object Technol. **11**(2), 1–6 (2012)
6. Brunsfeld, M.: Tree-sitter. https://github.com/tree-sitter/tree-sitter (Accessed 1 Jan 2023)
7. Campbell, L.R., et al.: Restricted growth function patterns and statistics. Adv. Appl. Math. **100**, 1–42 (2018)
8. Chen, Y.T., Huang, C.Y., Yang, T.H.: Using multi-pattern clustering methods to improve software maintenance quality. IET Softw. **17**(1), 1–22 (2023)
9. Chung, F.R.: Spectral graph theory, vol. 92. American Mathematical Soc. (1997)
10. Corradini, A., König, B., Nolte, D.: Specifying graph languages with type graphs. J. Logical Algebraic Methods Programm/ **104**, 176–200 (2019). https://doi.org/10.1016/j.jlamp.2019.01.005, https://www.sciencedirect.com/science/article/pii/S235222081730233X
11. Devroye, L.: Sample-based non-uniform random variate generation. In: Proceedings of the 18th Conference on Winter Simulation, pp. 260–265 (1986)
12. Fiedler, M.: A property of eigenvectors of nonnegative symmetric matrices and its application to graph theory. Czechoslov. Math. J. **25**(4), 619–633 (1975)
13. Fiedler, M.: Laplacian of graphs and algebraic connectivity. Banach Center Publ. **1**(25), 57–70 (1989)
14. GitHub: Github advanced search (2023). https://github.com/search/advanced, (Accessed 1 Jan 23)
15. GitHub: Octoverse 2022: 10 years of tracking open source (2023). https://github.blog/2022-11-17-octoverse-2022-10-years-of-tracking-open-source/, (Accessed 1 Jan 23)
16. Gupta, N., Kumar, S., Gupta, V., Vijh, S.: Novel automatic approach using modified differential evaluation to software module clustering problem. SN Comput. Sci. **4**(6), 816 (2023)
17. Harman, M., Swift, S., Mahdavi, K., Beyer, H.: An empirical study of the robustness of two module clustering fitness functions, In: genetic and Evolutionary Computation Conference; Conference date: 25–06-2005 Through 29–06-2005, pp. 1029–1036. Assoc Computing Machinery (2005),
18. Kang, Y., Xie, W., Wang, X., Wang, H., Wang, X., Li, J.: Mopisde: a collaborative multi-objective information-sharing de algorithm for software clustering. Expert Syst. Appli., 120207 (2023)
19. Khan, M.Z., et al.: A novel approach to automate complex software modularization using a fact extraction system. J. Mathem. 2022 (2022)

20. Harper, L.H.: Stirling behaviour is asymptotically normal. Annals Math. Stat. **3**(2), 410–414 (1967)

21. Kramer, O.: Iterated local search. In: A Brief Introduction to Continuous Evolutionary Optimization. SAST, pp. 45–54. Springer, Cham (2014). https://doi.org/10.1007/978-3-319-03422-5_5

22. Lu, K.: Practical program modularization with type-based dependence analysis. In: 2023 IEEE Symposium on Security and Privacy (SP), pp. 1256–1270. IEEE (2023)

23. Mancoridis, S., Mitchell, B.S., Chen, Y., Gansner, E.R.: Bunch: a clustering tool for the recovery and maintenance of software system structures. In: Proceedings IEEE International Conference on Software Maintenance-1999 (ICSM 1999). Software Maintenance for Business Change'(Cat. No. 99CB36360), pp. 50–59. IEEE (1999)

24. Mancoridis, S., Mitchell, B.S., Rorres, C., Chen, Y., Gansner, E.R.: Using automatic clustering to produce high-level system organizations of source code. In: Proceedings. 6th International Workshop on Program Comprehension, IWPC 1998 (Cat. No. 98TB100242), pp. 45–52. IEEE (1998)

25. Maramazi, F., Odebode, A., Mann, A., Swift, S., Arzoky, M.: Intelligent systems and applications.In: Proceedings of the 2024 Intelligent Systems Conference (Intellisys) vol. 1. LNNS, vol. 822, pp. 470. Springer (2024)

26. Mitchell, B.S., Mancoridis, S.: Clustering module dependency graphs of software systems using the bunch tool. Nat. Sci. Found., Alexandria, VA, USA, Tech. Rep (1998)

27. Prajapati, A., Parashar, A., Rathee, A.: Multi-dimensional information-driven many-objective software remodularization approach. Front. Comp. Sci. **17**(3), 173209 (2023)

28. Ramalhinho-Lourenço, H., Martin, O.C., Stützle, T.: Iterated local search (2000)

29. Rand, W.M.: Objective criteria for the evaluation of clustering methods. J. Am. Stat. Assoc. **66**(336), 846–850 (1971)

30. Savić, M., Rakić, G., Budimac, Z., Ivanović, M.: A language-independent approach to the extraction of dependencies between source code entities. Inf. Softw. Technol. **56**(10), 1268–1288 (2014)

31. SciTools: Understand: The software developer's multi-tool (2023). https://scitools.com/, (Accessed 10 Nov 2023)

32. Tan, A.J.J., Chong, C.Y., Aleti, A.: Closing the loop for software remodularisation-rearrange: an effort estimation approach for software clustering-based remodularisation. arXiv preprint arXiv:2303.06283 (2023)

33. Temme, N.M.: Asymptotic estimates of stirling numbers. Stud. Appl. Math. **89**(3), 233–243 (1993)

34. Tucker, A., Swift, S., Liu, X.: Variable grouping in multivariate time series via correlation. IEEE Trans. Syst. Man Cybernet. Part B (Cybernet). **31**(2), 235–245 (2001)

35. Weiss, K., Banse, C.: A language-independent analysis platform for source code. arXiv preprint arXiv:2203.08424 (2022)

36. Weisstein, E.W.: Stirling number of the second kind (2002). https://mathworld.wolfram.com/

37. Yang, K., Wang, J., Fang, Z., Wu, P., Song, Z.: Enhancing software modularization via semantic outliers filtration and label propagation. Inf. Softw. Technol. **145**, 106818 (2022)

A Novel Two-Level Clustering-Based Differential Evolution Algorithm for Training Neural Networks

Seyed Jalaleddin Mousavirad[1]([✉]), Diego Oliva[2], Gerald Schaefer[3],
Mahshid Helali Moghadam[4], and Mohammed El-Abd[5]

[1] Mid Sweden University, Sundsvall, Sweden
jalalmousavirad@gmail.com
[2] Depto. de Ingeniería Electro -Fotónica, Universidad de Guadalajara, CUCEI,
Guadalajara, Mexico
[3] Department of Computer Science, Loughborough University, Loughborough, UK
[4] Mälardalens University, Västerås, Sweden
[5] College of Engineering and Applied Sciences, American University of Kuwait,
Salmiya, Kuwait

Abstract. Determining appropriate weights and biases for feed-forward neural networks is a critical task. Despite the prevalence of gradient-based methods for training, these approaches suffer from sensitivity to initial values and susceptibility to local optima. To address these challenges, we introduce a novel two-level clustering-based differential evolution approach, C2L-DE, to identify the initial seed for a gradient-based algorithm. In the initial phase, clustering is employed to detect some regions in the search space. Population updates are then executed based on the information available within each region. A new central point is proposed in the subsequent phase, leveraging cluster centres for incorporation into the population. Our C2L-DE algorithm is compared against several recent DE-based neural network training algorithms, and is shown to yield favourable performance.

Keywords: Differential evolution · clustering · neural network training · regularisation

1 Introduction

Feed-forward neural networks (FFNNs) are a widely adopted artificial neural network (ANN) architecture employed in diverse classification and regression problems [11,27]. Comprising basic components known as neurons and connections linking them, FFNNs allow the flow of information from the input layer through hidden layers, ultimately reaching the output layer. Each connection is characterised by a weight that signifies its strength. The training process in FFNNs aims to determine optimal weights that minimise the error between actual and predicted outputs. Although gradient-based approaches such as the

S. Smith et al. (Eds.): EvoApplications 2024, LNCS 14634, pp. 259–272, 2024.
https://doi.org/10.1007/978-3-031-56852-7_17

back-propagation (BP) algorithm are prevalent, they tend towards local optima and thus provide sub-optimal results [30].

Population-centric metaheuristic (PCM) algorithms, such as differential evolution (DE) [40] and particle swarm optimisation (PSO) [38], provide a useful alternative to address the challenges encountered by traditional algorithms. Evolutionary algorithms (EAs) are group of PCMs that has been widely applied in the training of FFNNs. [37] compares (BP) with a genetic algorithm (GA) for FFNN training, concluding that the latter excels in terms of effectiveness. [13] uses a modified GA for rapidly training FFNNs, demonstrating superior efficiency compared to conventional GA-based training algorithms. [7] proposes a hybrid approach combining GA and BP for determining the weights in FFNNs, outperforming both GA and BP individually.

Swarm intelligence algorithms form another group of PCMs. [34] combines PSO with the Levenberg-Marquardt (LM) algorithm to achieve faster convergence. [39] introduces an opposition PSO-based training method and evaluates it on various clinical datasets. [31] proposes a comprehensive learning strategy integrated with PSO and LM as a local search algorithm for neural network training. Various other PCM algorithms have been applied for FFNN training, including the imperialist competitive algorithm (ICA) [8,22], the firefly algorithm (FA) [17], the grey wolf optimiser (GWO) [2,19], and Lévy flight distribution [3,35], among others.

Differential evolution is a well-established PCM renowned for its outstanding performance in addressing complex optimisation problems [5,10,20,21]. It comprises three primary operators: mutation, crossover, and selection. Mutation facilitates the exchange of information among different individuals, crossover integrates a mutant vector with a target vector, and selection chooses superior individuals from old and new individuals into a new population.

DE has also been widely employed for FFNN training. [12] introduces a DE-based training algorithm, showcasing its ability to outperform gradient-based methods. [28] incorporates opposition-based learning into DE, demonstrating good performance across various classification problems. [32] employs an improved DE algorithm that incorporates opposition-based learning and a region-based strategy, while [36] proposes a centroid-based differential evolution algorithm with composite trial vector generation strategies and control parameters to optimise the weights and biases in FFNNs. In [24], a clustering-based DE approach for neural network training is employed.

In a recent enhancement to DE, [29] introduces a methodology involving centre-based sampling at the population level of DE, with the centre of the entire population incorporated as a new individual. Integrating the centre point is shown to effectively guide the population towards improved individuals. On the other hand, [4] indicates that cluster centres in a population are viable candidates in the search space to move towards. Building upon these two concepts, in this paper, we propose a novel two-level clustering-based differential evolution algorithm, C2L-DE, for training FFNNs. At the first level, the clustering algorithm works like a multi-parent crossover to update the population. In contrast, at the second level, the central point of population clustering is injected as a new individual into the population.

The main characteristics of C2L-DE are:

- a clustering strategy is employed at the first level to update the population;
- clustering is used to introduce a new individual into the population at the second level;
- a regularisation term is incorporated into the objective function to enhance generalisation;
- the weights and biases determined by C2L-DE are fed into the Levenberg-Marquardt algorithm as the initial seed.

The remainder of the paper is organised as follows: Section 2 gives an overview of some essential concepts. Section 3 presents our proposed approach, detailing the fundamental components of C2L-DE and explaining its overall structure. In Sect. 4, the performance of C2L-DE is assessed across various benchmark problems, while Sect. 5 concludes the paper.

2   Background

2.1   Differential Evolution

Differential evolution (DE) [40] is a straightforward yet highly effective PCM algorithm widely recognised for excellent performance in addressing complex optimisation problems [10,41]. DE begins with N_P individuals randomly generated from a uniform distribution. To update the population, three primary operators are employed: mutation, crossover, and selection.

The mutation operator produces a mutant vector, $v_i = (v_{i,1}, v_{i,2}, ..., v_{i,D})$, defined as

$$v_i = x_{r1} + F(x_{r2} - x_{r3}),\tag{1}$$

where x_{r1}, x_{r2}, and x_{r3} are three distinct randomly chosen individuals from the current population, and F represents a scale factor.

Crossover is responsible for incorporating the mutant vector into the target vector. For binomial crossover, this is performed as

$$u_{i,j} = \begin{cases} v_{i,j} & \text{if rand}(0,1) \leq CR \text{ or } j == j_{rand} \\ x_{i,j} & \text{otherwise} \end{cases},\tag{2}$$

where CR denotes the crossover rate, j_{rand} is a random number ranging from 1 to N_P, and $i = 1, ..., NP$, $j = 1, ..., D$.

Selection identifies the superior individual from the trial and target vectors, ensuring the progression of more promising solutions in the population.

The iterative process enhances the algorithm's ability to effectively explore and exploit the search space.

2.2   Pattern Clustering

The fundamental aim of clustering is to arrange a collection of patterns so that the members within each group share more similarities than those in different groups. Mathematically, clustering involves defining a set P consisting of N d-dimensional patterns, denoted as $P = \{p_1, p_2, \cdots, p_N\}$. The k-means algorithm [16] is the most widely adopted clustering algorithm and proceeds in the following steps:

1. Randomly initialise the cluster centres;
2. In the allocation step, assign each pattern to its nearest cluster centre (e.g., using Euclidean distance);
3. In the update step, recalculate the position of each cluster centre as the centroid of its assigned patterns;
4. Repeat steps 2 and 3 until convergence or a predefined stopping criterion is met.

2.3   Feed-Forward Neural Networks

FFNNs, a widely employed class of ANNs, are trained in a supervised manner to handle pattern recognition problems [1,33]. The typical architecture of an FFNN consists of three types of layers: an input layer, one or more hidden layers, and an output layer. Each node in these layers incorporates an activation function that defines how the weighted sum of inputs transforms into the output. The connections between layers are assigned weights, indicating the strength between the respective nodes. Weights are critical for FFNN performance, making determining suitable weight values one of the most vital and challenging aspects of FFNNs. Among various approaches, gradient descent-based methods form the most widely adopted technique for this training process.

3   Proposed C2L-DE Algorithm

Our proposed C2L-DE algorithm leverages clustering at two distinct levels. At the first level, specific individuals are substituted with cluster centres, while at the second level, the central point of a cluster centre is introduced as a new individual into the population. Additionally, our proposed algorithm incorporates a regularisation-based objective function to enhance the generalisation capabilities of the algorithm.

3.1   First-Level Clustering

At the first level, C2L-DE employs a clustering algorithm to construct areas in search space using the k-means algorithm. Determining the number of clusters is accomplished by selecting a random number within the range of 2 to $\sqrt{N_P}$. The resulting cluster centres are analogous to a multi-parent crossover, representing the cumulative solutions within a cluster.

C2L-DE's population update strategy involves adopting a generic population-based algorithm (GPBA) [6]. This approach aligns with a GPBA methodology and encompasses the folloingw steps:

- **Diversity selection:** individuals are randomly chosen from the current population, mirroring the initialisation of points in the k-means algorithm;
- **Clustered generation:** k-means is applied to generate m individuals (set A). Each cluster centre determined through this process corresponds to a new individual;
- **Individual substitution:** from the current population, m individuals (set B) are (randomly) selected for substitution;
- **Elite update:** the best m individuals from the combined set $A \cup B$ are selected as $\bar{B}$, and the new population is formed as $(P - B) \cup \bar{B}$.

This population update procedure integrates elements from clustering algorithms and population-based strategies, ensuring an effective and dynamic approach in C2L-DE.

It is important to note that C2L-DE does not employ the clustering algorithm in each iteration. Instead, following [5, 32], clustering is applied periodically based on a clustering period.

3.2 Second-Level Clustering

DE-centre-p [29] is a centre-based DE algorithm where an individual, determined by the central point defined as the centre of the N best individuals, is introduced as a new member of the population. The population is then divided into two parts, one a set of individuals that undergo positional updates through standard mutation and crossover operations, and one that is an individual exclusively devoted to preserving the centre of the N best individuals. On the other hand, [4] suggests that cluster centres within a population represent promising candidates in search space, in particular for directional movement. Consequently, at the second level of our proposed clustering scheme, we introduce a novel approach for incorporating a new individual into the population based on cluster centres.

Following the initial clustering phase, the N most promising areas are identified using a one-step k-means algorithm. The value of N is not fixed and is randomly chosen between 2 and $\sqrt{N_P}$. Cluster centres serve as representatives for each cluster. Subsequently, the central point of these cluster centres is selected as a new individual, obtained as

$$\overrightarrow{x_{centre}} = \frac{\overrightarrow{x_{c1}} + ... \overrightarrow{x_{ci}} + ... + \overrightarrow{x_{cN}}}{N}, \tag{3}$$

where x_{ci} is the i-th cluster centre. While DE-centre-p injects this individual into the population with a fixed location, in C2L-DE we dynamically select this location based on the objective function. In other words, this new individual replaces the worst individual and endeavours to substitute the least favourable solution with the central point derived from several promising candidates within the population. Figure 1 illustrates the process creating a new individual based on the central point of cluster centres.

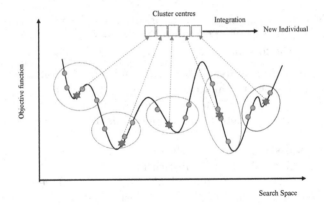

Fig. 1. Clustering at the second level. Circle-shaped points show individuals in the population, while star-shapesd points indicate cluster centres.

3.3 Encoding Strategy

Our approach uses a real-valued encoding scheme to represent individuals. Each solution is described by a vector comprising connection weights and bias values. The encoding length directly correlates with the problem's complexity, reflecting the total number of connection weights and biases that require optimisation.

3.4 Objective Function

We use an objective function for FFNN training that incorporates a regularisation term and is calculated as

$$f = \frac{100}{P} \sum_{p=1}^{P} \xi(x_p) + \frac{\lambda}{2m} \sum ||W||^2, \tag{4}$$

with

$$\xi(\vec{p}) = \begin{cases} 1 & \text{if } \vec{o_p} \neq \vec{d_p} \\ 0 & \text{otherwise} \end{cases}, \tag{5}$$

where d_p and o_p are the actual and predicted outputs, respectively, and m is the total number of samples. The regularisation parameter, λ, serves as a hyper-parameter, penalising large values of weights and biases. If λ is excessively large, numerous weights will approach zero, simplifying the FFNN and making it prone to underfitting. Conversely, if λ is too small, the regularisation term's influence diminishes. An optimal choice of λ is crucial as it helps control the weights, preventing overfitting while maintaining the model's performance.

3.5 Levenberg-Marquardt Algorithm

We use the weights obtained by C2L-DE as an initial seed to the Levenberg-Marquardt (LM) algorithm [15,18]. LM aims to optimise the objective function by adjusting the network weights using an update rule defined as

$$w_{t+1} = w_t - (J_t^T J_t + \mu I)^{-1} J_t^k E_t, \tag{6}$$

with

$$E_t = \sum_{i=1}^{N} (d_i - y_i)^2, \tag{7}$$

where J is the Jacobian matrix of the error vector E_t, J^T is its transpose, I is the identity matrix with dimensions matching the Hessian $J^p J$, N is the number of training samples, and μ is a damping factor adjusted during the optimisation process. $J^k E$ indicates the gradient of the error function E.

It is worth noting that the LM algorithm converges faster compared to other algorithms, such as BP or back-propagation with momentum [9, 14].

3.6   C2L-DE Algorithm

Algorithm 1 presents our proposed C2L-DE algorithm in pseudo-code form. C2L-DE first creates an initial population and evaluates the objective function of each individual. The algorithm then iteratively performs mutation, crossover, and selection operations. Periodically, it undergoes the two levels of clustering. At the first level, based on the clustering period (C_P), a combination of k-means clustering and random selection is employed to update the population, while at the second level, k-means is employed to identify cluster centres and create a new individual as the average of these centres. This new individual then replaces the worst individual in the population. The algorithm iterates until the maximum number of function evaluations is reached. It is worth noting that we utilise a one-step k-means algorithm due to its O(1) complexity, ensuring no change in the overall complexity.

Upon completion, the best individual, $\vec{x}^*$, is identified. If the maximum number of function evaluations is surpassed, the algorithm proceeds to the secondary phase. It initialises ω as the best individual and resets the iteration count. It then iteratively computes the Jacobian, the approximated Hessian, and the error, updating the weights using the Levenberg-Marquardt algorithm. This process continues until a specified maximum number of iterations is reached.

4   Experimental Results

To evaluate the effectiveness of the proposed C2L-DE algorithm, we conduct a set of experiments on diverse datasets from the UCI machine learning repository[1], namely:

- *Iris*: a well-known classification dataset with 150 samples, 4 features, and 3 classes;

---

[1] https://archive.ics.uci.edu/ml/index.php.

Algorithm 1: C2L-DE algorithm

---

1 **Initialisation:**;
2 Initialise N_{pop}, $NFE_{\max}$, $iter_{\max}$, J_r, C_P, λ;
3 $NFE = 0$, $iter = 1$;

4 **while** $NFE \leq NFE_{max}$ **do**
5   Generate initial population Pop using uniformly distributed random numbers;
6   Calculate objective function of each individual in Pop using Eq. (4);
7       $NFE = N_{\text{pop}}$;
8   **foreach** *individual* **do**
9   Perform mutation operation;
10   Perform crossover operator;
11   Calculate objective function using Eq. (4);
12   Perform selection operation;
13   **end**
14      $NFE = NFE + N_{\text{pop}}$;
        // First-level Clustering
15   **if** $rem(iter, C_P) == 0$ **then**
16   Randomly generate k as a random number between 2 and $\sqrt{N_P}$;
17   Conduct a single step of k-means clustering and designate the cluster centres as set A;
18   Randomly pick k individuals from the current population and designate them as set B;
19   From the union of sets A and B, select the best k individuals and denote them as $\bar{B}$;
20   Choose the new population as $(Pop - B) \cup \bar{B}$;
21   **end**
        // Second-level Clustering
22   Randomly generate k as a random number between 2 and $\sqrt{N_P}$;
23   Conduct a single step of k-means clustering;
24   Select N cluster centre solutions as $\vec{x_{c1}}, \vec{x_{c2}}, ..., \vec{x_{cN}}$;
25      $x_{new} = \frac{\vec{x_{c1}} + \vec{x_{c2}} + ... + \vec{x_{cN}}}{N}$;
26      $x_{worst} \leftarrow x_{new}$;
27 **end**
28 $\vec{x^*} \leftarrow$ the best individual in pop
29 $iter = iter + 1$;
30 **if** $NFE > NFE_{max}$ **then**
31   Initialise ω as $\vec{x^*}$ (i.e. the best individual in the current population);
32   Set the current iteration $iter$ to 0;
33   **while** $iter < iter_{max}$ **do**
34   Compute the Jacobian J, the approximated Hessian $J^T J$, and the error E_t;
35   Update weights using Eq. (7);
36   Recalculate E_t;
37   **if** $iter < iter_{max}$ **then**
38   Increment $iter$ by 1;
39   **end**
40   **end**
41 **end**

- *Breast Cancer*: comprising 699 samples, 9 features, and 2 classes;
- *Liver*: a binary clinical dataset from BUPA Medical Research Ltd., with 345 instances and 7 features;
- *Pima*: a challenging clinical classification dataset featuring 768 samples, 2 classes, and 8 features;
- *Seed*: an agricultural dataset with seven geometrical features of wheat kernels, containing 210 samples divided into 3 categories.
- *Vertebral*: A clinical dataset incorporating biomechanical features, categorized into 3 classes with 310 samples.

Here we do not focus on determining the optimal FFNN architecture, but adopt the approach from [23,25], setting the number of neurons in the single hidden layer to $2 \times N + 1$, where N is the number of inputs. For evaluation, we employ 10-fold cross-validation.

C2L-DE is benchmarked against a number of state-of-the-art and recently proposed DE-based trainers, including standard DE, QODE, RDE-OP, Reg-IDE, and Cen-CODE. The number of function evaluations for all PCMs is fixed at 25,000 [26]. The population size for all PCMs is set to 50. For C2L-DE, the crossover probability, scaling factor, and jumping rate are set to 0.9, 0.5, and 0.3, respectively, and the clustering period and regularisation parameter are also chosen as 10 and 0.1, respectively. For the remaining algorithms, we use the default parameters as per the cited publications.

The obtained results on the Iris dataset, presented in Table 1, reveal valuable insights into the performance of different DE algorithms. Our proposed C2L-DE algorithm stands out prominently, achieving the joint highest mean fitness value of 99.33 (along with Reg-IDE), showcasing the effectiveness of C2L-DE in converging towards optimal solutions. In addition, the low standard deviation of 2.10 indicates the robustness of C2L-DE across multiple runs. In contrast, standard DE and other comparative algorithms such as QODE and Cen-CODE exhibit lower mean fitness values and higher standard deviations.

Table 1. Experimental results on Iris dataset.

	mean	std.dev	rank
DE	92.00	5.26	6
QODE	95.33	6.32	5
RDE-OP	96.67	6.48	4
Reg-IDE	99.33	2.11	1.5
Cen-CODE	98.00	3.22	3
C2L-DE	99.33	2.10	1.5

The results on the Breast Cancer dataset are given in Table 2. From there, we can see that all algorithms except DE provide a similar mean accuracy. C2L-DE

Table 2. Experimental results on Breast Cancer dataset.

	mean	std.dev	rank
DE	97.36	2.06	6
QODE	98.10	0.99	5
RDE-OP	98.82	1.67	1
Reg-IDE	98.39	2.24	3
Cen-CODE	98.38	1.61	4
C2L-DE	98.53	1.64	2

is second ranked with a mean fitness value of 98.53, demonstrating its competitive performance.

Table 3 shows the results on the Liver dataset. C2L-DE is top ranked with a mean fitness value of 77.64, highlighting its superior performance. QODE and Reg-IDE also exhibit competitive mean fitness values of 76.82 and 76.26, respectively, resulting in the second and third ranks.

Table 3. Experimental results on Liver dataset.

	mean	std.dev	rank
DE	67.81	8.21	6
QODE	76.82	9.46	2
RDE-OP	75.63	6.45	4
Reg-IDE	76.26	4.03	3
Cen-CODE	75.10	6.66	5
C2L-DE	77.64	5.83	1

Table 4 presents the results on Pima dataset. C2L-DE is again top ranked here, with a mean fitness of 81.50. Reg-IDE is second ranked with a mean fitness value of 80.60, followed by RDE-OP (80.21) and QODE (79.55).

The results on the Seed dataset, given in Table 5, also show our C2L-DE algorithms as the top-performing approach, achieving a mean accuracy of 93.80. Cen-CODE follows with a mean accuracy of 82.38, while algorithms like DE, QODE, and RDE-OP perform less effectively.

The experimental results on the Vertebral dataset, reported in Table 6, reveal QODE as the top-performing algorithm. However, QODE generally does not achieve satisfactory results on the other datasets. C2L-DE follows closely with a mean accuracy of 87.74, while DE and Cen-CODE exhibit lower performance.

The obtained results across multiple datasets clearly demonstrate the superior performance of our proposed C2L-DE algorithm compared to the other methods, while also proving it to be a robust method.

Table 4. Experimental results on Pima dataset.

	mean	std.dev	rank
DE	76.94	4.97	6
QODE	79.55	4.94	4
RDE-OP	80.21	5.73	3
Reg-IDE	80.60	4.15	2
Cen-CODE	77.99	4.12	5
C2L-DE	81.50	5.34	1

Table 5. Experimental results on Seed dataset.

	mean	std.dev	rank
DE	70.00	11.01	4
QODE	67.62	3.01	5
RDE-OP	67.62	4.92	5
Reg-IDE	80.60	4.15	3
Cen-CODE	82.38	8.1	2
C2L-DE	93.80	5.96	1

Table 6. Experimental results on Vertebral dataset.

	mean	std.dev	rank
DE	85.16	5.31	5.5
QODE	88.39	8.76	1
RDE-OP	86.77	4.42	3.5
Reg-IDE	86.77	5.37	3.5
Cen-CODE	85.16	6.48	5.5
C2L-DE	87.74	6.23	2

5 Conclusions

In this paper, we have presented the C2L-DE algorithm as a novel effective
solution for the complex task of determining optimal weights and biases in
feed-forward neural networks. Traditional gradient-based methods, while widely
employed, encounter challenges such as sensitivity to the initial values and susceptibility to local optima. Our two-level clustering-based differential evolution
approach addresses these issues by introducing a dynamic and informed population update strategy. In the initial phase, clustering identifies diverse regions
within the search space, guiding population updates based on localised information. Subsequently, a central point derived from cluster centres, is introduced

as a new individual into the population. A comparative analysis against several recent DE training algorithms confirms the promising performance of C2L-DE.

In future work, we intend to extend the application of our algorithm to other ANN-related tasks, such as neural architecture search. Additionally, C2L-DE holds potential for hyperparameter optimisation, showcasing its versatility and adaptability in various aspects of neural network optimisation.

References

1. Abrishami, N., Sepaskhah, A.R., Shahrokhnia, M.H.: Estimating wheat and maize daily evapotranspiration using artificial neural network. Theor. Appl. Climatol. **135**(3), 945–958 (2018). https://doi.org/10.1007/s00704-018-2418-4
2. Amirsadri, S., Mousavirad, S.J., Ebrahimpour-Komleh, H.: A levy flight-based grey wolf optimizer combined with back-propagation algorithm for neural network training. Neural Comput. Appl. **30**(12), 3707–3720 (2017). https://doi.org/10.1007/s00521-017-2952-5
3. Bojnordi, E., Mousavirad, S.J., Pedram, M., et al.: Improving the generalisation ability of neural networks using a Lévy flight distribution algorithm for classification problems. New Gener. Comput. **41**(2), 225–242 (2023). https://doi.org/10.1007/s00354-023-00214-5
4. Bojnordi,E., Mousavirad,S.J., Schaefer, G., Korovin, I.: MCS-HMS: a multi-cluster selection strategy for the human mental search algorithm. In: IEEE Symposium Series on Computational Intelligence, pp. 1–6, 2021
5. Cai, Z., Gong, W., Ling, C.X., Zhang, H.: A clustering-based differential evolution for global optimization. Appl. Soft Comput. **11**(1), 1363–1379 (2011)
6. Deb, K.: A population-based algorithm-generator for real-parameter optimization. Soft. Comput. **9**(4), 236–253 (2005)
7. Ding, S., Chunyang, S., Junzhao, Yu.: An optimizing BP neural network algorithm based on genetic algorithm. Artif. Intell. Rev. **36**(2), 153–162 (2011)
8. Duan, H., Huang, L.: Imperialist competitive algorithm optimized artificial neural networks for UCAV global path planning. Neurocomputing **125**, 166–171 (2014)
9. El-Bakry, M.Y., El-Dahshan, E.-S.A., El-Hamied, E.F.A.: Charged particle pseudo-rapidity distributions for Pb-Pb and Au-Au collisions using neural network model. Ukrainian J. Phys. **58**(8), 709–709 (2013)
10. Fister, I., Fister, D., Deb, S., Mlakar, U., Brest, J.: Post hoc analysis of sport performance with differential evolution. Neural Comput. Appl. **32**, 1–10 (2018)
11. Hosaka, T.: Bankruptcy prediction using imaged financial ratios and convolutional neural networks. Expert Syst. Appl. **117**, 287–299 (2019)
12. Ilonen, J., Kamarainen, J.-K., Lampinen, J.: Differential evolution training algorithm for feed-forward neural networks. Neural Process. Lett. **17**(1), 93–105 (2003)
13. Kim, D., Kim, H., Chung, D.: A modified genetic algorithm for fast training neural networks. In: International Symposium on Neural Networks, pp. 660–665 (2005)
14. Lera, G., Pinzolas, M.: Neighborhood based Levenberg-Marquardt algorithm for neural network training. IEEE Trans. Neural Networks **13**(5), 1200–1203 (2002)
15. Levenberg, K.: A method for the solution of certain non-linear problems in least squares. Q. Appl. Math. **2**(2), 164–168 (1944)
16. MacQueen, J.: Some methods for classification and analysis of multivariate observations. In: 5th Berkeley Symposium on Mathematical Statistics and Probability, pp. 281–297 (1967)

17. Mandal, S., Saha, G., Pal, R.K.: Neural network training using firefly algorithm. Glob. J. Adv. Eng. Sci. **1**(1), 7–11 (2015)
18. Marquardt, D.W.: An algorithm for least-squares estimation of nonlinear parameters. J. Soc. Ind. Appl. Math. **11**(2), 431–441 (1963)
19. Mirjalili, S.: How effective is the Grey Wolf optimizer in training multi-layer perceptrons. Appl. Intell. **43**(1), 150–161 (2015). https://doi.org/10.1007/s10489-014-0645-7
20. Moravvej, S.V., Mousavirad, S.J., Oliva, D., Schaefer, G., Sobhaninia, Z.: An improved DE algorithm to optimise the learning process of a BERT-based plagiarism detection model. In: IEEE Congress on Evolutionary Computation, pp. 1–7 (2022)
21. Mousavirad, S.J., Bidgoli, A.A., Rahnamayan, S.: Tackling deceptive optimization problems using opposition-based DE with center-based Latin hypercube initialization. In: 14th International Conference on Computer Science and Education (2019)
22. Mousavirad, S.J., Bidgoli, A.A., Ebrahimpour-Komleh, H., G.S.: A memetic imperialist competitive algorithm with chaotic maps for multi-layer neural network training. Int. J. Bio-Inspired Comput. **14**(4), 227–236 (2019)
23. Mousavirad, S.J., Bidgoli, A.A., Ebrahimpour-Komleh, H., Schaefer, G., Korovin, I.: An effective hybrid approach for optimising the learning process of multi-layer neural networks. In: International Symposium on Neural Networks, pp. 309–317 (2019)
24. Mousavirad, S.J., Gandomi, A.H., Homayoun, H.: A clustering-based differential evolution boosted by a regularisation-based objective function and a local refinement for neural network training. In: IEEE Congress on Evolutionary Computation, pp. 1–8 (2022)
25. Mousavirad, S.J., Jalali, S.M.J., Sajad, A., Abbas, K., Schaefer, G., Nahavandi, S.: Neural network training using a biogeography-based learning strategy. In: International Conference on Neural Information Processing (2020)
26. Mousavirad, S.J., Oliva, D., Hinojosa, S., Schaefer, G.: Differential evolution-based neural network training incorporating a centroid-based strategy and dynamic opposition-based learning. In: IEEE Congress on Evolutionary Computation, pp. 1233–1240 (2021)
27. Mousavirad, S.J., Rahmani, R., Dolatabadi, N.: A transfer learning based artificial neural network in geometrical design of textured surfaces for tribological applications. Surf. Topogr. Metrol. Prop. **11**(2), 025001 (2023)
28. Mousavirad, S.J., Rahnamayan, S.: Evolving feedforward neural networks using a quasi-opposition-based differential evolution for data classification. In: IEEE Symposium Series on Computational Intelligence (2020)
29. Mousavirad, S.J., Rahnamayan, S.: A novel center-based differential evolution algorithm. In: Congress on Evolutionary Computation (2020)
30. Mousavirad, S.J., Schaefer, G., Jalali, S.M.J., Korovin, I.: A benchmark of recent population-based metaheuristic algorithms for multi-layer neural network training. In: Genetic and Evolutionary Computation Conference Companion, pp. 1402–1408 (2020)
31. Mousavirad, S.J., Schaefer, G., Korovin, I.: An effective approach for neural network training based on comprehensive learning. In: International Conference on Pattern Recognition (2020)
32. Mousavirad, S.J., Schaefer, G., Korovin, I., Oliva, D.: RDE-OP: a region-based differential evolution algorithm incorporation opposition-based learning for optimising the learning process of multi-layer neural networks. In: Castillo, P.A., Jiménez

Laredo, J.L. (eds.) EvoApplications 2021. LNCS, vol. 12694, pp. 407–420. Springer, Cham (2021). https://doi.org/10.1007/978-3-030-72699-7_26

33. Munkhdalai, L., Lee, J.Y., Ryu, K.H.: A hybrid credit scoring model using neural networks and logistic regression. In: Pan, J.-S., Li, J., Tsai, P.-W., Jain, L.C. (eds.) Advances in Intelligent Information Hiding and Multimedia Signal Processing. SIST, vol. 156, pp. 251–258. Springer, Singapore (2020). https://doi.org/10.1007/978-981-13-9714-1_27

34. Nawi, N.M., khan, A., Rehman, M.Z., Aziz, M.A., Herawan, T., Abawajy, J.H.: An accelerated particle swarm optimization based Levenberg Marquardt back propagation algorithm. In: Loo, C.K., Yap, K.S., Wong, K.W., Teoh, A., Huang, K. (eds.) ICONIP 2014. LNCS, vol. 8835, pp. 245–253. Springer, Cham (2014). https://doi.org/10.1007/978-3-319-12640-1_30

35. Pedram, M., Mousavirad, S.J., Schaefer, G.: Training neural networks with Lévy flight distribution algorithm. In: 7th International Conference on Harmony Search, Soft Computing and Applications, pp. 93–103 (2022)

36. Rahmani, S., Mousavirad, S.J., El-Abd, M., Schaefer, G., Oliva, D.: Centroid-based differential evolution with composite trial vector generation strategies for neural network training. In: Correia, J., Smith, S., Qaddoura, R. (eds.) International Conference on the Applications of Evolutionary Computation, vol. 13989, pp. 608–622. Springer, Cham (2023). https://doi.org/10.1007/978-3-031-30229-9_39

37. Sexton, R.S., Gupta, J.N.D.: Comparative evaluation of genetic algorithm and backpropagation for training neural networks. Inform. Sci. **129**(1–4), 45–59 (2000)

38. Shi, Y., Eberhart, R.: A modified particle swarm optimizer. In: IEEE International Conference on Evolutionary Computation, pp. 69–73 (1998)

39. Si, T., Dutta, R.: Partial opposition-based particle swarm optimizer in artificial neural network training for medical data classification. Int. J. Inform. Technol. Decis. Making **18**(5), 1717–1750 (2019)

40. Storn, R., Price, K.: Differential evolution-a simple and efficient heuristic for global optimization over continuous spaces. J. Global Optim. **11**(4), 341–359 (1997)

41. Wang, X., et al.: Massive expansion and differential evolution of small heat shock proteins with wheat (triticum aestivum l.) polyploidization. Sci. Rep. **7**(1), 1–12 (2017)

Iterated Beam Search for Wildland Fire Suppression

Gustavo Delazeri$^{(\boxtimes)}$ and Marcus Ritt

Universidade Federal do Rio Grande do Sul, Porto Alegre, Brazil
{gustavo.delazeri,marcus.ritt}@inf.ufrgs.br

Abstract. Wildfires cause significant damage costs globally, and it is likely that they are becoming more damaging due to climate change. Here we study methods for fire suppression, after a breakout of fire. In our model, we have a grid graph $G = (V, A)$ that represents the discretization of a terrain into cells and an ignition node $s \in V$ from which the fire spreads to other nodes. The spread of the fire is defined by the arc weights, which can be used to model important factors such as wind direction and vegetation type. At various points in time, one or more fire suppression resources become available to be applied to nodes in the graph that are not yet burned. Applying a resource to a node $v \in V$ adds a delay to the outgoing edges of v, which causes a local slowdown in fire propagation. The goal is to find an allocation of resources to the nodes of the graph such that the total burned area at a target time is minimized. In this work, we propose a heuristic algorithm based on beam search to tackle this problem. Our computational experiments show that our approach is able to consistently find the optimal solution to almost all instances used in literature, but in considerably less time than previous approaches.

Keywords: wildfire suppression · heuristic search · beam search

1 Introduction

Wildfires are estimated to have caused global damage costs of about USD 69 billion in 2018–2023 [8, 12]. Their frequency and damage are likely to increase with climate change, with longer wildfire seasons, larger affected areas, and new locations of occurrence. They are at the same time harder to handle, since they coincide more frequently with dry air [6, 14]. Although deaths from wildfires are rare in comparison to other natural disasters, they destroy ecosystems, threaten homes, livelihoods, technical infrastructure such as railways and the electricity grid, and lead to a reversal of carbon capture [6, 13]. An increased frequency of wildfires demands a comprehensive and urgent response, and governments around the world already are investing in wildfire research with the goal of understanding its causes and how damages can be mitigated [3].

According to [10], the operations research community has been studying wildfire management since the early 1960s, and [7] is one of the first works dealing

S. Smith et al. (Eds.): EvoApplications 2024, LNCS 14634, pp. 273–286, 2024.
https://doi.org/10.1007/978-3-031-56852-7_18

with the application of operations research techniques to forest fire problems. Since then, we can find in the literature a variety of mathematical models that aim to capture decisions related to the process of preventing and suppressing a wildfire, such as the coordination of fire crews, the deployment of aerial fire-fighting assets and the routing of vehicles to transport firefighters and other equipment. To give some examples, in 1995 [5] proposed the firefighter problem, which is defined on a graph where fire spreads from an ignition node to adjacent nodes in sequential time steps. At each time instant, a certain number of fire suppression resources is available and can be deployed to unburned nodes. Applying a resource to a node prevents the fire from spreading through its outgoing edges to adjacent nodes, and the goal is to stop the fire in the minimum amount of time steps. [2] proposed a more realistic mixed-integer linear programming model that integrates fire spread behavior and the placement of suppression resources. The landscape is represented by a graph and the model comprises control variables, to decide which nodes will receive fire suppression resources, and response variables, which define fire spread paths, fire arrival times, and fire intensity for all the nodes. The goal is to minimize the total value of the burned area together with operational costs.

In this work, we consider a problem first defined in [1] and [11]. Similarly to [2], we are given a graph representing a landscape, an ignition node and some fire suppression resources spread over time. The goal is to allocate the fire resources to the nodes in order to minimize the burned area at some target time instant. In this context, our main contribution is a heuristic algorithm based on iterated beam search that achieves better results than previous approaches in a fraction of the time.

To close this section, we give an overview of what follows. In Sect. 2, we formally define the problem. Section 3 goes over the algorithmic approaches to this problem that can be found in the literature. Section 4 provides a series of definitions that will be used to explain our algorithm, which is presented in Sect. 5. In Sect. 6, we conduct some computational experiments to study the performance of our algorithm. Section 7 concludes the work and proposes new research directions.

2   Problem Description

Fire propagation is modeled by a directed graph $G = (V, A)$ with travel times t_a on arcs $a \in A$, which model the time required for fire to propagate from a node to a neighboring node. A directed graph permits to model different fire travel times in opposite directions, which can occur due to factors like wind and terrain slope. Given an ignition node $s \in V$, the travel times define a shortest-path tree rooted at s in which each node $v \in V$ has an associated fire arrival time a_v and a predecessor p_v. Now assume we have k fire suppression resources which can be allocated to nodes $v \in V$, and each resource adds a delay Δ to the outgoing arcs of v. Each resource $i \in [k]$ is available at time r_i, and can only be allocated to a node v if $a_v \geq r_i$, i.e. if v is not burned yet[1]. We also assume that each node can receive at most one resource. Finally, we have a time horizon H and are interested in nodes that do not burn until H.

The allocation of resources to nodes can be represented by an injective function $\Lambda : [k] \rightarrow V$. By definition, such an allocation changes the travel times t, but it can also change the topology and the arrival times of the shortest-path tree. As a result, given an allocation of resources Λ, we denote the resulting fire propagation times by t^Λ, the fire arrival times by a^Λ, and the predecessor relation by p^Λ. The problem, then, is to find a feasible allocation of resources Λ that minimizes the number of burned nodes at time instant H, i.e.

$$b = \sum_{v \in V} [a_v^\Lambda \leq H].$$

3   Related Work

The problem we are interested in was first proposed as a mixed-integer linear programming (MIP) model by [1]. In [11], the authors propose a set of representative instances for this model and an iterated local search to solve them. The authors compare the performance of the local search with the performance of a commercial solver on the mathematical formulation of [1]. In computational experiments, they show that the heuristic achieves good results in a reasonable amount of time for all instances, while the solver needs more time to produce results and, for some large instances, fails to produce a feasible solution within the time limit of 2 h.

[4] extend the work of [1] and [11] by proposing a better MIP formulation of the problem, an exact algorithm using logic-based Benders decomposition, and a simple greedy heuristic used to warm-start the exact algorithm. In computational experiments, they show that the exact algorithm and a commercial solver using the new MIP model can solve all the instances proposed by [11] in a few seconds. In light of that, they propose new instances consisting of 20×20 grids and a larger optimization horizon. In another round of computational experiments, they compare the performance of the solver, the iterated local search of [11] and the proposed exact algorithm considering a time limit of 2 h. The solver was not able to prove the optimality of any instance, failed to produce a feasible solution in some cases and had the overall worst performance regarding solution quality. The iterated local search was able to find the optimal solution of some instances, but in most of the time it stayed behind the exact algorithm, which was able to find and prove the optimality of all instances.

4   Preliminaries

Consider a grid graph $G = (V, A)$. The immediate neighborhood of a node $v \in V$, denoted as $\mathbb{N}(v)$, encompasses nodes reachable through outgoing arcs of v. Sim-

---

[1] We use $[n]$ to denote a set containing the first n natural numbers, i.e. $[n] = \{1, \ldots, n\}$.

ilarly, the extended neighborhood $N^*(v)$ includes nodes reachable via outgoing arcs as well as diagonal connections from v. Time instants where resources become available are represented by a sequence of times $T = (t_1, t_2, \dots)$, in ascending order. We denote by $\alpha(t) = \min_{i > 0 | t_i > t} t_i$ the first time instant after t when new resources become available, with $\alpha(t) = H$ if no further resources become available after t. Finally, for each time instant $t \in [0, H]$, $R_t \subseteq [k]$ is a set containing the resources that become available at time t.

When an allocation of resources Λ assigns a resource to a node $v \in V$, we say that v is *protected*. We denote by $P^\Lambda \subseteq V$ the set of nodes protected by Λ. If $|P^\Lambda| = k$ we say that Λ is a *complete allocation*. Conversely, if $|P^\Lambda| < k$ we say that Λ is *partial*. The special allocation that does not protect any node is denoted by Λ_0, i.e. $P^{\Lambda_0} = \emptyset$. Finally, given an allocation Λ and a time instant $t \in [0, H]$, we define $B_t^\Lambda = \{v \in V \mid a_v^\Lambda < t\}$ as the set of nodes that are burned at t. Note that our goal is to find an allocation Λ such that $|B_H^\Lambda|$ is minimized.

5 Proposed Algorithm

Beam search is a graph search algorithm that visits nodes in a breadth-first manner until a target node is reached. Starting from the root node, beam search keeps a list of β nodes and, at each level of the search tree, nodes in the list are expanded η times. In the literature, β is known as the beam width and η as the ramification factor. A heuristic function is then used to rank the $\beta\eta$ expansions, and the best β nodes are selected to continue to the next iteration. Beam search has been extensively used to tackle optimization problems [9]. In the context of our problem, each interior node of the search tree represents a partial allocation of resources, and leaf nodes are complete allocations. The root node is Λ_0, and for each $t \in T$, we expand the current set of allocations by applying the resources in R_t. The best leaf node is returned by the algorithm.

5.1 Beam Search

Algorithm 1 gives a high level view of our approach. In line 1, we create a set $\mathcal{A}$ containing only Λ_0, which will represent the current state of the search tree. For each time instant $t \in T$, we use the function Step to expand each node in $\mathcal{A}$, and we store all the expansions of the current level in the set E. In line 4, we use a heuristic function to select the β best allocations to continue to the next iteration, and in line 6 we return the best leaf node in the search tree. We explain how to expand a given allocation in Sect. 5.2. We will next define the heuristic function used to prune the search tree.

We propose two heuristic functions to evaluate a partial allocation of resources Λ. The first one, which we call h_1, is equal to the number of burned nodes at time instant H.

$$h_1(\Lambda) = \sum_{v \in V} [a_v^\Lambda \leq H]$$

Algorithm 1: BeamSearch

Data: Fire perimeter size z.

Result: An allocation of resources Λ.

1   $\mathcal{A} \leftarrow \{\Lambda_0\}$

2   **for** $t \in T$ **do**

3      $E \leftarrow \bigcup_{\Lambda \in \mathcal{A}} \text{Step}(\Lambda, t, z)$

4      $\mathcal{A} \leftarrow \text{prune}(E, t)$

5   **end**

6   **return** $\arg\min_{\Lambda \in \mathcal{A}} |B_H^\Lambda|$

Heuristic h_1 can be quite uninformative in the first few time instants, especially when the delay Δ is low and the optimization horizon H is large. In such situations, it is likely that the first few resources available cannot save any nodes, hence a comparison between two allocations is uninformative. In light of that, we propose a second heuristic, called h_2, which aims to measure how much delay an allocation Λ introduces in the network.

$$h_2(\Lambda) = \sum_{v \in V} \max\{H - a_v^\Lambda, 0\}$$

As we will see in the experimental section, we can obtain better results by starting with h_2 as the guiding heuristic and then switching to h_1 at some point in time. We call the time instant at which we start using h_1 *transition instant*, and we denote it by $\hat{t}$. It is better to define the transition instant relative to the velocity with which the fire propagates. To this end, we define the *free burning time* of an instance as the time instant at which the last node is burned assuming that no node is protected by a resource, i.e. the free burning time equals $\max_{v \in V} a_v^{\Lambda_0}$.

We can now specify the transition instant as a percentage of the free burning time, and we denote this percentage by $\hat{p}$. To give an example, if the free burning time of an instance is 10 and $\hat{p} = 0.5$, we have that the transition instant $\hat{t}$ is equal to 5.

In summary, if $t < \hat{t}$ we prune the search tree by selecting the β best partial allocations in E using h_2. If $t \geq \hat{t}$, we use h_1. In Sect. 6.3, we study how the transition instant affects the performance of our algorithm.

5.2   Expanding an Allocation of Resources

We now consider the problem of generating the expansions of a given allocation in the search tree, as is done in line 3 of Algorithm 1. As a first step, we will develop a procedure to create a single expansion (Algorithm 2) and later we will embed it into Algorithm 3, which implements the function Step, called in line 3 of Algorithm 1.

Algorithm 2: Expand

Data: A partial allocation of resources Λ, a time instant t, the fire
perimeter size z.

Result: An expansion of Λ using the resources in R_t.

1 $F \leftarrow F_t^\Lambda(z)$

2 $N \leftarrow F \cap \bigcup\limits_{v \in P^\Lambda} N^*(v)$

3 **for** $i \in R_t$ **do**

4   **if** $N \neq \emptyset$ and $p < \text{rand}(0, 1)$ **then**

5 | $v \leftarrow$ Randomly pick an element of N

6   **else**

7 | $v \leftarrow$ Randomly pick an element of F

8   **end**

9      $\Lambda_i \leftarrow v$

10      $F \leftarrow F \setminus \{v\}$

11      $N \leftarrow (N \cup N^*(v)) \cap F$

12 **end**

13 **return** Λ

Given an allocation of resources Λ and a time instant $t \in T$, we have a set
of candidate nodes $C = V \setminus (B_t^\Lambda \cup P^\Lambda)$ which can receive a resource and $|R_t|$
resources available. Our goal is to select a subset of C of size $|R_t|$ to apply the
resources in R_t. Algorithm 2 is based on two key observations about which nodes
tend to receive a resource first in high quality solutions:

1. Nodes that are close to burned nodes;
2. Nodes that are neighbors of protected nodes.

Motivated by the first observation, we define the notion of *fire perimeter*, i.e. a
set of nodes that are close to the current set of burned nodes. Since the arcs
of an instance represent fire velocity instead of physical distance, our notion of
closeness must be based on fire arrival time. With that in mind, we define the
fire perimeter at a time instant t as

$$F_t^\Lambda(z) = \{v \in C \mid t \leq a_v^\Lambda \leq f(t, z)\}$$

for some non-negative integer z, where $f : T \times \mathbb{N}_0 \to [0, H]$ is[2]

$$f(t, z) = \left(\alpha^{\lceil (z+1)/2 \rceil}(t) + \alpha^{\lceil (z+2)/2 \rceil}(t)\right)/2.$$

where $\alpha(t)$ is the earliest time after t in which new resources get available, as
defined at the end of Sect. 4.

Intuitively, the fire perimeter at time instant t is the set of unprotected
nodes whose fire arrival time is between t and some other time instant t', where

---

[2] We write $\alpha^n(t)$ for the composition of α with itself n times, e.g. $\alpha^2(t) = \alpha(\alpha(t))$.

$t' = f(t, z)$ for some non-negative integer z. Increasing z will increase t', which in turn may increase the size of $F_t^\Lambda(z)$. As a result, z controls the size of the fire perimeter. We clarify this notion with an example. Suppose we have resources at time instants 10, 20, and 30, and the optimization horizon H is equal to 60. Assume that the current time instant is 10 and no resources were deployed yet, i.e. the current allocation is Λ_0. In this scenario, $F_{10}^{\Lambda_0}(0)$, $F_{10}^{\Lambda_0}(1)$, $F_{10}^{\Lambda_0}(2)$, $F_{10}^{\Lambda_0}(3)$ contain the nodes that will burn between time instant 10 and $f(10, 0) = 20$, $f(10, 1) = 25$, $f(10, 2) = 30$, and $f(10, 3) = 45$, respectively. Figure 1 illustrates the example.

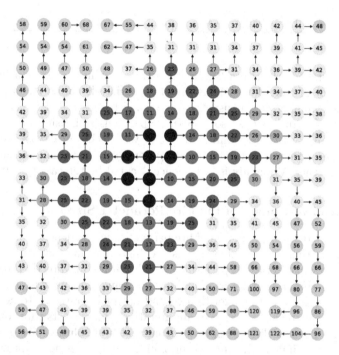

Fig. 1. A simple illustration of our definition of fire perimeter. In the example, we have resources at time instants 10, 20, and 30, and the optimization horizon H is equal to 60. We are at time instant 10 and no resources were deployed yet, i.e. the current allocation is Λ_0. The set $B_{10}^{\Lambda_0}$ is represented by the black colored nodes, $F_{10}^{\Lambda_0}(0)$ by purple nodes, $F_{10}^{\Lambda_0}(1)$ by red nodes, $F_{10}^{\Lambda_0}(2)$ by orange nodes, and $F_{10}^{\Lambda}(3)$ by yellow nodes. Note that $F_{10}^{\Lambda_0}(0) \subset F_{10}^{\Lambda_0}(1) \subset F_{10}^{\Lambda_0}(2) \subset F_{10}^{\Lambda_0}(3)$. (Color figure online)

In summary, Algorithm 2 considers only nodes in $F_t^\Lambda(z)$ instead of exploring all nodes in C (line 1). Similarly, and motivated by the second observation, we define a set N with all the nodes in $F_t^\Lambda(z)$ that have a neighbor in P^Λ (line 2). Using the sets $F_t^\Lambda(z)$ and N, Algorithm 2 proceeds as follows: for each resource $i \in R_t$, with probability p we select a node from N to be protected, and with probability $1 - p$ we select any node from F (lines 3 to 11).

Algorithm 3: Step

    Data: A partial allocation of resources Λ, a time instant t, the fire
          perimeter size z.

    Result: A set with at most η expansions of Λ using the resources in R_t.

1 $E \leftarrow \emptyset$

2 **repeat** $c|F_t^\Lambda(z)|$ **times**

3      $\Lambda' \leftarrow \text{Expand}(\Lambda, t, z)$

4      $E \leftarrow E \cup \{\Lambda'\}$

5 **end**

6 $E \leftarrow \text{sort}(E, t)$

7 **return** First η expansions in E

We now embed Algorithm 2 into Algorithm 3, which gives us a procedure to create a set of candidate expansions of a partial allocation Λ. In lines 2 to 5 we create a number of expansions proportional to the size of $F_t^\Lambda(z)$, for some constant $c \in \mathbb{Z}_+$. In line 6 we sort all the expansions in E using some heuristic function. Similarly to Algorithm 1, if $t < \hat{t}$, we use h_2, otherwise we use h_1. Finally, in line 7 we return the first η allocations in E. Note that in line 4 we do not check whether Λ' already is in E, hence it could be the case that $|E| < \eta$.

5.3 Dynamical Update of the Fire Perimeter Size

In Sect. 5.2, we defined the notion of fire perimeter, which depends on an integer constant z. Setting z to a value that is too high may increase running times, since the number of iterations performed by Algorithm 3 is directly proportional to the size of the fire perimeter. On the other hand, setting z to a value that is too low may impede the algorithm to find optimal solutions. To account for that, we propose to start with $z = 0$ and iteratively increase its value once a call to Algorithm 1 is not able to improve the current best solution. We observed in preliminary experiments that increasing z indefinitely does not improve performance and slows downs the algorithm in some cases, so we propose to define a maximum value for z and, once this value is reached, we cycle back to $z = 0$. Line 8 of Algorithm 4 illustrates that. Note that, for a given choice of z_{max}, the maximum value of z is $z_{max} - 1$.

6 Experimental Evaluation

In this section we present some computational experiments. All the experiments were done on a platform with a 3.5 GHz AMD Ryzen 9 3900X 12-Core processor, 32 GB of main memory, and Ubuntu Linux 20.04 LTS. Our algorithm was implemented in C++ and compiled with GCC 9.4 with maximum optimization. Our implementation and detailed computational data is available at https://github.com/gutodelazeri/Iterated-Beam-Search.

Algorithm 4: Main Algorithm

Result: An allocation of resources Λ^*.

1 $\Lambda^* \leftarrow \Lambda_0$
2 $z \leftarrow 0$
3 **while** Termination criteria not met **do**
4      $\Lambda \leftarrow$ BeamSearch(z)
5   **if** $|B_H^\Lambda| < |B_H^{\Lambda^*}|$ **then**
6          $\Lambda^* \leftarrow \Lambda$
7   **else**
8          $z \leftarrow (z+1) \mod z_{max}$
9   **end**
10 **end**
11 **return** Λ^*

Table 1. Instances used in the experiments.

Group	Resources per time instant						H	Δ
	10	20	30	40	50	60		
LA	3	3	3	3	0	0	70	50
LB	3	3	3	3	3	3	70	30

6.1   Test Instances

In this work we consider the set of instances proposed by [4]. This set consists of 16 instances, where each instance is a 20×20 grid graph. In all instances, the ignition node is at a central location in the graph and the optimization horizon is 70. The 16 instances are divided into two groups of 8 instances each, based on the magnitude of the delay caused by a resource and the quantity of resources released at each time instant. The optimal solution of all 16 instances is known, so in the sections below we report algorithm performance in terms of the absolute deviation from the optimal objective value[3]. Table 1 summarizes the instance set.

In this set of instances, edge weights attempt to model the fire propagation influenced by wind direction. In practice, the weight of each edge is sampled from a uniform distribution, and the range of values in this distribution depends on the direction to which the edge points. For further information, readers can refer to Table 5 in [4].

---

[3] In [4], the optimality of instance LB7 could not be proved. By executing their method with a time limit of 3 h we were able to find the optimal solution.

Table 2. Description of parameter values.

Parameter	Value	Description
β	50	Beam width
η	70	Ramification factor
c	30	See Algorithm 3
p	0.5	See Algorithm 2
z_{max}	3	See Algorithm 4

Table 3. Performance of beam search using different transition instants, as a function of a percentage $\hat{p}$ of the free burning time. We denote by δ the absolute difference between the obtained solution and the optimal solution, and columns $\bar{\delta}$ and σ_δ show the average and the standard deviation of δ across the 320 executions (16 instances and 20 replications). Similarly, we denote by ttb the time in seconds required to find the best (not necessarily optimal) solution. Columns $\overline{ttb}$ and σ_{ttb} give the average and the standard deviation of ttb. Lastly, column "Opt." has the percentage all 320 executions in which the optimal solution was found.

$\hat{p}$	$\hat{t}$	$\bar{\delta}$	σ_δ	$\overline{ttb}$	σ_{ttb}	Opt. (%)
0.1	6.9	1.41	3.59	61.18	105.06	75
0.2	13.8	0.15	0.35	52.23	110.73	85
0.3	20.7	0.12	0.33	26.68	46.40	88
0.4	27.6	0.12	0.33	26.62	46.24	88
0.5	34.5	0.18	0.45	43.38	82.68	85

6.2 Parameter Values

In all the experiments below, our algorithm uses the same set of parameter values, which are specified in Table 2. As stated in Sects. 5.1 and 5.2, the guiding heuristic used depends on the transition instant. In the next section, we conduct an experiment to find the best transition instant for the instances we are considering.

6.3 Transition Instant

In this section, we analyse how the transition instant affects the performance of our algorithm. Recall that in Sect. 5.1 we defined the transition instant as a percentage $\hat{p}$ of the free burning time of an instance. In this experiment, for each value of $\hat{p} \in \{0.1, 0.2, 0.3, 0.4, 0.5\}$ we run the beam search algorithm 20 times with different seed values on each of the 16 instances. The termination criterion was a maximum running time of 600 s. For the set of instances we are considering, the free burning time is always equal to 69, so for any value of $\hat{p}$ the transition instant is the same for all 16 instances. For each run, we collected the best objective value obtained and the time to find the best solution. Table 3 summarizes the results.

For the instances we are considering, a transition instant of 6.9 means that only the heuristic h_1 is used. As the first row shows, this is the worst version of our algorithm. When the transition instant is 13.8, we use h_2 when $t = 10$ and h_1 otherwise. As the second row shows, this version obtains better results when compared to using h_1 only. When the transition instant is between 20 and 30, as is the case of rows three and four, heuristic h_2 is used when $t = 10$ and when $t = 20$. The table shows that this is the best version of our algorithm. This version was able to find the optimal solution in 88% of the 320 executions, and obtained an average absolute gap of $\delta = 0.12$.

6.4   Comparison with the Literature

In this section we compare the best version of our algorithm found in the last experiment (BS) against the logic-based Benders decomposition of [4] (LBBD) and the iterated local search of [11] (ILS). We use the implementation of LBBD and ILS provided by [4] and run them in the same computational environment as BS. Following the protocol of Sect. 6.3, all three algorithms were executed 20 times on each of the 16 instances. The termination criterion for LBBD and BS was a maximum running time of 600 s. The termination criterion of ILS was a maximum number of iterations in stagnation, as specified in [11]. Table 4 shows the results.

As we can see, BS solves to optimality 14 out of the 16 instances in all 20 replications, while LBBD does so for 9 instances and ILS for only one instance. We can also see that BS obtains the smallest average absolute gap δ. In [4], LBBD was compared to ILS given a time limit 7200 s. Here we can see that, even with a fraction of the time limit, LBBD still beats ILS by a significant margin. Regarding the average time to find the best solution, we can see that BS obtains the smallest one in all instances. Considering that BS finds an optimal solution in most of the executions, this shows the efficiency of our algorithm.

To close this section, we analyse the performance profile of the three algorithms over the 320 executions. Figure 2 shows the percentage of the 320 executions that found an optimal solution within a particular interval of time. As we can see, within just 100 s our algorithm finds an optimal solution in about 80% of all executions, while LBBD does so for around 40% and ILS for around 20%. Within 300 s, the curves of BS and ILS stagnate. This is not true for LBBD, since it explores the search space systematically. Finally, within 600 s LBBD finds an optimal solution in about 70% of all executions and ILS in about 20%. As we saw in the last section, BS is able to find an optimal solution in 88% of all executions.

Table 4. Comparison between BS, ILS, and LBBD. We denote by δ the absolute difference between the obtained solution and the optimal solution, and the first six columns show the average and the standard deviation of δ over the 320 executions. Similarly, we denote by ttb the time in seconds required to find the best (not necessarily optimal) solution, and the last three columns show the average ttb of each algorithm. The values in these columns are expressed in terms of the average ttb of BS. For example, by looking at the first row we can see that ILS takes, on average, 50 s more than BS to arrive at the best solution.

	$\bar{\delta}$			σ_δ			$\overline{ttb}$		
	BS	ILS	LBBD	BS	ILS	LBBD	BS	ILS	LBBD
LA0	0	1.45	0.00	0	2.28	0.00	0	50.59	37.39
LA1	0	5.65	0.00	0	4.85	0.00	0	61.25	5.72
LA2	0	3.05	0.00	0	3.17	0.00	0	0.52	13.57
LA3	0	10.65	0.00	0	6.27	0.00	0	53.38	52.64
LA4	0	0.00	0.00	0	0.00	0.00	0	49.03	66.41
LA5	0	1.85	0.00	0	1.46	0.00	0	83.55	64.96
LA6	0	8.05	0.00	0	6.68	0.00	0	91.25	100.63
LA7	0	6.40	0.00	0	5.08	0.00	0	42.70	154.64
LB0	1	5.75	0.55	0	5.34	1.10	0	126.58	174.00
LB1	0	12.00	0.00	0	4.67	0.00	0	45.62	104.94
LB2	1	5.00	1.10	0	4.77	0.79	0	97.23	164.39
LB3	0	10.30	3.50	0	4.05	2.42	0	115.61	206.35
LB4	0	9.65	7.05	0	6.13	4.97	0	118.18	262.09
LB5	0	14.40	3.85	0	3.39	3.10	0	82.33	266.53
LB6	0	15.00	8.60	0	5.34	3.90	0	106.62	340.72
LB7	0	9.20	6.35	0	5.52	3.25	0	71.82	242.15
Avg	0.12	7.40	1.94	0	4.31	1.22	0	74.77	141.07

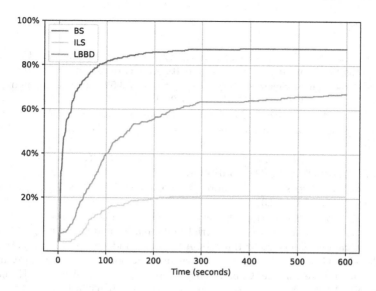

Fig. 2. Performance profile for the three algorithms, considering all 320 executions (16 instances and 20 replications). The x-axis shows time in seconds and the y-axis shows the percentage of the 320 runs in which the algorithm found the optimal solution within that time.

7   Conclusions and Future Work

In this work we proposed a heuristic algorithm for a problem related to wildfire suppression. The goal was to allocate fire suppression resources to regions of a landscape represented by a graph in order to minimize the total burned area. Our algorithm is a beam search guided by two heuristic functions to evaluate partial solutions and some heuristic rules on how to better expand the search tree at each level. In computational experiments, we showed that we can obtain better results by starting with one of the heuristic functions and then switching to the other at some point in time. Using these findings, we compared our approach to previous works in the literature. Our results indicate that the beam search algorithm can consistently find the optimal solution of most instances in considerably less time than alternative algorithms.

As future work, we would like to test our algorithm in more challenging instances, both in terms of grid size and the degree of irregularity of the landscapes. It would also be interesting to extend our approach to take into account different objective functions, like operational costs and the cost of the burned area.

Acknowledgments. M. R. acknowledges support from CNPq (grant 437859/2018-5), Coordenação de Aperfeiçoamento de Pessoal de Nível Superior, Brasil (CAPES), Finance Code 001, and CYTED (Grant P318RT0165).

References

1. Alvelos, F.: Mixed integer programming models for fire fighting. In: Gervasi, O., et al. (eds.) Computational Science and Its Applications - ICCSA 2018. Lecture Notes in Computer Science(), vol. 10961, pp. 637–652. Springer, Cham (2018). https://doi.org/10.1007/978-3-319-95165-2_45
2. Belval, E.J., Wei, Y., Bevers, M.: A mixed integer program to model spatial wildfire behavior and suppression placement decisions. Can. J. For. Res. **45**(4), 384–393 (2015)
3. Dimitropoulos, S.: Fighting fire with science. Nature **576**(7786), 328–328 (2019). https://doi.org/10.1038/d41586-019-03747-2
4. Harris, M.G., Forbes, M.A., Taimre, T.: Logic-based benders decomposition for wildfire suppression (2023)
5. Hartnell, B.L.: Firefighter! an application of domination. In: Proceedings of 25th Manitoba Conference on Combinatorial Mathematics and Computing (1995)
6. IPCC: Climate Change 2022: Impacts, Adaptation and Vulnerability. Contribution of Working Group II to the Sixth Assessment Report of the Intergovernmental Panel on Climate Change. Cambridge University Press, Cambridge, UK and New York, NY, USA (2022). https://doi.org/10.1017/9781009325844
7. Jewell, W.S.: Forest fire problems—a progress report. Oper. Res. **11**(5), 678–692 (1963)
8. Joint Economic Committee, U.S. Senate: Climate-exacerbated wildfires cost the U.S. between $394 to $893 billion each year in economic costs and damages (2023). https://web.archive.org/web/20231114011935/https://www.jec.senate.gov/public/index.cfm/democrats/reports?id=E31AF93E-34C7-4C35-A416-533FF796369B. Accessed 13 Nov 2023
9. Lowerre, B.: The harpy speech recognition system. Ph.D. thesis, CMU (1976)
10. Martell, D.L.: A review of operational research studies in forest fire management. Can. J. For. Res. **12**(2), 119–140 (1982)
11. Mendes, A.B., e Alvelos, F.P.: Iterated local search for the placement of wildland fire suppression resources. Eur. J. Oper. Res. **304**(3), 887–900 (2023)
12. Munich Re: Wildfires and bushfires - Climate change increasing wildfire risk (2023). https://www.munichre.com/en/risks/natural-disasters/wildfires.html. Accessed 19 Jan 2024
13. Reuters: Death toll from Hawaii wildfires drops to 97 (2023). https://www.reuters.com/world/us/death-toll-hawaii-wildfires-drops-97-missing-is-now-31-hawaii-governor-2023-09-15. Accessed 19 Jan 2024
14. United Nations: As wildfires increase, integrated strategies for forests, climate and sustainability are ever more urgent (2023). https://www.un.org/en/un-chronicle/wildfires-increase-integrated-strategies-forests-climate-and-sustainability-are-ever-0. Accessed 19 Jan 2024

A New Angle: On Evolving Rotation Symmetric Boolean Functions

Claude Carlet[1,2], Marko Durasevic[3], Bruno Gasperov[3],
Domagoj Jakobovic[3(✉)], Luca Mariot[4], and Stjepan Picek[5]

[1] Department of Mathematics, Université Paris 8, 2 Rue de la Liberté,
93526 Saint-DenisCedex, France
[2] University of Bergen, Bergen, Norway
[3] University of Zagreb Faculty of Electrical Engineering and Computing,
Zagreb, Croatia
{marko.durasevic,bruno.gasperov,domagoj.jakobovic}@fer.hr
[4] Semantics, Cybersecurity & Services Group, University of Twente Drienerlolaan 5,
7522 NB Enschede, The Netherlands
l.mariot@utwente.nl
[5] Digital Security Group, Radboud University, PO Box 9010,
Nijmegen, The Netherlands
stjepan.picek@ru.nl

Abstract. Rotation symmetric Boolean functions represent an interesting class of Boolean functions as they are relatively rare compared to general Boolean functions. At the same time, the functions in this class can have excellent cryptographic properties, making them interesting for various practical applications. The usage of metaheuristics to construct rotation symmetric Boolean functions is a direction that has been explored for almost twenty years. Despite that, there are very few results considering evolutionary computation methods. This paper uses several evolutionary algorithms to evolve rotation symmetric Boolean functions with different properties. Despite using generic metaheuristics, we obtain results that are competitive with prior work relying on customized heuristics. Surprisingly, we find that bitstring and floating point encodings work better than the tree encoding. Moreover, evolving highly nonlinear general Boolean functions is easier than rotation symmetric ones.

Keywords: rotation symmetry · Boolean functions · metaheuristics · nonlinearity

1 Introduction

Boolean functions are mathematical objects with various applications, including cryptography [17], combinatorics [27], coding theory [14,20], sequences [20], telecommunications [22], and computational complexity theory [1]. Naturally, for Boolean functions to be useful across various applications, they must fulfill various properties, such as being balanced and exhibiting high nonlinearity. Finding

S. Smith et al. (Eds.): EvoApplications 2024, LNCS 14634, pp. 287–302, 2024.
https://doi.org/10.1007/978-3-031-56852-7_19

Boolean functions with specific properties can be rather difficult, which is why the research community has been actively investigating the design of Boolean functions for nearly 50 years. In that respect, approaches to constructing Boolean functions can be divided into algebraic construction and various search techniques.[1]. Within search techniques, the most common division is into random search and metaheuristics. Unfortunately, sometimes even those approaches are not sufficient due to the vast number of Boolean functions of n inputs, which is equal to 2^{2^n} (see Table 1). Clearly, for $n = 6$, an exhaustive search already becomes impossible.[2] In such cases, it might be beneficial to focus on special classes of Boolean functions that are smaller and, thus, more amenable to search and enumeration but still large enough to contain many interesting functions. One such class is rotation symmetric Boolean functions - those functions that are invariant under cyclic shifts of the input coordinates. These functions have played a pivotal role in surpassing the quadratic bound, as discussed later.

The initial motivation for studying rotation symmetric Boolean functions can be traced back to the reason above: this class is significantly smaller than the class of general Boolean functions while still containing a large number of interesting functions. Moreover, such functions have a nice structure and allow for a compact representation [18]. We provide comparisons of class sizes for general Boolean functions, bent functions, and rotation symmetric functions in Table 1. Finally, Boolean functions in the class of rotation symmetric Boolean functions can have very good cryptographic properties. For instance, Kavut et al. found Boolean functions in 9 variables with nonlinearity 241 [12]. This achievement resolved an almost three-decade-old open problem and was accomplished using heuristics.

Table 1. The number of Boolean functions. Note that there is no known bound on the number of bent rotation symmetric (RS) functions.

n criterion	4	5	6	7	8	9	10	11	12	13	14	15	16
# general	2^{16}	2^{32}	2^{64}	2^{128}	2^{256}	2^{512}	2^{1024}	2^{2048}	2^{4096}	2^{8192}	2^{16384}	2^{32768}	2^{65536}
# bent	896	–	5425430528	–	$2^{106.3}$	–	2^{638}	–	2^{2510}	–	2^{9908}	–	2^{39203}
# RS	2^6	2^8	2^{14}	2^{20}	2^{36}	2^{60}	2^{108}	2^{188}	2^{352}	2^{632}	2^{1182}	2^{2192}	2^{4116}

Unfortunately, despite belonging to a much smaller class, the space of rotation symmetric Boolean functions still becomes too large for exhaustive search already for $n = 9$. This motivates the need to investigate diverse metaheuristic techniques and the construction of rotation symmetric Boolean functions.

---

[1] Some works also combine theory and search techniques, e.g., [12,25].

[2] One could still assume the hybrid mode where one 1) considers the equivalences that preserve the parameters of interest, 2) classify the functions under these equivalences, and 3) study each representative.

Multiple works leverage evolutionary algorithms to construct Boolean functions with specific properties, commonly focusing on properties like balancedness and nonlinearity, which we also consider in this work. However, most of these studies do not consider rotation symmetric Boolean functions but remain confined to the general classes of balanced, highly nonlinear functions or bent functions. The literature on rotation symmetric Boolean functions and metaheuristics is much more sparse. Despite this scarcity of research, significant findings were made already more than 15 years ago [12]. Even more time ago, Patterson and Wiedemann also dealt with rotation symmetric functions (whose name was introduced later) [?]. On the other hand, the first work considering evolutionary algorithms in this context appeared only in 2022 [30].

This paper investigates how various evolutionary algorithms can construct rotation symmetric Boolean functions, including both bent and balanced functions. We consider three solution encodings: bitstring, tree, and floating point, and two fitness functions. To the best of our knowledge, we are the first to investigate tree and floating-point encodings for this problem. The tree encoding represents an especially intriguing option, as state-of-the-art results indicate its superior performance over bitstring (see, e.g., [5]). As far as we know, no prior work has applied evolutionary algorithms to construct bent rotation symmetric Boolean functions. Our main findings are:

- We found rotation symmetric Boolean functions for every tested dimension. At the same time, genetic programming (GP) that evolves general (i.e., not confined to the rotation symmetric class of functions) Boolean functions finds functions with the same or higher nonlinearity. Therefore, we cannot conclude that finding a rotation symmetric Boolean function is simpler due to the smaller search space.
- While tree encoding is considered the best approach for general Boolean functions, we observe that both bitstring and floating point encoding perform better for rotation symmetric functions. This is because the latter two encodings significantly reduce the search space due to efficient encoding, which is not the case for GP (tree encoding).
- While the best results in related works are reported with customized heuristics, we reached the same (or even better) values with general metaheuristics. As such, we question whether developing custom heuristics is as beneficial as developing more powerful fitness functions.

2   Background

Let us denote positive integers with n and m: $n, m \in \mathbb{N}^+$. We denote the Galois (finite) field with two elements as $\mathbb{F}_2$ and the Galois field with 2^n elements by $\mathbb{F}_{2^n}$. An (n, m)-function represents a mapping F from $\mathbb{F}_2^n$ to $\mathbb{F}_2^m$.

When $m = 1$, the function f is called a Boolean function (in n inputs/variables). We endow the vector space $\mathbb{F}_2^n$ with the structure of that field, since for every n, there exists a field $\mathbb{F}_{2^n}$ of order 2^n that is an n-dimensional vector space. The usual inner product of a and b equals $a \cdot b = \bigoplus_{i=1}^{n} a_i b_i$ in $\mathbb{F}_2^n$.

2.1 Boolean Function Representations

The simplest way to uniquely represent a Boolean function f on $\mathbb{F}_2^n$ is by its truth table (TT). The truth table of a Boolean function f is the list of pairs of function inputs (in $\mathbb{F}_2^n$) and function values, with the size of the value vector being 2^n. The value vector is the binary vector composed of all $f(x), x \in \mathbb{F}_2^n$, with a certain order selected on F_2^n. Usually, as seen in, e.g., [3], one uses a vector $(f(0), \ldots, f(1))$ that contains the function values of f, ordered lexicographically. While the truth table representation is simple and "human-readable", little can be deduced from it except the Hamming weight.

The Walsh-Hadamard transform W_f is a unique representation of a Boolean function f that measures the correlation between $f(x)$ and the linear functions $a \cdot x$, see, e.g., [3]:[3]

$$W_f(a) = \sum_{x \in \mathbb{F}_2^n} (-1)^{f(x)+a \cdot x}. \tag{1}$$

The Walsh-Hadamard transform is very useful as many Boolean function properties can be evaluated through it. Since the complexity of calculating the Walsh-Hadamard transform with a naive approach equals 2^{2n}, it is common to employ a more efficient method called the fast Walsh-Hadamard transform, where the complexity is reduced to $n2^n$.

2.2 Boolean Function Properties and Bounds

Balancedness. A Boolean function f is called balanced if it takes the value one exactly the same number of times (2^{n-1}) as the value zero when the input ranges over $\mathbb{F}_2^n$.

Nonlinearity. The minimum Hamming distance between a Boolean function f and all affine functions, i.e., the functions with the algebraic degree[4] at most 1 (in the same number of variables as f), is called the nonlinearity of f. The nonlinearity nl_f of a Boolean function f can be easily calculated from the Walsh-Hadamard coefficients, see, e.g., [3]:

$$nl_f = 2^{n-1} - \frac{1}{2} \max_{a \in \mathbb{F}_2^n} |W_f(a)|. \tag{2}$$

The Parseval relation $\sum_{a \in \mathbb{F}_2^n} W_f(a)^2 = 2^{2n}$ implies that the nonlinearity of any n-variable Boolean function is bounded above by the so-called covering radius bound:

$$nl_f \leq 2^{n-1} - 2^{\frac{n}{2}-1}. \tag{3}$$

---

[3] Note that the sum is calculated in $\mathbb{Z}$.

[4] The algebraic degree deg_f of a Boolean function f is defined as the number of variables in the largest product term of the function's algebraic normal form having a non-zero coefficient, see, e.g., [16]. The algebraic normal form is a unique representation where an n variable Boolean function can be considered to be a multivariate polynomial over $\mathbb{F}_2$.

Eq. (3) cannot be tight when n is odd. For n odd, a slightly better bound is $2\lfloor 2^{n-2} - 2^{\frac{n}{2}-2} \rfloor$ [8]. We will consider Boolean functions that approach the covering radius bound as highly nonlinear. We show the values for the covering radius bound for each n in Table 2.

Bent Boolean Functions. The functions whose nonlinearity equals the maximal value $2^{n-1} - 2^{n/2-1}$ are referred to as bent, and they exist only for n even, see, e.g., [3]. Bent Boolean functions are a very active research topic with applications in, e.g., coding theory [14] and telecommunications [20]. They are also commonly discussed in cryptography but are not used since they are not balanced (despite being maximally nonlinear). Bent Boolean functions are rare, and we know the exact numbers of bent Boolean functions for $n \leq 8$ only. The numbers of Boolean functions (or upper bound values) are given in Table 1.

2.3   Rotation Symmetric Boolean Functions

A Boolean function over $\mathbb{F}_2^n$ is called rotation symmetric (RS) if invariant under any cyclic shift of input coordinates. Stated differently, it is invariant under a primitive cyclic shift, for instance:

$$(x_0, x_1, \ldots, x_{n-1}) \rightarrow (x_{n-1}, x_0, x_1, \ldots, x_{n-2}).$$

Since the above expression holds, the number of rotation symmetric Boolean functions will be less than the number of Boolean functions, as the output value remains the same for certain input values. Let us provide a small example of a rotation symmetric Boolean function when $n = 3$. We obtain the following partitions:

$$\{(0,0,0)\} \tag{4}$$
$$\{(0,0,1), (0,1,0), (1,0,0)\}$$
$$\{(0,1,1,), (1,1,0), (1,0,1)\}$$
$$\{(1,1,1)\}$$

Stănică and Maitra use the Burnside lemma to show that the number of rotation symmetric Boolean functions equals 2^{g_n}, where g_n equals [29]:

$$g_n = \frac{1}{n} \sum_{t|n} \phi(t) 2^{\frac{n}{t}}, \tag{5}$$

where ϕ is the Euler phi function.

Bent rotation symmetric functions are maximally nonlinear and invariant under any cyclic shift of input coordinates. Rotation symmetric bent functions are much rarer than general bent functions [18]. The motivation for considering bent rotation symmetric Boolean functions stems from the fact that such functions can have a simple structure (leading to new bent functions, e.g., Niho bent functions) and representation. Moreover, it is possible to compute them efficiently. However, there are some drawbacks, the most notable being that no new bent function has ever been found among rotation symmetric functions, as all those found belong to the well-known general classes of bent functions [18]. We provide results on the upper bounds of nonlinearity and the best-known nonlinearities in Table 2. More information about Boolean functions can be found in, e.g., [3,16].

Table 2. Nonlinearities of Boolean functions. Note that the bound equals the covering radius bound when n is even. Moreover, the best-known nonlinearities when the function is imbalanced and n is even are obtained for bent functions. The best-known results are taken from [3].

n													
condition	4	5	6	7	8	9	10	11	12	13	14	15	16
$2\lfloor 2^{n-2} - 2^{\frac{n}{2}-2}\rfloor$	6	12	28	58	120	244	496	1000	2016	4050	8128	16292	32640
balanced													
best-known nl_f	4	12	26	56	116	240	492	992	2010	4036	8120	16272	NA
imbalanced													
best-known nl_f	6	12	28	56	120	242	496	996	2016	4040	8128	16276	32640

3   Related Work

The research community has been active in evolving Boolean functions with specific cryptographic properties for almost 30 years [19]. While many settings have been tried, the most used solution encodings are the bitstring encoding and the tree encoding [5]. As far as we know, Fuller et al. were the first to consider evolving bent Boolean functions [6]. The authors started with a low-order Boolean function of input size n and then generated bent functions of higher algebraic order by iteratively adding ANF terms and checking whether the resulting function is bent. Yang et al. used evolutionary algorithms to evolve bent Boolean functions [31]. They used the trace representation of Boolean functions. Radek and Vaclav used Cartesian Genetic Programming to evolve bent Boolean functions up to 16 inputs [9]. To achieve this goal, the authors used various parallelization techniques. Picek and Jakobovic used GP to evolve algebraic constructions, which were then used to construct bent Boolean functions [23]. The authors showcased that the approach is highly efficient and provided results for up to 24 inputs, marking the first time that EC successfully constructed such large bent Boolean functions. Husa and Dobai employed linear GP to evolve bent Boolean functions, reporting superior results compared to related works, as they managed to evolve bent Boolean functions up to 24 inputs [10].

Stănică et al. used simulated annealing to evolve rotation symmetric Boolean functions [28]. By reducing the search space in this manner, the authors could construct 9-variable plateaued functions with nonlinearity 240 (among other properties). Kavut et al. utilized a steepest descent-like iterative algorithm to discover highly nonlinear Boolean functions [12]. The authors found imbalanced Boolean functions in 9 variables with a nonlinearity of 241. This represented a significant breakthrough, as the question of whether such functions existed had remained unanswered for nearly three decades. Moreover, the authors found Boolean functions in 10 variables with nonlinearity 492. Kavut and Yucel used a steepest-descent-like iterative algorithm to construct imbalanced Boolean functions in 9 variables with nonlinearity 242 [13] where the authors considered the generalized rotation symmetric Boolean functions. Liu and Youssef used simulated annealing to construct balanced rotation symmetric Boolean functions with nonlinearity equal to 488 [15]. Wang et al. employed genetic algorithms (GAs) to construct rotation symmetric Boolean functions [30]. The authors reported constructing balanced, highly nonlinear rotation symmetric functions.

4   Experimental Settings

4.1   Representations

Bitstring Encoding. The most widely used method for encoding a Boolean function is the bitstring representation [5]. The bitstring represents the truth table of the function with which the algorithm works directly. For a general Boolean function with n inputs, the truth table is encoded as a bit string with a length of 2^n. In the case of rotation symmetric Boolean functions, the number of truth table entries that need to be encoded is considerably smaller. For instance, for a 3-variable function, instead of $2^3 = 8$ bits, we only need to encode 4 bits, which is equal to the number of partitions in the example in the previous section (see Eq. (4)). The number of distinct bits that need to be encoded, corresponding to the genotype length, is shown in Table 3 for a given number of variables.

Table 3. The number of the encoding bits (genotype size) for rotation symmetric Boolean functions

variables	1	2	3	4	5	6	7	8	9	10	11	12	13	14	15	16
g_n		2	3	4	6	8	14	20	36	60	108	188	352	632	1182	2192	4116

In each evaluation, the bitstring genotype is first decoded into the full Boolean truth table, and the desired property is calculated. Although the bitstring representation usually performs worse than other encodings [5], especially for a larger number of variables, this might not be the case here due to the largely reduced genotype size.

The corresponding variation operators we use are the simple bit mutation, which inverts a randomly selected bit, and the shuffle mutation, which shuffles the bits within a randomly selected substring. For the crossover operators, we use the one-point crossover, which combines a new solution from the first part of one parent and the second part of the other parent with a randomly selected breakpoint. The second operator is the uniform crossover that randomly selects one bit from both parents at each position in the child bitstring that is copied. Each time the evolutionary algorithm invokes a crossover or mutation operation, one of the previously described operators is randomly selected.

Floating Point Encoding. The second approach we use for representing a Boolean function is the floating point genotype, defined as a vector of continuous variables. With this representation, one needs to define the translation of a vector of floating point numbers into the corresponding genotype, which is then translated into a full truth table (binary values). The idea behind this translation is that each continuous variable (a real number) of the floating point genotype represents a subsequence of bits in the genotype. All the real values in the floating point vector are constrained to the interval $[0, 1]$. If the genotype size is g_n, the number of bits represented by a single continuous variable of the floating point vector can vary and is defined as:

$$decode = \frac{g_n}{dimension},\qquad(6)$$

where the parameter *dimension* denotes the floating point vector size (number of real values). This parameter can be modified as long as the genotype size is divisible by this value. The first step of the translation is to convert each floating point number to an integer value. Since each real value must represent *decode* bits, the size of the interval decoding to the same integer value is given as:

$$interval = \frac{1}{decode}.\qquad(7)$$

To obtain a distinct integer value for a given real number, every element d_i of the floating point vector is divided by the calculated interval size, generating a sequence of integer values:

$$int\_value_i = \left\lfloor \frac{d_i}{interval} \right\rfloor.\qquad(8)$$

The final translation step consists of decoding the integer values to a binary string that can be used for evaluation. As an example, consider a genotype of 8 bits. Suppose we want to represent it with 4 real values; in this case, each real value encodes 2 bits from the truth table. A string of two bits may have 4 distinct combinations. Therefore, a single real value must be decoded into an integer value from 0 to 3. Since each real value is constrained to $[0, 1]$, the corresponding integer value is obtained by dividing the real value by $2^{-2} = 0.25$ and truncating it to the nearest smaller integer. Finally, the integer values are translated into the sequence of bits they encode.

Tree Encoding. In the third approach, we use tree-based GP to evolve a function in the symbolic form using a tree representation. The terminal set includes a given number of Boolean variables, $x_0, x_1, \ldots, x_{n-1}$. The function set consists of several Boolean primitives that can be used to represent any Boolean function. In our experiments, we use the following function set: OR, XOR, AND, AND2, XNOR, IF, and function NOT that takes a single argument. The function AND2 behaves the same as the function AND but with the second input inverted. The function IF takes three arguments and returns the second one if the first one evaluates to true and the third one otherwise. The output of the root node is the output value of the Boolean function. The corresponding truth table of the function $f : \mathbb{F}_2^n \to \mathbb{F}_2$ is determined by evaluating the tree over all possible 2^n assignments of the input variables at the leaf nodes. The genetic operators used in our experiments with tree-based GP are simple tree crossover, uniform crossover, size fair, one-point, and context preserving crossover [26] (selected at random), and subtree mutation.

Since GP, in this manner, evolves any Boolean function, and not solely rotation symmetric ones, we do not use the GP-derived truth table directly. Instead, it is treated as the bitstring genotype, the same as in the previous two representations, and decoded into a rotation symmetric function. This allows GP to use fewer variables than n since the genotype size is considerably smaller than the resulting truth table; for instance, for $n = 8$, the genotype size $g_n = 36$ (instead of 256), and GP will need to use only 6 variables to produce a bitstring of at least the required size. Unfortunately, since the genotype size (see Table 3) is not a power of 2, a part of the GP-produced bitstring (e.g., of size 64 with six variables) will not be used in any way. More importantly, there is no direct translation between the truth table of the GP-produced Boolean function, with fewer variables, and the actual rotation symmetric function being decoded and optimized, which may prove detrimental to the GP.

4.2   Fitness Functions

In our experiments, we optimize two different types of Boolean functions: 1) maximally nonlinear (bent) functions and 2) balanced, highly nonlinear functions. The first fitness function maximizes the nonlinearity value, nl_f, but is designed to consider the whole Walsh-Hadamard spectrum and not only its extreme value (see Eq. (2)). More specifically, we count the number of occurrences of the maximal absolute value in the spectrum, denoted as $\#max\_values$. Since higher nonlinearity corresponds to a *lower* maximal absolute value, we aim for as few occurrences of the maximal value as possible to make it easier for the algorithm to reach the next nonlinearity value. With this in mind, the fitness function is defined as:

$$fitness_1 : nl_f + \frac{2^n - \#max\_values}{2^n}. \tag{9}$$

The second term never reaches the value of 1 since, in that case, we effectively reach the next nonlinearity level.

With the second criterion, we aim to find balanced, highly nonlinear functions. We use a two-stage objective function in which a bonus equal to the previous fitness value is awarded only to a balanced function; otherwise, the objective value is only the balancedness penalty. The balancedness penalty BAL is the difference up to the balancedness (i.e., the number of bits to be changed to make the function balanced). This difference is included in the objective function with a negative sign to act as a penalty in maximization scenarios. The delta function $\delta_{BAL,0}$ assumes the value one when $BAL = 0$ and is zero otherwise.

$$fitness_2 : -BAL + \delta_{BAL,0} \cdot (nl_f + \frac{2^n - \#max\_values}{2^n}). \tag{10}$$

5   Experimental Results

Regarding bitstring (denoted as TT) and tree encoding (denoted as GP), we employ the same evolutionary algorithm: a steady-state selection with a 3-tournament elimination operator. In each iteration of the algorithm, three individuals are chosen at random from the population for the tournament, and the worst one in terms of fitness value is eliminated. The two remaining individuals in the tournament are used with the crossover operator to generate a new child individual, which then undergoes mutation with individual mutation probability $p_{mut} = 0.5$. Finally, the mutated child takes the place of the eliminated individual in the population. The population size in all experiments was 500, and the termination criteria were set to 10^6 evaluations. Each experiment was repeated for 30 runs. We consider Boolean function sizes from 8 to 16 inputs, as with less, finding rotation symmetric functions is easy and well within reach of an exhaustive search (see Table 3). The floating point representation can be used with any continuous optimization algorithm, which increases its versatility. In our experiments, we used the following optimization algorithms: Artificial Bee Colony (ABC) [11], Clonal Selection Algorithm (CLONALG) [2], CMA-ES [7], Differential Evolution (DE) [21], Optimization Immune Algorithm (OPTIA) [4], and a GA-based algorithm with steady-state selection (GA-SST), as described above. The implementation of all the algorithms and their default parameter settings are available in the ECF software framework.[5]

5.1   General Vs. Rotation Symmetric Functions

To facilitate easier comparison with related work, we also provide results for general balanced, highly nonlinear functions and general bent functions, along with the corresponding rotation symmetric ones (Tables 4 and 5). The results for general Boolean functions were reproduced with GP since, in that scenario, existing research points to GP as the most efficient approach [5,24]. Observe that in the case of balanced functions, the results are better for general functions than for rotation symmetric ones. Our results (the general ones) are also competitive

---

[5] Evolutionary Computation Framework, http://solve.fer.hr/ECF/.

with the best-known nonlinearities up to $n = 12$ and for $n = 14$ (see Table 2). The nonlinearities when using rotation symmetric functions are the same as the best-known ones only for $n = 8, 9$.

The results are slightly different for imbalanced functions (as we do not manage to obtain bent functions in all the cases). For small sizes (up to $n = 12$), the results for general functions are better than for rotation symmetric functions, but for $n = 14, 16$, the opposite is true. We suspect this happens due to the large search space size for such n values, where GP is known to face issues for such large Boolean functions [5]. The general results are competitive with the best-known nonlinearities up to $n = 12$, while the rotation symmetric ones are competitive for $n = 8$ only. We note that for general functions, we do not reach bent ones for $n = 14, 16$; for rotation symmetric ones, bent functions are reached only for $n = 8$.

Table 4. General (30 runs with GP) and rotation symmetric balanced Boolean functions, the best-obtained nonlinearities.

Size									
	8	9	10	11	12	13	14	15	16
general	116	240	492	992	2000	4032	8120	16256	32608
rot sym	116	240	488	988	1992	4012	8058	16186	32456

Table 5. General (30 runs with GP) and rotation symmetric imbalanced Boolean functions, the best-obtained nonlinearities.

Size					
	8	10	12	14	16
general	120	496	2016	7994	32332
rot sym	120	488	1992	8062	32468

5.2 Rotation Symmetric Balanced, Highly Nonlinear Boolean Functions

We provide results for balanced rotation symmetric functions in Table 6 and Fig. 1. Interestingly, the best results for most sizes are attained by the TT representation, except $n = 14$ and $n = 16$, for which the FP-SST representation provides the best results. When FP encoding is used, one can vary the number of bits a single FP value will represent (*decode*, Eq. 6). In our preliminary experiments, the best results were obtained with a relatively small *decode* (i.e., with one FP value representing a small number of bits), consequently resulting in a larger number of FP variables. This analysis is not included for brevity, but all FP-based algorithms used the same optimized setting with $decode = 3$.

Table 6. Median of nonlinearity values obtained for balanced Boolean functions for different numbers of variables. The N.F. entry denotes that the algorithm could not obtain a balanced Boolean function.

Representation	Size								
	8	9	10	11	12	13	14	15	16
TT	116.94	240.61	484.99	985	1988	4009	8049	16179	32435
GP	116.72	236.97	480.99	981	1976	3993	8032	16143	32394
FP-ABC	116.69	236.95	480.99	981	1977	3992	8033	16147	32406
FP-CLONALG	116.88	239.73	484.98	985	1988	4005	8036	16137	32385
FP-CMAES	116.81	236.95	480.99	977	1971	3983	8014	16113	N.F
FP-DE	116.80	236.93	480.98	977	1969	3969	7954	N.F	N.F
FP-OPTIA	115.83	237.94	484.98	985	1981	3988	8019	16117	32362
FP-SST	116.88	240.59	484.98	985	1987	4005	8053	16169	32443

5.3 Rotation Symmetric Bent Boolean Functions

We provide results for bent (thus, imbalanced) rotation symmetric functions in Table 7 and Fig. 2. TT provides superior results mainly because of the greatly reduced search space size compared to general Boolean functions. FP-SST is among the best, likely because our implementation includes a variety of floating-point crossover and mutation operators. Notice that GP provides worse results than TT because there is no semantic link between the GP genotype and the resulting decoded rotation symmetric Boolean function. Among the FP-based algorithms, CMAES and DE exhibit surprisingly unsatisfactory performance, not even managing to find balanced functions for larger n values. We note that the results for rotation symmetric functions are better than general Boolean results for imbalanced nonlinear functions for sizes 14 and 16, possibly again because of the reduced search size in the rotation symmetric encoding.

Finally, we compare our results with the two most relevant related works. Kavut et al. considered rotation symmetric functions in sizes 9 to 11 [12]. For $n = 9$, the best nonlinearity for a balanced function equals 240, the same as we achieve. For $n = 10$, Kavut et al. reported nonlinearity equal to 488 and 492, but the functions are imbalanced in both cases. We reach balanced functions with nonlinearity 488. For $n = 11$, Kavut et al. reported a nonlinearity of 988 for the balanced function and 992 for the imbalanced function; we also reach the nonlinearity of 988 for balanced functions. Later, Kavut et al. applied affine transformation and changed imbalanced functions into balanced ones, but the resulting functions are not rotation symmetric anymore, prohibiting direct comparison. Moreover, to reach such results, they utilized custom heuristics.

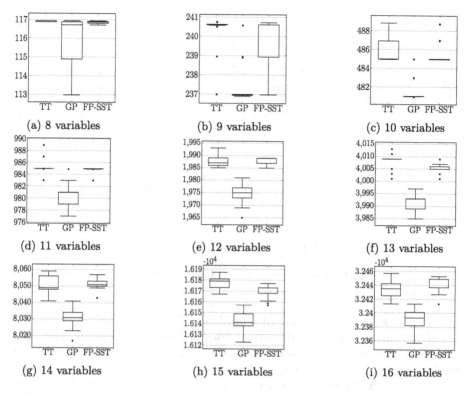

Fig. 1. Box plots for nonlinearity values obtained for balanced Boolean functions

Table 7. Median of nonlinearity values obtained for bent Boolean functions for a different number of variables.

Representation	Size				
	8	10	12	14	16
TT	120.00	488.71	1990.97	8056.99	32455.50
GP	120.00	484.88	1979.99	8038.00	32411.50
FP-ABC	119.53	484.41	1980.00	8037.00	32410.00
FP-CLONALG	120.00	487.89	1990.98	8045.00	32414.00
FP-CMAES	118.78	483.96	1976.00	8025.50	32382.50
FP-DE	120.00	482.98	1974.99	8007.50	32348.00
FP-OPTIA	119.53	486.93	1987.98	8036.50	32398.50
FP-SST	120.00	487.90	1990.96	8056.00	32458.50

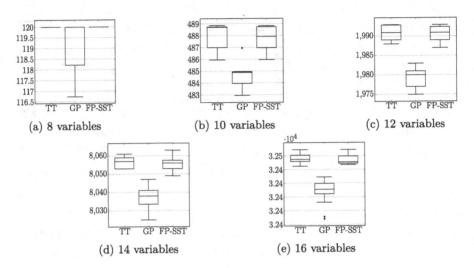

Fig. 2. Box plots for nonlinearity values obtained for bent Boolean functions

Wang et al. used a custom version of the GA for their experiments and considered only balanced rotation symmetric functions [30]. More precisely, they used "vanilla" GA, followed by two custom algorithm modifications where good results were reached only for those modified algorithms. For $n = 8$, they reached nonlinearity 116, the same as we. For $n = 10$, they obtained a nonlinearity of 488, which is again the same as we achieve. Finally, for $n = 12$, they reported a nonlinearity of 1996 but only provided an example with nonlinearity 1992, which is the same as our best result.

6   Conclusions and Future Work

This paper explores the difficulty of evolving rotation symmetric Boolean functions. While this class of Boolean functions is much smaller than general Boolean functions, we did not observe the problem to be simpler. Nevertheless, the obtained results are good and rival the related works even though they use customized heuristics while we use generic metaheuristics. We observe that tree encoding is not the best for evolving rotation symmetric functions, but bitstring and floating point work much better (differing from the situation when evolving general Boolean functions). The reason is that the reduction of the search space for bitstring and floating points is significant, while for tree encoding, we can reduce it only marginally. For future work, we consider two directions to be especially interesting. First, considering (bent) rotation symmetric Boolean functions, it would be interesting to see whether constructions of such functions could be found following the approach from [23]. Next, while this work considers rotation symmetric Boolean functions, it would be interesting to consider vectorial rotation symmetric functions (rotation symmetric S-boxes).

References

1. Arora, S., Barak, B.: Computational Complexity: A Modern Approach, 1st edn. Cambridge University Press, USA (2009)
2. Brownlee, J., et al.: Clonal selection algorithms. Swinburne University of Technology, Australia, Complex Intelligent Systems Laboratory (2007)
3. Carlet, C.: Boolean Functions for Cryptography and Coding Theory. Cambridge University Press, Cambridge (2021). https://doi.org/10.1017/9781108606806
4. Cutello, V., Nicosia, G., Pavone, M.: Real coded clonal selection algorithm for unconstrained global optimization using a hybrid inversely proportional hypermutation operator. In: Proceedings of the 2006 ACM symposium on Applied computing, pp. 950–954 (2006)
5. Djurasevic, M., Jakobovic, D., Mariot, L., Picek, S.: A survey of metaheuristic algorithms for the design of cryptographic Boolean functions. Crypt. Commun. **15**(6), 1171–1197 (2023). https://doi.org/10.1007/s12095-023-00662-2
6. Fuller, J., Dawson, E., Millan, W.: Evolutionary generation of bent functions for cryptography. In: Proceedings of the IEEE Congress on Evolutionary Computation, CEC 2003, Canberra, Australia, 8–12 December 2003, pp. 1655–1661. IEEE (2003)
7. Hansen, N., Müller, S.D., Koumoutsakos, P.: Reducing the time complexity of the derandomized evolution strategy with covariance matrix adaptation (CMA-ES). Evol. Comput. **11**(1), 1–18 (2003)
8. dong Hou, X.: On the norm and covering radius of the first-order reed-muller codes. IEEE Trans. Inform. Theory **43**(3), 1025–1027 (1997). https://doi.org/10.1109/18.568715
9. Bartz-Beielstein, T., Branke, J., Filipič, B., Smith, J. (eds.): PPSN 2014. LNCS, vol. 8672. Springer, Cham (2014). https://doi.org/10.1007/978-3-319-10762-2
10. Husa, J., Dobai, R.: Designing bent Boolean functions with parallelized linear genetic programming. In: Proceedings of the Genetic and Evolutionary Computation Conference Companion, p. 1825–1832. GECCO 2017, Association for Computing Machinery, New York, NY, USA (2017). https://doi.org/10.1145/3067695.3084220
11. Karaboga, D., Gorkemli, B., Ozturk, C., Karaboga, N.: A comprehensive survey: artificial bee colony (ABC) algorithm and applications. Artif. Intell. Rev. **42**, 21–57 (2014)
12. Kavut, S., Maitra, S., Yucel, M.D.: Search for Boolean functions with excellent profiles in the rotation symmetric class. IEEE Trans. Inf. Theory **53**(5), 1743–1751 (2007). https://doi.org/10.1109/TIT.2007.894696
13. Kavut, S., Yücel, M.D.: 9-variable boolean functions with nonlinearity 242 in the generalized rotation symmetric class. Inform. Comput. **208**(4), 341–350 (2010). https://doi.org/10.1016/j.ic.2009.12.002, https://www.sciencedirect.com/science/article/pii/S0890540109002454
14. Kerdock, A.: A class of low-rate nonlinear binary codes. Inf. Control **20**(2), 182–187 (1972)
15. Liu, W.M., Youssef, A.: On the existence of (10, 2, 7, 488) resilient functions. IEEE Trans. Inf. Theory **55**(1), 411–412 (2009). https://doi.org/10.1109/TIT.2008.2008140
16. MacWilliams, F.J., Sloane, N.J.A.: The Theory of Error-Correcting Codes. Elsevier, Amsterdam, North Holland (1977). ISBN 978-0-444-85193-2

17. Méaux, P., Journault, A., Standaert, F.-X., Carlet, C.: Towards stream ciphers for efficient FHE with low-noise ciphertexts. In: Fischlin, M., Coron, J.-S. (eds.) EUROCRYPT 2016. LNCS, vol. 9665, pp. 311–343. Springer, Heidelberg (2016). https://doi.org/10.1007/978-3-662-49890-3_13

18. Mesnager, S.: Bent Functions. Springer, Cham (2016). https://doi.org/10.1007/978-3-319-32595-8

19. Han, Y., Okamoto, T., Qing, S. (eds.): ICICS 1997. LNCS, vol. 1334. Springer, Heidelberg (1997). https://doi.org/10.1007/BFb0028456

20. Olsen, J., Scholtz, R., Welch, L.: Bent-function sequences. IEEE Trans. Inf. Theory **28**(6), 858–864 (1982)

21. Pant, M., Zaheer, H., Garcia-Hernandez, L., Abraham, A., et al.: Differential evolution: a review of more than two decades of research. Eng. Appl. Artif. Intell. **90**, 103479 (2020)

22. Paterson, K.: On codes with low peak-to-average power ratio for multicode CDMA. Inf. Theory IEEE Trans. **50**, 550–559 (2004)

23. Picek, S., Jakobovic, D.: Evolving algebraic constructions for designing bent Boolean functions. In: Proceedings of the Genetic and Evolutionary Computation Conference 2016, pp. 781–788. GECCO 2016, Association for Computing Machinery, New York, NY, USA (2016). https://doi.org/10.1145/2908812.2908915

24. Picek, S., Jakobovic, D.: Evolutionary computation and machine learning in security. In: Proceedings of the Genetic and Evolutionary Computation Conference Companion, pp. 1572–1601. GECCO 2022, Association for Computing Machinery, New York, NY, USA (2022). https://doi.org/10.1145/3520304.3534087

25. Picek, S., Marchiori, E., Batina, L., Jakobovic, D.: Combining evolutionary computation and algebraic constructions to find cryptography-relevant Boolean functions. In: Bartz-Beielstein, T., Branke, J., Filipič, B., Smith, J. (eds.) PPSN 2014. LNCS, vol. 8672, pp. 822–831. Springer, Cham (2014). https://doi.org/10.1007/978-3-319-10762-2_81

26. Poli, R., Langdon, W.B., McPhee, N.F.: A Field Guide to Genetic Programming. lulu.com (2008)

27. Rothaus, O.: On "bent" functions. J. Comb. Theory Ser. A **20**(3), 300–305 (1976)

28. Stănică, P., Maitra, S., Clark, J.A.: Results on rotation symmetric bent and correlation immune Boolean functions. In: Roy, B., Meier, W. (eds.) FSE 2004. LNCS, vol. 3017, pp. 161–177. Springer, Heidelberg (2004). https://doi.org/10.1007/978-3-540-25937-4_11

29. Stănică, P., Maitra, S.: Rotation symmetric Boolean functions-count and cryptographic properties. Discrete Appl. Math. **156**(10), 1567–1580 (2008). https://doi.org/10.1016/j.dam.2007.04.029, https://www.sciencedirect.com/science/article/pii/S0166218X07001734

30. Wang, Y., Gao, G., Yuan, Q.: Searching for cryptographically significant rotation symmetric Boolean functions by designing heuristic algorithms. Secur. Commun. Netw. **2022**, 1–6 (2022). https://doi.org/10.1155/2022/8188533

31. Yang, M., Meng, Q., Zhang, H.: Evolutionary design of trace form bent functions. Cryptology ePrint Archive, Paper 2005/322 (2005). https://eprint.iacr.org/2005/322

Analysis of Evolutionary Computation Methods: Theory, Empirics, and Real-World Applications

On the Potential of Multi-objective Automated Algorithm Configuration on Multi-modal Multi-objective Optimisation Problems

Oliver Ludger Preuß[1]([✉])[ID], Jeroen Rook[2][ID], and Heike Trautmann[1,2][ID]

[1] Machine Learning and Optimisation, Paderborn University, Paderborn, Germany
{oliver.preuss,heike.trautmann}@uni-paderborn.de
[2] Data Management and Biometrics, University of Twente, Enschede,
The Netherlands
j.g.rook@utwente.nl

Abstract. The complexity of Multi-Objective (MO) continuous optimisation problems arises from a combination of different characteristics, such as the level of multi-modality. Earlier studies revealed that there is a conflict between solver convergence in objective space and solution set diversity in the decision space, which is especially important in the multi-modal setting. We build on top of this observation and investigate this trade-off in a multi-objective manner by using multi-objective automated algorithm configuration (MO-AAC) on evolutionary multi-objective algorithms (EMOA). Our results show that MO-AAC is able to find configurations that outperform the default configuration as well as configurations found by single-objective AAC in regards to objective space convergence and diversity in decision space, leading to new recommendations for high-performing default settings.

Keywords: Automated Algorithm Configuration · Multi-Objective Optimisation · Multimodality · Evolutionary Computation

1 Introduction

Multi-objective optimisation (MOO) aims at solving multi-objective optimisation problems (MOPs). The goal is to find a set of solutions that form an optimal trade-off between multiple conflicting objectives, i.e. the Pareto set (PS) in decision and the Pareto front (PF) in objective space, respectively. Popular algorithms to solve MOPs, are evolutionary multi-objective optimisation algorithms (EMOA) [7] which are inspired by concepts of variation and selection from natural evolution. However, EMOAs usually only find an approximation of the true PF [12], and their performance is sensitive to underlying parameter settings. For respective optimal configuration, automated algorithm configuration (AAC) [14,22] is promising and will be shown to be highly effective.

A challenge that often needs to be faced in MOO is multi-modality, i.e. the presence of multiple local and global optima. The multi-global case, which we will

S. Smith et al. (Eds.): EvoApplications 2024, LNCS 14634, pp. 305–321, 2024.
https://doi.org/10.1007/978-3-031-56852-7_20

focus on, is characterised by different solutions in decision space corresponding to the same point in objective space [12] (Fig. 1).

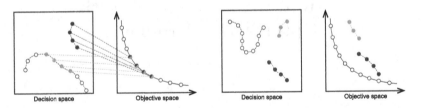

Fig. 1. Multi-modal MOPs: multi-global case (left), multi-modal case (right). (Fig. taken from [13])

Thereby, it is not sufficient to solely focus on convergence of EMOAs towards the PF but also to find at best all solutions in the decision space that map to the PF. This trade-off strongly depends on the configuration used for the chosen EMOA. Rook et al. [22] showed that configurations favouring convergence in objective space negatively affect diversity in decision space and vice versa leading to a multi-objective algorithm configuration problem. We will thus simultaneously optimise for both performance criteria, reflected by the *Dominated Hypervolume* (HV, [30]) and the *Solow-Polasky measure* (SP, [26,27]), using multi-objective automated algorithm configuration (MO-AAC, [21]).

Three different research questions (RQs) will be answered: 1) How competitive are EMOAs configured for both convergence towards the PF and diversity in decision space compared to EMOAs solely configured for a respective single performance objective? 2) How configurable are EMOAs in terms of versatility and competitiveness? And 3) How does the trade-off between both performance criteria look like when configuring for both simultaneously?

We will specifically show that in sum, this study has led to the proposition of new default configurations for all examined EMOAs, which have been found to enhance their efficacy pertaining to both performance indices concurrently. Thereby, the EMOA Omni-Optimizer [8] experimentally outperforms competing EMOAs regarding versatility and competitiveness.

Section 2 provides details on MOO and respective performance measures, followed by an overview of MO-AAC approaches in Sect. 3. We will then discuss experimental results in Sect. 4 and conclude with Sect. 5 including an outlook on future research perspectives.

2   Multi-objective (Evolutionary) Optimisation

The goal of (continuous) MOO is to simultaneously optimise multiple (conflicting) objective functions $f_i : \mathcal{X} \rightarrow \mathbb{R}, i \in [m] := \{1, ..., m\}, m \geq 2$; w.l.o.g. minimisation of all objectives is assumed. Ranking of solutions in the multi-objective setting poses a challenge since multiple objectives usually are conflicting. Thus the notion of *(Pareto-)dominance* needs to be introduced. Given two solutions $\mathbf{x}, \mathbf{y} \in \mathcal{X}$, we say that $\mathbf{x}$ *(Pareto-)dominates* $\mathbf{y}$, denoted as $\mathbf{x} \prec \mathbf{y}$, iff $f_i(x) \leq f_i(y) \ \forall_{i \in [m]}$, and $f_j(x) < f_j(y) \ \exists_{j \in [m]}$. As a consequence, the optimal result of a MOP is commonly not a single solution but rather a set of solutions where every solution is not dominated by any other solution. This set of optimal trade-off solutions – also called Pareto set (PS) – can be described as $\mathcal{X}^* = \{\mathbf{x} \in \mathcal{X} \mid \not\exists \, \mathbf{x}' \in \mathcal{X} : \mathbf{x}' \prec \mathbf{x}\}$. The projection of the PS in the objective space is referred to as Pareto front (PF).

Traditional EMOAs, e.g. NSGA-II [9] and SMS-EMOA [3], approximate the PF w.r.t. convergence and diversity of solutions in objective space [2]. *Dominated Hypervolume* (HV) [30] is a widely used Pareto-compliant metric which measures the area enclosed by a set of non-dominated solutions in objective space and an anti-optimal reference point r (see Fig. 2b).

Multi-modal MOPs yield new challenges in terms of locally efficient sets, ridges and basin structures [11]. Especially in the multi-global case, this can negatively impact EMOAs' performance as they tend to get stuck in global basins in the decision space that only partially cover the PS, as was demonstrated in earlier studies [22]. We use the *Solow-Polasky measure* (SP) [26] to measure the extent of coverage of all global basins. SP was designed to measure the diversity of species in biology. It was later adopted by [27] in the context of *Evolutionary Diversity Optimization*. It measures the pairwise distances of points in the decision space to assess its diversity. SP is defined as $SP(P) = \sum_{1 \leq i, j \leq \mu} M_{ij}^{-1} \in [1, \mu]$, where $P = \{P_1, \ldots, P_\mu\}$ is a population of μ individuals, and M^{-1} is the Moore-Penrose generalised inverse matrix of M with $M_{i,j} = \exp(-d(P_i, P_j))$ where d is the (Euclidean) distance between two individuals. If the solution set is spread out over the decision space, SP will be higher compared to the case when the points are clustered and do not cover the whole space (Fig. 2a).

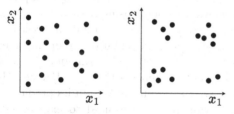 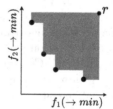

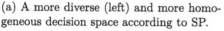

(a) A more diverse (left) and more homogeneous decision space according to SP.

(b) The area calculated by HV.

Fig. 2. Examples of how SP characterises diversity and HV convergence.

3   Multi-objective Automated Algorithm Configuration

Algorithm configuration (AC) in general aims to find the best configuration θ^* from a parameter space Θ of an algorithm A on a given problem instance set Π, such that $c(\theta^*, \Pi) = \text{opt}_{\theta \in \Theta} c(\theta, \Pi)$. I.e., the configuration that yields the best overall performance of a quality metric c. In the context of meta-heuristics AC is often considered as an offline and generative hyper-heuristic [6]. Typically, $c(\theta, \Pi)$ is obtained by aggregating the quality on all instances $\pi \in \Pi$ for a fixed parameter configuration $\theta \in \Theta$, e.g. by taking the mean $c(\theta, \Pi) = \frac{1}{|\Pi|} \sum_{\pi \in \Pi} c(\theta, \pi)$. The search for θ^* is *de facto* an optimisation problem but has some unique characteristics. Expensive function evaluations, mixed-type parameter/decision spaces, and aggregated performance observations result into that *off-the-shelf* evolutionary algorithms are inefficient in solving the AC problem. Methods like irace [18], ParamILS [15], and SMAC [16] are problem specific frameworks that perform single-objective (SO) automated algorithm configuration (AAC) [14]. An extensive overview of AAC algorithms can be found in [25].

Since in this study two EMOA performance objectives are of interest, we extend the AC formulation to the multi-objective setting by using the notion of Pareto dominance. The goal is now to find a set of configurations that are the optimal trade-off configurations, i.e., the PF of multiple quality metrics simultaneously, i.e., $c^i : \Theta \to \mathbb{R}, i \in [m], m \geq 2$. The final incumbent in the MO case is thus a set of optimal trade-off configurations, i.e., the PF, which can be described as $\Theta^* = \{\theta \in \Theta \mid \nexists \theta' \in \Theta : \theta' \prec \theta\}$ and is analogous to PF definition for MOO. Also for the multi-objective AC (MO-AC) scenarios, tailored MO-AAC approaches exist, such are MO-ParamILS [4] and MO-SMAC [21][1]. For this paper, we decided to use SMAC and MO-SMAC for the AAC and MO-AAC scenarios, respectively. These configurators support mixed-type and nested parameters spaces, which the others do not have. They also support logarithmically scaled parameters, which could taken into account when sampling new configurations.

SMAC is a model-based configurator and is internally alternating between two phases; the Bayesian Optimisation (BO) phase and the intensification phase. In the BO phase, promising configurations are found using a surrogate model trained on previous algorithm runs. The surrogate model is a random forest that takes the configuration and instance features as input and uses the quality metric as output label. Configurations are found by performing random and local searches, which are then ranked based on how promising they are. Because the predictions of the surrogate model also return the uncertainty in its prediction, the Expected Improvement (EI) acquisition function is used to express how promising a configuration is. The intensification phase validates the proposed configurations to see if they are actually better than the believed-to-be-best configuration, i.e., the incumbent. This is efficiently done by first running the challenging configuration on only one instance and comparing its performance against the performance of the incumbent on that instance. If the challenger is worse, it is rejected, and the comparison stops. Otherwise, the challenger is run on more instances, and the comparison is made again on their aggregated performance. If the challenger is still better than the incumbent after the challenger

---

[1] Source code: https://github.com/jeroenrook/SMAC3/tree/mosmac-anon.

ran on all the instances the incumbent ran on, the challenger is accepted as the new incumbent.

MO-SMAC has the same working principle as SMAC. However, there are several differences. First, the incumbent is now a set of non-dominated configurations. Secondly, the BO-phase has a surrogate model for each objective and the predicted HV improvement (PHVI) acquisition function combines the predictions of these models to score configurations on how much they will improve the incumbent. Thirdly, comparisons are made based on Pareto dominance relations during intensification. As long as the challenger is not dominated by the incumbent configuration closest to the challenger, the validation continues. Because the number of problem instances the incumbent evaluates on increases during a configuration run, the incumbent's size is limited to 8 to ensure this progression. When a new configuration is added to the incumbent and the limited is exceeded, the configuration with the lowest crowding distance [9] to the others is removed.

4   Experiments

MO-AAC experiments were conducted to automatically configure EMOAs on a complementary set of multi-modal MOPs for simultaneously showing convergence in objective and diversity in decision space. Thereby, we will specifically address the research questions posed in the introduction.

4.1   Experimental Setup

The experimental setup is aligned with the experiments of [22] to build on top of their work and ensure comparability. Seven different EMOAs are considered, the first four are MOEA/D [20], NSGA-II [9], Omni-Optimizer [8], and SMS-EMOA [3]. These are classical EMOAs that intrinsically focus on convergence towards the PF and thus may not be able to find diverse solutions in decision space according to SP. An exception here might be Omni-Optimizer designed to also favor a diverse decision space. The remaining three EMOAs utilize gradient information of a MOP. HIGA-MO [28] focuses on the HV gradient while MOGSA [11] and MOLE [23] use a gradient to utilize landscape characteristics to move along local structures and preserve different solutions in the decision space.

The set of problem instances includes all problems of ZDT [30], DTLZ [10], and MMF [29] except ZDT5 and MMF13, and instances f_{46}, f_{47}, and f_{50} from bi-objective BBOB [5]. This results in a total of 33 instances. All instances are bi-objective and have a 2-dimensional decision space. The EMOA population size μ was fixed to 100, however, MOLE and MOGSA can return a larger solution set than 100, as they do not have a population. In the rare case they return more than 2 000 non-dominated points, 2 000 points were randomly sampled without replacement to keep the SP computation – which relies on the matrix inversion of dense matrices – possible.

Table 1. EMOA configuration spaces.

MOEA/D	Default	Range
T	20	[10, 40]
Tr	8	[4, 20]
aggregation	wt	{wt, awt, pbi}
archive	0	{0, 1}
decomp	SLD	{SLD, Uniform}
∇p	1	[0.1, 1]
neighbor	λ	{ λ, x}
nr	1	[1, 10]
update	norm	{norm, best, restrict}
Variation level 1		
method	sbx	{bin, diff, poly, sbx }
diffmut	rand	{rand, mean, wgi }
sbx η	20	[1, 100]
sbx pc	1	[1, 10]
Variation level 2		
method	poly	{bin, diff, poly, sbx}
diffmut	rand	{rand, mean, wgi}
sbx η	20	[1, 100]
sbx pc	1	[1, 10]
Variation level 2		
method	off	{bin, diff, poly, sbx, off}
diffmut	rand	{rand, mean, wgi}
sbx η	20	[1, 100]
sbx pc	1	[1, 10]

MOLE	Default	Range
descent parameters		
armijo_factor	1^{-4}	$[1^{-5}, 0.1]$
direction_min	1^{-8}	[0, 1]
histroy_size	100	$[1, 2^{32} - 1]$
max_iter_descent	1000	$[1, 2^{32} - 1]$
scale_factor	2	[1.1, 5]
step_min	1^{-6}	$[1^{-8}, 0.01]$
step_max	0.1	$[1^{-4}, 0.01]$
explore parameters		
angle_max	45	[10, 90]
scale_factor	2	[1.1, 5]
step_min	1^{-4}	$[1^{-6}, 100]$
step_max	0.1	[0.1, 100]
refine parameters		
after_nstarts	100	[1, 100]
hv_target	2^{-5}	$[1^{-6}, 0.1]$
other parameters		
epsilon_gradient	1^{-8}	[0, 1]
max_local_sets	1000	$[0, 2^{32} - 1]$

SMS-EMOA	Default	Range
mutator	poly	{gauss, poly, uni}
mutPoly_eta	10	[0, 100]
mutPoly_p	0.2	[0, 1]
mutGauss_eta	1	[0, 1]
mutGauss_sdev	0.05	[0, 1]
recombinator	sbx	{cross, int, sbx}
recSBX_eta	5	[0, 10]
recSBX_p	1	[0, 1]

NSGA-II	Default	Range
mutator	poly	{gauss, poly, uni}
mutPoly_eta	10	[0, 100]
mutPoly_p	0.2	[0, 1]
mutGauss_eta	1	[0, 1]
mutGauss_sdev	0.05	[0, 1]
recombinator	sbx	{cross, int, sbx}
recSBX_eta	5	[0, 10]
recSBX_p	1	[0, 1]

HIGA-MO	Default	Range
dominated_steer	NDS	{M[1, 6], NDS}
sampling	uni	{uni, LHS, Grid}
step_size	1^{-3}	$[1^{-9}, 1]$

Omni-Optimizer	Default	Range
delta	1^{-3}	[0, 1]
eta_cross	20	[5, 20]
eta_mut	20	[5, 20]
init	random	{random, LHS}
mate	norm	{norm, restrict}
p_cross	0.6	[0.6, 1]
p_mut	0.1	[0, 1]
space_niching	0.5	[0, 1]

MOGSA	Default	Range
exploration_step	0.2	[0, 1]
ls_method	both	{bi, mo-ls, both}
max_no_basins	50	[1, 2000]
max_no_basins_ls	500	[1, 1000]
prec_angle	1^{-4}	[0, 0.01]
prec_grad	1^{-6}	[0, 0.01]
prec_norm	1^{-6}	[0, 0.01]
scale_step	0.5	[0, 1]

For each problem instance, a reference set was empirically approximated by combining all function evaluations while running all considered EMOAs 10 times with an evaluation budget of 100 000. If the reference point for calculating the HV of an instance was unknown, it was obtained by taking the maximum function values of the obtained reference sets increased with a small constant to account for solutions on the extremes [1]. To ensure comparability of HV values across different instances and enable aggregation, the HV was normalised by dividing by the HV of the reference set of the specific instance.

SO-SMAC configures for SP and HV separately, whereas MO-SMAC configures for both simultaneously. Experiments of [22] with SO-SMAC were reproduced as MO-SMAC is built on top of a different SMAC version [17] as was originally used by the authors. Considering the 7 different EMOAs, a total of 21 different configuration scenarios were performed. Each configuration scenario had a termination criterion of 250 algorithm calls. All other (MO-)SMAC parameters were set to default. Each of the EMOAs had a function call limit of 20 000 within configuration. A preliminary study showed that most algorithms showed sufficient convergence given this budget. Table 1 lists the EMOA configuration spaces. We did not configure for population size to prevent the configurators from finding configurations where the population size equals the number of function evaluations, which yields a high diversity in decision space but does not actually run the respective EMOA beyond the initialisation of the population.

For computational reasons we used 10-fold cross-validation (CV) instead of leave-one-out CV as in [22]. In each fold, 10 separate configuration runs were performed to account for the stochastic behaviour of (MO-)SMAC. Out of these runs, the incumbent solution for SMAC was selected based on the one that yielded the best average performance on the training instances. For MO-SMAC all configurations from the sets of incumbents out of the 10 configuration runs were combined, and only the overall non-dominated configurations were selected as the final incumbent. Again, this was based on their aggregated mean performance on the instances in the training partition of the fold. On top of the CV folds, additional configuration scenarios were conducted where the configurators ran on all instances. The resulting configurations were evaluated on the test instances of the respective folds for the CV scenarios and on all the instances for the other scenarios. Each evaluation of a configuration on an instance was based on 25 independent runs that were each seeded differently. To run all the experiments, approximately 5 000 CPU hours were needed on the HPC cluster PALMA II of the University of Münster. An overview of the experimental setup can be found in Fig. 3[2].

4.2 Results

RQs were addressed based on the test instances of the CV results, apart from 3*b* which relies on configuring on all instances in total. Figure 4 provides an overview of the specific aspects the RQs investigate. On the left side RQ 2b focuses on the problem space whereas all other RQs focus on the performance measure space depicted on the right side.

In addition – to enhance clarity – each experimental result is labelled to describe the origin and abstraction level they originated from. The shape of the label describes if the origin is based on the CV results (●) or on runs

---

[2] Experimental code can be found at https://github.com/jeroenrook/MMMOO-moconfig-exp.

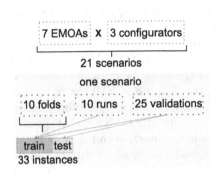

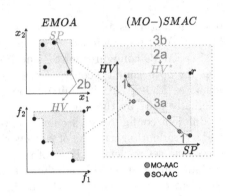

Fig. 3. Overview of the experimental setup: scenarios, folds, configuration runs, and validation.

Fig. 4. Overview of RQs (orange, see Sect. 4.2) and related spaces (left: function-space, right: indicator-space). (Color figure online)

with all instances (■). The colour describes the abstraction level where green (■) corresponds to decision-space (SP), blue (■) to objective-space (HV), and yellow (□) to indicator-space. As an example, ● describes that the results from the CV are used and that we look at their objective-space performance.

R1. How competitive are EMOAs configured for both convergence towards the PF and diversity in decision space compared to EMOAs configured for a respective single objective? ○ Rook et al. [22] showed the potential of AAC on multi-modal MOPs by configuring the EMOAs separately for SP and HV. We deem it important that MO-AAC is able to find comparable configurations to SO-AAC. This shows the competitiveness of MO-AAC with SO-AAC even though MO-AAC has a more dispersed task. More specifically, Fig. 5 compares the best-performing non-dominated MO configurations for SP and HV over the 10 runs out of every fold with the respective SO configuration. Interestingly, all solvers improve the SO-AAC solution for SP. Here, MOGSA and MOLE have the highest increase, while HIGA-MO and NSGA-II have the lowest increase. For HV, MOGSA and MOLE have the highest increase again, while all other solvers show little to no substantial improvement.

R2. How Configurable are EMOAs? Specifically, we investigate *versatility* and *competitiveness*. Versatility measures the adaption capability of an algorithm to different requirements, e.g., the ability to compromise between HV and SP gradually.

R2a. Which EMOA is Most Versatile? ○ To answer this question, we now consider the bi-objective space spanned by SP and HV based on the configurations found with MO-SMAC. As HV in general measures quality and spread of solutions in objective (here: performance) space, it is well-suited to express the desired EMOA's versatility, i.e. $HV^* := HV_{HV,SP}$, where the non-dominated configurations, on basis of the training instances, of the combined 10

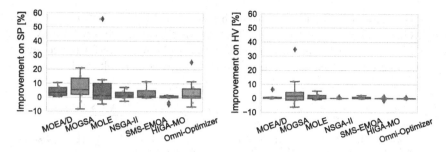

Fig. 5. Relative improvement of the best MO configuration for SP (left) and HV (right) compared to the SO configuration, respectively.

configurator runs are used as input. HV^* is calculated over the mean performance over each fold's test instances and validation seed (25), resulting in 250 scores per algorithm. Based on these HV^*'s the algorithms were ranked.

The average over all these rankings is shown in a Critical Difference (CD) diagram, displayed in Fig. 6, where a lower ranking indicates better versatility. The figure displays the CD as a black line when EMOAs are statistically tied. It is based on a Nemenyi test [19] with $\alpha = 0.1$, resulting in a CD of 0.52. The plot shows that Omni-Optimizer has the highest HV^* overall, followed by MOLE, and are therefore considered the most versatile EMOAs. HIGA-MO, MOEA/D, and MOGSA achieve average ranking and are statistically tied. The worst HV^* was achieved by SMS-EMOA and NSGA-II which are also tied with each other.

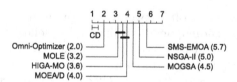

Fig. 6. Rankings based on HV^* of the configurations.

R2b. Which EMOA is Most Competitive? ◑ ● Based on the mean performance over all non-dominated MO-AAC configurations of each EMOA, we now rank them for SP and HV separately. The mean of SP and HV over the test instances of all folds (250) is considered. By this an overall performance value for all found non-dominated configurations is provided and EMOA rankings are similarly calculated as in R2a with also a critical difference of 0.52.

For SP, Omni-Optimizer outperforms all other algorithms. MOLE, HIGA-MO, and MOEA/D achieve average ranking. Also, HIGA-MO is statistically tied with MOLE and MOEA/D. The worst performing algorithms for SP are MOGSA, NSGA-II, and SMS-EMOA. The rankings for HV look vastly different than the SP rankings. Here, NSGA-II and SMS-EMOA are the best performing algorithms, while they are the worst performing regarding SP. Omni-Optimizer is now the third best algorithm, although it is tied with SMS-EMOA. MOEA/D and MOLE follow in the third and fourth position and are statistically tied. HIGA-MO and MOGSA are tied as well and have the worst performance regarding HV. These rankings closely resemble the rankings of Rook et al. [22].

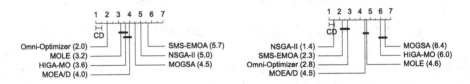

Fig. 7. Performance rankings for SP (left,) and HV (right).

The rankings for versatility (Fig. 6) and competitiveness (Fig. 7) with regard to SP are qualitatively similar. Here the algorithms that are designed to keep a diverse decision space (Omni-Optimizer and MOLE) achieve the best performance. The HV rankings, conversely, are dissimilar to the versatility ranks. Here, algorithms that intrinsically focus on convergence in the objective space (NSGA-II and SMS-EMOA) achieve the highest performance. Omni-Optimizer also achieves good performance with regards to HV competitiveness as it statistically ties with SMS-EMOA which ranks second. It is expected that there will be trade-offs between algorithms with regard to SP and HV, which makes Omni-Optimizer's performance as the best algorithm with regard to SP while achieving a good ranking for HV noteworthy.

R3. How does the trade-off between the convergence towards the PF and diversity in decision space look like? In the following, we will explicitly investigate the extent and characteristics of the trade-off between HV and SP of the non-dominated solutions generated by MO-AAC.

R3a What is the Extent of the Trade-Off? ●● Figure 8 visualises the relative loss in SP and HV of the two extreme solutions on the trade-off surface of the non-dominated MO-AAC configurations. Those were identified based on the results on the validation instances, i.e. the median of the 25 repetitions per instance and applying the arithmetic mean afterwards. For those configurations we calculate the (relative) loss as follows: Denoting the best MO-AAC configurations regarding HV and SP as C^{HV} and C^{SP}, respectively, relative losses in HV and SP result as $HV(C^{HV})/HV(C^{SP}) - 1$ and $SP(C^{SP})/SP(C^{HV}) - 1$, deliberately using only a conceptual notation here leaving out the EMOAs in between. More specifically, the relative loss is the percentage of how one indicator worsens when using the best configuration for the other indicator compared to its own best configuration.

The more extreme the trade-off behaviour, the more versatile an EMOA is in general since there are specific configurations that perform especially well on SP or HV. SP shows, in general, a higher loss than HV across all EMOAs reflecting a higher parameter sensitivity in this regard. MOEA/D shows the highest loss in

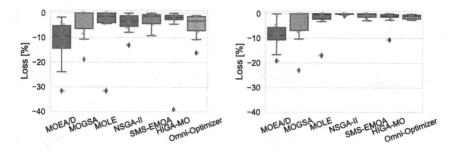

Fig. 8. Relative loss for SP (left) and HV (right) on MO-AAC configurations.

both SP and HV, followed by MOGSA. Omni-Optimizer and NSGA-II have the second highest loss of SP when considering the mean loss. This is also reflected in Figs. 9 and 10.

R3b. How do the Actual Configurations Differ on this Trade-off? ☐ We now focus on the actual parameter settings resulting from the MO-AAC experiments. Relying on CV confirming generalisation capability of results, we consider MO-AAC on all instances for further analysis, i.e. for each EMOA all incumbent configurations of the 10 conducted configuration runs are chosen. Table 2 shows that all EMOAs, except Omni-Optimizer, obtained solely unique configurations reflecting both stochastic behaviour as well as exploration capability of MO-AAC. However, only few non-dominated configurations are resulting after all.

Table 2. Number of MO-AAC configurations found.

Algorithm	# configs	unique configs	non-dominated
MOLE	19	19	1
MOGSA	17	17	2
NSGA-II	27	27	4
HIGA-MO	17	17	3
MOEA/D	29	29	3
Omni-Optimizer	42	40	4
SMS-EMOA	23	23	3

Figure 9 shows the performance of all configurations referred to in Table 2, complemented by the default EMOA configurations and the individual SO configurations for SP and HV. Almost all of the non-dominated configurations dominate the default configuration for every EMOA, confirming the potential of AAC in general. When considering the SO configurations, 8 out of 14 are not dominated by others when comparing to the configurations obtained with MO-AAC. Thus, they would be part of the non-dominated front if found by MO-AAC. This gives reason that MO-AAC is, in theory, also able to find these configurations. This is actually the case for MOEA/D, where the SO configuration for SP is one found by MO-AAC as well. An exception here is MOGSA, where the SO config-

uration is the only non-dominated configuration overall and even dominates the
SO configuration for HV.

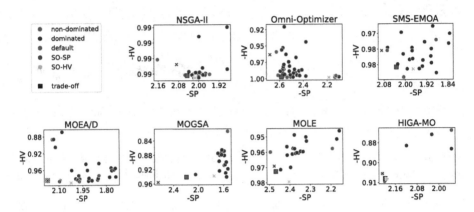

Fig. 9. MO-AAC, SO-AAC and default configurations for every EMOA separately,
trained and validated on all instances. The best trade-off solution reflects the carefully
picked best compromise between both objectives.

Aiming to provide recommendations for the overall best performing algorithm
when optimally configured, Fig. 10 combines all non-dominated (including best
hand-picked trade-off) MO-AAC configurations from the sub-figures of Fig. 9
relating to R1 in a visual way reflecting EMOA competitiveness. The optimally
configured Omni-Optimizer is the overall most performant EMOA followed by
NSGA-II with also a few configurations on the overall trade-off surface, however,
falling substantially behind in SP compared to a small loss in HV only. This is
unsurprising since Omni-Optimizer is an extension of NSGA-II to favor a diverse
decision space. SMS-EMOA and MOEA/D are second in line and comparable,
however, clearly dominated overall confirming results of R2b, while HIGA-MO
and MOGSA rank worst.

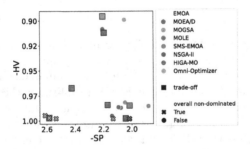

Fig. 10. Combined non-dominated MO-AAC configurations for all EMOAs, trained
and validated on all instances.

Figure 11 shows the actual parameter values of the two most competitive algorithms Omni-Optimizer and NSGA-II. The new recommended default configurations corresponds to the selected best trade-off solution.

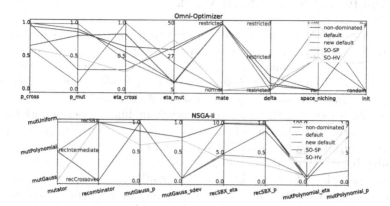

Fig. 11. Summary of all non-dominated, default, SO, and proposed new default configuration parameters for Omni-Optimizer and NSGA-II corresponding to Fig. 9. All configurations were found by training and testing on all instances.

4.3   Summary and Discussion of Results

Altogether, the results give interesting insights into the trade-off between HV and SP in the extremes using MO-AAC. MO-AAC is capable of finding configurations that have competitive performance with SO-AAC (Fig. 5). This is underlined by Fig. 9, where in most cases, the SO-AAC and MO-AAC configurations form a non-dominated front when considered together. Thus, MO-AAC is able to find trade-off configurations that lie in between the SO-AAC configurations but is also able to find configurations that overlap or even dominate SO-AAC configurations.

The second insight is the dominant performance of Omni-Optimizer. It achieved the best average ranking regarding HV^* of the non-dominated MO-AAC configurations. A contributing factor here is the ability of Omni-Optimizer to find multiple (non-dominated), and specifically high performing configurations (Table 2). Second, Omni-Optimizer substantially outperforms the other EMOAs regarding performance w.r.t. SP and achieves competitive performance w.r.t. to HV. Here it is tied with SMS-EMOA and only NSGA-II achieves a better ranking. This is unsurprising since Omni-Optimizer is an extension of NSGA-II, focusing on preserving diversity in decision space. Both aspects are also reflected in Figs. 9 and 10.

Moreover, experimental results allow us to recommend new default configurations for each EMOA – under similar problem dimensionalities – dominating the previous defaults for both HV and SP. Figure 11 displays the new default configurations for Omni-Optimizer and NSGA-II. All other configurations can be found with the experimental code.

5   Conclusion and Future Work

In this context, various EMOAs were automatically configured to simultaneously generate diverse solutions in the decision space and foster convergence towards the Pareto front. These configurations were obtained using the model-based AAC framework (MO-)SMAC which demonstrated its high performance and potential for solving multi-objective configuration tasks. As test instances, a set of multi-modal multi-objective optimisation instances from different benchmark function collections were utilised. Omni-Optimizer is shown to outperform competing EMOAs regarding the trade-off behaviour w.r.t. HV and SP when optimally configured by MO-AAC, it shows high versatility and competitiveness. NSGA-II shows comparable performance but smaller versatility in configuring for diversity in decision space. Overall, we were able to recommend new default configurations in similar scenario conditions for all considered EMOAs which resulted in higher performance regarding both performance criteria simultaneously.

The experimental results are promising in that there is a lot to gain by further pursuing this line of research, fostering algorithm design, understanding of algorithm behaviour, and automated configuration. A straightforward extension of this work will be investigating scalability and generalisability of results, both in terms of enlarging the dimensionality of the decision space, and possibly also objective space. Here, the normalisation of SP needs to be looked at as it becomes difficult to compare different instances in higher and differing dimensionalities. Parallel to this, the effect of the EMOA's evaluation budgets can be investigated. Also, other benchmark sets, such as recently provided in [24], can be included. Experimental results can then be further detailed and more specifically analysed, focusing on specific benchmark sets and problem characteristics.

Moreover, MO configuration studies internally solve specific MOO problems themselves, and it is crucial to analyse the characteristics of the resulting optimisation landscapes in order to get a detailed understanding of problem hardness and structural properties such as multi-modality. This will further help in understanding the performance differences of different configurators applied to the underlying scenario. Respective experiments will thus include a comparison with MO-ParamILS [4] and potentially specific racing approaches as well. Robustness of resulting EMOA configurations is also an issue and could even be integrated as a third performance criterion into the MO configuration scenario. However, different notions of robustness exist which could be explored first in this regard.

Additionally, automated configurators do have parameters themselves which should be analysed further regarding parameter importance and sensitivity of results regarding the chosen settings within SO- and MO-SMAC. However, one has to be careful not to end up in a 'vicious circle' of meta-configuring configurators.

References

1. Afsar, B., Fieldsend, J.E., Guerreiro, A.P., Miettinen, K., Rojas Gonzalez, S., Sato, H.: Many-Objective Quality Measures. In: Brockhoff, D., Emmerich, M., Naujoks, B., Purshouse, R. (eds.) Many-Criteria Optimization and Decision Analysis. Natural Computing Series, pp. 113–148. Springer, Cham (2023). https://doi.org/10.1007/978-3-031-25263-1_5

2. Audet, C., et al.: Performance indicators in multiobjective optimization. Eur. J. Oper. Res. **292**(2), 397–422 (2021). issn: 0377–2217. https://doi.org/10.1016/j.ejor.2020.11.016

3. Beume, N., Naujoks, B., Emmerich, M.: SMS-EMOA: multiobjective selection based on dominated hypervolume. Eur. J. Oper. Res. **181**(3), 1653–1669 (2007). issn: 03772217. https://doi.org/10.1016/j.ejor.2006.08.008

4. Blot, A., Hoos, H.H., Jourdan, L., Kessaci-Marmion, M.É., Trautmann, H.: MO-ParamILS: a multi-objective automatic algorithm configuration framework. In: Festa, P., Sellmann, M., Vanschoren, J. (eds.) LION 2016. LNCS, vol. 10079, pp. 32–47. Springer, Cham (2016). https://doi.org/10.1007/978-3-319-50349-3_3

5. Brockhoff, D., et al.: Using Well-Understood Single-Objective Functions in Multiobjective Black-Box Optimization Test Suites (2016). https://doi.org/10.48550/ARXIV.1604.00359

6. Burke, E.K., et al.: Hyper-heuristics: a survey of the state of the art. J. Oper. Res. Soc. **64**(12), 695–1724 (2013). issn: 0160–5682, 1476–9360. https://doi.org/10.1057/jors.2013.71

7. Coello, C.A.C., Lamont, G.B., Van Veldhuisen, D.A.: Evolutionary algorithms for solving multi-objective problems. 2nd ed. Genetic and Evolutionary Computation Series. Springer, New York (2007). isbn: 978-0-387-36797-2

8. Deb, K., Tiwari, S.: Omni-optimizer: a procedure for single and multi-objective optimization. In: Coello Coello, C.A., Hernández Aguirre, A., Zitzler, E. (eds.) EMO 2005. LNCS, vol. 3410, pp. 47–61. Springer, Heidelberg (2005). https://doi.org/10.1007/978-3-540-31880-4_4

9. Deb, K., et al.: A fast and elitist multiobjective genetic algorithm: NSGA-II. IEEE Trans. Evol. Comput. **6**(2), 182–197 (2002). issn: 1089778X. https://doi.org/10.1109/4235.996017

10. Deb, K., et al.: Scalable test problems for evolutionary multiobjective optimization. In: Evolutionary Multiobjective Optimization, pp. 105–145. Springer, London (2005). isbn: 978-1-85233-787-2. https://doi.org/10.1007/1-84628-137-7_6

11. Grimme, C., Kerschke, P., Trautmann, H.: Multimodality in multi-objective optimization – more boon than bane? In: Deb, K., Goodman, E., Coello Coello, C.A., Klamroth, K., Miettinen, K., Mostaghim, S., Reed, P. (eds.) EMO 2019. LNCS, vol. 11411, pp. 126–138. Springer, Cham (2019). https://doi.org/10.1007/978-3-030-12598-1_11

12. Grimme, C., et al.: Peeking beyond peaks: challenges and research potentials of continuous multimodal multi-objective optimization. In: Computers and Operations Research 136, 105489 (2021). issn: 03050548. https://doi.org/10.1016/j.cor.2021.105489

13. Heins, J., et al.: BBE: basin-based evaluation of multimodal multiobjective optimization problems. In: Parallel Problem Solving from Nature - PPSN XVII, vol. 13398. Cham: Springer (2022), pp. 192–206. isbn: 978-3-031-14714-2. https://doi.org/10.1007/978-3-031-14714-2_14

14. Hoos, H.H.: Automated algorithm configuration and parameter tuning. In: Hamadi, Y., Monfroy, E., Saubion, F. (eds.) Autonomous Search, pp. 37–71. Springer, Heidelberg (2011). https://doi.org/10.1007/978-3-642-21434-9_3

15. Hutter, F., et al.: ParamILS: an automatic algorithm configuration framework. J. Artif. Intell. Res. **36**, 267–306 (2009). issn: 1076–9757. https://doi.org/10.1613/jair.2861

16. Hutter, F., Hoos, H.H., Leyton-Brown, K.: Sequential model-based optimization for general algorithm configuration. In: Coello, C.A.C. (ed.) LION 2011. LNCS, vol. 6683, pp. 507–523. Springer, Heidelberg (2011). https://doi.org/10.1007/978-3-642-25566-3_40

17. Lindauer, M., et al.: SMAC3: a versatile bayesian optimization package for hyperparameter optimization. JMLR **23**(54), 1–9 (2022). http://jmlr.org/papers/v23/21-0888.html

18. López-Ibáñez, M., et al.: The irace package: iterated racing for automatic algorithm configuration. Oper. Res. Perspectives **3**, 43–58 (2016). issn: 22147160. https://doi.org/10.1016/j.orp.2016.09.002

19. Nemenyi, P.B.: Distribution-free Multiple Comparisons. Ph.D. thesis. Princeton University (1963)

20. Zhang, Q., Li, H.: MOEA/D: a multiobjective evolutionary algorithm based on decomposition. IEEE Trans. Evol. Comput. **11**(6), pp. 712–731 (2007). issn: 1941–0026, 1089–778X. https://doi.org/10.1109/TEVC.2007.892759

21. Rook, J., et al.: MO-SMAC: multi-objective sequential model-based algorithm configuration. In: Manuscript Under Review, pp. 1–8 (2024)

22. Rook, J., et al.: On the potential of automated algorithm configuration on multimodal multi-objective optimization problems. In: Proceedings of the Genetic and Evolutionary Computation Conference Companion, pp. 356–359. ACM, Boston, July 2022. isbn: 978-1-4503-9268-6. https://doi.org/10.1145/3520304.3528998

23. Schäpermeier, L., Grimme, C., Kerschke, P.: MOLE: digging tunnels through multimodal multi-objective landscapes. In: Proceedings of the Genetic and Evolutionary Computation Conference. Boston Massachusetts: ACM, July 2022, pp. 592–600. isbn: 978-1-4503-9237-2. https://doi.org/10.1145/3512290.3528793

24. Schäpermeier, L., et al.: Peak-a-boo! generating multi-objective multiple peaks benchmark problems with precise pareto sets. In: Evolutionary Multi-Criterion Optimization, vol. 13970, pp. 291–304. Springer, Cham (2023). https://doi.org/10.1007/978-3-031-27250-9. isbn:978-3-031-27250-9_21

25. Schede, E., et al.: A survey of methods for automated algorithm configuration. J. Artif. Intell. Res. **75**, 425–487 (2022). issn: 1076–9757. https://doi.org/10.1613/jair.1.13676

26. Solow, A.R., Polasky, S.: Measuring biological diversity. In: Environ. Ecol. Stat. **1**(2), 95–103 (1994). issn: 1352–8505, 1573–3009. https://doi.org/10.1007/BF02426650

27. Ulrich, T., Thiele, L.: Maximizing population diversity in single-objective optimization. In: Proceedings of the 13th Annual Conference on Genetic and Evolutionary Computation. Dublin Ireland: ACM, July 2011, pp. 641–648. isbn: 978-1-4503-0557-0. https://doi.org/10.1145/2001576.2001665

28. Wang, H., Deutz, A., Bäck, T., Emmerich, M.: Hypervolume indicator gradient ascent multi-objective optimization. In: Trautmann, H., Rudolph, G., Klamroth, K., Schütze, O., Wiecek, M., Jin, Y., Grimme, C. (eds.) EMO 2017. LNCS, vol. 10173, pp. 654–669. Springer, Cham (2017). https://doi.org/10.1007/978-3-319-54157-0_44

29. Yue, C., et al.: A novel scalable test problem suite for multimodal multiobjective optimization. Swarm Evol. Comput. **48**, 62–71 (2019). issn: 22106502. https://doi.org/10.1016/j.swevo.2019.03.011

30. Zitzler, E., et al.: Performance assessment of multiobjective optimizers: an analysis and review. IEEE Trans. Evol. Comput. **7**(2), 117–132 (2003). issn: 1089–778X. https://doi.org/10.1109/TEVC.2003.810758

A Simple Statistical Test Against Origin-Biased Metaheuristics

Aidan Walden and Maxim Buzdalov[✉]

Aberystwyth University, Aberystwyth, UK
mbuzdalov@gmail.com

Abstract. One of the strong points of evolutionary algorithms and other similar metaheuristics is their robustness, which means that their performance is consistent across large varieties of problem settings. In particular, such algorithms avoid preferring one solution to another unless the optimized function gives enough reasons for doing that. This property is formally captured as invariance with regards to certain transformations of the search space and the problem definition, such as translation or rotation.

The lack of some basic invariance properties in some recently proposed "nature-inspired" algorithms, together with the deliberate misuse of commonly used benchmark functions, can present them as excellent optimizers, which they are not. One particular class of such algorithms, origin-biased metaheuristics, are good at finding an optimum at the origin and are much worse for any other purpose.

This paper presents a statistical testing procedure which can help to reveal such algorithms and to illustrate the negative aspects of their behavior. A case study involving 15 different algorithms shows that this test successfully detects most origin-biased algorithms.

Keywords: Biased algorithms · Statistical tests · Nature-inspired algorithms

1 Introduction

An implicit assumption behind good designs of algorithms intended for black-box search and optimization is that such algorithms shall perform equally well for problems that are identical up to certain natural transformations. This assumption is very useful for reasoning about these algorithms: once they are shown to work well on one problem from the class of identical problems, all other problems are covered as well. This also contributes to the robustness of the algorithms, because the algorithm essentially no longer depends on problem features considered unimportant.

For pseudo-Boolean optimization problems, such natural transformations include inverting bit values and exchanging bit indices in the problem definition. For instance, the problem of finding a bit string maximizing the number of ones is very similar to maximizing the number of zeros, so it is fair to expect from

© The Author(s), under exclusive license to Springer Nature Switzerland AG 2024
S. Smith et al. (Eds.): EvoApplications 2024, LNCS 14634, pp. 322–337, 2024.
https://doi.org/10.1007/978-3-031-56852-7_21

a good algorithm to have identical performance on these two problems. Formally, this concept is captured in a definition of *unbiased algorithms* [13], and, conceptually, easily extends to arbitrary discrete search spaces [23]. As a side effect, this enables better evaluation of the complexity of optimization problems [5].

In the case of continuous optimization, there are also some formal concepts that are helpful in evaluating how robust or symmetric an algorithm is, typically expressed in form of *invariants*. For instance, a problem of minimizing $f(\boldsymbol{x}) = x_1 x_2$ with constraints $-1 \leq x_1, x_2 \leq 1$ is sufficiently similar to a problem of minimizing $f(\boldsymbol{y}) = y_1 y_2 - y_1 - y_2 + 1$ with constraints $0 \leq y_1, y_2 \leq 2$, because the latter can be transformed into the former by a substitution $x_1 := y_1 - 1$ and $x_2 := y_2 - 1$. One may expect that any algorithm that is claimed to be a black-box optimizer should have identical performance on these two problems, which is captured in a notion of *translational invariance*. Similarly, one can consider multiplying variables and constraints by constants different from zero (*scaling invariance*), including mirroring of the search space, and even rotations of the search space: an algorithm which does not make any difference between such problems can be called *affine invariant*.

Necessary Invariance and Benchmark Suites. While rotation invariance can be difficult to achieve in high-performing algorithms, for which the complexity of CMA-ES [9] can be a good example, translation and scaling invariance properties are considered to be a necessary feature of any real-valued black-box optimizer. To support this claim, we may cite the problem definition of the CEC-2014 competition on single-objective real-parameter numeric optimization [14]. Here, all test functions, most of which have their optimum at the origin, such as the "bent cigar" function $f_2(\boldsymbol{x}) = x_1^2 + 10^6 \sum_{i=2}^{D} x_i^2$, are required to be shifted by a vector $\boldsymbol{o}$, which is provided for each function in a separate file and is unknown to the optimizer, and they are also required to be scaled. Similar requirements are present in all such competitions nowadays.

Problems from such competitions are typically used for benchmarking for many subsequent years after the competition is over. Outside of the strict competition framework, however, it is difficult to control whether these problems are used correctly, and, in particular, whether the necessary transformations are applied to the functions. This made it possible to propose algorithms that favor optima at the origin and claim that they are superior by providing comparisons on benchmarks such as CEC-2014 without applying the mandatory shift vectors. In some of the cases, a series of algorithm designs and flaws in the use of benchmarks by the same authors hints, in our opinion, at malicious intents.

Origin-Biased Algorithms. An example of such an algorithm is the sine-cosine algorithm [17], which features the following update rule:

$$x_{ij} = x_{ij} + r_1 \sin(r_2) \cdot |r_3 y_j - x_{ij}|,$$

where x_i and y are individuals (so x_{ij} and y_j are their j-th decision variables), r_1 is some positive constant depending on the current iteration, r_2 is uniformly sampled from $[0; 2\pi)$, and r_3 is uniformly sampled from $[0; 2)$. Due to the presence of r_3, which is not identical to one, this update rule does not preserve translational invariance, as $|r_3(y_j + o_j) - (x_{ij} + o_j)|$ is not identical to $|r_3 y_j - x_{ij}|$ for an arbitrary shift component o_j. This algorithm has already been criticized for this design pattern and its performance consequences [2]. In particular, the change to x_{ij} will more likely be small when x_{ij} and y_j are closer to the origin, which nicely emulates adaptive step size when the optimum is at the origin too, but does an improper job otherwise.

Translational invariance can easily be satisfied when a new individual is either sampled uniformly from the constrained search space or created as a linear combination of other individuals

$$x_{\text{new}} = \alpha_1 x_1 + \alpha_2 x_2 + \ldots + \alpha_n x_n,$$

such that $\sum_i \alpha_i$ sums up to exactly one, and is obviously violated when this sum is not identical to one. However, algorithm descriptions can be vague, and the degree to which finding an optimum at the origin is easier (or harder) than at a significant distance from the origin, cannot be easily inferred from the provided update equations. As a result, an experimental tool is needed to identifying origin-biased algorithms and demonstrate the difference in convergence speeds.

Our Contribution. We propose a simply-defined benchmark problem with two identical and symmetrically located global optima (and without any additional local optima), one of which is at the origin. Using this problem, we investigate the sequences of best known solutions over several iterations across a number of independent runs, and analyse them with non-parametric statistical tests: whenever a test detects a statistically significant difference in behavior around these global optima, we may suspect that the algorithm in question is origin-biased.

We also perform an experimental evaluation of the proposed technique using 15 algorithms, ranging from the classic ones to modern efficient optimizers, and also including a variety of the so-called "nature-inspired" metaheuristics, many of which are origin-biased. We show that this procedure is indeed able to distinguish certain classes of algorithms and produce a clear evidence that they work worse when the optimum is not at the origin.

Related Work. A number of recent works, such as [1,4,24], complain about a flurry of "novel" metaheuristic algorithms that use questionable metaphors in questionable way, often solely to make an impression of something new. Some works [3] explicitly point out that many of these algorithms are copies of the already existing algorithms, although our paper indicates that some of these claims are not exactly true (and these algorithms are something even worse).

A research which uses statistical machinery to reveal potentially unwanted dependencies of the performance of algorithms on the properties of the problem being solved has been recently conducted in [28], where the authors investigate the concept of *structural bias*. This elaborate suite of experiments along with statistical tests can determine different possible kinds of biases, such as, for instance, the absence of rotational invariance in some classical variations of differential evolution [25], as well as the degree of biasedness. However, they seem to be focused more on fine effects found within the classic algorithms. What is more, it is too easy to produce various falsely positive conclusions about the presence of bias when constraint optimization is considered and constraints are not modified in a correct way along with the rest of the problem definition, a pitfall that can be found in subsequent work [27] of the same authors.

The effects of the use of linear recombination operators that violate translational invariance have been extensively studied in [26] with an obvious conclusion that such operators do not contibute positively if the optimum is away from the origin. Yet another recent work [12] also explores the bias towards origin as its main topic, however, it mainly concentrates on the final optimization outcomes, and does not attempt to measure the effect of the bias.

Structure of the Paper. In Sect. 2, we are going to describe the statistical procedure aimed at detecting origin-biased algorithms. The experimental evaluation of this procedure is detailed in Sect. 3. Section 4 concludes the paper.

2   Proposed Testing Procedure

In this section we describe the testing procedure along with some rationale behind it. The main idea of the procedure is to run an algorithm on a perfectly symmetric optimization problem with two different local optima, so that a perfectly unbiased algorithm can take either direction from the start contribute. We perform N runs of the algorithm on this problem for a limited number of iterations T and record, for each iteration, the best individual and its fitness. Ideally, for each fixed iteration, such record has equal chances of being closer to either of the local optima, and the fitness measures should not be dependent on which optimum is chosen. So if a difference in behavior is detected between the runs that are closer to either of the optima, the algorithm is likely biased.

We chose a variation of the so-called Sphere function, which amounts to sum of squares of variables. The number of decision variables D is set to 2, and the function is as follows:

$$f_T(\boldsymbol{x}) = \min\{x_1^2 + x_2^2, \quad (x_1 - 10)^2 + (x_2 + 10)^2\}$$
$$\text{subject to} -5 \leq x_1 \leq 15$$
$$-15 \leq x_2 \leq 5.$$

The values are chosen pretty much arbitrarily, such that the search space is perfectly symmetrical. We also found it useful to avoid having the matching variable values in the global optima, so that if an algorithm is biased with regards to one particular variable, we do not miss it. The problem is illustrated in Fig. 1.

The next question is what exactly to measure. Our first idea was to simply count how many runs of the algorithm ended up in either of the optima, $O_0 =$

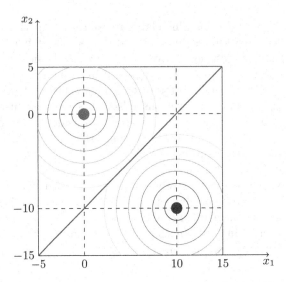

Fig. 1. An illustration of the optimization problem $f_T(x) = \min\{x_1^2 + x_2^2, (x_1 - 10)^2 + (x_2 + 10)^2\}$, a part of the testing procedure

$(0,0)$ or $O_1 = (10, -10)$, and if the algorithm converges to O_0 significantly more or less often than to O_1, we declare it origin-biased. However, the preliminary experiments showed that the algorithms available for evaluation are typically not biased in this regard, and for those which are, a very large number of runs is required to obtain any statistical significance.

The second idea was to measure the convergence speed among those runs which tend to either O_0 or O_1, and to compare whether these speeds are statistically different. To achieve this, we record the best individuals for each iteration in each run, which gives a set of values $x_1^{(1)}, \ldots, x_T^{(1)}, x_1^{(2)}, \ldots, x_T^{(2)}, \ldots, x_T^{(N)}$. Then, for each iteration number $1 \leq t \leq T$, we collect those which are closer to each of the optima as follows:

$$V_0 = \{x_t^{(i)} \mid ||x_t^{(i)} - O_0|| < ||x_t^{(i)} - O_1||\}$$
$$V_1 = \{x_t^{(i)} \mid ||x_t^{(i)} - O_0|| \geq ||x_t^{(i)} - O_1||\}.$$

The sets $F_0 = \{f_T(x) \mid x \in V_0\}$ and $F_1 = \{f_T(x) \mid x \in V_1\}$ represent the corresponding fitness values.

Our null hypothesis is that the algorithm in question is unbiased, which means that F_0 and F_1 are sampled from the same distribution. Since fitness values in such problems are far from being normally distributed, we use the non-parametric Wilcoxon rank sum test, also known as the Mann-Whitney U test [15, 29]. This way, we obtain a sequence of p-values, one per each iteration.

Using this sequence, we can investigate whether the bias is present, for which we may use the minimum p-value observed: the smaller the value, the more prominent the bias. If there is a bias, we can also see in which iterations it is the most prominent.

Before comparing these values with any thresholds, we need to perform a correction for multiple comparisons, as the adjacent p-values may be arbitrarily correlated. We use the Bonferroni correction [6], as it is the most conservative one. This correction can be interpreted as division of the threshold by the number of comparisons, which is T in our case.

The next section applies this idea to a number of existing algorithms and investigates the results.

3 Experiments

We implemented our experiments atop a Python framework that implements many algorithms we suspect to be biased: EvoloPy [8]. The list of the authors suggests that at least half of the algorithms with mostly the same authors have an implementation which was approved by the actual authors. We feel that this is important, so that we cannot be blamed for intentionally implementing the algorithms in a wrong way and claiming that they are biased.

As an example of an algorithm which is known to be unbiased we also use the reference Python implementation of CMA-ES [9], also maintained by the main author of the algorithm and his collaborators. We implemented thin wrappers around the corresponding frameworks, so that we can collect the intermediate best values $x_t^{(i)}$ mentioned above, and performed the analysis of the obtained traces. We used the number of runs $N = 10^3$ and the number of iterations $T = 100$. Also, we set identical population size $P = 50$ for all the algorithms.

Our experiments are available for reproduction on GitHub in the following repository: https://github.com/AidanWalden1/origin-based-stat-test.

The total list of all algorithms we tested is as follows. Except for the parameters mentioned above, all implementations are the defaults from their respective libraries.

1. Genetic algorithm [11];
2. Differential evolution [25];
3. Particle swarm optimization [7];
4. Cuckoo search [33];
5. CMA-ES [9] with and without boundary checking;
6. Bat algorithm [32];
7. Firefly algorithm [30,31];
8. Salp swarm algorithm [18];
9. Moth flame optimizer [16];
10. Multiverse optimizer [20];
11. Grey wolf optimizer [21];
12. Sine-cosine algorithm [17];
13. Harris hawks optimizer [10];
14. Whale optimization algorithm [19];
15. The Jaya algorithm [22].

For each algorithm the corresponding figure on the following pages contains two plots:

- The plot of p-values against the iteration number, with two additional lines corresponding to the significance levels $p = 0.05$ (uncorrected threshold) and $p = 0.0005$ (corrected threshold).
- The quartiles of the fitness values. In red, we plot the fitness values of the points that are closer to $O_0 = (0,0)$, and in blue the ones that are closer to $O_1 = (10, -10)$. If the algorithm is unbiased, these two should nearly coincide.

For the sake of brevity, we group similar algorithms and discuss them jointly in the following subsections.

3.1   Algorithms that Pass the Test

The following algorithms passed the test and are mathematically confirmed to have no origin bias: differential evolution, genetic algorithm, particle swarm optimization, cuckoo search, bat algorithm and firefly algorithm.

Of these, differential evolution, genetic algorithm, particle swarm optimization are the classical algorithms. For cuckoo search, it is known that is technically a fast evolution strategy [34] with Lévy distribution chosen for the heavy-tailed step size distribution. Similarly, bat algorithm and firefly algorithm are forms of particle swarm optimization per [3].

Figures 2–7 show that none of these algorithms obtained sufficient statistical significance for detecting a bias. The convergence plots, partitioned with regards to the optimum being converged to, also look very similar with only negligible differences. This confirms, to the degree possible using p-value-based statistical testing, that these algorithms are all unbiased with regards to translational invariance.

3.2   CMA-ES

The CMA-ES algorithm [9] is widely recognized as an unbiased algorithm that has very good performance in many applications, where maintaining rotational invariance, along with translational and scaling invariance, is one of the crucial enabling features to achieve this performance.

Our first move was to call CMA-ES in the same way as other algorithms by also specifying the box constraints. The CMA-ES algorithm does not natively include support for constrained optimization, and there are multiple ways to introduce this feature to this algorithm. In Fig. 8 the default constraint handling mechanism was enabled, which, surprisingly, resulted in occasionally significant bias detected in the first 50 iterations, and a vastly significant bias in the last 50 iterations.

In an attempt to clarify the matter, we also ran CMA-ES without any constraint-handling. The results are presented in Fig. 9. Here we see that the first 50 iterations can indeed be treated as having no bias, but the last 50 iterations again result in a high bias.

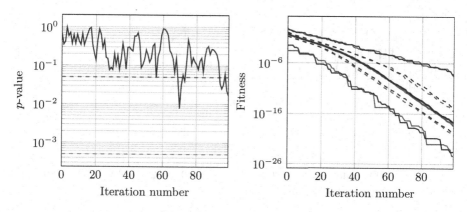

Fig. 2. Differential evolution [25], min p-value: 0.0077, corrected: 0.77

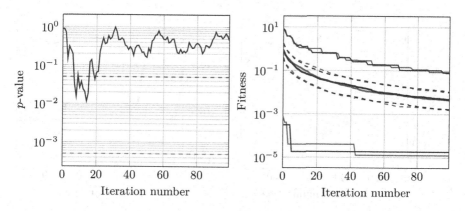

Fig. 3. Genetic algorithm [11], min p-value: 0.012, corrected: 1

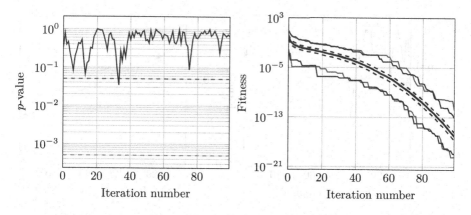

Fig. 4. Particle swarm optimization [7], min p-value: 0.034, corrected: 1

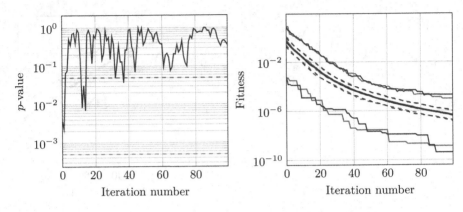

Fig. 5. Cuckoo search [33], min p-value: 0.0022, corrected: 0.22

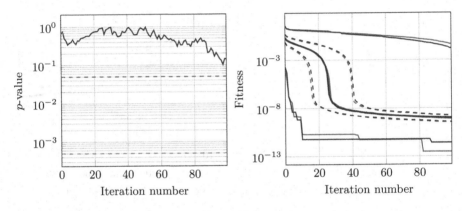

Fig. 6. Bat algorithm [32], min p-value: 0.1, corrected: 1

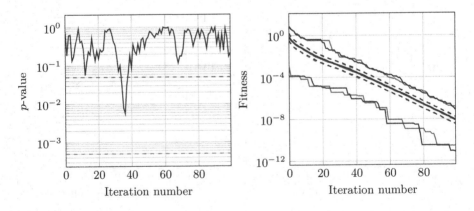

Fig. 7. Firefly algorithm [30,31], min p-value: 0.0056, corrected: 0.56

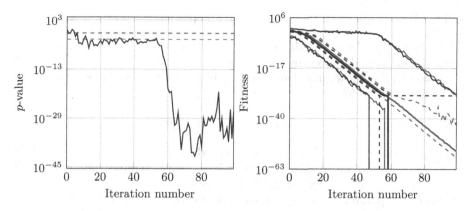

Fig. 8. CMA-ES [9] with constraint handling. Minimum p-value in first 50 iterations: $6.85 \cdot 10^{-7}$, corrected: $6.85 \cdot 10^{-5}$. Minimum p-value in last 50 iterations: $4.67 \cdot 10^{-42}$

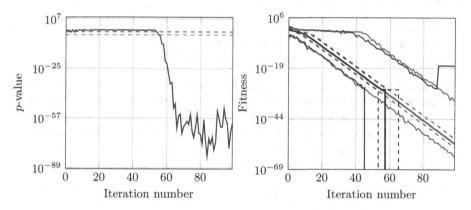

Fig. 9. CMA-ES [9] without constraint handling. Minimum p-value in first 50 iterations: 0.024, corrected: 1. Minimum p-value in last 50 iterations: $2.66 \cdot 10^{-82}$

However, by closely looking at the convergence plots, and also by remembering the IEEE 754 standard for floating-point arithmetics in computers, the effect appearing in the last 50 iterations can be in fact explained. While converging to $O_0 = (0,0)$, the algorithm can operate with extremely small numbers in a good precision, such as $5.301599682053987 \cdot 10^{-52}$, whereas while converging to $O_1 = (10, -10)$, this precision is not reachable, because in machine precision $10 + 5.301599682053987 \cdot 10^{-52}$ results in 10. As a result, the algorithm has to either stay at distances of order 10^{-17} from the optimum, or jump straight into it. This is why the blue curves in Figs. 8–9 diverge around iteration 50.

The mild bias in the constrained version of CMA-ES can, on the other hand, only be explained by some bias introduced by the constraint-handling code. This resembles some of the effects reported in [27], and requires further investigation.

3.3   Algorithms that Pass the Test but are Biased Differently

The salp swarm algorithm, the multiverse optimizer and the moth flame optimizer demonstrated no bias in our test, as shown in Figs. 10–12.

However, the salp swarm algorithm and the multiverse optimizer both contain an operator of the form:

$$x_i = x_j + \alpha(l + \beta(u - l)),$$

where α is chosen such that $|\alpha| \in (c_1, c_2)$, where c_1 is a positive constant, is a function of the number of iterations and the sign is sampled uniformly from $\{-1, +1\}$, and β is sampled uniformly from $[0; 1]$.

Such an operator is clearly biased with regards to constraints: if the constrains are centered ($l = -u$), this operator is unbiased, otherwise it is not. We performed an additional test following the same methodology as above, but on a problem defined as follows:

– minimize $f(x_1, x_2) = (x_1 - 99)^2 + (x_2 - 99)^2$;
– first constraint set: $-100 \le x_1, x_2 \le 100$;
– second constraint set: $98 \le x_1, x_2 \le 298$.

Two constraint sets are essentially mirrored with regards to the optimum, and since the function itself is symmetric, the problems are isomorphic and there should be no performance difference. Still, the minimum observed p-value obtained for this problem is $1.01 \cdot 10^{-162}$ for the multiverse optimizer and $1.9 \cdot 10^{-151}$ for the salp swarm algorithm. So, they are both clearly biased.

On the other hand, the moth flame optimizer uses the following update rule:

$$x_{ij} = y_{ij} + |x_{ij} - y_{ij}| \cdot \exp(t) \cdot \cos(2\pi t),$$

where x_i and y_i are individuals, and t is sampled uniformly from $[a; 1]$ where a linearly decreases from -1 to -2 over time. The expectation of $\exp(t) \cos(2\pi t)$ is not zero for most a, so there is clearly a directional bias. Our simple test framework, however, is apparently insufficient to detect this bias experimentally.

3.4   Algorithms that Fail the Test

The grey wolf algorithm, the sine-cosine algorithm, the Harris hawks optimization, the whale optimization algorithm and the Jaya algorithm clearly fail the developed test, as illustrated in Figs. 13–17. In all these cases, we see bias rising sharply from first iterations, and the convergence speeds are vastly different for different optima O_0 and O_1. All these algorithms clearly converge to the origin much faster than to the point distant from it, as illustrated by the red convergence plots getting down way quicker than the blue ones. This is clearly an indication that such algorithms do not fulfil their claims about performance stated in their respective papers.

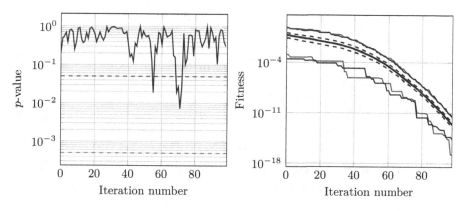

Fig. 10. Salp swarm algorithm [18], min p-value: 0.0073, corrected: 0.73

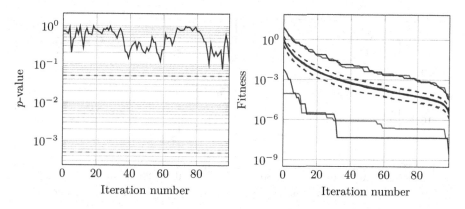

Fig. 11. Multiverse optimizer [20], min p-value: 0.076, corrected: 1

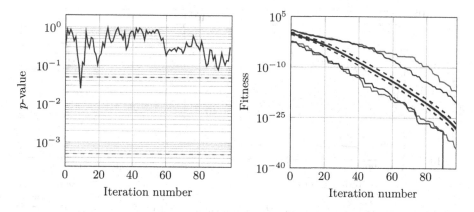

Fig. 12. Moth flame optimizer [16], min p-value: 0.025, corrected: 1

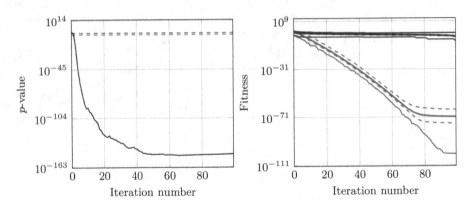

Fig. 13. Grey wolf optimizer [21] (Color figure online), min p-value: $1.21 \cdot 10^{-149}$, corrected: $1.21 \cdot 10^{-147}$

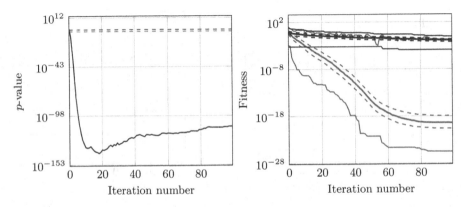

Fig. 14. Sine-cosine algorithm [17], min p-value: $1.10 \cdot 10^{-140}$, corrected: $1.10 \cdot 10^{-138}$

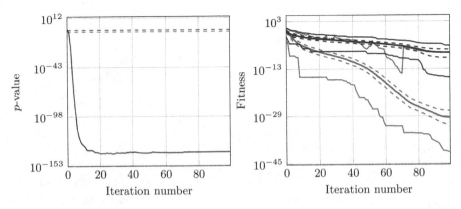

Fig. 15. Harris hawks optimization [10], min p-value: $2.2 \cdot 10^{-140}$, corrected: $2.2 \cdot 10^{-138}$

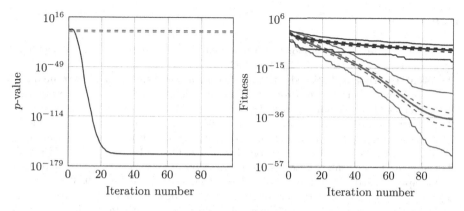

Fig. 16. Whale optimization algorithm [19], min p-value: $3.3 \cdot 10^{-164}$, corrected: $3.3 \cdot 10^{-162}$

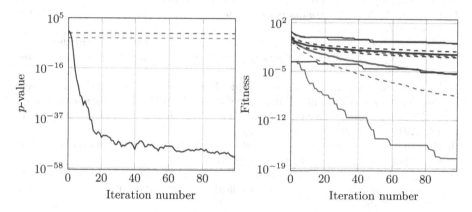

Fig. 17. Jaya [22], min p-value: $5.32 \cdot 10^{-54}$, corrected: $5.32 \cdot 10^{-52}$

4 Conclusion

We presented a simple and effective testing procedure based on statistical tests, that can detect the presence of a bias towards an origin in an optimizer claiming to be black-box. In our experimental study, it detected a significant proportion of algorithms that can be confirmed by code inspection to be origin-biased.

The case of CMA-ES though indicated that, for algorithms that converge really quickly, using the computer floating-point arithmetics following the IEEE 754 standard can introduce the effect which can also be interpreted as a bias, however, the actual optimizer cannot be blamed. Perhaps using different kind of floating-point arithmetics consistently can alleviate this issue.

We hope that this paper can contribute to putting an end to the recent domination of poorly-designed metaheuristics, which in turn would ensure a brighter future to the domain of evolutionary computation.

References

1. Aranha, C., et al.: Metaphor-based metaheuristics, a call for action: the elephant in the room. Swarm Intell. **16**, 1–6 (2022)
2. Askari, Q., Younas, I., Saeed, M.: Critical evaluation of sine cosine algorithm and a few recommendations. In: Proceedings of Genetic and Evolutionary Computation Conference Companion, pp. 319–320 (2020)
3. Camacho-Villalón, C.L., Dorigo, M., Stützle, T.: Exposing the grey wolf, moth-flame, whale, firefly, bat, and antlion algorithms: six misleading optimization techniques inspired by bestial metaphors. Int. Trans. Oper. Res. **30**, 2945–2971 (2023)
4. Campelo, F., Aranha, C.: Ec bestiary: A bestiary of evolutionary, swarm and other metaphor-based algorithms (2018). https://zenodo.org/records/1293352
5. Doerr, C.: Complexity theory for discrete black-box optimization heuristics. In: Doerr, B., Neumann, F. (eds.) Theory of Evolutionary Computation: Recent Developments in Discrete Optimization, pp. 133–212. Springer International Publishing, Cham (2020). https://doi.org/10.1007/978-3-030-29414-4_3
6. Dunn, O.J.: Multiple comparisons among means. J. Am. Stat. Assoc. **56**(293), 52–64 (1961)
7. Eberhart, R.C., Kennedy, J.: A new optimizer using particle swarm theory. In: Proceedings of the Sixth International Symposium on Micro Machine and Human Science, pp. 39–43 (1995)
8. Faris, H., Aljarah, I., Mirjalili, S., Castillo, P.A., Merelo, J.J.G.: Evolopy: an open-source nature-inspired optimization framework in python. In: Proceedings of International Joint Conference on Computational Intelligence. vol. 1 (ECTA), pp. 171–177 (2016)
9. Hansen, N., Ostermeier, A.: Completely derandomized self-adaptation in evolution strategies. Evol. Comput. **9**, 159–195 (2001)
10. Heidari, A.A., Mirjalili, S., Faris, H., Aljarah, I., Mafarja, M., Chen, H.: Harris hawks optimization: algorithm and applications. Futur. Gener. Comput. Syst. **97**, 849–872 (2019)
11. Holland, J.H.: Adaptation in Natural and Artificial Systems. University of Michigan (1975)
12. Kudela, J.: A critical problem in benchmarking and analysis of evolutionary computation methods. Nature Mach. Intell. **4**, 1238–1245 (2022)
13. Lehre, P.K., Witt, C.: Black-box search by unbiased variation. Algorithmica **64**, 623–642 (2012)
14. Liang, J.J., Qu, B.Y., Suganthan, P.N.: Problem definitions and evaluation criteria for the CEC 2014 special session and competition on single objective real-parameter numerical optimization. Tech. Rep. 201311, Nanyang Technological University (2013)
15. Mann, H.B., Whitney, D.R.: On a test of whether one of two random variables is stochastically larger than the other. Ann. Math. Stat. **18**(1), 50–60 (1947)
16. Mirjalili, S.: Moth-flame optimization algorithm: a novel nature-inspired heuristic paradigm. Knowledge-based Systems **89**, 229–249
17. Mirjalili, S.: SCA: a sine cosine algorithm for solving optimization problems. Knowl.-Based Syst. **96**, 120–133 (2016)
18. Mirjalili, S., Gandomi, A.H., Mirjalili, S.Z., Saremi, S., Faris, H., Mirjalili, S.M.: Salp swarm algorithm: a bio-inspired optimizer for engineering design problems. Adv. Eng. Softw. **114**, 163–191 (2017)

19. Mirjalili, S., Lewis, A.: The whale optimization algorithm. Adv. Eng. Softw. **95**, 51–67 (2016)
20. Mirjalili, S., Mirjalili, S.M., Hatamlou, A.: Multi-verse optimizer: a nature-inspired algorithm for global optimization. Neural Comput. Appl. **27**, 495–513 (2016)
21. Mirjalili, S., Mirjalili, S.M., Lewis, A.: Grey wolf optimizer. Adv. Eng. Softw. **69**, 46–61 (2014)
22. Rao, R.V., Saroj, A.: A self-adaptive multi-population based Jaya algorithm for engineering optimization. Swarm Evol. Comput. **37**, 1–26 (2017)
23. Rowe, J., Vose, M.: Unbiased black box search algorithms. In: Proceedings of Genetic and Evolutionary Computation Conference, pp. 2035–2042 (2011)
24. Sörensen, K.: Metaheuristics-the metaphor exposed. Int. Trans. Oper. Res. **22**, 3–18 (2015)
25. Storn, R., Price, K.: Differential evolution - a simple and efficient heuristic for global optimization over continuous spaces. J. Global Optim. **11**(4), 341–359 (1997)
26. Tsai, H.C.: Potential bias when creating a differential-vector movement algorithm. Appl. Soft Comput. **113**, Part A, 107925 (2021)
27. Vermetten, D., Caraffini, F., van Stein, B., Kononova, A.V.: Using structural bias to analyse the behaviour of modular CMA-ES. In: Proceedings of Genetic and Evolutionary Computation Companion, pp. 1674–1682 (2022)
28. Vermetten, D., van Stein, B., Caraffini, F., Minku, L.L., Kononova, A.V.: BIAS: a toolbox for benchmarking structural bias in the continuous domain. IEEE Trans. Evol. Comput. **26**(6), 1380–1393 (2022)
29. Wilcoxon, F.: Individual comparisons by ranking methods. Biometrics Bull. **1**(6), 80–83 (1945)
30. Yang, X.S.: Firefly algorithms for multimodal optimization. In: Stochastic Algorithms: Foundations and Applications, pp. 169–178. No. 5792 in Lecture Notes in Computer Science (2009)
31. Yang, X.S.: Firefly algorithm, stochastic test functions and design optimisation. Int. J. Bio-Inspired Comput. **2**(2), 78–84 (2010)
32. Yang, X.S.: A new metaheuristic bat-inspired algorithm. In: Nature inspired cooperative strategies for optimization, pp. 65–74. No. 284 in Studies in Computational Intelligence (2010)
33. Yang, X.S., Deb, S.: Cuckoo search via Lévy flights. In: Proceedings of World Congress on Nature and Biologically Inspired Computing, pp. 210–214 (2009)
34. Yao, X., Liu, Y.: Fast evolution strategies. In: Proceedings of International Conference on Evolutionary Programming, pp. 149–161. No. 1213 in Lecture Notes in Computer Science (1997)

Computational Intelligence
for Sustainability

Optimizing Urban Infrastructure for E-Scooter Mobility

Diego Daniel Pedroza-Perez$^{(\boxtimes)}$ ⓘ, Jamal Toutouh ⓘ, and Gabriel Luque ⓘ

ITIS Software, Universidad de Málaga, Málaga, Spain
{pedroza,jamal,gluque}@uma.es

Abstract. This paper addresses the optimization of urban infrastructure for e-scooter mobility through a multi-criteria approach. The proposed problem considers redesigning road infrastructure to integrate e-scooters into a city's multimodal transportation system. The objectives involve improving cycle lane coverage for e-scooters while minimizing installation costs. A parallel multi-objective evolutionary algorithm is introduced to solve this problem, applied to a real-world instance based on Málaga city data. The results showcase the algorithm's effectiveness in exploring the Pareto front, offering diverse trade-off solutions. Key solutions are analyzed, highlighting different zones with varying trade-offs between travel time improvement and installation costs. Visualization of proposed infrastructure changes illustrates significant reductions in travel time and enhanced multimodality. Computational efficiency analysis indicates successful parallelization, achieving substantial speedup and high efficiency with up to 32 processing elements.

Keywords: E-scooter mobility · Urban infrastructure · Parallel multi-objective optimization

1 Introduction

In recent years, micromobility transport means, such as bikes, e-scooters, and other electric micromobility devices, have been adopted to urban transportation by citizens as a sustainable alternative to internal combustion engine vehicles. For instance, 136 million shared micromobility trips were taken in 2019 in the United States [31]. This widespread use of micromobility devices has presented new challenges for urban and transportation planners, as they must now deal with a large influx of vehicles with varying sizes, mobility patterns, security requirements, and technologies on city streets [27].

This research is partially funded by the Universidad de Málaga (UMA); under grant PID 2020-116727RB-I00 (HUmove) funded by MCIN/AEI/10.13039/501100011033; under grant number PRE2021-100645 by MCIN/AEI/10.13039/501100011033 and by the FSE+; and TAILOR ICT-48 Network (No 952215) funded by EU Horizon 2020 research and innovation programme.The authors thank the Supercomputing and Bioinformatics center at the UMA for their computer resources and assistance.

S. Smith et al. (Eds.): EvoApplications 2024, LNCS 14634, pp. 341–357, 2024.
https://doi.org/10.1007/978-3-031-56852-7_22

Due to the novelty of e-scooters in urban transportation, few attempts have been made for appropriate specific micromobility network planning. Different issues regarding e-scooters have been discussed in recent literature [11], such as parking spot location [35] or injury analysis and prevention [11]. One advantage of e-scooters is that they are compact and easy to carry, store, and park. This feature makes it possible to integrate e-scooters as a viable option in multimodal transportation solutions [16, 32]; e.g., people can use an e-scooter for a part of their trip, then switch to walking or public transport.

This paper aims to integrate e-scooters as part of a multimodal transportation solution in cities. One solution to effectively include e-scooter mobility in cities is setting up specific road infrastructure, such as cycle lanes [2], to be shared by both bikes and e-scooters. Studies show that adding cycle lanes to roads with two or more lanes does not significantly impact traffic [15, 24]. Sustainable urban micromobility requires road redesign, including the installation of cycle lanes in road segments. This poses logistical challenges similar to other location-based public services in smart cities [6, 9, 21, 30].

Existing literature falls short in investigating the redesign of road infrastructure to enhance e-scooter transportation, with limited emphasis on cyclists. One study suggested employing geographic information system tools and the network robustness index (NRI) to assess the travel-time impact for drivers caused by implementing separated bike lanes in Toronto. The NRI computations were utilized to pinpoint suitable locations for these lanes, factoring in driver travel-time impacts [5]. Another study developed a comprehensive optimization framework using real data from a bike-sharing system. This framework formalizes the bike lane planning problem, incorporating cyclists' utility functions. The proposed integer optimization model maximizes utility by considering cyclists' route choices, accounting for bike trip coverage and lane continuity [19]. Notably, these studies overlook the economic cost associated with installing new bike lanes, a crucial consideration for the execution of such infrastructure projects.

This research proposes a novel problem named Multiobjective Urban Road Infrastructure Redesign for E-Scooters Integration (URIReI), a multi-criteria optimization problem that considers redesigning road infrastructure to integrate e-scooter mobility considering two relevant objectives: the cycle lane infrastructure coverage and the installation cost. A parallel multi-objective evolutionary algorithm is applied to address URIReI considering a real-world instance based on data of Málaga (Spain). The main contributions of this article are: *a)* defining and formulating URIReI, a new realistic multi-criteria optimization problem for redesigning the road infrastructure for e-scooters on a city scale, considering the cycle lanes network coverage and installation costs; *b)* proposing a parallel multi-objective evolutionary algorithm (pMoEA) to tackle URIReI; and *c)* defining a realistic instance using real-world data to address the optimization problem.

The rest of the article is organized as follows: Sect. 2 describes the URIReI problem. Section 3 presents the pMoEA applied in the experimentation. Sections 4 and 5 report the experimental setup and results. Finally, Sect. 6 presents the conclusions and formulates the main lines for future work.

2   The Multiobjective Urban Road Infrastructure Redesign for E-Scooters Integration Problem

The problem considered in this paper aims to select the best location for new cycle lanes in the current roads in a city to improve (multimodal) e-scooter trips' quality of service (QoS) by minimizing travel times and simultaneously minimizing the costs of modifying the roads (i.e., installing cycle lanes).

The QoS of a given solution (tentative city roads design) is evaluated regarding the reduction of travel times. This reduction is measured in terms of the proportion of time required to carry out the trips compared to the current (actual) design of the city in order to make it a minimization problem. Thus, the QoS assigned to solutions that do not improve anything is 1, and the QoS of a solution that reduces travel times by 33% is 0.66. In this version of the problem, the cost of installing cycle lanes depends on the length of the road and a constant that represents the monetary costs per unit length.

The two objectives of this problem (i.e., improving QoS and reducing cycle lane installation costs) are in conflict. While providing cycle lanes on every road in a city would offer optimal quality of service, it would also come with a steep economic cost and impede regular road traffic. To aid decision-makers in these matters, the main objective of this research is to define and address the proposed optimization problem to find new road designs that effectively balance the trade-offs between these competing objectives.

2.1   Problem Description

In this section, we will present the mathematical formulation of the problem. First, we will describe our system's input elements, including the map and routes. Then, we will move on to model our optimization problem.

To represent the map of a city, we use a directed graph, denoted as $M = (V, E)$. Here, E represents the edges (road segments of the city), while V represents the vertices (points of interest in the city, mainly road intersections). We further divide the edges into two distinct sets: $E_1 \subseteq E$ and $E_2 \subseteq E$, with $E_1 \cap E_2 = \emptyset$ and $E_1 \cup E_2 = E$. The set E_1 corresponds to road segments suitable for new infrastructure installation, specifically cycle lanes. The set E_2 denotes segments where installing such infrastructure is either unnecessary (due to existing e-scooter infrastructure) or unfeasible due to specific characteristics.

We associate each edge with two functions, d and t. The function $d : E \to \mathbb{R}$ represents the length of the road segment, while the function $t : E \times \{0, 1\} \to \mathbb{R}$ accounts for the reduction in travel time based on whether infrastructure is added. The latter is defined in Eq. 1.

$$t(e, m) = \begin{cases} 1 & \text{if } m = 0 \\ \frac{t_{\text{cycle\_lane}}(e)}{t_{\text{current}}(e)} & \text{otherwise} \end{cases} \quad (1)$$

In this model, t_{current} represents the time it currently takes to travel the road segment, while $t_{\text{cycle\_lane}}$ is the time required to cross it if a cycle lane is added.

For edges in E_2, where no infrastructure can be added, the function evaluates to 1, indicating no improvement in travel time. Similarly, for edges in E_1 where infrastructure can be added but is not ($m = 0$), the function also evaluates to 1. However, if the infrastructure is added ($m = 1$) for edges in E_1, the function evaluates to a value less than 1, indicating a potential reduction in travel time.

We also have a set of routes of interest: $R = \{r_i \mid r_i = (v_{o_i}, v_{d_i}), v_{o_i} \in V, v_{d_i} \in V\}$, where each route is represented by its origin (v_{o_i}) and destination (v_{d_i}).

In this optimization problem, given a city map $M = (V, E_1 \cup E_2)$ and a set of routes R, we are looking for a binary vector $\vec{x} = \{x_1, \ldots, x_{|E_1|}\}$, where each x_i is set to 1 if it is proposed to install a cycle lane on $e_i \in E_1$, or 0 if the road remains unchanged. This aims to minimize expressions 2 and 3.

$$f_1(\vec{x}) = \frac{1}{|R|} \cdot \sum_{r \in R} \frac{1}{|path(r, \vec{x})|} \cdot \left(\sum_{\substack{e \in path(r, \vec{x}) \\ e \in E_1}} t(e, x_e) + \sum_{\substack{e \in path(r, \vec{x}) \\ e \in E_2}} t(e, 0) \right) \quad (2)$$

$$f_2(\vec{x}) = \sum_{r \in R} \left(\sum_{\substack{e \in path(r, \vec{x}) \\ e \in E_1}} x_e \cdot d(e) \cdot C \right) \quad (3)$$

where $path(r, \vec{x})$ identifies the edges of the optimal path for route r using the infrastructure proposed by $\vec{x}$. The constant C signifies the cost per unit length required to install a cycle lane. Therefore, f_1 measures the average reduction in travel time for completing all routes using the proposed infrastructure (QoS), while f_2 represents the cost of implementing that infrastructure.

3   Parallel Multi-objective Evolutionary Algorithm

Addressing multi-criteria decision-making is key for improving sustainability in cities. Non-dominated Sorting Genetic Algorithm, version II (NSGA-II) [8], has been frequently used to solve the underlying multi-objective optimization problems, achieving highly competitive results [6, 21, 25, 29, 30, 34]. Since the evaluation of the individuals (tentative solutions) requires considerable computational costs (about 3 min for each individual), we applied a parallel master-slave version. This section summarizes NSGA-II, its operators and parallel implementation.

3.1   The NSGA-II Algorithm Applied to URIReI

NSGA-II employs a non-dominated elitist ordering strategy to enhance convergence speed [8]. It implements a crowding technique to preserve solution diversity and utilizes a fitness assignment method that considers dominance ranks and crowding distance values. Algorithm 1 shows the pseudo-code of NSGA-II. This paper used a NSGA-II that employs a file that saves all the non-dominated solutions of all evaluations.

Algorithm 1. Pseudo-code of the NSGA-II algorithm

1: $t \leftarrow 0$
2: offspring $\leftarrow \emptyset$
3: $\leftarrow$ **initialize**($P(0)$)
4: **while** not stopping_criterion **do**
5:     **evaluate**($P(t)$)
6:     R $\leftarrow P(t) \cup$ offspring
7:     fronts $\leftarrow$ **non-dominated sorting**(R))
8:     $P(t+1) \leftarrow \emptyset$; $i \leftarrow 1$
9:     **while** $|P(t+1)| + |\text{fronts}(i)| \leq N$ **do**
10:         **crowding distance**(fronts(i))
11:         $P(t+1) \leftarrow P(t+1) \cup$ fronts(i)
12:         $i \leftarrow i+1$
13:     **end while**
14:     **sorting by distance** (fronts(i))
15:     $P(t+1) \leftarrow P(t+1) \cup$ fronts(i)[1:(N - $|P(t+1)|$)]
16:     selected $\leftarrow$ **selection**($P(t+1)$)
17:     offspring $\leftarrow$ **evolutionary operators**(selected)
18:     $t \leftarrow t + 1$
19: **end while**
20: **return** computed Pareto front

3.2   Evolutionary Operators

Solution Encoding. Solutions are encoded as a binary vector $\vec{x}$ of the size of the search space, i.e., $\vec{x} = \{x_1, \ldots, x_{|E_1|}\}$. Each position in the vector represents a possible inclusion of a cycle lane in a specific road segment on the map. If $x_i=1$, a cycle lane is installed in the road segment represented by i. If $x_i=0$, no cycle lane is installed. Figure 1 illustrates an example of solution encoding of a scenario with 7 road segments (locations candidates) in which the road segments 2, 4, 5 and 6 have installed a cycle lane.

Initialization. The population is randomly initialized. For each position (x_i) in the solution vector of an individual, the initialization vector randomly assigns a binary value. The values 0 and 1 have the same probability of being selected. This process is repeated for each individual in the initial population.

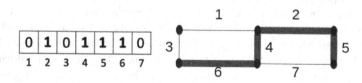

Fig. 1. Example of solution encoding of a scenario with 7 road segments.

Selection, Replacement, and Fitness Assignment. NSGA-II uses the $(\mu+\lambda)$ evolution model. The selection process is based on dominance through binary tournament selection. Those that are non-dominated are evaluated using crowding distance. Fitness assignment is performed by considering Pareto dominance rank, valid results, and crowding distance value.

Recombination and Mutation Operators. Through the evolutionary process, recombination and mutation operators are applied to create the offspring. A two-point crossover is used as a recombination operator between two individuals. This crossover operator, applied with a crossover probability of p_C, selects two random positions from the solution vectors representing the parents. It then exchanges the bits within those positions to create two new solutions. After the crossover operation, a multi-bitflip mutation is applied to the newly created solutions. This operator flips each bit's value with a mutation probability of p_M.

3.3   Parallel Master-Slave Implementation

By distributing tasks such as population splitting or fitness function evaluation across multiple processing elements, parallel implementations of EAs (pEAs) have shown the capability to achieve high-quality results within reasonable execution times [13]. The multiprocessor parallel NSGA-II proposed in this work aligns with the master-slave model, as classified by Alba et al. [1].

Given that the evaluation of the fitness functions of this problem demands a more significant computational time cost than the application of variation operators, it has been identified as a prime candidate for parallelization. Consequently, our master-slave pEA is structured hierarchically, with a master process performing the evolutionary search and controlling a group of slave processes that evaluate the fitness function.

4   Experimental Setup

This section describes the problem instance addressed in this study, the metrics used to evaluate our approach, and the execution platform. We also report the results of the experiments performed to select the parameters of our algorithm.

4.1   Problem Instance

Málaga, in Spain, was chosen as the case study for this research. Málaga has e-scooter regulations regarding the speed and roads that can be used [23,26]. The urban infrastructure graph of the city was obtained using real-world open data from Open Street Map [12]. The edges of this graph, which resemble the roads and streets, have attributes that correspond to additional information, such as the type, the length, or whether they have cycle lanes. In this work, we are interested only in cycle lanes since previous studies [7,33] suggest this structure is a safer option for e-scooter users.

The search space consisted of roads with two or more lanes, not motorways. Figure 2 is a coloured-painted map of Málaga. It displays the potential cycle lanes in green and in blue existing infrastructure. Furthermore, multimodal options are considered: subway, e-scooters, walking or a combination.

The experiment was conducted using real-world open data to test our algorithm. The graph was retrieved from Open Street Map and the official website of Málaga subway[1]. Points of interest and other additional data were extracted from Junta Andalusia website [17] and from the Málaga Open Data portal [20].

Fig. 2. A zoomed version of the map of Málaga without rural areas. The points in red are the subway stations (Color figure online)

The origin points are the centre of mass of the 11 districts of Málaga. The population density of each district's census segment was clustered to obtain each district's centre of mass. Routes to educational institutions have been taken into account, given that the primary users of this mode of transportation are typically individuals aged 16 and over [28]. A total of 558 routes had to be evaluated.

Installing dedicated cycle lanes in the real world incurs various expenses [22]. We use 480,000 euros per kilometer as reported in one of Spain's latest projects [4].

4.2   Evaluated Metrics

In this subsection, we will describe the metrics employed to analyze the results of our approaches from different perspectives.

Multiobjective Optimization Metrics. Relevant MO metrics are applied to assess the search capabilities of the proposed approach. Regarding the quality of the computed solutions (proximity to the Pareto front), generational distance (gd) and inverted generational distance (igd) are computed [14]. The *spread* metric is applied to evaluate the dispersion of non-dominated solutions found in the search [18], and the number of non-dominated solutions ($\#nds$) computed is evaluated. Finally, the combined relative hypervolume (rhv) metric is applied to analyze the coverage and dominance of the search space [36]. For those MO metrics that requires the real Pareto front (which is unknown for the real problem solved as case study), an approximation was computed by gathering all the non-dominated solutions found in all the executions performed.

---

[1] Málaga Subway website: www.metromalaga.es.

Computational Efficiency. The assessment of parallel algorithm performance commonly relies on speedup and efficiency. Speedup (s_m) measures the degree to which a parallel algorithm outpaces its corresponding sequential counterpart. It is evaluated as the ratio of execution times between the sequential algorithm (T_1) and the parallel version executed on m processing elements (T_m). Efficiency (e_m) is the normalized measure of speedup regarding the number of processing elements used in executing a parallel algorithm, facilitates the comparison of algorithms running on potentially dissimilar computing platforms.

Problem Related Metrics. To assess the effectiveness of the proposed solution, we utilize two key metrics: the total travel time required to complete all routes using the suggested infrastructure and the cost associated with implementing said infrastructure. Our calculation of travel time involves the utilization of Eq. 2, in which the optimal route $(path(r, \vec{x}))$ is determined by the Dijkstra algorithm. It is important to note that this process can be quite time-consuming due to the extensive search space and multitude of analyzed routes. We also summarize the cost of the proposed infrastructure through Eq. 3.

4.3 Parameter Settings

A set of parametric setting experiments was performed to determine the best parameter values for the proposed NSGA-II, which applies the operators described in Sect. 3.2. The population size ($\#pop$) and the maximum number of generations ($\#gen$) were calibrated in preliminary experiments. Due to the computational time cost required to evaluate the fitness of the individuals and limited access to computational resources, we have performed a limited number of fitness evaluations per independent run. Thus, after a preliminary analysis, the configuration with $\#pop=32$ and $\#gen=50$ provided a good exploration pattern. Candidate values for p_C and p_M were $p_C \in \{0.5, 0.7, 0.9\}$ and $p_M \in \{0.30, 0.10, 0.05, 0.01, 0.001\}$. Each configuration was assessed on 15 independent runs by using the rhv quality metric. Figure 3 summarizes the experimental results by showing the boxplot with the distribution of rhv results computed using each configuration. The most competitive competitive configuration was $p_C=0.9$ and $p_M=0.01$. The mean computational time cost was 108.61 min.

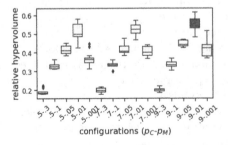
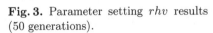
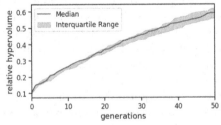

Fig. 3. Parameter setting rhv results (50 generations).

Fig. 4. Median of rhv for the 15 runs (50 generations).

After analyzing the convergence of the chosen configuration in terms of rhv (see to Fig. 4), we observed that the algorithm was unable to converge after 50 generations. In order to improve this aspect, we conducted twice as many fitness evaluations, as the computational time cost was still acceptable (expected to be twice as long). Thus, we repeated all parameterization experiments by conducting 100 generations. Boxplot with the distribution of rhv in Fig. 5 summarizes the experimental results. As expected, the results were significantly improved by doubling the number of generations. The configuration with p_C=0.9 and p_M=0.01 was the most competitive, as was the case with $\#gen$=50.

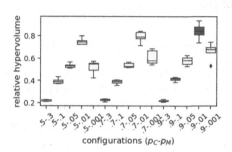

Fig. 5. Parameter setting rhv results (100 generations).

4.4 Execution Platform and Implementation Details

The parallel NSGA-II was implemented[2] in Python version 3.10.12. Our experiments were run using Slurm into the Picasso supercomputer of the University of Málaga that has a computation cluster of 126 X SD530 nodes: 52 cores (Intel Xeon Gold 6230R @ 2.10 GHz), 192 GB of RAM. InfiniBand HDR100 network. 950 GB of local scratch disks. Slurm is workload management software that allows the distribution of jobs into several resources.

The software required for the computations has been developed using Python. The code was implemented using specific software libraries such as Numpy, scikit-learn, and Pandas. OSMnx [3] and NetworkX were used to compute the routes and paths. Finally, DEAP [10] and multiprocessing library were employed to implement the NSGA-II and to parallel the implementation.

5 Experimental Evaluation

This section reports the experimental evaluation of the proposed approach addressing the problem over the presented instance on 30 independent runs. This analysis considers the multiobjective optimization metrics, the quality and the cost of the proposed solutions, and the computational efficiency.

---

[2] Github of the project: https://github.com/pedrozad/e-scooter-way.

5.1   Multi-objective Optimization Evaluation

Table 1 presents the main descriptive statistics of the results of the studied multi-objective optimization metrics. Since the Kruskal-Wallis statistical test rejected the null hypothesis that the results follow a normal distribution, median and interquartile range (iqr) are reported as quality and dispersion estimators. ↑ indicates that higher values of the reported metric are better, ↓ stands for lower values are better. Although we do not compare our approach with other methods, the authors consider it necessary to report the primary metrics used in multiobjective optimization because it may be helpful for further comparisons and reproducibility of the work.

Table 1. Results of MO metrics for the studied instance.

	minumum	median	iqr	maximum
rhv ↑	0.582	0.738	0.145	0.901
gd ($\times 10^{-3}$) ↓	0.194	0.557	0.478	1.088
igd ($\times 10^{-3}$) ↑	0.414	0.835	0.599	1.627
$spread$ ↑	0.059	0.198	0.088	0.369
#nds ↑	22	54	24	72

The computed relative hypervolume values range from 0.582 to 0.901, indicating that the algorithm successfully explores a substantial portion of the Pareto front. The median rhv of 0.738 suggests a consistent performance across independent runs, with higher values signifying a better coverage of the approximated problem Pareto front. The interquartile range (iqr) is 0.145, which underlines the reduced variability in the attained hypervolume values.

The generational distance values are scaled by 10^{-3} for readability. The gd results range from 0.194 to 1.088. Lower gd values are desirable as they indicate solutions closer to the Pareto front. The median gd value (0.557) suggests a good algorithm convergence toward the Pareto front. The inverted generational distance is also scaled by 10^{-3}. The igd results range from 0.414 to 1.627. Higher igd values are favourable, indicating solutions closer to the Pareto front. The median igd of 0.835 suggests effective convergence.

The spread metric assesses the distribution of non-dominated solutions along the Pareto front. The spread metric ranges from 0.059 to 0.369. Higher spread values are desirable as they indicate a more evenly distributed set of solutions. The median spread of 0.198 suggests a reasonable dispersion of solutions, while the iqr of 0.088 reflects variability in the spread across independent runs.

The number of non-dominated solutions ranges from 22 to 72 across runs. A higher number of non-dominated solutions implies a more diverse set of trade-off solutions. For initial populations of 32 individuals, the median #nds of 54 indicates a substantial number of high-quality solutions. The iqr of 24 highlights variability in the algorithm's ability to discover non-dominated solutions.

Summarizing, the proposed pNSGA-II performs commendably in addressing the multi-objective optimization problem for the studied instance. The consistency in achieving high rhv and the presence of a considerable number of non-dominated solutions underscore the algorithm's effectiveness in exploring the Pareto front. Variability in generational distance, inverted generational distance, and spread metrics across independent runs (i.e., iqr values) may indicate sensitivity to specific problem characteristics.

5.2 Solution Analysis and Interpretation

In the previous section, we discussed the numerical performance of our algorithmic approach. Now, we will shift our focus to evaluate the quality of the solutions in the problem domain. This section explores the trade-off between enhancing travel time and minimizing the installation cost of cycle lanes.

Our analysis begins with the presentation of Table 2, which showcases the key characteristics of essential solutions. The table highlights the extremes derived from the Pareto front, such as a solution that minimizes Quality of Service (es_T), as well as a solution that minimizes cost (es_C). Furthermore, it features the closest solution to the ideal vector (cs_{IV}). The table also includes an alternative scenario that covers the entire search space with cycle lanes (ALL) and the base case of our study (BASE), which only uses the current existing infrastructure.

Table 2. Descriptive indicators for some key solutions. Cost in millions of euros

	time (routes)				time (improvement)		cost
	min	mean	std	max	hours	%	(millions of €)
cs_{IV}	46.45	1543.68	0.39	3532.08	31.34	11.58	263.82
es_T	46.45	1539.21	0.39	3506.16	32.03	11.84	272.64
es_C	53.33	1583.98	0.38	3532.08	25.09	9.27	256.35
ALL	46.45	1393.05	0.40	3439.45	54.69	20.21	562.24
BASE	73.88	1745.87	0.38	3562.66	–	–	0.00

The table presents various solutions that offer a detailed analysis of the trade-offs between travel time improvement and cycle lane installation costs. The ALL scenario represents the upper limits in both travel time reduction (20.21%) and associated costs (€562.24 millions), covering the entire search space with cycle lanes. Although this scenario showcases the potential for substantial improvements in travel time, it requires a considerably higher economic investment compared to the base case (BASE), which serves as the reference.

Further analysis of the extremes reveals that the cs_{IV} solution, the closest to the ideal vector, strikes a balance with a significant 11.58% improvement in travel time while maintaining a relatively moderate cost increase. Conversely, the extreme solution minimizing Quality of Service (es_T) achieves the highest travel time reduction (11.84%) at a slightly increased cost, emphasizing the potential benefits of prioritizing travel time over cost considerations. On the other hand, the extreme solution minimizing cost (es_C) is a cost-efficient option with a 9.27% reduction in travel time. This solution reflects the scenario where minimizing installation costs takes precedence, showcasing the inherent trade-offs between economic efficiency and travel time optimization.

Although the cost of these solutions seems high (exceeding €250 millions) for only an 11–12% improvement (31–32 h), it's important to note that this cost-benefit analysis assumes a one-time use of each route. These cycle lanes would be utilized repeatedly over multiple years, amplifying the overall impact. Apart from the quantitative metrics, the investment in cycle lanes also yields substantial social benefits by facilitating mobility, enhancing the safety of alternative transportation users, and contributing to improvements in public health and the environment. The provision of such infrastructure also encourages the adoption of sustainable transportation alternatives, potentially reducing reliance on more pollutant modes of transport like cars.

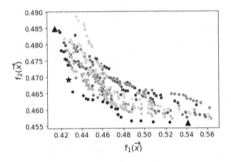

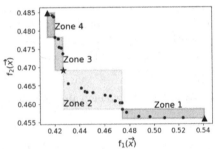

Fig. 6. All non-dominated solutions computed by each run.

Fig. 7. Non-dominated solutions computed for all runs and evaluated zones.

After analyzing some key solutions, we show all the non-dominated solutions computed in Fig. 6. The identified non-dominated solutions are marked by black dots, serving as an approximate representation of the problem's Pareto front. We use a black triangle ▲ to denote the extreme solutions minimizing Quality of Service (es_T) and minimizing cost (es_C). The black star ★ represents the closest solution to the ideal vector (cs_{IV}). It is worth noting that the objectives in the figure are normalized using the improvements in travel time and cost of the ALL and BASE solutions as reference points. This normalization provides a more comprehensive understanding of the non-dominated solutions. In terms of solution quality, numerous options provide considerable time improvements, exceeding fifty percent of the gains of the ALL solution, without incurring similarly high costs (less than fifty percent of the cost).

Figure 7 shows the non-dominated solutions computed in the whole experimentation. Besides, it includes four shaded zones identified, representing different trade-offs between travel time improvement and e-scooter lane installation costs. Table 3 summarizes the main results in terms of travel time and deployment cost for each zone. In **Zone 1**, the solutions offer significant enhancements in Quality of Service with minimal infrastructure costs. Investing in electric scooter lane infrastructure in this zone may be considered a cost-effective option for improving urban mobility system efficiency. In **Zone 2**, time improvements persist but with a gradual cost increase. These solutions strike a balance between travel time efficiency and economic efficiency. With 25 solutions in this zone, decision-makers have a range of options for achieving a reasonable compromise between travel time improvements and cost-effectiveness.

Zones 3 and **4** experience steeper cost increments in exchange for more substantial improvements in travel time efficiency. Although these solutions require a more significant investment, they offer the most substantial reductions in travel time. In environments where enhancing the transportation system's efficiency is a priority, and there is a broader budget, solutions from these zones could be considered for more strategic implementation.

Table 3. Pareto front zones summary.

	number of solutions	time (seconds)				cost (millions of euros)			
		min	mean	std	max	min	mean	std	max
Zone 1	7	1560.33	1569.86	8.72	1583.98	256.35	256.82	0.54	257.81
Zone 2	25	1543.68	1550.83	4.99	1560.26	258.76	260.81	1.38	263.82
Zone 3	7	1541.33	1542.88	0.87	1543.61	266.33	267.30	1.50	268.85
Zone 4	9	1539.21	1540.31	0.84	1541.05	272.01	272.30	0.28	272.65

Finally, Fig. 8 illustrates a visual representation of a route (depicted in yellow) that connects a start point to a destination. The blue lines show the cycle lanes installed. Left side of this figure shows the current configuration in Málaga city and the right side presents the proposed infrastructure installation based on cs_{IV} solution. This visualization highlights the significant reduction in travel time (approximately 15%), and the promotion of multimodality in the trip. The route computed when using the cs_{IV} infrastructure allows for smooth transitions, incorporating the use of the metro (represented by red dots). Using the public transportation optimizes travel time because the proposed new road infrastructure facilitates the integration of various modes of transportation within the city, enhancing overall mobility and providing a clear example of the practical benefits of the proposed urban redesign.

Fig. 8. Representation of a route (orange line) using the current infrastructure (left) and the infrastructure proposed by the cs_{IV} solution (right). (Color figure online)

5.3 Computational Efficiency

An additional ten independent runs of the selected parameterization for the experimental analysis have been performed with 4, 8, and 16 processing elements to evaluate the computational efficiency. Table 4 reports the mean, normalized standard deviation (std%), and maximum (max) of the computational efficiency metrics assessed for the pNSGA-II across the evaluated numbers of processing units. These values were presented due to the results follow a normal distribution.

As the number of processing elements increases, a notable reduction in mean computational time is observed, indicating improved efficiency in solving the optimization problem. The speedup metric quantifies the algorithm's acceleration with respect to the sequential version. A substantial increase in speedup is evident as the number of processing elements grows. This behaviour indicates successful parallelization, with the algorithm achieving a speedup of up to 29.496 on 32 processing elements. Efficiency shows a decrease with the increasing number of processing elements. Nonetheless, the algorithm maintains high-efficiency levels, surpassing 90% even with 32 processing elements.

Table 4. Speedup and efficiency regarding to the number of processing units.

#procs.	comp. time (minutes)			speedup			efficiency		
	mean	std%	max	mean	std%	max	mean	std%	max
4	1358.994	0.035	1463.916	3.924	0.006	3.957	0.981	0.006	0.989
8	736.572	0.049	801.928	7.712	0.012	7.814	0.964	0.012	0.977
16	395.819	0.087	468.358	15.329	0.008	15.477	0.958	0.008	0.967
32	213.234	0.067	240.290	29.496	0.053	30.531	0.922	0.053	0.954

The results show that the pNSGA-II efficiently uses additional processing elements, significantly reducing computational time and notable speedup. The observed efficiency levels highlight the algorithm's scalability and effectiveness in leveraging parallelization for addressing the analyzed optimzation problem.

6 Conclusions and Future Work

Our study focuses on optimizing urban infrastructure for e-scooter mobility. We propose a multi-criteria optimization problem, called URIReI, that addresses the challenge of integrating e-scooters into multimodal transportation systems by redesigning road infrastructure. To solve this problem, we introduce a parallel version of the well-known NSGA-II.

Our experimental evaluation was based on real-world data from the city of Málaga, Spain. The proposed approach was highly effective, consistently exploring the Pareto front and providing decision-makers with diverse non-dominated solutions. These solutions represent trade-offs between travel time improvement and infrastructure installation costs, offering flexibility in urban planning decisions. We analyzed distinct zones, each with its characteristics, aiding decision-makers in choosing solutions aligned with its goals and budget constraints.

Furthermore, we found that our algorithm significantly reduced computational time, achieving notable speedup while maintaining high-efficiency levels even with a larger number of processing elements.

Future work should involve testing alternative algorithms to validate the approach's robustness. Also, incorporating additional modes of transportation, such as buses, could enhance the model's realism. We also plan to consider the cycle lanes connectivity and assessing the impact on overall urban transportation networks. Finally, exploring the proposed infrastructure changes' economic, environmental, and social effects would contribute to a more holistic evaluation of the solutions. This multi-faceted analysis would enrich decision-making and foster sustainable and socially responsible urban development.

References

1. Alba, E., Luque, G., Nesmachnow, S.: Parallel metaheuristics: recent advances and new trends. Int. Trans. Oper. Res. **20**(1), 1–48 (2013)
2. Bai, L., Liu, P., Chan, C.Y., Li, Z.: Estimating level of service of mid-block bicycle lanes considering mixed traffic flow. Transp. Res. Part A: Policy Pract. **101**, 203–217 (2017)
3. Boeing, G.: OSMnx: New methods for acquiring, constructing, analyzing, and visualizing complex street networks (Sep 2017). https://doi.org/10.1016/j.compenvurbsys.2017.05.004, http://dx.doi.org/10.1016/j.compenvurbsys.2017.05.004
4. Buczyński, A.: The costs of cycling infrastructure (2021). https://ecf.com/system/files/The_Costs_of_Cycling_Infrastructure_Factsheet.pdf, online; accessed 01 November 2023
5. Burke, C.M., Scott, D.M.: Identifying "sensible locations" for separated bike lanes on a congested urban road network: A toronto case study. Prof. Geogr. **70**(4), 541–551 (2018)
6. Cintrano, C., Toutouh, J.: Multiobjective electric vehicle charging station locations in a city scale area: malaga study case. In: Jiménez Laredo, J.L., Hidalgo, J.I., Babaagba, K.O. (eds.) Appl. Evol. Comput., pp. 584–600. Springer International Publishing, Cham (2022)

7. Cloud, C., Heß, S., Kasinger, J.: Shared e-scooter services and road safety: Evidence from six european countries. Europ. Econ. Rev. **160**, 104593 (2023). https://doi.org/10.1016/j.euroecorev.2023.104593, https://www.sciencedirect.com/science/article/pii/S0014292123002210

8. Deb, K.: Multi-Objective Optimization using Evolutionary Algorithms. John Wiley & Sons (2001)

9. Fabbiani, E., Nesmachnow, S., Toutouh, J., Tchernykh, A., Avetisyan, A., Radchenko, G.: Analysis of mobility patterns for public transportation and bus stops relocation. Program. Comput. Softw. **44**(6), 508–525 (2018)

10. Fortin, F.A., De Rainville, F.M., Gardner, M.A., Parizeau, M., Gagné, C.: DEAP: Evolutionary algorithms made easy. J. Mach. Learn. Res. **13**, 2171–2175 (Jul 2012)

11. Gössling, S.: Integrating e-scooters in urban transportation: problems, policies, and the prospect of system change. Transp. Res. Part D: Transp. Environ. **79**, 102230 (2020)

12. Haklay, M., Weber, P.: Openstreetmap: user-generated street maps. IEEE Pervasive Comput. **7**(4), 12–18 (2008). https://doi.org/10.1109/MPRV.2008.80

13. Harada, T., Alba, E.: Parallel genetic algorithms: a useful survey. ACM Comput. Surv. (CSUR) **53**(4), 1–39 (2020)

14. Ishibuchi, H., Masuda, H., Tanigaki, Y., Nojima, Y.: Modified distance calculation in generational distance and inverted generational distance. In: Gaspar-Cunha, A., Henggeler Antunes, C., Coello, C.C. (eds.) Evolutionary Multi-Criterion Optimization: 8th International Conference, EMO 2015, Guimarães, Portugal, March 29 –April 1, 2015. Proceedings, Part II, pp. 110–125. Springer International Publishing, Cham (2015). https://doi.org/10.1007/978-3-319-15892-1_8

15. Jaffe, E.: When adding bike lanes actually reduces traffic delays. https://www.bloomberg.com/news/articles/2014-09-05/when-adding-bike-lanes-actually-reduces-traffic-delays. Accessed on November 1st (2023)

16. Javadiansr, M., Davatgari, A., Rahimi, E., Mohammadi, M., Mohammadian, A., Auld, J.: Coupling shared e-scooters and public transit: a spatial and temporal analysis. Transp. Lett. pp. 1–18 (2023)

17. Junta Andalusia Open Data Portal: Directorio de centros docentes no universitarios de andalucía - portal de datos abiertos. https://www.juntadeandalucia.es/datosabiertos/portal/dataset/directorio-de-centros-docentes-de-andalucia

18. Li, M., Zheng, J.: Spread assessment for evolutionary multi-objective optimization. In: Ehrgott, M., Fonseca, C.M., Gandibleux, X., Hao, J.-K., Sevaux, M. (eds.) EMO 2009. LNCS, vol. 5467, pp. 216–230. Springer, Heidelberg (2009). https://doi.org/10.1007/978-3-642-01020-0_20

19. Liu, S., Shen, Z.J.M., Ji, X.: Urban bike lane planning with bike trajectories: Models, algorithms, and a real-world case study. Manufact. Serv. Oper. Manage. **24**(5), 2500–2515 (2022)

20. Malaga Open Data Portal: Sistema de Información Cartográfica - Sección censal. https://datosabiertos.malaga.eu/dataset/sistema-de-informacion-cartografica-seccion-censal

21. Massobrio, R., Toutouh, J., Nesmachnow, S., Alba, E.: Infrastructure deployment in vehicular communication networks using a parallel multiobjective evolutionary algorithm. Int. J. Intell. Syst. **32**(8), 801–829 (2017)

22. Meng, L.: Political economy and cycling infrastructure investment. Transp. Res. Interdiscip. Perspect. **14**, 100618 (2022). https://doi.org/10.1016/j.trip.2022.100618, https://www.sciencedirect.com/science/article/pii/S259019822200080X

23. Málaga: Boletín oficial de la provincia de málaga. edicto 252/2021 - published from january 11 to 19, 2021 (01 2021). http://www.bopmalaga.es/cve.php?cve=20210119-00252-2021

24. Nanayakkara, P.K., Langenheim, N., Moser, I., White, M.: Do safe bike lanes really slow down cars? a simulation-based approach to investigate the effect of retrofitting safe cycling lanes on vehicular traffic. Int. J. Environ. Res. Public Health 19(7), 3818 (2022)

25. Nesmachnow, S., Rossit, D.G., Toutouh, J.: Comparison of multiobjective evolutionary algorithms for prioritized urban waste collection in Montevideo. Uruguay. Electron. Notes Discr. Math. 69, 93–100 (2018)

26. Official State Gazette Agency: Official state gazette nᵒ 297 of november 11, 2020 (boe) (11 2020). https://www.boe.es/eli/es/rd/2020/11/10/970

27. Olabi, A., et al.: Micromobility: progress, benefits, challenges, policy and regulations, energy sources and storage, and its role in achieving sustainable development goals. Int. J. Thermofluids 100292 (2023). https://doi.org/10.1016/j.ijft.2023.100292

28. Pedroza-Perez, D.D., Toutouh, J., Luque, G.: E-scooters routes potential: open data analysis in current infrastructure. malaga case. In: Dorronsoro, B., Chicano, F., Danoy, G., Talbi, E.-G. (eds.) Optimization and Learning: 6th International Conference, OLA 2023, Malaga, Spain, May 3–5, 2023, Proceedings, pp. 380–392. Springer Nature Switzerland, Cham (2023). https://doi.org/10.1007/978-3-031-34020-8_29

29. Péres, M., Ruiz, G., Nesmachnow, S., Olivera, A.C.: Multiobjective evolutionary optimization of traffic flow and pollution in Montevideo. Uruguay. Appl. Soft Comput. 70, 472–485 (2018)

30. Rossit, D.G., Toutouh, J., Nesmachnow, S.: Exact and heuristic approaches for multi-objective garbage accumulation points location in real scenarios. Waste Manage. 105, 467–481 (2020)

31. Shaheen, S., Cohen, A.: 12. shared micromobility: policy and practices in the united states. In: A Modern Guide to the Urban Sharing Economy, chap. 12, pp. 166–180. Edward Elgar Publishing (2021)

32. Sherriff, G., Lomas, M., Blazejewski, L., Larrington-Spencer, H.: A micromobility buffet: e-scooters in the context of multimodal spaces and practices in greater manchester. Active Travel Stud. 3(1) (2023)

33. Tian, D., Ryan, A.D., Craig, C.M., Sievert, K., Morris, N.L.: Characteristics and risk factors for electric scooter-related crashes and injury crashes among scooter riders: A two-phase survey study. Int. J. Environ. Res. Public Health 19(16) (2022)

34. Toutouh, J., Rossit, D., Nesmachnow, S.: Soft computing methods for multiobjective location of garbage accumulation points in smart cities. Ann. Math. Artif. Intell. 88(1), 105–131 (2020)

35. Zakhem, M., Smith-Colin, J.: Micromobility implementation challenges and opportunities: analysis of e-scooter parking and high-use corridors. Transp. Res. Part D: Transp. Environ. 101, 103082 (2021)

36. Zitzler, E., Thiele, L.: Multiobjective optimization using evolutionary algorithms — a comparative case study. In: Eiben, A.E., Bäck, T., Schoenauer, M., Schwefel, H.-P. (eds.) Parallel Problem Solving from Nature — PPSN V, pp. 292–301. Springer Berlin Heidelberg, Berlin, Heidelberg (1998). https://doi.org/10.1007/BFb0056872

Evolutionary Computation in Edge, Fog, and Cloud Computing

Simple Efficient Evolutionary Ensemble Learning on Network Intrusion Detection Benchmarks

Zhilei Zhou, Nur Zincir-Heywood(ID), and Malcolm I. Heywood(✉)(ID)

Faculty of Computer Science, Dalhousie University, Nova Scotia, Canada
{ZhileiZhou,nzincirh,mheywood}@dal.ca

Abstract. Training and deploying genetic programming (GP) classifiers for intrusion detection tasks on the one hand remains a challenge (high cardinality and high class imbalance). On the other hand, GP solutions can also be particularly 'lightweight' from a deployment perspective, enabling detectors to be deployed 'at the edge' without specialized hardware support. We compare state-of-the-art ensemble learning solutions from GP and XGBoost on three examples of intrusion detection tasks with 250,000 to 700,000 training records, 8 to 115 features and 2 to 23 classes. XGBoost provides the most accurate solutions, but at two orders of magnitude higher complexity. Training time for the preferred GP ensemble is in the order of minutes, but the combination of simplicity and specificity is such that the resulting solutions are more informative and discriminatory. Thus, as the number of features increases and/or classes increase, the resulting ensembles are composed from particularly simple trees that associate specific features with specific behaviours.

Keywords: Boosting · Bagging · Stacking · Evolutionary Ensemble Learning · Intrusion Detection

1 Introduction

Intrusion detection tasks are typically either described by flow data collected across networks or log files summarizing the operation of servers appearing on a network. In this work, we assume the former flow based data, which is to say, the packets transferred across a network are described in terms of source–destination statistics, i.e. a flow. Several tools are available for constructing flows (e.g. Argos, Tranalizer, WireShark). The intrusion detection task is then addressed using some form of (supervised) machine learning algorithm. However, challenges appear on account of the high degree of class imbalance and high cardinality of the data sets. In this work, we revisit the task of constructing machine learning solutions to the flow based network intrusion detection problem, but with the additional objective of engineering features that discriminate

Research enabled by NSERC Discovery Grant RGPIN-2020-04438.

between different behaviours appearing in the data. Moreover, the resulting simple solutions are then able to operate in real-time on very modest computing platforms, i.e. an IoT/edge scenario.[1]

Previous research has deployed genetic programming (GP) to the network intrusion detection task by addressing issues such as data set cardinality/imbalance [3,20] or multi-class classification [1,16]. However, since these works were performed there have both been advances to the datasets used to capture properties of the intrusion detection task and evolutionary ensemble learners. The latter development implies that multiple (GP) classifiers participate in providing a label [10]. One of the central questions when developing ensembles is how to construct a suitably diverse set of base models [4,15]. That is to say, if the classifiers participating in an ensemble are trained on the same data, their behaviours will likely be correlated, rendering their combination in an ensemble ineffective. With this in mind, different methods have been proposed for 'perturbing' the training conditions, i.e. constructing models from different subsets of features (e.g. Random Forests), re-weighting/sampling the training partition (e.g. Bagging or Boosting [4]) and/or randomizing the training procedure (e.g. stochastic weight/population initialization). In addition, ensemble methods do not necessarily return solutions that are *informative* from an end user perspective. That is to say, if multiple models have to be applied simultaneously to collectively produce a label, then it becomes increasingly difficult to determine the basis for decisions.

In this work, we revisit the network intrusion detection problem through ensemble learning using two evolutionary ensemble learning frameworks: BStacGP (Sect. 2) and Symbolic Bid-Based GP (Sect. 3). BStacGP explicitly constructs a 'stack' of classifiers using a boosting process that returns a residual dataset after adding each classifier. The cardinality of the training partition therefore decreases as each new tree is added to the stack. Symbolic Bid-Based GP on the other hand employs a competitive coevolutionary relationship between a population of teams (candidate ensembles) and a data subset (i.e. bagging through coevolution). A 'winner takes all' model of aggregation is assumed, so each label is associated with a single program, but programs in themselves might be complex. In addition we compare solutions to those returned using XGBoost and Decision Trees. Our interest is to assess the relative performance versus solution complexity against well known baselines for ensemble learning and single model classification.

Section 4 reports on the benchmarking study performed across three datasets (CTU-13, Kitsune and KDD-99) where these are representative of recent and historically relevant intrusion detection benchmarks. As such the types of feature, attack and normal data vary. Moreover, the number of classes requiring detection range from 2 to 23 and the class distribution might be relatively balanced to extremely imbalanced (<0.001%). BStacGP and XGBoost are most consistently

---

[1] This is distinct but complementary to assuming that intrusion detection can be performed at some centralized cloud based resource using more computationally expensive paradigms, such as deep learning.

able to provide solutions across the different datasets, but only BStacGP is additionally able to return simple solutions. Simplicity in this case provides more clarity with respect to how features are used and supports execution on IoT/edge platforms.

2 BStacGP Framework

BStacGP represents a process for evolving a stack of predictors (Sect. 2.1) and a process for 'navigating' the resulting stack (Sect. 2.2). Previous research indicated that good scaling with cardinality was possible [21], but the impact of class imbalance is unknown.

2.1 Stack Construction

The BStacGP framework is summarized by Algorithm 1. Unlike the majority of frameworks for ensemble learning, BStacGP incrementally constructs a 'stack' of GP classifiers. The classifier maps an input, X_p, to a number line divided into a discrete number of bins, Fig. 1. As long as inputs mapped to the same bin have the same class label, then the bin is said to be *pure*.[2] If on the other hand, multiple inputs with different labels are mapped to the same bin, then the bin is *ambiguous*. Any other bins are considered *empty*. Such a mapping is independent from the number of classes involved, i.e. the number of bins is significantly more than the number of classes and mappings are rewarded for maximizing bin purity.

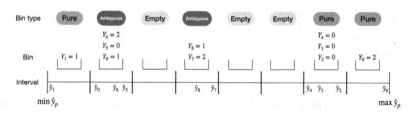

Fig. 1. Pictorial relation between program outputs ($\hat{y}_p$), bins, labels ($0 \leq Y_p < C$) and bin type. Each bin spans an equal interval ($\frac{\max \hat{y}_p - \min \hat{y}_p}{\texttt{MaxBins}-1}$). Each input, X_p, is mapped to a single scalar position on the program's output $\hat{y}_p$ for which there is a known class label, Y_p. Depending on the distribution of Y_p in the same bin, a bin is said to be pure, ambiguous or empty

Having mapped the training partition to an individual's number line, the fitness function is designed to capture the properties of the bins making up the resulting histogram distribution (as defined by the mapping). Specifically, for each bin, the following purity/impurity metric is estimated,

---

[2] Up to β instances from other classes accepted before pure considered ambiguous.

$$bin(i,c) = \frac{Count(i,c)}{S(i) \times Inst(c)} \tag{1}$$

where $Count(i,c)$ is the number of class 'c' records mapped to bin 'i', $S(i)$ is the number of records mapped to interval 'i', and $Inst(c)$ is the number of records from class 'c' appearing in the training partition. Fitness (f_{gini}) is now incrementally defined as per the following formulation,[3]

$$f_{gini} \leftarrow \sum_{i,c}^{B,C} (bin(i,c))^2 \times Inst(c); \forall i \in B, c \in C \tag{2}$$

where B and C are the number of bins and classes respectively.

Algorithm 1. BStacGP framework

1: $Ensemble \leftarrow \emptyset$
2: **while** !$MaxBoost$ **do**
3:     initialize(**Pop**)   ▷ Initialize a Pop of single node 'trees'
4:     $Champ \leftarrow \emptyset$
5:     **repeat**
6:         Fitness ← Evaluate($\langle X, Y \rangle$,**Pop**)   ▷ Evaluate using GiniIndex
7:         $Ranked$ ← Sort(**Pop**, GiniFitness)   ▷ Rank Pop
8:         **PPool** ← Top($Ranked$, **Pop**, %Gap)   ▷ Drop worst %Gap from Pop
9:         $Champ$ ← TestHistogram(**PPool**) ▷ Identify Champion from parent pool
10:        **Offspring** ← Variation(**PPool**, %Gap)
11:        **Pop** ← **PPool** ∪ **Offspring**
12:    **until** $Champ \neq \emptyset$ OR !$MaxGen$
13:    $\langle X', Y' \rangle$ ← MarkPure($Champ$)   ▷ Identify correctly labelled
14:    $\langle X, Y \rangle$ ← Residual($\langle X', Y' \rangle$)   ▷ Return residual data partition
15:    $Ensemble \leftarrow Ensemble \cup Champ$ ▷ Update ensemble complement
16: **end while**

Steps 7 and 8 rank the population relative to this fitness function and drop the worst %Gap individuals, i.e. a breeder. The resulting parent pool is tested for a champion. A champion is defined as an individual with fitness better than the last individual added to the stack and with at least one pure bin. Should such an individual exist, then the inner loop exits, otherwise the loop continues with the creation of offspring and their addition to the population.

The outer loop of Algorithm 1 completes by identifying the training records that the champion successfully mapped to pure bins. These are removed from the training partition, and the 'residual' training partition identified. Such a process incrementally decreases the cardinality of the training partition and helps to focus the next round of evolution on what the ensemble *cannot* correctly classify.

---

[3] Motivated by information theoretic formulations employed in decision tree methods, e.g. Chapter 8 in [6].

2.2   Ensemble Querying Post Training

During training, programs are incrementally added to the ensemble using specific 'splits' of the data. Let a first-in-first-out list, L, reflect the order in which (champion) programs were added to the ensemble. Given a set of test records, $\langle X^t, Y^t \rangle$, a record is presented to the first program from the list, $L_{i=0}$. Program execution again maps the record to a bin. We now query the bin type (as established during training), as follows:

- **pure bin:** the bin's label is returned as the class prediction, y'_t, of this test record. If $y'_t = Y^t$ then a correct classification results, otherwise the predicted class was incorrect.
- **ambiguous bin:** increment the program list pointer (i.e. select the next program L_i) and repeat execution–bin querying for program, $i = i + 1$.
- **empty bin:** identify the nearest bin that is either ambiguous or pure. Interpret as above.

Such a process deploys the ensemble sequentially, with the ambiguous bin category causing the next program from the list to be referenced. For each mapping to an ambiguous bin, the next program from the ensemble list is selected, L_i. Mapping to a pure bin returns the label associated with the bin during training. If the last program fails to provide a label (record still mapped to an ambiguous bin), then a default class can be returned (e.g. most frequent, most costly).

3   Symbolic Bid Based GP

Symbolic Bid Based GP (hereafter SBB) constructs an ensemble of classifiers using a symbiotic relationship between learners and teams [11]. In addition, a data subset is assumed for decoupling the cost of fitness evaluation from training partition cardinality using a competitive coevolutionary formulation (a form of boosting). SBB has previously demonstrated its effectiveness under a range of imbalanced multi-class [14], high dimensional [5] and streaming classification tasks [13]. In the following we summarize the symbiotic and competitive coevolutionary components respectively.

3.1   Symbiotic Model

The symbiotic framework assumes that two populations are maintained: a team population $(\mathcal{T})$ and a learner population $(\mathcal{L})$. The team population attempts to discover good combinations of Learners to appear in a team, whereas the learner population represents the source of programs to appear in Teams. Learners are defined in terms of a program, p, and an action, a, where actions are initialized from the set of class labels, $\mathcal{A}$. Each member of the team population identifies a subset of Learners to appear in a Team such that,

- the complement of Learners are unique.[4]
- at least two Learners appear per Team.[5]
- there are at least two different actions sampled by the Learners appearing in the *same* Team.[6]

Evaluation of a team implies that the programs from all learners associated with the same team are executed on the current training exemplar. Whichever program has the maximum output wins the right to suggest its action (label) resulting in a binary outcome for each interaction between Team, t_i, and training record p_k, or

$$G(t_i, p_k) \leftarrow \begin{cases} 1, \text{ if team } t_i \text{ classifies record } p_k \\ 0 \text{ otherwise} \end{cases} \tag{3}$$

Once all teams are evaluated across all data records appearing in the data subset, fitness sharing is applied,

$$f_i = \sum_k \left(\frac{G(t_i, p_k)}{1 + \sum_j G(t_j, p_k)} \right)^2 \tag{4}$$

Such a function helps to maintain the diversity of the team population with the objective of letting the more specialist teams exist long enough to be subsumed into the teams labeling the 'low hanging fruit'. The other element to diversity maintenance takes the form of the competitive coevolutionary relationship between team and data subset (Sect. 3.2). Moreover, maintaining diversity across a population provides the basis for independent cycles of evolution in which new solutions describe their actions in terms of previously evolved teams. This results in a hierarchy of teams or stacking [12,19].

After all the teams have their fitness evaluated on the current data subset, they are ranked and the bottom $G_T\%$ of the population are deleted leaving a parent pool from which $G_T\%$ children are composed. Any Learners from the learner population that are not indexed by the parent pool are also deleted. Variation operators are applied hierarchically to compose new teams/learners by first cloning $G_T\%$ teams (selected uniformly) and learners in order to avoid disrupting working relationships in the respective team and learner parent pools. Further details of variation can be found in the earlier works [5,14].

3.2 Competitive Coevolution

SBB coevolves a 'point population' ($\mathcal{P}$), the content of which defines the data subset from which fitness evaluation is performed. The point population is stratified such that each class is represented equally with $G_P\%$ of the point population replaced at each generation. Although new training records are sampled uniformly from the original training partition to replace the $G_P\%$ records deleted

---

[4] Otherwise a trivial redundancy appears.

[5] A single learner would only be able to suggest a single class.

[6] All data labeled as the same class.

at each generation, records are prioritized to remain in the data subset using the concept of distinctions [7]. Record p_k is said to form a distinction between two teams, t_i and t_j if $G(t_i, p_k) > G(t_j, p_k)$. This results in an $\mathcal{P} \times (\mathcal{T}^2 - \mathcal{T})$ matrix of distinctions. Fitness sharing can also be applied to the resulting matrix of distinctions [5,7,14]. Records are then prioritized for retention in the point population that, for example, identify a single team as classifying a record than no other team can classify.[7]

4 Results

4.1 Datasets and Parameterization

Benchmarking is performed across two recent large intrusion detection datasets CTU-13 [8] and the Kitsune network attack dataset [17] and the historically relevant KDD-99 dataset. Summary dataset statistics appear in Table 1.

The **CTU-13 dataset**[8] contains seven different types of Botnet traffic and is described using 8 features derived from flows provided by the Argos Netflow generator. The task is posed as separating normal from Botnet, i.e. anomaly detection. This results in a relatively balanced distribution of attack versus normal, but under a low dimensional feature space. This means that feature engineering will likely be necessary in order to provide effective anomaly detectors.

The **Kitsune dataset**[9] describes eight different types of attack, including Man-in-the-Middle, Denial-of-Service and probing/scanning. Moreover, the devices used to collect the data are representative of IoT applications (e.g. web cams and baby monitors). Each flow is described in terms of 115 features with the goal of the classifier to identify each type of attack (as well as normal traffic). The high dimensionality is assumed to provide the basis for better classifier accuracy. With this in mind, the goal of the classifier is to identify each attack instance as well as normal. The class distribution is imbalanced with 4 classes appearing in less than 0.5% of the data.

Finally, the **KDD-99 dataset**[10] has been deployed in multiple ways over the years, e.g. anomaly only, distinguish between the 5 attack types and normal. In this work, the full 23 class formulation is assumed, where this represents a particularly challenging task scenario as individual attack types should be identified and the high degree of class imbalance is present.

Algorithm parameterization is performed using a grid search. In the case of BStacGP and SBB the resulting parameterizations are reported in Tables 2 and 3 respectively. In all cases 30 runs are performed. Test data is employed post training and does not inform parameter choices. XGBoost [2] and Decision Trees [18] are also subject to task specific parameter tuning, but space restrictions preclude an enumeration of the parameters.

---

[7] Put another way, without a specific training record, the significance of a Team would be lost.

[8] https://www.stratosphereips.org/datasets-ctu13.

[9] https://archive.ics.uci.edu/dataset/516/kitsune+network+attack+dataset.

[10] https://www.openml.org/search?type=data&sort=runs&id=1113&status=active.

Table 1. Dataset properties. #Train and #Test are the cardinality for training and test partitions respectively. D is the number of features and C the number of classes. Class Distribution reflects the (approx.) distribution of each class. Largest single class corresponds to 'normal'. All the rest are different instances of an attack

Dataset	CTU-13	Kistume	KDD-99
#Train	560,792	735,612	247,010
#Test	240,340	315,263	247,010
Dimension (D)	8	115	41
Labels (C)	2	9	23
Class Distribution (%)	55/ 45	75.8/ 6.7 /5.4/ 4.3/ 4.3/ 2/ remaining with <0.5%	48.1/ 33.8/ 16.7/ 0.4/ 0.3/ 0.2/ 0.17/ 0.17/ 0.16/ 0.04/ 0.04 remaining with <0.001%

Table 2. BStacGP parameters. MaxBoost through to Gap are defined in Algorithm 1. MaxBins appears in Fig. 1. β is the bin purity threshold

Dataset	CTU13	Kitsune	KDD99
MaxBoost	10	100	200
MaxGen	2	1	10
Pop	300	1000	300
Gap	30%	30%	30%
MaxBins	100,000	30	100
β	2	2	2

4.2   Benchmarking Comparison

Table 4 summarizes performance on the test partition using the Balanced Accuracy and macro F1-score metrics [9] (averaged across 30 champions identified from training runs). In addition, we also capture multiple statistics to quantify the complexity of a solution. We note that XGBoost provides the most accurate solutions, but also typically the most complex. Solutions based on the Decision Tree method will always have the least number of trees (1), but are not necessarily the simplest solution (high Tree Depth and/or Total number of nodes). Solutions identified by SBB typically concentrate on classifying the most frequently occurring classes (difference between Balanced Accuracy and macro F1-score) and are therefore generally not competitive.

Under the balanced two class CTU13 dataset, decision tree solutions were the most complex, whereas they were the simplest under the 23 class KDD99 dataset. We also consider the impact of attempting to reduce the decision tree

Table 3. SBB parameters. MaxGen/Team/Length represent max. generations/team size/num. nodes per tree. #levels is the max. depth of the hierarchy. Gap is the number of individuals replaced in the 'point' and 'team' populations respectively

Dataset	CTU13	Kitsune	KDD99
#levels	10	5	10
MaxGen		10,000	
MaxTeam		10	
MaxLength		96	
Pop. Size		200	
Gap		20 (10)	

and XGBoost solutions to match the complexity of BStacGP.[11] These results are summarized in the last two columns in Table 4. It is apparent that reducing the complexity of decision tree or XGBoost (to that of BStacGP) has a negative impact on macro F1-score.

Table 5 reports the typical training time for constructing classifiers using each approach. The two GP approaches are the slowest, particularly with respect to the CTU13 and KDD99 datasets. On the largest cardinality dataset (Kitsune), however, BStacGP runtimes were in the same order of magnitude of runtime as recorded for XGBoost. Moreover, BStacGP runtimes were very consistent across all three datasets. Decision tree training was always the fastest (only constructs a single classifier), but appeared to slow considerably on the Kitsune dataset. SBB was by far the slowest and most variable.

4.3   BStacGP Behavioural Properties

XGBoost deploys all the ensemble to label every data record. This makes it difficult to learn from the resulting model (e.g. identify the most discriminatory features) as well as costing more computationally. Decision trees need only use part of a tree to make a decision, but as the tree depth increases, it also becomes increasingly difficult to provide knowledge transfer. BStacGP on the other hand explicitly organizes trees hierarchically as a stack. A tree either provides a label or declares a data point as ambiguous. Only ambiguous records are forwarded to the next tree (Sect. 2.2). This means that only a single tree is responsible for making each prediction.

CTU13 is formulated as a two class, low dimensional task for which feature engineering appears to have taken place, Fig. 2. There are only two trees in the stack, with a comparatively high complexity (Table 4). No particular preference to predicting normal/attack appears across the layers, and both trees appear to have a similar complexity. Conversely, Fig. 3 illustrates how a BStacGP

---

[11] BStacGP average rank of 1.33 for total number of nodes, versus an average rank of 2.67 and 3.67 for decision tree and XGBoost.

Table 4. Test performance. Bal acc. is the balanced accuracy metric as applied to all classes. F1-score is the multi-class ratio of precision to recall with equal class weighting. # Trees are the number of trees appearing in a solution. Tree Depth is the max. tree depth. Total # nodes is the total operator count across all components of a solution. †denotes the best test classification performance of the simplest solutions. Bold are the best classification performance ignoring model complexity

CTU13

Model	BStacGP	SBB	DT	XGboost	simple DT	simple XGboost
Bal. acc.	93.1†	80.9	96.0	**96.0**	92.3	88.8
F1-score	93.1†	81.1	96.1	**96.0**	92.1	88.8
#Trees	2	3	1	100	1	5
Tree Depth	5.43	4	49	6	15	4
Total #nodes	**86.2**	184	44,659	9,834	99	147

Kistume

Model	BStacGP	SBB	DT	XGboost	simple DT	simple XGboost
Bal. acc.	90.7	91.1	99.8	**99.9**	89.5	96.2†
F1-score	92.0†	49.1	99.6	**99.9**	84.2	80.2
#Trees	28.5	9	1	1,000	1	20
Tree Depth	1	4	35	3.42	11	2
Total #nodes	**85.5**	383.3	555	15,250	89	136

KDD-99

Model	BStacGP	SBB	DT	XGboost	simple DT	simple XGboost
Bal. acc.	73.5	68.4	68.0†	**85.8**	–	63.5
F1-score	71.5	44.7	69.9†	**81.0**	–	62.0
#Trees	101	16	1	2,300	–	69
Tree Depth	2.59	4	34	1.92	–	3.14
Total #nodes	817.8	857.5	**405**	13,790	–	855

Table 5. Training times (approx). All times on a common Intel i7012799H computing platform. SBB, BStacGP and Decision Trees perform training using a *single thread* whereas XGBoost employed 20

Dataset	CTU13	Kitsune	KDD99
Decision Tree	≈1 s	5.2 min	≈1 s
XGBoost	14 s	10 min	≈1 s
BStacGP	5–6 min	13 min	10 min
SBB	30 min to 45 h		

solution decomposes the Kitsune test partition. There are 53 trees in this particular champion (x-axis). In the case of normal (subplot (a)) predictions are made cumulatively across the entire stack whereas for SSDP flood (subplot (e)) most predictions are made early on, particularly level 0. Fuzzing operations are detected by classifiers at levels 3 and 27 in particular (subplot (b)). The majority of man-in-the-middle attacks (ARM MitM and Video Injection) are also associated with specific levels of the stack (level 4 and 10 respectively). Conversely, the SYS DoS attack is identified much later in the stack (subplot (f)).

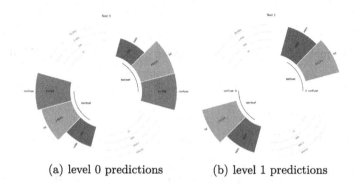

(a) level 0 predictions   (b) level 1 predictions

Fig. 2. BStacGP trees under CTU13 test partition. Hit, Miss and Confuse are the counts of correct, incorrect and ambiguous bin associations respectively

Figure 4 (top row) illustrates the specific predictions on Kitsune corresponding to particular levels of the BStacGP solution from Fig. 3. Moreover, all the trees comprising this stack take the form of decision tree stumps (a single arithmetic operation with two features) as follows: level 0: $x_{62} - x_{27}$; level 3: $x_{62} + x_{58}$; level 10: $x_{62} + x_9$; and level 48: $x_6 - x_{27}$. Figure 4 (bottom row) illustrates the frequency with which different features are indexed across *all* trees for 3 example classes. Thus, feature 62 is frequent and common to all, but features 27, 58 and 9 (or 20) are frequent and specific to each class.

Figure 5 provides a summary of test behaviour from a BStacGP solution over 6 classes from the KDD99 dataset. There are 117 levels to this stack. It is again clear that specific trees contribute specific properties, albeit with a collection of trees distributed across the stack classifying each class. However, the 'Smurf DoS' class represents an exception in which all 140,000 instances are correctly labeled by a single tree (subplot (b)). Moreover, the consistently low miss rates indicates again that the BStacGP classifiers are incrementally 'picking off' very specific behaviours from particular classes. Thus, the Neptune DoS attack appears to be labeled by 4 trees from the stack (subplot (c)), whereas ≈75% of Warez r2l attacks are detected by 2 trees (subplot (d)). Similarly, ≈90% of the Satan and ipsweep probe behaviours are detected by 3 to 4 trees (subplots (e) and (f)). Given that these preferences also appear under the training condition, specific trees and features can again be associated with specific attacks.

Figure 6 (top row) illustrates specific predictions on KDD99 corresponding to particular levels of the BStacGP solution from Fig. 5. It is again apparent that trees with very specific properties/functions result. Moreover, these trees are also relatively simple. For example:

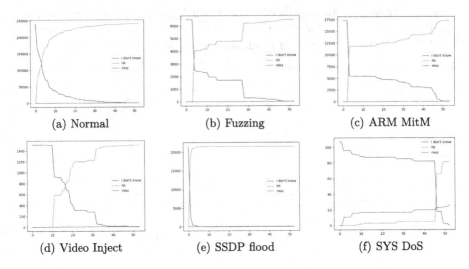

Fig. 3. Operation of BStacGP under Kitsune test partition. x-axis is the stack level, y-axis is the number of the class labeled correctly (orange), incorrectly (green) or 'I don't know' (blue). Only the latter are forwarded to the next stack level for labelling

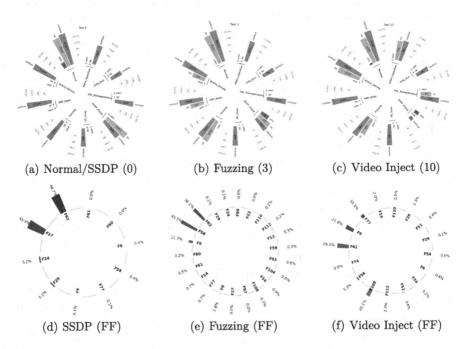

Fig. 4. Example classifier performance under Kitsune test partition. (X) indicates stack level. Bars represent number of miss/correct/confuse predictions per class. (FF) indicates feature frequency across the entire class

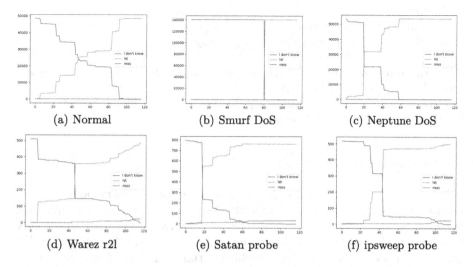

Fig. 5. Operation of BStacGP under KDD-99 test partition. x-axis is the stack level, y-axis is the number of the class labeled correctly (orange), incorrectly (green) or 'I don't know' (blue). Only the latter are forwarded to the next stack level for labelling. 6 of 23 classes shown due to space restrictions

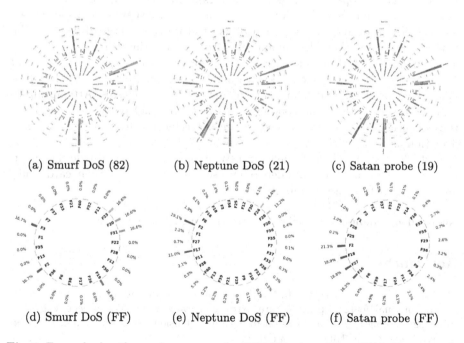

Fig. 6. Example classifier performance under KDD99 test partition. (X) indicates stack level. Bars represent number of miss/correct/confuse predictions per class. (FF) indicates feature frequency across the entire class

- Smurf DoS (level 82 tree): $x_{31} + x_{19} + (x_2 \times x_3) - x_{20} - x_{23}$
- Neptune DoS (level 21 tree): $(x_{17} - x_2) \times (x_{28} - x_{24}) + x_{11} + x_{17} - x_2$
- Satan probe (level 19 tree): $x_{18} + (x_2 + (x_{27} - x_{16}))$

Fig. 6 (bottom row) illustrates the frequency with which specific features are utilized across all trees for 3 classes. Feature 2 is common to all, but other features appear to be class specific, e.g. feature 3, 19, 20, 23, 31 (Smurf), 17 24, 28 (Neptune) or 16, 18, 27 (Satan).

5   Conclusion

IoT or edge computing devices increasingly represent the entry point for network borne attacks. Such devices typically have a limited computing capability. However, as the first point of entry they also represent the first opportunity to minimize the impact of malicious behaviour. We conduct a benchmarking study in which we build candidate (machine learning) detectors from three rather different datasets representing different perspectives on the intrusion detection task (balanced anomaly detection, imbalanced high dimensional feature space over 9 classes or imbalanced high class count over a medium number of features). XGBoost provided the best classification accuracy, but also represented solutions that are opaque (thousands of trees). Conversely, the preferred evolutionary ensemble (BStacGP) provided increasingly transparent solutions as the number of classes and/or dimension of the feature space used to describe the task increased. In effect, the amount of feature engineering appears to decrease as the number of features increases, making for trees that often only consist of 1 to 5 nodes. Moreover, such trees are highly discriminatory in that they provide labels for classifying significant amounts of specific classes. Attempting to tune XGBoost to the same complexity as BStacGP had particularly negative consequences for classification performance, precluding their use for intrusion detection on IoT devices.

Future research will continue to investigate new ways for constructing evolutionary ensemble methods in order to both scale to higher dimension/cardinality and develop better interpretability.

Acknowledgements. Circos plot constructed using http://circos.ca.

References

1. Badran, K.M.S., Rockett, P.I.: Multi-class pattern classification using single, multidimensional feature-space feature extraction evolved by multi-objective genetic programming and its application to network intrusion detection. Genet. Program Evolvable Mach. **13**(1), 33–63 (2012)

2. Chen, T., Guestrin, C.: XGBoost: a scalable tree boosting system. In: Proceedings of the ACM SIGKDD International Conference on Knowledge Discovery and Data Mining, pp. 785–794. ACM (2016)

3. Curry, R., Lichodzijewski, P., Heywood, M.I.: Scaling genetic programming to large datasets using hierarchical dynamic subset selection. IEEE Trans. Syst., Man, Cybernet.s - Part B **37**(4), 1065–1073 (2007)

4. Dietterich, T.G.: An experimental comparison of three methods for constructing ensembles of decision trees: bagging, boosting, and randomization. Mach. Learn. **40**(2), 139–157 (2000)

5. Doucette, J.A., McIntyre, A.R., Lichodzijewski, P., Heywood, M.I.: Symbiotic coevolutionary genetic programming: a benchmarking study under large attribute spaces. Genet. Program Evolvable Mach. **13**(1), 71–101 (2012)

6. Duda, R.O., Hart, P.E., Stork, D.G.: Pattern Classification. Wiley and Sons, 2nd edn. (2001)

7. Ficici, S.G., Pollack, J.B.: Pareto optimality in coevolutionary learning. In: Kelemen, J., Sosík, P. (eds.) Advances in Artificial Life, pp. 316–325. Springer Berlin Heidelberg, Berlin, Heidelberg (2001). https://doi.org/10.1007/3-540-44811-X_34

8. García, S., Grill, M., Stiborek, J., Zunino, A.: An empirical comparison of botnet detection methods. Comput. Secur. **45**, 100–123 (2014)

9. Grandini, M., Bagli, E., Visani, G.: Metrics for multi-class classification: an overview. CoRR abs/2008.05756 (2020). https://arxiv.org/abs/2008.05756

10. Heywood, M.I.: Evolutionary ensemble learning. In: Banzhaf, W., Machado, P., Zhang, M. (eds.) Handbook of Evolutionary Machine Learning, pp. 205–243. Springer Nature Singapore, Singapore (2024). https://doi.org/10.1007/978-981-99-3814-8_8

11. Heywood, M.I., Lichodzijewski, P.: Symbiogenesis as a mechanism for building complex adaptive systems: a review. In: Di Chio, C., et al. (eds.) Applications of Evolutionary Computation, pp. 51–60. Springer Berlin Heidelberg, Berlin, Heidelberg (2010). https://doi.org/10.1007/978-3-642-12239-2_6

12. Kelly, S., Lichodzijewski, P., Heywood, M.I.: On run time libraries and hierarchical symbiosis. In: Proceedings of the IEEE Congress on Evolutionary Computation, pp. 1–8. IEEE (2012)

13. Khanchi, S., Vahdat, A., Heywood, M.I., Zincir-Heywood, A.N.: On botnet detection with genetic programming under streaming data label budgets and class imbalance. Swarm Evol. Comput. **39**, 123–140 (2018)

14. Lichodzijewski, P., Heywood, M.I.: Managing team-based problem solving with symbiotic bid-based genetic programming. In: Proceedings of the Genetic and Evolutionary Computation Conference, pp. 363–370. ACM (2008)

15. Ma, S., Ji, C.: Performance and efficiency: recent advances in supervised learning. Proc. IEEE **87**(9), 1519–1535 (1999)

16. McIntyre, A.R., Heywood, M.I.: Classification as clustering: a pareto cooperative-competitive GP approach. Evol. Comput. **19**(1), 137–166 (2011)

17. Mirsky, Y., Doitshman, T., Elovici, Y., Shabtai, A.: Kitsune: an ensemble of autoencoders for online network intrusion detection. In: Annual Network and Distributed System Security Symposium. The Internet Society (2018)

18. Quinlan, J.R.: C4.5: Programs for Machine Learning. Morgan Kaufmann (1993)

19. Smith, R.J., Heywood, M.I.: Coevolving deep hierarchies of programs to solve complex tasks. In: Proceedings of the Genetic and Evolutionary Computation Conference, pp. 1009–1016. ACM (2017)

20. Song, D., Heywood, M.I., Zincir-Heywood, A.N.: Training genetic programming on half a million patterns: an example from anomaly detection. IEEE Trans. Evol. Comput. **9**(3), 225–239 (2005)
21. Zhou, Z., et al.: A boosting approach to constructing an ensemble stack. In: Pappa, G., Giacobini, M., Vasicek, Z. (eds.) Genetic Programming: 26th European Conference, EuroGP 2023, Held as Part of EvoStar 2023, Brno, Czech Republic, April 12–14, 2023, Proceedings, pp. 133–148. Springer Nature Switzerland, Cham (2023). https://doi.org/10.1007/978-3-031-29573-7_9

Evolutionary Computation Meets Stream Processing

Vincenzo Gulisano[1] and Eric Medvet[2]([✉])

[1] Department of Computer Science and Engineering, Chalmers University of
Technology, Gothenburg, Sweden
`vincenzo.gulisano@chalmers.se`
[2] Department of Engineering and Architecture, University of Trieste, Trieste, Italy
`emedvet@units.it`

Abstract. Evolutionary computation (EC) has a great potential of
exploiting parallelization, a feature often underemphasized when describing evolutionary algorithms (EAs). In this paper, we show that the
paradigm of stream processing (SP) can be used to express EAs in a
way that allows the immediate exploitation of parallel and distributed
computing, not at the expense of the agnosticity of the EAs with respect
to the application domain. We introduce the first formal framework for
EC based on SP and describe several building blocks tailored to EC.
Then, we experimentally validate our framework and show that (a) it
can be used to express common EAs, (b) it scales when deployed on
real-world stream processing engines (SPEs), and (c) it facilitates the
design of EA modifications which would require a larger effort with traditional implementation.

Keywords: Parallellization · Design of EAs · Distributed computing

1 Introduction

Artificial intelligence (AI) has witnessed significant growth in recent decades,
propelled by advancements in hardware and the establishment of de-facto standard frameworks with rich application programming interfaces (APIs). These
frameworks play a pivotal role in decoupling aspects such as efficient implementation and interfacing with diverse hardware platforms. Despite this overall
progress, not all AI-related techniques have advanced uniformly. EC, the focus
of this paper, has seen the emergence of numerous fragmented and specialized
frameworks with limited customizability and in need of structural modifications
to address crucial aspects like scalability and parallel/distributed execution [2].

To make a significant step towards enhancing the adaptability and performance of EC, we present here a novel approach to overcome these limitations by
leveraging a state-of-the-art computing paradigm named SP, widely adopted in
the IoT-to-Cloud continuum [6,11,15,24]. SP allows to describe forms of computation which occur over streams of items that flow over time. By connecting

S. Smith et al. (Eds.): EvoApplications 2024, LNCS 14634, pp. 377–393, 2024.
https://doi.org/10.1007/978-3-031-56852-7_24

simple (or complex) SP building blocks, either stateless of stateful, and organizing them in graphs called queries, one may describe complex workflows in an elegant way. Moreover, and more importantly, SP is more than a scientifically mature field [4,30,34]: there exist many widespread software frameworks which are used in real-world production-level applications [17] and nicely couple with different kinds of distributed computing systems [16,29]. On the other side, most of the significant EAs used in EC are population-based and iterative. In practice, this means that several candidate solutions exist during the execution of an EA and they are modified over time by re-iteratively applying a small set of operations (e.g., selection, variation, evaluation), combined in simple or more complex ways. The potential link between SP and EC is hence clear: candidate solutions, i.e., individuals in the EC jargon, are the items that can be processed by SP blocks in EC-aware queries that reflect the overall working principles of the EA.

In this work, we lay down this link and propose the first formal framework based on the SP paradigm for EC. We describe a number of SP blocks, called operators, tailored to EC which may be used to define different EAs, without imposing limitations on the kind of entities the EA can work on. In fact, one of the strong points which has favored the usage of EC in very different domains is its capability to work with different kinds of solutions (e.g., numerical formulae [18], Boolean functions for cryptography [31], security policies [20], robotic controllers [27]) and hence on different domains [5]. With most of these solution kinds, simply resorting on GPU-based parallelization would not be enough. In contrast, SP facilitates seamless adaptation of existing EAs to different domains while decoupling the support for efficient parallel/distributed execution on larger and more heterogeneous computing systems.

We validate our proposal experimentally, by using our SP framework to implement two EAs and applying them to five problems with two kinds of solutions (bitstrings and mathematical formulae). We show that the implementations of stream-based EAs are effective and we provide evidence of how simultaneous executions (jobs) can be customized without requiring alterations to the underlying implementation. We also showcase the advantages of decoupling EA definition and execution in SP with advanced intra- and inter-job customization options such as merging of populations from different jobs at runtime, to trade-off fitting performance and job completion time.

To our knowledge, this is the first attempt to port EAs into the realm of SP. However, the power of SP has already been harnessed for improving other AI-related workflows [22,23], mostly based on machine learning (ML). Conversely, AI and ML proved useful to tune and optimize SP tasks. Actually, most of the existing literature (e.g., [33]) focuses on the latter case, which is not relevant to our work. Concerning the first case, i.e., SP for AI, we note that Apache Flink provides a set of ML APIs [8], which, however, do not support any EA. As stated in [3], the integration of ML within stream processing is still at its early stage, with many systems that support efficient data distribution across ML jobs but that, under the hood, still rely on RPC calls to external frameworks for carrying out the actual learning processes.

For what concerns the parallelization of EAs [1], many previous works targeted specific hardware platforms, with an increasing interest in the last decade in those based on GPUs [19,28]. However, despite the EC community acknowledges that scalability and exploitation of large computing infrastructures are key goals [14], the SP paradigm has not yet been applied to EC. Nevertheless, modern EC software [9,21] is often designed to exploit concurrency.

2 Preliminaries: Stream Processing

2.1 Definitions

A *stream* S is an unbounded sequence of tuples defined over the attributes $A(S) = \{\tau, a_1, \ldots, a_p\}$, where each *attribute* a has a domain $V(a)$. τ is a special attribute called *timestamp* defined in a time domain $V(\tau) = \mathbb{T} \cup \mathbb{R}$, $\mathbb{T}$ being the time domain. A *tuple* t of a stream S is composed of $|A(S)| = p + 1$ attribute values, with each value $v(a, t) \in V(a)$, with $a \in A(S)$.

For the sake of brevity, we write $V(A(S))$ for the set of all possible tuples defined on the attributes $A(S)$ of a stream S, i.e., $V(A(S)) = \mathbb{T} \times V(a_1) \times \cdots \times V(a_p)$ with $A(S) = \{\tau, a_1, \ldots, a_p\}$. Moreover, we define $A_\tau(S) = A(S) \setminus \{\tau\}$ and, accordingly, we write $v(A_\tau(S), t)$ for the part of the tuple of t consisting of all the attributes without the timestamp, i.e., $v(A_\tau(S), t) = \langle v(a_1, t), \ldots, v(a_p, t) \rangle \in V(A_\tau(S))$.

In the literature [12], several models build on different assumptions about the ordering of tuples within a stream S based on their τ attribute. For generality, we do not impose a total order on τ, but only assume there can exist substreams $S_k \in S$ so that $\forall t_i, t_j \in S, k(v(A_\tau(S), t_i)) = k(v(A_\tau(S), t_j)) \implies v(\tau, t_i) \leq v(\tau, t_j)$, where $k(v(A\tau(S), t)$ is an arbitrary function applied on any attribute value of tuple t, implies t_i is observed before t_j in S.

A *stream processing query* (or simply query) is a directed graph where nodes are either sources, operators, or sinks and edges are streams. A query meets the following criteria: (a) sources have no incoming streams, (b) sinks have no outgoing streams, (c) and operators have at least one incoming and one outgoing stream. Sources generate tuples over time. Sinks consume tuples. Operators process tuples from input streams and produce tuples to output streams, not necessarily resulting in each tuple in the input stream becoming a tuple in the output stream. We describe operators in the next section.

2.2 Operators

Operators are either stateless or stateful. Stateless operators do not maintain a state that evolves based on the tuples they process, while stateful operators do. Since a comprehensive overview of common operators found in SPEs [12] is not within the scope of this contribution, we present next the ones we consider in our work.

Merger. This is a stateless operator with n input streams $S_{in,1}, \ldots, S_{in,n}$ and one output stream S_{out}, such that $A(S_{in,1}) = \cdots = A(S_{in,n}) = A(S_{out})$. We denote by M a Merger operator.

Merger outputs each input tuple of each input stream on the output stream, keeping unmodified the timestamp and each attribute.

Delayer. This is a stateless operator with one input stream S_{in} and one output stream S_{out}, such that $A(S_{in}) = A(S_{out})$ and is defined by a delay value $\delta \in \mathbb{T}$. We denote by $D[\delta]$ a Delayer operator with its parameter.

Delayer outputs each input tuple t_{in} on the output stream as a tuple t_{out}, increasing the timestamp by δ, i.e., $v(\tau, t_{out}) = v(\tau, t_{in}) + \delta$.

FlatMap. This is a stateless operator with one input stream S_{in} and n output streams $S_{out,1}, \ldots, S_{out,n}$ and is defined by n functions $f_1, \ldots, f_n$, each processing an input tuple into a bag of tuples of the i-th output stream. Formally, $f_i : V(A_\tau(S_{in})) \to \mathcal{P}^*(V(A_\tau(S_{out,i})))$. We denote by $FM[f_1, \ldots, f_n]$ a FlatMap operator with its parameters.

Intuitively, FlatMap maps one incoming tuple to zero or more output streams. Formally, for each input tuple t_{in} and each f_i, $FM[f_1, \ldots, f_n]$ first computes the bag $T_{out,i} = f_i(v(A_\tau(S_{in}), t_{in}))$, then it outputs one tuple $t_{out,i}$ for each element in $T_{out,i}$ to the i-th output stream, setting $v(\tau, t_{out,i}) = v(\tau, t_{in})$.

Aggregate. This is a stateful operator with one input stream S_{in} and one output stream S_{out}. Aggregate is defined by: a key function $f_{key} : V(A_\tau(S_{in})) \to K$, with K being a discrete set of keys, that takes a tuple of S_{in} and returns a key; an output function $f_{out} : \mathcal{P}^*(V(A_\tau(S_{in}))) \to \mathcal{P}^*(V(A_\tau(S_{out})))$ that takes a bag of tuples of S_{in} and produces a bag of tuples of S_{out}; a window size $w_s \in \mathbb{R}^+$; and a window advance $w_a \in \mathbb{R}^+$. We denote by $A[f_{key}, f_{out}, w_s, w_a]$ an Aggregate operator with its parameters.

Intuitively, Aggregate groups incoming tuples in sets (called *instances*), based on their key and timestamp, and transforms instances in outgoing tuples; the instances constitute the state of the operator: they are initially empty and are updated based on incoming tuples. Formally, $A[f_{key}, f_{out}, w_s, w_a]$ works as follows for each input tuple t_{in} of S_{in}: (1) it computes the key $k = f_{key}(v(A_\tau(S_{in}), t_{in}))$; (2) it computes the *epochs* $e_1, \ldots, e_h \in \mathbb{N}$ such that $v(\tau, t_{in}) \in [e_i w_a, e_i w_a + w_s[$; (3) it adds t_{in} to each instance I_{k,e_i} associated with k and e_i. Finally, based on the assumption that tuples in S_k are timestamp sorted (see Sect. 2.1), for each $I_{k',e}$ such that $ew_a + w_s < v(\tau, t_{in})$ and $k' = k$, (1) it computes $T_{out} = f_{out}(I_{k',e})$ and (2) it outputs one tuple t_{out} for each element of T_{out} with $v(\tau, t_{out}) = \max_{t \in I_{k',e}} v(\tau, t)$. After having set the tuples of T_{out}, Aggregate removes the corresponding instance $I_{k',e}$ from the state.

Multiplexer. This is a specialized $FM[f_1, \ldots, f_n]$ with one input stream S_{in} and n output stream $S_{out,1}, \ldots, S_{out,n}$, such that $A(S_{in}) = A(S_{out,1}) = \cdots =$

$A(S_{\text{out},n})$. All the $f_i : V(A(S_{\text{in}})) \rightarrow \mathcal{P}^*(V(A(S_{\text{out}})))$ functions are the same: they return a one-element-bag containing the input. Hence, this operator forwards each input tuple on each output stream. We denote by X a Multiplexer.

Filter. This is a specialized FM$[f_1]$ with one input stream S_{in} and one output stream S_{out}, such that $A(S_{\text{in}}) = A(S_{\text{out}})$. The only $f_1 : V(A(S_{\text{in}})) \rightarrow \mathcal{P}^*(V(A(S_{\text{out}})))$ applies a predicate $\pi : V(A(S_{\text{in}})) \rightarrow \{\text{true}, \text{false}\}$ to the input $v(A(S_{\text{in}}), t_{\text{in}})$ and returns an empty bag if $\pi(v(A(S_{\text{in}}), t_{\text{in}}))$ is false or a one-element-bag containing only the input otherwise. We denote by F$[\pi]$ a Filter with its parameter.

3   EAs as Queries

We consider optimization problems defined by a search space P and a fitness function $q : P \rightarrow \mathbb{R}$. We assume, without loss of generality, that q has to be minimized, i.e., the goal is to find $p^* = \arg\min_{p \in P} q(P)$.

We employ an EA for solving the optimization problem. We do not enforce any specific constraint on the EA. We only assume it is iterative and population-based, i.e., that it evolves a population (formally, a bag) of individuals iteratively until some predefined termination criterion is met. We call *individual* a triplet given by an *genotype* $g \in G$, a *phenotype* $p \in P \cup \varnothing$, which is a candidate solution to the optimization problem, and its fitness, which can be either $q(p)$ or $\varnothing$; for both the phenotype and the fitness, $\varnothing$ represents the case when they are not yet been evaluated for g. We call genotype-phenotype *mapping* a function $\phi : G \rightarrow P$ that allows to obtain a phenotype $p = \phi(g)$ from a genotype g: in EC terms, ϕ (together with its domain G and co-domain P) defines the representation of solutions. We denote by $\mathcal{I} = G \times P \times \mathbb{R}$ the set of all possible individuals for a problem defined over P and tackled with a $\phi : G \rightarrow P$ representation.

An EA has some parameters and is, in general, stochastic. We call *job* an execution of an EA with some predefined parameter values: the outcome of a job is one solution p^* corresponding to the best individual, i.e., the one with the best fitness in the population at the last iteration of the EA.

In the following sections, we describe how to use stream processing to describe EAs, namely, how to express EAs as queries. To this aim, we introduce a number of operators, defined as specializations of the FlatMap and Aggregate operators described in Sect. 2.2, with names and functionalities which are familiar to the EC community.

In a query representing an EA there are three kinds of streams: (a) streams S_J of jobs, where $A(S_J) = \{\text{jobId}, \dots\}$, $V(\text{jobId}) = \mathbb{N}$, and the other attributes describe possible other parameters of the EA; (b) streams S_I of individuals, where $A(S_I) = \{\text{jobId}, \text{individual}\}$ and $V(\text{individual}) = \mathcal{I}$; (c) streams S_{I*} of bags of individuals, where $A(S_{I*}) = \{\text{jobId}, \text{individuals}\}$ and $V(\text{individuals})$ is $\mathcal{P}^*(\mathcal{I})$.

Finally, in a query representing an EA the time domain $\mathbb{T}$ is $\mathbb{N}$ and $v(\tau, t)$ represents the iteration of "birth" of an individual.

IndividualFactory. This is a specialized FM[f_1] with one input stream of jobs $S_{J,\text{in}}$ and one output stream of individuals $S_{I,\text{out}}$. The only $f_1 : \mathcal{I} \to \mathcal{P}^*(\mathcal{I})$ takes a job and returns a bag of n individuals where only the genotype is set, according to the parameters of the input job. In EC terms, f_1 represents the population initialization procedure. We denote by IF[n] an IndividualFactory operator with its parameter.

FitnessEvaluator. This is a specialized FM[f_1] with one input stream of individuals $S_{I,\text{in}}$ and one output stream of individuals $S_{I,\text{out}}$. The only $f_1 : \mathcal{I} \to \mathcal{P}^*(\mathcal{I})$ "fills" the phenotype and fitness of the individual, if they are $\varnothing$. Note that, f_1 always outputs bags of one element, i.e., $\forall i \in \mathcal{I}, |f_1(i)| = 1$. We denote by FE a FitnessEvaluator operator.

GeneticOperator. This is a specialized FM[f_1] with one input stream of bags of individuals $S_{I^*,\text{in}}$ and one output stream of individuals $S_{I,\text{out}}$. The only $f_1 : \mathcal{I} \to \mathcal{P}^*(\mathcal{I})$ takes the input individuals and applies a genetic operator $o : G^* \to G$ to their genotypes: the resulting genotype g is set as the genotype of the output individual with $\phi(g)$ as phenotype and $\varnothing$ as fitness. As for FE, here f_1 always outputs bags of one element. We denote by GO[o] a GeneticOperator with its parameter.

Selector. This is a specialized A[$f_{\text{key}}, f_{\text{out}}, 1, 1$] (i.e., with $w_s = w_a = 1$) with one input stream of individuals $S_{I,\text{in}}$ and one output stream of bags of individuals $S_{I^*,\text{out}}$. The key function f_{key} returns the jobId of the individual, hence instances contain individuals of the same iteration and of the same job— this is the default behavior for the Selector operator; later in the paper, we explore different alternatives. The output function f_{out} works as follows: let $\Sigma : \mathcal{P}^*(V(A_\tau(S_I))) \to \mathcal{P}^*(V(A_\tau(S_I)))$ be a stochastic function that takes a bag of individuals and returns a subbag of those individuals, then, given an instance I, f_{out} applies Σ to I for n_{sel} times, hence obtaining n_{sel} bags. We denote by S[Σ, n_{sel}] a Selector with its parameters.

IndividualWrapper. This is a specialized FM[f_1] with one input stream of individuals $S_{I,\text{in}}$ and one output stream of bags of individuals $S_{I^*,\text{out}}$. The only $f_1 : \mathcal{I} \to \mathcal{P}^*(\mathcal{P}^*(\mathcal{I}))$ takes an individual and returns a bag containing a one-element-bag containing that individual. Hence, this operator "wraps" input individuals in bags. We denote by IW a IndividualWrapper.

IndividualUnwrapper. This is a specialized FM[f_1] with one input stream of bags of individuals $S_{I^*,\text{in}}$ and one output stream of individuals $S_{I,\text{out}}$. The only $f_1 : \mathcal{P}^*(\mathcal{I}) \to \mathcal{P}^*(\mathcal{I})$ is the identity. Hence, this operator "unwraps" an input bag of individuals to its elements, which are sent on the output stream. We denote by IU a IndividualUnwrapper.

3.1   Example: Genetic Algorithm (GA) Query

We here show an example of an EA expressed as a query.

We consider the case of a rather standard genetic algorithm (GA), mostly agnostic with respect to the solution representation ϕ—in Sect. 4 we discuss the experiments we performed with a bitstring representation and a tree-based representation for mathematical formulae.

This EA works as follows. Initially, it builds a population of n_{pop} individuals, according to a representation-specific procedure. Then it iterates the following steps. First, it builds an offspring of n_{pop} individuals by generating $0.8n_{pop}$ individuals with crossover followed by mutation and the remaining $0.2n_{pop}$ with just mutation—in both cases, it selects parents with tournament selection. Then it merges the parents with the offspring and selects the best n_{pop} individuals that will constitute the population at the next iteration. The EA keeps iterating for n_{iter} times. Figure 1 shows the query corresponding to this EA.

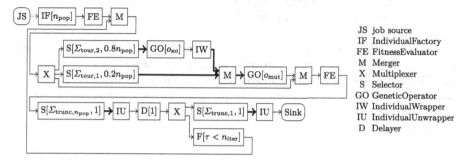

JS	job source
IF	IndividualFactory
FE	FitnessEvaluator
M	Merger
X	Multiplexer
S	Selector
GO	GeneticOperator
IW	IndividualWrapper
IU	IndividualUnwrapper
D	Delayer

Fig. 1. The query for a standard GA. Arrow types indicate the stream types: dotted ⤍ for streams of jobs S_J, solid → for streams of individuals S_I, solid thick ⇥ for streams of bags of individuals S_{I*}.

$\Sigma_{tour,n}$ represents tournament selection: given a bag of individuals, it repeats n times the following steps: it first selects n_{tour} individuals—n_{tour} being a parameter of tournament selection—from the bag (randomly with repetition), then it selects the best individual in the subbag; the output is a bag of n individuals, consistently with the parameters needed by the $S[\Sigma, n_{sel}]$ operator. Note that in the query for this EA $\Sigma_{tour,n}$ is used two times, once for generating $n_{sel} = 0.8n_{pop}$ bags of $n = 2$ individuals (that will be the parents of a new individual built with crossover followed by mutation), once for generating $n_{sel} = 0.2n_{pop}$ bags of $n = 1$ individual (that will be the parent of a new individual built with mutation).

$\Sigma_{trunc,n}$ represents truncation selection: given a bag of individuals, it returns the n best individuals in the subbag—$\Sigma_{trunc,n}$, differently from $\Sigma_{tour,n}$, is hence deterministic. There are two operators in the query based on this selection function: $S[\Sigma_{trunc,n_{pop}}, 1]$ takes input individuals of one iteration (recall that S is an Aggregate with $w_s = w_a = 1$ and that τ is the iteration number) and outputs one bag of the n_{pop} best ones. $S[\Sigma_{trunc,1}, 1]$ just outputs the best individual of each iteration: this one is then unwrapped and sent to the sink.

The operators $D[1]$ and $F[\tau < n_{iter}]$ govern the iterations of the EA. The former increases the iteration number; the latter stops sending back individuals to the first Multiplexer when n_{iter} iterations occurred: that is, it acts as a termination criterion.

The job source JS and the Sink represent the "start" and "end" of the evolutionary optimization. Namely, we assume the source emits one job tuple upon some user action as, e.g., the submission of an optimization task to the SPE running the query representing the EA. On the other end, the arrival of one individual (which is the best individual at each iteration) to the Sink might trigger the storing or logging of the individual (i.e., the solution and its fitness) for later analysis.

3.2 Example: Random Walk (RW) Query

Here we show an example of a query corresponding to another, much simpler EA. This EA is a form of random walk (RW) where the population is constituted by one single individual.

In detail, RW works as follows. Initially, it builds the first individual according to a representation-specific procedure. Then, at each iteration, it mutates the individual, used as parent, and compares the obtained offspring against the parent. If the offspring is better than the parent, it keeps it as the parent for the next iteration; otherwise, it keeps the parent. The EA keeps iterating for n_{iter} times. Figure 2 shows the query corresponding to RW.

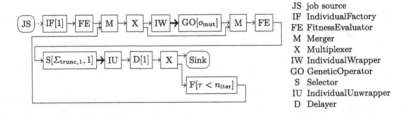

Fig. 2. The query for RW. Arrow types indicate the stream types, as in Fig. 1.

3.3 Streaming-Based Implementation Details

For ease of exposition, Sects. 3.1 and 3.2 discuss the steps performed by the operators while processing the individuals of a single job. One of the key advantages of stream processing, though, is the possibility of leveraging task/operator pipelining and data parallelism while processing the individuals of one or more jobs in a concurrent, parallel, and distributed fashion. In this section, we provide further insights about how stream processing achieves this while efficiently handling the tuples flowing through operators.

Watermarks and Result Production. In Sect. 2.2, we stated that the A operator (i.e., the stateful operator specialized by the Selector operator S—see Sect. 3) produces a result for $I_{k,e}$ upon the reception of a tuple $t_{\text{in}} \in S_{\text{in}}$ so that $ew_a + w_s < v(\tau, t_{\text{in}})$ and $f_{\text{key}}(v(A_\tau(S_{\text{in}}), t_{\text{in}})) = k$.

In the examples from Sects. 3.1 and 3.2, we note all the individuals of a job belonging to the first iteration are eventually fed to the S operator. For such individuals to be fed to f_{out}, nonetheless, S needs to receive at least a tuple with the timestamp and key required to trigger the invocation of f_{out}. Under the hood, such triggering is based on special tuples called *watermarks* [12] that only carry a timestamp and a key. In SPEs, watermarks are automatically generated, forwarded, and processed by operators to trigger results production while correctly enforcing operators' semantics.

Concurrent, Parallel, and Distributed EA Job Execution. Two important aspects must be taken into account when multiple EA jobs are carried out at the same time by a given query. First, due to the asynchronous analysis of streaming operators, individuals of one or more jobs could be processed at different paces, thus advancing their iterations at different rates. Second, depending on the f_{key} used by the S operator, two bags of individuals from the same or different jobs that are not jointly processed at the i-th iteration could later be expected to be jointly processed at the j-th iteration, with $j > i$ (this holds true also for their offspring individuals).

Accounting for the first aspect alone, the watermarks described in the previous section are sufficient to enforce correct semantics even when intra- or inter-job iterations advance at different paces, since the watermark triggering the invocation of f_{out} on such individuals is received after all such individuals. To handle the second aspect, though, watermarks from different keys need to be merged, using only the minimum of the latest received values [12] as a trigger for result production. By doing this, being P_i and P_i' two bags of individuals associated to two different keys k, k' at the i-th iteration, and so that individuals in P_i and P_i' (or their offspring) are to be jointly processed in bag P_j at the j-th iteration, with $j > i$, the watermark forwarded after invoking f_{out} on P_i cannot prematurely trigger the invocation of f_{out} on P_j before P_i' individuals/offspring are added to P_j (and vice-versa). This mechanism is transparently handled by SPEs, as we also exemplify in Sect. 4.3.

Query Optimizations. SPEs usually compile the queries defined by users (referred to as *logical*) into *physical* queries by automatically applying optimizations such as operator *chaining* and parallelization to boost performance while preserving the semantics of the logical query [10,25]. Such a conversion is also applied to the queries considered in this work. During such conversion, UW operators are chained with their upstream peers. An S followed by a UW, for instance, performs directly UW's unwrapping upon the production of a bag of individuals, thus avoiding unnecessary tuple communication costs between S and

UW. IW wrapping, moreover, is transparently handled by the stream connecting IW to its preceding/subsequent operator, avoiding also in this case extra
communication overheads to/from the IW operator itself.

4   Experimental Evaluation

Our experiments aim at answering the following research questions: (RQ1) does
an EA implemented as an SP query deliver the same search effectiveness of
its "classic" implementation? (RQ2) does an EA query scale, in terms of search
efficiency, with the degree of parallelism available to the SPE? (RQ3) is it possible
to express new EAs conveniently in the form of queries?

We performed the experiments considering the EAs presented in Sects. 3.1
and 3.2, each one tailored to two different representations and used to solve two
different problems. Unless otherwise specified, we set $n_{\text{pop}} = 100$, $n_{\text{tour}} = 5$, and
$n_{\text{iter}} = 100$ for GA and $n_{\text{iter}} = 10\,000$ for RW. Note that this way, both EAs
generate $10\,000$ new individuals for each job. For easing the comprehension, in
the following we present the results of the experiments in terms of the number
n_{evals} of fitness function evaluations, rather than of n_{iter}.

Problems and Representations. Concerning the problems, we considered
two cases. First, we considered the classic one-max (OM) problem, in which the
goal is to find an ℓ-long bitstring of ones: we took $\ell \in \{100, 1000\}$. The fitness
function q gives the rate of bits in the string set to zero, to be minimized. For
OM, $G = P = \{0, 1\}^\ell$, i.e., genotypes and phenotypes are bitstrings and the
mapping function ϕ is the identity. We used as genetic operator o_{mut} the bitflip
mutation and as o_{xo} the uniform crossover followed by a bitflip mutation, both
mutations with probability 0.01. We recall that RW uses only o_{mut}, while GA
uses both genetic operators. Finally, when building the initial population for
GA and the initial genotype for RW, we simply sampled ℓ bits for each genotype
with uniform probability.

Second, we considered symbolic regression (SR), in which the goal is to find
a mathematical expression which minimizes the prediction error on a dataset
$\{x^{(i)}, y^{(i)}\}_{i=1}^{i=n}$, with $x \in \mathbb{R}^p$ and $y \in \mathbb{R}$. In particular, we considered the datasets
corresponding to three popular benchmarks [36]: *Keijzer-6*, where $y = \sum_{j=1}^{j=\lfloor x_1 \rfloor} \frac{1}{j}$
and the dataset contains $n = 50$ observations with x_1 evenly spaced in $[1, 50]$;
Nguyen-7, where $y = \ln(x_1+1)+\ln(x_1^2+1)$ and the dataset contains $n = 20$ points
with x_1 randomly distributed in $[0, 2]$; *Pagie-1*, where $y = \frac{1}{1+x_1^{-4}} + \frac{1}{1+x_2^{-4}}$ and the
dataset contains $n = 625$ points with both x_1 and x_2 evenly spaced in $[-5, 5]$.
In the three cases, the fitness function q is given by the mean squared error
(MSE) of a candidate solution on the dataset—we remark that we did not use
linear scaling while evaluating individuals [35]. For SR we adopted a tree-based
representation for the individuals—when coupled with GA, they correspond to a
standard form of genetic programming (GP). In detail, the genotype space G is
the set of all the trees with a depth in $[3, 8]$ in which the non-terminal nodes are

$\bullet + \bullet$, $\bullet - \bullet$, $\bullet \times \bullet$, $\bullet / \bullet$, or $\ln^* \bullet$ (where $\bullet$ represents child nodes and $/$*, $\ln^*$ are the protected versions of the corresponding operations) and terminal nodes are the problem independent variables x_i or the constants 0.1, 1, 10. The phenotype space P is the set of mathematical expressions corresponding to the trees in G. With this representation, we used the standard tree mutation and standard tree crossover as genetic operators.

Implementation and Baseline. We implemented the queries corresponding to GA and RW (and all the stream operators adopted by them) using Flink 1.15.2 [7], a popular and well-established SPE also offered by Cloud providers such as AWS. For the experiment related to (RQ1), we used JGEA [21] as the classic implementation of the two EAs. We remark that both implementations are based on Java.

We run the experiments on an Intel Xeon E5-2637 v4 @ 3.50 GHz (4 cores, 8 threads) server with 64 GB of RAM with Ubuntu 18.04. In general, the two implementations exhibited similar performance in terms of running time: however, we remark that a thorough comparison of the computational efficiency of JGEA against our prototypical EA queries was not a goal of this study.

4.1 (RQ1): Equivalence of Search Effectiveness

We performed 30 jobs, i.e., evolutionary runs, for each combination of problem, EA, and implementation (i.e., query or classic). We report the results in Figs. 3 and 4 for OM and in Figs. 5 and 6 for SR. Namely, Figs. 3 and 5 show the fitness $q(p^*)$ of the best individual during the evolution, while Figs. 4 and 6 detail the distribution of the best fitness at two stages of the evolution, at $n_{\text{evals}} = 1000$ and at $n_{\text{evals}} = 10\,000$, i.e., at the end of the evolution.

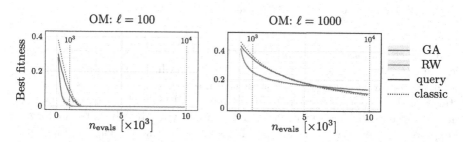

Fig. 3. Best fitness (median and interq. range) during the evolution.

By observing the figures, we note that the general trend of the lines (for Figs. 3 and 5) is similar for the two implementations, regardless of the problem and of the EA. This finding suggests that an EA implemented as a query is "functionally" similar to its counterpart implemented in a classic way: solutions found with a query are as good as those found with a classic implementation.

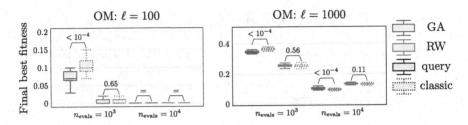

Fig. 4. Distribution of the best fitness $n_{\text{evals}} = 10^3$ and 10^4. Over the pairs of boxplots, p-value of the Wilcoxon test: $=$ means that all the samples are 0.

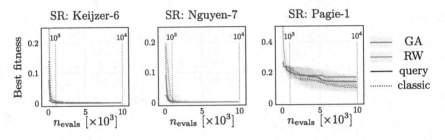

Fig. 5. Best fitness (median and interq. range) during the evolution.

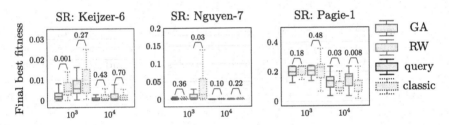

Fig. 6. Distr. of the best fitness at $n_{\text{evals}} = 10^3$ and 10^4; p-values as in Fig. 4.

For each combination of problem and EA, we performed a statistical significance test (Wilcoxon signed rank, after having verified the proper hypotheses) with the null hypothesis of equality of the means of the best fitness, at the two stages of the evolution corresponding to $n_{\text{evals}} = 1000$ and $n_{\text{evals}} = 10\,000$. Figures 4 and 6 report the p-values (and the underlying distributions, in form of boxplots): in most of the cases the differences are not statistically significant.

4.2 (RQ2): Scalability

SPEs are highly optimized to exploit parallelism. We wanted to verify that this capability holds also when the SPE is executing a query corresponding to an EA. For this experiment, we considered the Pagie-1 problem and GA, because they correspond to the most computationally intensive combination. We submitted to the query a number n_{jobs} of concurrent jobs (with different random seeds), with

$n_{\text{jobs}} \in \{1, 3, \ldots, 29\}$, and we measured the average completion time (wall time) for each job. We repeated the experiment four times setting Flink parallelism to 1, 2, 4, or 8 (with 4 and 8 being the parallelism degrees for which all cores are used). Figure 7 presents the salient results of this experiment, i.e., the average wall time vs. n_{jobs} for different parallelism degrees.

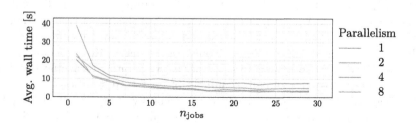

Fig. 7. Average job completion time (wall time) when performing n_{jobs} concurrent jobs using different degrees of parallelism.

It can be seen that, provided a large enough number of jobs (here, ≈ 10) is submitted to the query, the SPE can keep the average wall time constant. Also, by comparing the different lines, one can observe that the greater the parallelism, the lower the average time: namely, it corresponds to 9.5 s, 7.0 s, 6.2 s, and 5.5 s for 1, 2, 4, 8 parallelism, respectively, for $n_{\text{jobs}} = 9$ and to 7.7 s, 4.9 s, 3.5 s, 3.0 s for $n_{\text{jobs}} = 29$. Note that, the gains of higher parallelism degrees are less pronounced as Flink's parallelization grows, especially once 5 or more jobs are run in parallel. This is expected since the 4 available cores are fully utilized for a parallelism degree higher than 4 or when 5 or more jobs run in parallel.

4.3 (RQ3): Expressive Power of Query-Based EAs

Quantifying the expressive power of queries for EAs, i.e., "counting" how many sound EAs can be expressed as queries is an hard task which is beyond the scope of this study. Nevertheless, we attempted to show concretely that a practitioner would easily be capable of modifying an EA by acting on the query.

Beside trivial modifications involving single operators (e.g., changing the selection function Σ of the S operators in Fig. 1 from tournament to roulette wheel), and simple modifications of the query (e.g., removing the "lowest" path exiting from the first X operator in Fig. 1, hence making the generational model without overlapping), we considered a modification showcasing the ease for rich customization enable by SP-based EA queries. In particular, we put ourselves in the perspective of a user who does not know how to set some hyperparameter of an EA for a given problem, considering GA applied to Pagie-1 and the set C of constants defining the genotype space of trees as hyperparameter.

A straightforward approach would be to consider n_p candidates for C and then run n_p jobs. Eventually, one candidate would turn out to be the best one, i.e., the one delivering the best solution. However, this approach would be computationally heavy, because all jobs would be executed entirely, including those corresponding to bad values of C. A smarter approach is to start n_p jobs in parallel with n_p candidate values for C and to merge them after a number n_{merge} of *bootstrap* iterations, on the assumption that in the merged population the individuals built with the best C will likely score better.

While merging jobs is not straightforwardly supported in a traditional EC software (e.g., in JGEA it would imply a rewriting of the full GA), it is trivial in the context of a streaming query, since all individuals (from any number of jobs) are being processed within the same instance of the query. More concretely, this can be achieved by customizing the f_{key} of the S operator, returning the same key for different jobs after the given number of bootstrap iterations.

We realized this modification and experimentally compared it against the expensive approach where all the candidate values are tested entirely—we remark, however, that our goal was not to propose a novel and effective meta-EA, but to show that realizing it is easy if it is an SP-based EA. Namely, we tested two cases: one ("at once") in which we started n_p jobs and merged all the jobs in a single job after n_{merge} iterations and one ("continuous") in which, starting at the n_{merge}-th iteration, we merged two random jobs at each generation until obtaining one single job.

Figure 8 shows the results of the comparison of this meta-EA against the trivial approach ("baseline") where n_p jobs are executed all together and entirely, all terminating after 100 iterations. We executed 10 times each of the five meta-EA and reported the best fitness across all their n_p. We set $n_p = 30$ and built the 30 candidates for C by randomly sampling 3 values in $[0, 10]$ for each one.

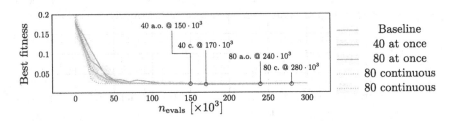

Fig. 8. Best fitness (median and interq. range) during the evolution w meta-EAs.

By observing Fig. 8, we see that the experiment highlights the differences between meta-EAs. In particular, all the variants of the smart approach are cheaper in terms of the overall number of fitness evaluations. Moreover, they also appear to be faster in convergence.

5   Concluding Remarks

We proposed a formal framework based on SP for EAs in the context of EC. We implemented two EAs in the form of SP queries, i.e., graphs of SP operators, and experimentally showed they are as effective as their non-SP counterparts, i.e., the EAs implemented in a traditional way. Also, we showed that using our framework, one may easily define new EAs or modify existing ones to explore new algorithmic possibilities or to streamline complex experimental procedures.

We believe our work may indicate a path towards a democratized EC. SP, a paradigm widely used in the IoT-to-Cloud continuum and offered as a service by, e.g., AWS, may enable the effective use of EAs beyond the communities of researchers and of specialists of specific implementations, without, remarkably, diminishing the wide applicability of EC to very different domains. Moreover, we think that by leveraging the existing literature aimed at characterizing [13] and explaining [26] the execution of SP queries, EC might stand out, in the broad family of AI, in consistency and explainability [32].

Acknowledgements. This study was carried out within the PNRR research activities of the consortium iNEST (Interconnected North-Est Innovation Ecosystem) funded by the European Union Next-GenerationEU (Piano Nazionale di Ripresa e Resilienza (PNRR) - Missione 4 Componente 2, Investimento 1.5 - D.D. 1058 23/06/2022, ECS_00000043), and by the Marie Skłodowska-Curie Doctoral Network project RELAX-DN, funded by the European Union under Horizon Europe 2021–2027 Framework Programme Grant Agreement number 101072456.

References

1. Alba, E., Luque, G., Nesmachnow, S.: Parallel metaheuristics: recent advances and new trends. Int. Trans. Oper. Res. **20**(1), 1–48 (2013)
2. Bartoli, A., Manzoni, L., Medvet, E.: Commentary on "Jaws 30", by W.B. Langdon. Genet. Program Evolvable Mach. **24**, 23 (2023). https://doi.org/10.1007/s10710-023-09471-1
3. Carbone, P., Fragkoulis, M., Kalavri, V., Katsifodimos, A.: Beyond analytics: the evolution of stream processing systems. In: Proceedings of the 2020 ACM SIGMOD International Conference on Management of Data, pp. 2651–2658 (2020)
4. Cardellini, V., Lo Presti, F., Nardelli, M., Russo, G.R.: Runtime adaptation of data stream processing systems: the state of the art. ACM Comput. Surv. **54**(11s), 1–36 (2022)
5. De Lorenzo, A., Bartoli, A., Castelli, M., Medvet, E., Xue, B.: Genetic programming in the twenty-first century: a bibliometric and content-based analysis from both sides of the fence. Genet. Program Evolvable Mach. **21**, 181–204 (2020)
6. Duvignau, R., Gulisano, V., Papatriantafilou, M., Savic, V.: Streaming piecewise linear approximation for efficient data management in edge computing. In: Proceedings of the 34th ACM/SIGAPP Symposium on Applied Computing (2019)
7. Flink: Apache Flink (2023). https://flink.apache.org. Accessed 27 Jan 2023
8. FlinkML: Apache Flink ML Documentation (2023). https://nightlies.apache.org/flink/flink-ml-docs-stable/. Accessed 14 Nov 2023
9. Fortin, F.A., De Rainville, F.M., Gardner, M.A.G., Parizeau, M., Gagné, C.: DEAP: evolutionary algorithms made easy. J. Mach. Learn. Res. **13**(1), 2171–2175 (2012)

10. Frasca, F., Gulisano, V., Mencagli, G., Palyvos-Giannas, D., Torquati, M.: Accelerating stream processing queries with congestion-aware scheduling and real-time Linux threads. In: Proceedings of the 20th ACM International Conference on Computing Frontiers, pp. 144–153 (2023)

11. Gulisano, V., Jimenez-Peris, R., Patino-Martinez, M., Soriente, C., Valduriez, P.: StreamCloud: an elastic and scalable data streaming system. IEEE Trans. Parallel Distrib. Syst. **23**(12), 2351–2365 (2012)

12. Gulisano, V., Palyvos-Giannas, D., Havers, B., Papatriantafilou, M.: The role of event-time order in data streaming analysis. In: Proceedings of the 14th ACM International Conference on Distributed and Event-Based Systems, DEBS 2020, pp. 214–217. Association for Computing Machinery, New York (2020). ISBN 9781450380287. https://doi.org/10.1145/3401025.3404088

13. Gulisano, V., Papadopoulos, A.V., Nikolakopoulos, Y., Papatriantafilou, M., Tsigas, P.: Performance modeling of stream joins. In: Proceedings of the 11th ACM International Conference on Distributed and Event-based Systems, pp. 191–202 (2017)

14. Harada, T., Alba, E.: Parallel genetic algorithms: a useful survey. ACM Comput. Surv. **53**(4), 1–39 (2020)

15. Havers, B., Duvignau, R., Najdataei, H., Gulisano, V., Koppisetty, A.C., Papatriantafilou, M.: DRIVEN: a framework for efficient data retrieval and clustering in vehicular networks. In: 2019 IEEE 35th International Conference on Data Engineering (ICDE), pp. 1850–1861. IEEE (2019)

16. Hummer, W., Satzger, B., Dustdar, S.: Elastic stream processing in the cloud. Wiley Interdisc. Rev. Data Min. Knowl. Disc. **3**(5), 333–345 (2013)

17. Isah, H., Abughofa, T., Mahfuz, S., Ajerla, D., Zulkernine, F., Khan, S.: A survey of distributed data stream processing frameworks. IEEE Access **7**, 154300–154316 (2019)

18. La Cava, W., et al.: Contemporary symbolic regression methods and their relative performance. arXiv preprint arXiv:2107.14351 (2021)

19. Maitre, O., Baumes, L.A., Lachiche, N., Corma, A., Collet, P.: Coarse grain parallelization of evolutionary algorithms on GPGPU cards with EASEA. In: Proceedings of the 11th Annual Conference on Genetic and Evolutionary Computation, pp. 1403–1410 (2009)

20. Medvet, E., Bartoli, A., Carminati, B., Ferrari, E.: Evolutionary inference of attribute-based access control policies. In: Gaspar-Cunha, A., Henggeler Antunes, C., Coello, C.C. (eds.) EMO 2015. LNCS, vol. 9018, pp. 351–365. Springer, Cham (2015). https://doi.org/10.1007/978-3-319-15934-8_24

21. Medvet, E., Nadizar, G., Manzoni, L.: JGEA: a modular java framework for experimenting with evolutionary computation. In: Proceedings of the Genetic and Evolutionary Computation Conference Companion, pp. 2009–2018 (2022)

22. Najdataei, H., Gulisano, V., Tsigas, P., Papatriantafilou, M.: pi-Lisco: parallel and incremental stream-based point-cloud clustering. In: Proceedings of the 37th ACM/SIGAPP Symposium on Applied Computing, pp. 460–469 (2022)

23. Najdataei, H., Nikolakopoulos, Y., Gulisano, V., Papatriantafilou, M.: Continuous and parallel LiDAR point-cloud clustering. In: 2018 IEEE 38th International Conference on Distributed Computing Systems (ICDCS), pp. 671–684. IEEE (2018)

24. Palyvos-Giannas, D., Havers, B., Papatriantafilou, M., Gulisano, V.: Ananke: a streaming framework for live forward provenance. Proc. VLDB Endow. **14**(3), 391–403 (2020)

25. Palyvos-Giannas, D., Mencagli, G., Papatriantafilou, M., Gulisano, V.: Lachesis: a middleware for customizing OS scheduling of stream processing queries. In: Proceedings of the 22nd International Middleware Conference, pp. 365–378 (2021)

26. Palyvos-Giannas, D., Tzompanaki, K., Papatriantafilou, M., Gulisano, V.: Erebus: explaining the outputs of data streaming queries. In: Very Large Data Base, vol. 16, pp. 230–242 (2023)

27. Pigozzi, F., Medvet, E.: Evolving modularity in soft robots through an embodied and self-organizing neural controller. Artif. Life **28**(3), 322–347 (2022)

28. Pospichal, P., Jaros, J., Schwarz, J.: Parallel genetic algorithm on the CUDA architecture. In: Di Chio, C., et al. (eds.) EvoApplications 2010. LNCS, vol. 6024, pp. 442–451. Springer, Heidelberg (2010). https://doi.org/10.1007/978-3-642-12239-2_46

29. Rathore, M.M., Son, H., Ahmad, A., Paul, A., Jeon, G.: Real-time big data stream processing using GPU with spark over Hadoop ecosystem. Int. J. Parallel Program. **46**(3), 630–646 (2017). https://doi.org/10.1007/s10766-017-0513-2

30. Röger, H., Mayer, R.: A comprehensive survey on parallelization and elasticity in stream processing. ACM Compu. Surv. (CSUR) **52**(2), 1–37 (2019)

31. Rovito, L., De Lorenzo, A., Manzoni, L.: Evolution of Walsh Transforms with genetic programming. In: Proceedings of the Companion Conference on Genetic and Evolutionary Computation, pp. 2386–2389 (2023)

32. Rudin, C.: Stop explaining black box machine learning models for high stakes decisions and use interpretable models instead. Nat. Mach. Intell. **1**(5), 206–215 (2019)

33. Russo, G.R., Cardellini, V., Presti, F.L.: Reinforcement learning based policies for elastic stream processing on heterogeneous resources. In: Proceedings of the 13th ACM International Conference on Distributed and Event-Based Systems, pp. 31–42 (2019)

34. Stephens, R.: A survey of stream processing. Acta Informatica **34**, 491–541 (1997)

35. Virgolin, M., Alderliesten, T., Bosman, P.A.: Linear scaling with and within semantic backpropagation-based genetic programming for symbolic regression. In: Proceedings of the Genetic and Evolutionary Computation Conference, pp. 1084–1092 (2019)

36. White, D.R., et al.: Better GP benchmarks: community survey results and proposals. Genet. Program Evolvable Mach. **14**, 3–29 (2013)

Evolutionary Computation in Image Analysis, Signal Processing and Pattern Recognition

Integrating Data Augmentation in Evolutionary Algorithms for Feature Selection: A Preliminary Study

Tiziana D'Alessandro[ID], Claudio De Stefano[ID], Francesco Fontanella[(✉)][ID], and Emanuele Nardone[ID]

Department of Electrical and Information Engineering (DIEI), University of Cassino and Southern Lazio, Via G. Di Biasio 43, 03043 Cassino, FR, Italy
{tiziana.dalessandro,destefano,fontanella,emanuele.nardone}@unicas.it

Abstract. In many machine learning applications, there are hundreds or even thousands of features available, and selecting the smallest subset of relevant features is a challenging task. More recently, researchers have investigated how data augmentation affects feature selection performance. Although evolutionary algorithms have been widely used for feature selection, no studies have investigated how data augmentation affects their performance on this challenging task. The study presented in this paper investigates how data augmentation affects the performance of evolutionary algorithms on feature selection problems. To this aim, we have tested Genetic Algorithms and Particle Swarm Optimization and compared their performance with two widely used feature selection algorithms. The experimental results confirmed that data augmentation is a promising tool for improving the performance of evolutionary algorithms for feature selection.

1 Introduction

In recent years, machine learning has seen a strong growth in the number of features used to represent the ever-increasing data collected. Unfortunately, in many applications, a part of those features may be redundant or irrelevant with respect to the target concept of the problem at hand. Those features may give rise to the curse of dimensionality phenomenon: when the dimensionality increases, the volume of the space increases so fast that the available data becomes sparse. This sparsity negatively affects the performance of the learned models and increases training time complexity. Those harmful features can be identified and then eliminated by using feature selection techniques. Feature selection refers to the process of finding the smallest subset of relevant features to use for the model construction. Feature selection is a search problem in the search space made of all the possible subsets of the available features. In addition to a search strategy, feature selection requires an evaluation function to estimate feature subset quality. Evaluation functions can be divided into two wide categories, namely filter and wrapper. Filter functions are based on statistics measures, whereas

S. Smith et al. (Eds.): EvoApplications 2024, LNCS 14634, pp. 397–412, 2024.
https://doi.org/10.1007/978-3-031-56852-7_25

wrapper functions use the performance achieved by a given classification algorithm trained on the subset of features to be evaluated [15]. As concerns the search strategy, the exponential nature of the search space (if N is the number of available features, the total number of possible solutions is 2^N) makes the exhaustive search impracticable in most real-world problems. For this reason, many search techniques have been applied to feature selection, such as complete search, greedy search, and heuristic search [21]. Unfortunately, the effectiveness of most of these approaches is limited by their high computational costs or by early stagnation in local optima.

Thanks to their global search ability, evolutionary computation (EC) techniques have been widely used as search tools in feature selection problems [1,22]. Furthermore, EC techniques do not need domain knowledge and do not make any assumptions about the search space, such as whether it is linearly or non-linearly separable and differentiable [5,6]. Among the EC-based approaches, Genetic Algorithms (GAs) have been widely used. GA binary vectors provide a natural and straightforward representation for feature subsets: the value 1 or 0 of the chromosome i-th element indicates whether the i-th feature is included or not. This allows GA-based algorithms to be used for feature selection without any modification [7]. More recently, also Particle Swarm Optimization (PSO) has been widely used. PSO has been inspired by the social behavior of birds and fish. In the PSO metaphor, a swarm of "particles" (the potential solutions) move through the search space by adjusting their velocities and positions. Particle dynamics is based on the knowledge learned so far in the search space by stochastically toward their own best-known position as well as the entire swarm's best-known position. For feature selection, particles can be either binary vectors, as in the GA algorithm mentioned above, or real-valued numbers [14]. In the second case, the i-th feature is selected only if the i-th value in the particle is larger than a given threshold θ [23].

Data augmentation involves creating new training data by applying various transformations to the original data [11], and improves model performance by increasing the diversity of training samples. Data augmentation is widely used in deep learning to create new training images using transformations like flipping, rotating, and scaling, among others [20]. Although less common, data augmentation is also used for tabular data. In this case, transformations include techniques like adding noise, interpolating data, or shuffling features [26]. Also, EC-based techniques have been used for data augmentation. In [18], the authors introduce a novel approach for finding effective data augmentation strategies to train deep neural networks in the medical imaging domain. In [16], the authors use an evolutionary algorithm to select the input samples generated by a generative adversarial network (GAN).

Recent studies have investigated the effectiveness of integrating data augmentation and feature selection. In [25], the authors use a generative adversarial network (GAN)-based technique and a hybrid feature selection method for small sample credit risk assessment with high dimensionality. In [4], the authors combine data augmentation and feature selection to build a data-driven system for automatic model recommendation in a computational physics problem. However, although EC-based approaches have been widely used for feature selection and, more recently, data augmentation, no studies have investigated how data augmentation affects the performance of evolutionary algorithms for feature selection.

In this paper, we try to fill the gap mentioned above by presenting a study in which we have tested the effectiveness of EC-based algorithms on augmented data. In particular, we tested GA and PSO as evolutionary algorithms. To investigate the performance of GA and PSO, we first compared their performance with the baseline results achieved without feature selection. Then, we compared their results with those achieved by two widely used feature selection algorithms. The first algorithm is recursive feature elimination (RFE) [13], whereas the second is a filter-based algorithm [19], both implemented in scikit-learn [17], used as ML backend library for the project. The experimental results confirmed that data augmentation is a promising approach for improving the effectiveness of EC-based algorithms for feature selection. The analysis of those results has allowed us to define a wide range of future research activities.

The remainder of the paper is organized as follows: Sect. 2 details the proposed approach, whereas Sect. 3 reports and discusses the experimental results. Finally, Sect. 4 is devoted to the conclusions and summarizes the future work inspired by the analysis of the experimental results.

2   The Proposed Approach

This study is aimed at investigating how data augmentation (DA) affects evolutionary algorithms in feature selection (FS) problems. Our approach was to explore how these techniques influence the performance of machine learning models, primarily in terms of accuracy and number of features selected. In the following subsections, we detail the procedure we implemented for data augmentation as well as the implementation of the GA and PSO algorithms considered in our study.

2.1   Data Augmentation

Data augmentation, a pivotal technique in machine learning, is essential for tackling imbalanced datasets and enhancing model generalization. Our algorithm (*Algorithm* 1), streamlines the generation of new samples by initially creating a set of random samples. The process involves a class-wise computation of the mean and standard deviation for each feature in the dataset. The augmentation process is iterative, continuing until the predetermined augmentation percentage is met. During this iterative process, each class undergoes a procedure where new samples are perturbed in a manner that adheres to class-wise constraints. Each perturbed sample is then evaluated to ensure it falls within $\pm 3\sigma$ of the class's feature distribution. Samples that meet this criterion are retained in the augmented dataset, while those that do not are discarded and replaced with

---

Algorithm 1: Data Augmentation with Random Noise.

Input: Dataset with N features and C classes; Percentage of augmentation P.
Output: Augmented dataset.
Load Dataset;
Generate randomly k new samples;
while P *not met* **do**
   | **foreach** *Class* **do**
   | | Calculate the mean and standard deviation of each feature class-wise;
   | | Pertubate samples adhering to class-wise constraints;
   | | Evaluate if a samples falls within $\pm 3\sigma$ of the class distribution;
   | | **if** *sample fits within the range* **then**
   | | | Retain the sample in the augmented dataset;
   | | **else**
   | | | Discard the sample and generate a new one;
   | | **end**
   | **end**
   | Check the performance of the augmented dataset with a classifier
   | (K-Nearest Neighbors);
   | Ensure class balance in the distribution of new instances;
end

---

newly generated samples. This approach meticulously maintains the integrity and distribution of the original dataset, introducing only slight modifications to random instances.

The augmentation's effectiveness is estimated using a K-Nearest Neighbors classifier, ensuring class balance and adjusting generated sample numbers per class as needed.

2.2 Evolutionary Algorithms for Feature Selection

PSO for Feature Selection. Particle Swarm Optimization (PSO) is an evolutionary computation technique that mimics the social behavior of swarms. In the context of feature selection for machine learning, PSO can be harnessed to identify an optimal subset of features that enhance the performance of a predictive model.

The PSO algorithm begins with the initialization of a population of particles. Each particle represents a candidate solution to the feature selection problem and is characterized by a position vector in the search space and a velocity vector that determines its movement through the search space. The position of a particle corresponds to a specific subset of features from the dataset.

The fitness of each particle is described in Sect. 2.3. A sigmoid function is applied to the particle's position to decide if its value exceeds 0.5, including or excluding each feature. This function transforms the feature's score to a probability, creating a binary decision process. A particle's position, therefore, translates to a particular selection of features.

Table 1. PSO parameters setting.

Parameter	Value
Swarm Size	30
Cognitive Coefficient (ϕ_1)	2.0
Social Coefficient (ϕ_2)	2.0
Number of Iterations	60
Particle Position Limits (p_{min}, p_{max})	$[-1.0, 1.0]$
Velocity Limits (s_{min}, s_{max})	$[-1.0, 1.0]$

The PSO algorithm iteratively updates the particles' velocities and positions. This update is influenced by the personal best position of each particle and the global best position found by the swarm, moderated by cognitive and social coefficients. These coefficients determine the relative influence of an individual particle's experience and the collective experience of the swarm, respectively.

Each particle's velocity is adjusted by considering the difference between its current position, its personal best, and the global best positions. The velocity update reflects a balance between exploring new areas in the search space and exploiting known good solutions. The particles' positions are then updated by adding the new velocity to the current position, ensuring that particles move towards regions of the search space with higher fitness values.

As the algorithm proceeds, particles converge towards a subset of features that provides a balance between a high accuracy score and a low complexity penalty. The algorithm terminates after a pre-defined number of iterations.

Upon completion, the algorithm returns the best feature subset found during the search (Table 1 shows the parameter values used).

GA for Feature Selection. In this section, we describe the second method for the feature selection process, using a Genetic Algorithm (GA). Traditional methods may lead to suboptimal solutions. Hence, we employed a GA to efficiently search for the best subset of features that could yield the highest prediction accuracy.

The GA begins with a population of potential solutions, each representing a different combination of features. These solutions are encoded as binary strings, where the presence of a feature in a given solution is marked as '1' and its absence as '0'. The GA evolves these solutions over several generations. In each generation, the quality of each solution is evaluated according to the fitness function detailed in the next subsection.

The evolution process involves selection, crossover, and mutation operations. Selection favors solutions with higher fitness, allowing them to pass their genes to subsequent generations. Crossover combines pairs of solutions to produce offspring that inherit features from both parents, while mutation introduces random changes to maintain genetic diversity within the population. The algorithm

Table 2. GA parameters setting.

Parameter	Value
Population Size	50
Crossover Probability	0.6
Mutation Probability	1/#features
Tournament size	3
Number of Generations	40
Elitism	Keep best

iteratively refines the population of feature subsets through these operations. The best-performing individual (feature subset) is identified at the end of the process. This subset provides the features that better discriminate the classes of the problem at hand by excluding irrelevant and noisy features.

The GA's effectiveness is measured by the quality of the final feature subset it identifies, as well as the consistency of its performance across multiple runs. The approach is flexible and can be adapted to various types of data and models. This method allows for an automated and intelligent search for the optimal feature space, potentially uncovering interactions and dependencies that are not immediately evident using other feature selection methods (Table 2 shows the parameter values used).

2.3 Fitness

We use a wrapper fitness function implemented using the DEAP [9] Python library and based on the K-Nearest Neighbors (KNN) classifier, using five-fold cross-validation as a model performance evaluation strategy. Furthermore, the defined fitness function also considers the number of selected features. Given an Individual I, its fitness $f(I)$ is computed as follows:

$$f(I) = \text{Acc} + \alpha \left(\frac{N_{\text{tot}} - N_{\text{sel}}}{N_{\text{tot}}} \right) \tag{1}$$

Acc is the accuracy achieved by KNN on the subset of features encoded by I, α is a penalty coefficient set to 0.01, N_{tot} is the total number of available features, whereas N_{sel} is the cardinality of the subset encoded by I.

3   Experimental Results

As mentioned in the Introduction, this research aims to investigate the effect of data augmentation on EC-based techniques for feature selection; for this reason, we performed several sets of experiments. We used six datasets originating from various application domains. We employed datasets with diverse characteristics regarding sample sizes, features, and classes. In detail, we considered Hand [2], a

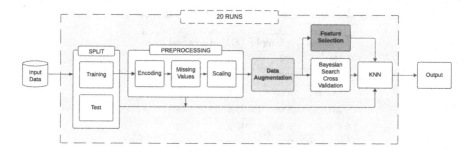

Fig. 1. Experimental workflow.

handwriting features dataset for Alzheimer's disease diagnosis; Isolet [3], known as the "Isolated Letter Speech Recognition" dataset contains features extracted from audio recordings, for speech and pattern recognition tasks; Mfeat1 and Mfeat2 are two subsets of the Multiple Features dataset [8], containing attributes associated with handwritten numerals. Other datasets are the Ozone [24] dataset, related to ozone levels for monitoring and studying climate and environment, and Toxicity [12], comprising toxic and non-toxic molecules specifically crafted for the functional domains of a protein playing a pivotal role in regulating circadian rhythms. For further details, please refer to Table 3.

The following subsections comprehensively describe each experiment, discussing the related results. For each dataset and experiment, we carried out twenty runs. Figure 1 shows the overall workflow of the experiments carried out. The first implementation refers to a baseline case, Sect. 3.1, where datasets were directly used for classification. Instead, four feature selection methods were exploited in the second experiment, Sect. 3.2. A third experiment, Sect. 3.3, involved the addition of a data augmentation module. In a fourth set of experiments, we investigated the behavior of GA and PSO during the evolution. Finally, to assess the effectiveness of GA and PSO, we compared the best results obtained using data augmentation with the baseline results as well as those achieved without data augmentation.

To assess the performance we used the accuracy, a common metric in machine learning that measures the overall correctness of a model's predictions. It was calculated as the ratio of correctly predicted instances over the total number of instances. As we perform 20 runs, we refer to the average accuracy and its standard deviation, which provide us with information about the stability of the experimental results. We also report the average number of selected features, as it is crucial in ML for efficiency, interpretability, and model performance.

Table 3. The datasets used in the experiments.

Dataset	#Samples	#Features	#Classes
Hand	174	90	2
Isolet	7797	617	26
Mfeat1	2000	216	10
Mfeat2	2000	64	10
Ozone	4748	72	2
Toxicity	171	1203	2

3.1 Baseline Experiment

The initial experimental setup served as a basic reference point for comparing its performance to the subsequent experiments, which were more time-consuming and resource-intensive. This comparison is helpful to determine whether the added complexity and effort in more advanced experiments lead to better performance. Figure 1 shows the final experimental workflow, where the baseline case is represented as the uncoloured section of the model. The system took a dataset, subjecting it to an initial preprocessing phase. This step is crucial for ensuring data quality and aligning the dataset with the standards expected by the subsequent classification algorithm. The procedure can be described through three operations: the encoding of categorical features, the handling of missing values, and a scaling transformation of all the features with the RobustScaler standardization technique. The first two operations were selectively applied to datasets where their application is meaningful and necessary. This step was the same for every experimental setting. Once the input dataset was processed, it was ready for the classification step. First, the dataset was split into training (80%) and test (20%) sets. Then, the training set was employed in a Bayesian search [10] to optimize the hyperparameters of the classification algorithm. The Bayesian search was implemented following a 5-fold cross-validation strategy. The chosen supervised ML algorithm is K-Nearest Neighbors (KNN), and after the hyperparameters' optimization, it was trained on the training set and tested on the test set. The final output was expressed in average and standard deviation accuracy, computed over the 20 runs. Table 4 shows for every dataset the obtained value of average accuracy and standard deviation.

Table 4. Baseline experiment results in terms of average accuracy and standard deviation computed over 20 runs.

Dataset	AVG	STD
Hand	58.8	7.1
Isolet	90.7	0.6
Mfeat1	95.7	1.1
Mfeat2	95.3	1.1
Ozone	94.1	3.8
Toxicity	67.4	5.5

3.2   Testing Feature Selection

The second experimental setting can be evinced from Fig. 1, considering the uncoloured section and the green box referring to the feature selection module. It is noticeable at first glance that the complexity is increased with respect to the baseline experiment, described in Sect. 3.1. The setup is similar to the previous one, with the addition of a feature selection module. Before the Bayesian search, we used the training set for feature selection (FS). We tested several FS techniques to understand which feature selection method was more convenient and to find a good trade-off between resources and performance. In particular, this experiment aims to compare common FS methods with those proposed in our approach Sect. 2.2, based on evolutionary algorithms inspired by the principles of biological evolution and natural selection. In detail, we considered two FS algorithms:

- Recursive Feature Elimination (RFE);
- SelectKBest (SKB).

RFE performs a greedy search to find the best-performing feature subset based on the backward elimination strategy [13]. Starting from the whole set of available features, the RFE algorithm iteratively creates models and determines the worst-performing feature at each iteration. Then, it builds the subsequent models with the remaining features until all of them are explored. Finally, features are ranked according to the order of their elimination. If the data contain N features, RFE evaluates N^2 subsets in the worst case. SelectKBest algorithm, instead, is a feature selection technique that identifies the most influential features in a dataset. It assigns scores to features based on a specified statistical test, which measures the strength of the relationship between each feature and the target variable. Common tests include ANOVA for regression and Chi-squared for classification. SelectKBest then ranks these features based on their scores. The parameter k determines the number of top-scoring features to retain. This process effectively reduces the feature space, enabling models to focus on the most relevant data, improving efficiency, and potentially enhancing model performance. The training set was partitioned to create a validation set comprising 20% of its samples. This division is crucial, as the validation accuracy is the scoring function for RFE and SKB obtained from the KNN algorithm.

Table 5 shows the result for the feature selection methods tested; for every dataset, the best results are in bold. From the table, we can observe that the best performance was achieved by PSO, which outperforms the others in four cases out of six. PSO outperformed the remaining algorithms both in terms of accuracy and the number of selected features. Though PSO was the best, another interesting trend from the table is that PSO and GA performance was comparable to the common and well-established FS method tested.

Table 5. Feature selection results in terms of average and standard deviation accuracy and average number of selected features computed over 20 runs.

Dataset(#features)	RFE		SKB		PSO		GA	
	Acc	#feat	Acc	#feat	Acc	#feat	Acc	#feat
Hand (90)	57.7 ± 7.5	45	58.0 ± 6.9	45	**58.8 ± 6.8**	**33.3**	58.4 ± 8.0	40.6
Isolet (617)	88.8 ± 1.4	308	88.3 ± 2.1	307	**92.1 ± 0.9**	**291.8**	91.5 ± 0.7	302.1
Mfeat1 (216)	95.0 ± 1.7	108	95.4 ± 1.6	109	**95.9 ± 0.9**	**90.9**	95.8 ± 1.1	103.3
Mfeat2 (64)	96.0 ± 1.0	32	**96.2 ± 0.9**	**31**	95.3 ± 1.0	31.5	95.3 ± 1.2	36.1
Ozone (72)	91.8 ± 3.2	36	92.3 ± 3.9	37	92.4 ± 3.3	**29.9**	**94.2 ±2.8**	32.6
Toxicity (1203)	64.8 ± 5.8	600	64.6 ± 6.6	601	**67.1 ± 7.5**	**509.9**	64.4 ± 8.7	597.9

3.3   Data Augmentation and Feature Selection

The third experiment shown in Fig. 1 was conducted with the experimental setting described in Sect. 3.2, with the addition of a data augmentation module (orange box). The workflow equals the previous one, so the input dataset was processed and split into training and test sets. This time, the training set was first augmented and then used for feature selection, bayesian search, and finally, for training the KNN algorithm. Using the data augmentation procedure detailed in Sect. 2.1, we tested three augmentation percentages, namely 10%, 20%, and 30%.

Table 6 shows the results obtained. To have a complete view of how the DA affects the performance, we also included the results achieved without DA (0%). From the table, it is evident that increasing the augmentation percentage generally leads to improved performance. There are exceptions to this trend, as indicated by the Ozone dataset when using the GA method, and the Isolet dataset in combination with RFE and SKB. However, for every dataset, the best performance was reached with the highest percentage of augmented data; in some cases, the increase in the accuracy reached the 5% from no data augmentation up to the 30% of augmented data, as for the case of the Isolet dataset with RFE. Besides this, the evolutionary FS methods outperformed the others for four datasets out of six, whereas for the remaining datasets, their performance was comparable to the other well-assessed FS methods. From the table, we can also observe that for all algorithms, data augmentation did not affect the number of features selected. Finally, comparing PSO and GA, we can see that the GA typically selected more features than the PSO.

3.4   Investigating the Behaviour of GA and PSO

In this set of experiments, we investigated the behavior of GA and PSO during the evolution. To this aim, we plotted the average fitness values as a function of the generation number. Figure 2 shows those plots for three out of the six datasets analyzed, namely Hand, Isolet, and Toxicity (the other ones showed similar trends). To investigate how data augmentation affects evolution, we plotted the average fitness for the three data augmentation percentages as well as

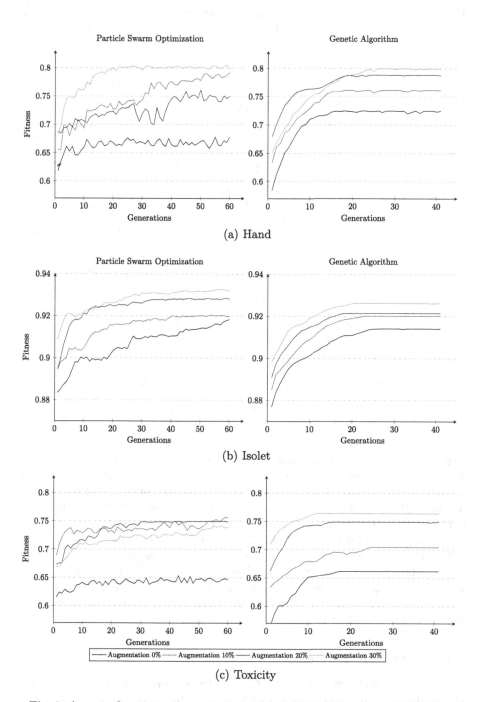

(a) Hand

(b) Isolet

(c) Toxicity

Fig. 2. Average fitness trend comparison: PSO on the left and GA on the right.

Table 6. Data augmentation and feature selection results in terms of average accuracy and standard deviation computed over 20 runs for every FS technique and DA percentage.

Dataset (#features)	DA (%)	RFE		SKB		PSO		GA	
		Acc	#feat	Acc	#feat	Acc	#feat	Acc	#feat
Hand (90)	0	57.7 ± 7.6	45	57.9 ± 6.9	45	58.8 ± 6.8	33.3	58.4 ± 8.1	40.6
	10	61.7 ± 7.0	45	60.1 ± 8.0	46	62.7 ± 7.4	38.0	62.8 ± 8.7	40.7
	20	65.6 ± 6.6	45	64.5 ± 8.5	44	66.3 ± 9.9	33.1	67.6 ± 6.8	40.1
	30	66.7 ± 7.2	44	64.0 ± 9.5	45	68.3 ± 6.2	34.3	**69.6 ± 5.7**	38.4
Isolet (617)	0	88.8 ± 1.4	308	88.3 ± 2.1	307	92.1 ± 0.9	291.8	91.5 ± 0.8	302.1
	10	82.2 ± 4.6	307	81.7 ± 3.6	308	92.0 ± 0.7	294.9	91.5 ± 1.6	301.5
	20	85.5 ± 3.8	308	85.9 ± 3.4	306	92.6 ± 0.7	295.6	92.4 ± 0.8	304.1
	30	88.9 ± 2.2	308	87.7 ± 2.5	308	**92.9 ± 0.9**	292.3	92.7 ± 0.7	299.4
Mfeat1 (216)	0	95.0 ± 1.7	108	95.4 ± 1.6	109	95.9 ± 0.9	90.9	95.9 ± 1.1	103.3
	10	95.5 ± 1.4	107	95.6 ± 0.9	109	96.2 ± 1.0	94.5	96.5 ± 0.8	101.2
	20	96.2 ± 0.9	108	96.3 ± 0.9	109	96.5 ± 0.8	89.5	96.5 ± 0.6	105.1
	30	96.5 ± 0.9	106	96.5 ± 0.9	109	96.7 ± 0.6	90.8	**96.9 ± 0.9**	104.6
Mfeat2 (64)	0	96.0 ± 1.0	32	96.2 ± 0.9	31	95.3 ± 1.1	31.5	95.3 ± 1.2	36.1
	10	96.0 ± 1.1	32	96.1 ± 1.4	33	95.4 ± 1.4	32.0	95.2 ± 1.4	35.6
	20	96.4 ± 0.9	32	95.6 ± 0.9	34	96.1 ± 1.0	32.4	96.2 ± 0.8	36.2
	30	96.1 ± 1.3	32	**96.5 ± 0.8**	32	95.7 ± 1.3	32.1	96.2 ± 1.0	34.7
Ozone (72)	0	91.8 ± 3.2	36	92.3 ± 3.9	37	92.4 ± 3.4	29.9	94.2 ± 2.9	32.6
	10	92.1 ± 3.1	37	93.3 ± 3.4	35	93.1 ± 2.4	30.2	92.8 ± 2.8	33.7
	20	93.0 ± 2.8	36	93.4 ± 3.2	35	94.1 ± 2.6	29.2	93.4 ± 2.5	33.7
	30	92.9 ± 2.4	36	93.2 ± 2.5	35	**94.8 ± 2.1**	30.0	94.6 ± 2.2	33.1
Toxicity (1203)	0	64.8 ± 5.8	600	64.6 ± 6.6	601	67.1 ± 7.5	509.9	64.4 ± 8.7	597.9
	10	64.5 ± 6.6	601	67.9 ± 7.5	602	66.1 ± 5.5	511.3	62.2 ± 4.2	593.1
	20	66.7 ± 9.5	601	66.9 ± 7.9	600	65.3 ± 6.3	502.7	63.4 ± 9.1	592.0
	30	**70.1 ± 8.9**	601	67.1 ± 9.5	600	69.6 ± 8.0	519.8	68.7 ± 8.0	593.3

that without augmentation. Looking at the plots, it is easy to notice that for each dataset the range of variation of the fitness is the same for the two algorithms and that in every case the line referring to the 0% of DA never exceeds the others, while the yellow line referring to a DA of 30% exceeds all the others: the only exceptions occur for some generations in the first part of the evolution, as shown in Fig. 2(a) for GA and in Fig. 2(b) for PSO. As expected, the increment of the fitness is directly linked to the increment of the augmented data. This is true in almost every case, except for PSO algorithm applied to Hand dataset, where the augmentation of the 10% outperforms the one of the 20%. Another interesting detail is that the fitness of the GA shows a more stable trend with respect to the one given by PSO.

3.5   Comparison Findings

Finally, to assess the effectiveness of GA and PSO, we compared the best results obtained using data augmentation with the baseline results (no FS) as well as those achieved without DA (see Table 7). To statistically validate the comparison results, we performed the non-parametric Wilcoxon rank-sum test ($\alpha = 0.05$).

Table 7. Comparison between baseline experiment and best performance achieved with FS and FS combined with DA.

Dataset (#features)	Baseline	Without DA			30% DA		
	Acc	FS	Acc	#feat	FS	Acc	#feat
Hand (90)	58.8 ± 7.1	PSO	58.8 ± 6.8	33.3	GA	**69.6 ± 5.7**	**38.45**
Isolet (617)	90.7 ± 0.6	PSO	92.1 ± 0.9	291.8	PSO	**92.9 ± 0.9**	**292.3**
Mfeat1 (216)	95.7 ± 1.1	PSO	95.9 ± 0.9	90.9	GA	**96.9 ± 0.9**	**104.6**
Mfeat2 (64)	95.3 ± 1.1	SKB	96.2* ± 0.9	31	SKB	96.5* ± 0.8	32
Ozone (72)	94.1* ± 3.8	GA	94.2* ± 2.8	32.6	PSO	94.8* ± 2.1	30
Toxicity (1203)	67.4* ± 5.5	PSO	67.1* ± 7.5	509.9	RFE	70.1* ± 8.9	601

The values in bold highlight for each dataset the results, which are significantly better with respect to the remaining ones, according to the Wilcoxon test. As concerns the results that do not present a statistically significant difference, the best two results are both starred.

From the table, we can see that the GA and PSO achieved the best statistically significant results on three out of the six datasets used (Hand, Isolet, and Mfeat1). It is interesting to note that for the Hand dataset, feature selection without data augmentation did not achieve any performance improvement, whereas data augmentation allowed us to achieve a large accuracy improvement (about 10%), but selecting more features A similar behavior, but with a smaller accuracy improvement, can be observed for Mfeat1, where data augmentation improved the accuracy by selecting more features.

Finally, from the table, we can also see that PSO performed better than GA on all three datasets, whereas data augmentation allowed GA to achieve the best performance on two out of the three datasets (Hand and Mfeat1), by selecting more features than those selected without data augmentation. These results suggest that data augmentation makes advantageous the GA property to select more features than PSO.

4   Conclusions and Future Work

In this research, we investigated the effect of data augmentation on the performance of evolutionary algorithms applied to feature selection problems. To this aim, we considered two effective and widely used evolutionary-based algorithms for feature selection, namely a genetic algorithm and a particle swarm optimization. They can explore a large search space of feature combinations thanks to their search efficiency. We tested these algorithms on six different datasets and considered different percentages of augmented data up to 30%. Finally, we also evaluated a baseline case where no feature selection and data augmentation were applied for comparison. For what concerns feature selection, we also tested two well-known techniques in machine learning: RFE and SKB. As expected, the developed experimental settings highlighted the importance of a feature selection technique and a data augmentation procedure.

The results obtained confirm that data augmentation always produces improvements in the performance of feature selection algorithms and that these improvements are directly proportional to the percentage of augmented data. In the case of the Hand dataset, the performance increment is much more significant compared to that of the other considered databases: this behavior may be due both to the fact that the baseline performances of the other databases, namely Isolet, Mfeat1, Mfeat2 and Ozone, are very high (over 90%) compared to those of Hand database (less than 60%), and to the fact that probably the available features are less redundant. This last point is also confirmed by the percentage of features selected for Hand database, which is lower than those of the above databases. The only exception is represented by the database Toxicity, which exhibits baseline performance rather low (less than 70% of accuracy), similar to Hand database, but does not allow the feature selection algorithms to achieve significant performance improvements. These results probably depend on the limited number of samples included in the toxicity database (171), but represented with a very high number of features (1203).

These interesting preliminary results paved the way for a wide range of future research activities. Indeed, we plan to extend our study by considering more datasets and evaluating different classification algorithms in both the final step and the fitness of the feature selection. We will also investigate different metrics for evaluating classification performance, e.g., F1-score, for imbalanced datasets. Future research activities will also investigate how data augmentation affects the performance of Multi-objective feature selection.

References

1. Cilia, N.D., De Stefano, C., Fontanella, F., Scotto di Freca, A.: Variable-length representation for EC-based feature selection in high-dimensional data. In: Kaufmann, P., Castillo, P.A. (eds.) EvoApplications 2019. LNCS, vol. 11454, pp. 325–340. Springer, Cham (2019). https://doi.org/10.1007/978-3-030-16692-2_22
2. Cilia, N.D., De Stefano, C., Fontanella, F., Molinara, M., Scotto Di Freca, A.: Handwriting analysis to support Alzheimer's disease diagnosis: a preliminary study. In: Vento, M., Percannella, G. (eds.) Computer Analysis of Images and Patterns: 18th International Conference, CAIP 2019, Salerno, Italy, September 3–5, 2019, Proceedings, Part II, pp. 143–151. Springer, Cham (2019). https://doi.org/10.1007/978-3-030-29891-3_13
3. Cole, R., Fanty, M.: ISOLET. UCI Machine Learning Repository (1994). https://doi.org/10.24432/C51G69
4. Daniel, T., Casenave, F., Akkari, N., Ryckelynck, D.: Data augmentation and feature selection for automatic model recommendation in computational physics. Math. Comput. Appl. **26**(1), 17 (2021). https://doi.org/10.3390/mca26010017

5. De Falco, I., Tarantino, E., Della Cioppa, A., Fontanella, F.: A novel grammar-based genetic programming approach to clustering. In: Proceedings of the 2005 ACM Symposium on Applied Computing, pp. 928–932 (2005)

6. De Falco, I., Tarantino, E., Cioppa, A.D., Fontanella, F.: An innovative approach to genetic programming—based clustering. In: Abraham, A., de Baets, B., Köppen, M., Nickolay, B. (eds.) Applied Soft Computing Technologies: The Challenge of Complexity, pp. 55–64. Springer, Heidelberg (2006). https://doi.org/10.1007/3-540-31662-0_4

7. De Stefano, C., Fontanella, F., Marrocco, C.: A GA-based feature selection algorithm for remote sensing images. In: Giacobini, M., et al. (eds.) Applications of Evolutionary Computing, pp. 285–294. Springer, Heidelberg (2008). https://doi.org/10.1007/978-3-540-78761-7_29

8. Duin, R.: Multiple Features. UCI Machine Learning Repository. https://doi.org/10.24432/C5HC70

9. Fortin, F.A., De Rainville, F.M., Gardner, M.A., Parizeau, M., Gagné, C.: DEAP: evolutionary algorithms made easy. J. Mach. Learn. Res. **13**, 2171–2175 (2012)

10. Frazier, P.I.: A tutorial on Bayesian optimization. arXiv preprint arXiv:1807.02811 (2018)

11. Goodfellow, I., Bengio, Y., Courville, A.: Deep Learning. Adaptive Computation and Machine Learning, MIT Press (2016)

12. Gul, S., et al.: Structure-based design and classifications of small molecules regulating the circadian rhythm period. Sci. Rep. **11**, 18510 (2021). https://api.semanticscholar.org/CorpusID:237546851

13. Guyon, I., Weston, J., Barnhill, S., Vapnik, V.: Gene selection for cancer classification using support vector machines. J. Mach. Learn. Res. **46**, 389–422 (2002)

14. Li, A.D., Xue, B., Zhang, M.: Multi-objective particle swarm optimization for key quality feature selection in complex manufacturing processes. Inf. Sci. **641**, 119062 (2023)

15. Li, J., et al.: Feature selection: a data perspective. ACM Comput. Surv. **50**(6), 1–45 (2017)

16. Mertes, S., Baird, A., Schiller, D., Schuller, B.W., André, E.: An evolutionary-based generative approach for audio data augmentation. In: 2020 IEEE 22nd International Workshop on Multimedia Signal Processing (MMSP), pp. 1–6 (2020)

17. Pedregosa, F., et al.: scikit-learn: machine learning in Python. J. Mach. Learn. Res. **12**, 2825–2830 (2011)

18. Pereira, S., Correia, J., Machado, P.: Evolving data augmentation strategies. In: Jiménez Laredo, J.L., Hidalgo, J.I., Babaagba, K.O. (eds.) EvoApplications 2022. LNCS, vol. 13224, pp. 337–351. Springer, Cham (2022). https://doi.org/10.1007/978-3-031-02462-7_22

19. Sánchez-Maroño, N., Alonso-Betanzos, A., Tombilla-Sanromán, M.: Filter methods for feature selection – a comparative study. In: Yin, H., Tino, P., Corchado, E., Byrne, W., Yao, X. (eds.) IDEAL 2007. LNCS, vol. 4881, pp. 178–187. Springer, Heidelberg (2007). https://doi.org/10.1007/978-3-540-77226-2_19

20. Shanmugamani, R., Moore, S.: Deep Learning for Computer Vision: Expert Techniques to Train Advanced Neural Networks Using TensorFlow and Keras. Packt Publishing (2018)

21. Venkatesh, B., Anuradha, J.: A review of feature selection and its methods. Cybern. Inf. Technol. **19**(1), 3–26 (2019)

22. Xue, B., Zhang, M., Browne, W.N., Yao, X.: A survey on evolutionary computation approaches to feature selection. IEEE Trans. Evol. Comput. **20**(4), 606–626 (2016)

23. Xue, B., Zhang, M., Browne, W.N.: Particle swarm optimisation for feature selection in classification: novel initialisation and updating mechanisms. Appl. Soft Comput. **18**, 261–276 (2014)
24. Zhang, K., Fan, W., Yuan, X.: Ozone level detection. UCI Machine Learning Repository (2008). https://doi.org/10.24432/C5NG6W
25. Zhang, X., Yu, L., Yin, H., Lai, K.K.: Integrating data augmentation and hybrid feature selection for small sample credit risk assessment with high dimensionality. Comput. Oper. Res. **146**, 105937 (2022)
26. Zheng, A., Casari, A.: Feature Engineering for Machine Learning: Principles and Techniques for Data Scientists, 1st edn. O'Reilly Media Inc. (2018)

Evolving Feature Extraction Models for Melanoma Detection: A Co-operative Co-evolution Approach

Taran Cyriac John[✉], Qurrat Ul Ain, Harith Al-Sahaf, and Mengjie Zhang

School of Engineering and Computer Science, Victoria University of Wellington,
P.O. Box 600, Wellington 6140, New Zealand
{taran.john,qurrat.ul.ain,harith.al-sahaf,mengjie.zhang}@ecs.vuw.ac.nz

Abstract. As global mortality rates rise alongside an increasing incidence of skin cancer, it becomes increasingly clear that the pursuit of an effective strategy to combat this challenge is gaining urgency. In traditional practices, the diagnosis of skin cancer predominantly depends on manual inspection of skin lesions. Despite its prevalent use, this approach is beset with several limitations, such as subjectivity, time constraints, and the invasive nature of biopsy procedures. Addressing these obstacles, the burgeoning field of Artificial Intelligence has been instrumental in advancing Computer Automated Diagnostic Systems (CADS) for skin cancer. A critical aspect of these systems is feature extraction, a process crucial for discerning and utilising key characteristics from raw image data, thereby bolstering the efficacy of CADS. This study introduces a feature extraction model that evolves automatically, leveraging the principles of genetic programming and cooperative coevolution. This method generates a ensemble of models that collaboratively work to extract discerning features from images of skin lesions. The model's effectiveness is evaluated using a publicly accessible dataset, whilst further analysis pertaining to interactions between the decomposition of image colour channels are explored. The findings indicate that the proposed method either matches or significantly surpasses the performance of established benchmarks and recent methodologies in this field, underscoring its potential in enhancing skin cancer diagnostic processes.

Keywords: Skin cancer · Genetic programming · Machine learning

1 Introduction

Skin cancer, a pervasive global health concern, is primarily categorized into non-melanoma skin cancer (NMSC) and cutaneous malignant melanoma (MM). NMSC, including Basal Cell Carcinoma (BCC) and Squamous Cell Carcinoma (SCC), accounts for the majority of cases but is less deadly compared to MM, which constitutes less than 5% of cases yet is responsible for about 65% of skin cancer-related deaths [1].

Currently, skin cancer diagnosis predominantly relies on manual examination of skin lesions, either through self-examination or by healthcare professionals,

© The Author(s), under exclusive license to Springer Nature Switzerland AG 2024
S. Smith et al. (Eds.): EvoApplications 2024, LNCS 14634, pp. 413–429, 2024.
https://doi.org/10.1007/978-3-031-56852-7_26

particularly dermatologists. Dermatologists use specific dermoscopic criteria for diagnosis, which involves pattern analysis, the 7-point checklist, the Menzies method, and comprehensive methodologies like the Colour, Architecture, Symmetry and Homogeneity (CASH) and Asymmetry, Border, Colour, and Dermoscopic structures (ABCD) [2–7].

However, several limitations are associated with these diagnostic approaches:

1. **Inherent Subjectivity**: Diagnostic conclusions can vary among specialists due to the subjective nature of manual evaluations, leading to potential discrepancies in treatment decisions and patient outcomes [8].
2. **Time Constraints**: The high demand for dermatological services often results in limited examination time, possibly leading to missed or delayed diagnoses with serious prognostic implications [9].
3. **Invasiveness of Biopsy Procedures**: When manual examination is inconclusive, biopsies are conducted. This procedure can cause discomfort or pain, adding to the patient's emotional and physical burden [10].
4. **Challenges in Early Detection**: Early stages of skin cancer, especially MM, are difficult to detect with the naked eye, often leading to delayed diagnoses and complicated treatment protocols [11].

In response to these challenges, the integration of Artificial Intelligence (AI) in healthcare, particularly Computer Automated Diagnostic Systems (CADS), is gaining traction. CADS, enhanced by the rapid advancement in AI and computational capabilities, holds the potential to improve the accuracy and efficiency of skin cancer detection, addressing the limitations of manual examinations [12]. Despite its potential, AI in the realm of skin cancer detection faces substantial challenges. The complexity of hand-crafting feature extraction (FE) algorithms introduces development intricacies and limits scalability. The uninterpretable nature of feature extraction in many AI models can deter medical professionals' trust in AI-based decisions.

The application of AI in skin cancer diagnosis, especially through CADS, offers a promising avenue to overcome the limitations of manual diagnostic methods. By enhancing accuracy and efficiency, AI could drastically contribute to better prognostic outcomes in skin cancer patients.

Genetic Programming (GP), is a technique known for evolving tree-structured programs to solve specific problems. Since the early 1990s [13], GP has seen successful applications in various fields, notably in image processing [14] and feature detection/extraction [15]. The concept of co-operative co-evolution (CC), often integrates with GP. CC tackles complex problems by breaking them down into simpler sub-problems or 'species' [16]. Each species evolves independently, with the performance of an individual evaluated based on its interaction with top performers from other species, known as the 'context vector'.

GP plays a significant role in AI-based CADS for detecting skin cancer. Ain et al. [17] utilised GP on dermoscopic images to develop a melanoma detection method, incorporating both domain-independent features, such as LBP descriptors, and domain-specific features from the 7-point dermatology checklist [17]. Their research demonstrated the efficacy of combining these feature types in classification, with GP models achieving high accuracy and interpretability in distinguishing between benign and malignant lesions. This work underscored the potential of pre-defined features like LBP and statistical wavelet decomposed images in achieving top-tier results. In a subsequent study, Ain et al. introduced a two-stage GP system (2SGP-W) for skin lesion analysis, which excelled in feature selection and construction, outperforming existing algorithms [18].

However, these hand-crafted, pre-defined features may overlook novel discriminatory attributes. Exploring ML, specifically GP, to develop an automatic FE algorithm is therefore promising. GP's ability to evolve feature extractors automatically offers numerous benefits over traditional methods like LBP or deep learning pipelines, including simultaneous feature extraction and selection. This dual functionality enhances model efficiency, reducing errors common in conventional methods with separately optimised components.

Support for the efficacy of GP in automatic image extraction can be found in the scientific literature. Specifically, in their work [19], Al-Sahaf et al. utilised GP to automatically evolve an LBP-like image, termed GP-criptor, using raw pixel values [19]. It is important to note the complexity pertaining to FE for skin-lesion images, marked by vast search spaces, instance diversity in both shape and appearance, and presence of artifacts and noise, necessitates innovative solutions. GP, with capacity to automatically navigate this intricate search space to evolve a solution, has the potential to effectively address these challenges.

1.1 Goals

The overall objective of this study is to leverage the capabilities of GP and CC, to automatically evolve a FE model capable of extracting discriminatory features pertaining to the malignance of a skin lesion instance. To fulfill this goal, the following objectives are pertinent:

- Evolving and aggregating multiple models for FE by performing colour decomposition on each skin lesion image in adherence to CC and GP principles;
- Crafting an appropriate fitness function incorporating both distance-based and wrapper-based measures to ascertain discriminatory features;
- Comparing the performance of the models synthesised by the proposed method to that of canonical as well as recently proposed FE methods; and
- Offering an insight into the characteristics of the proposed method through providing further analysis of the evolutionary process and results produced by experimentation.

2   Literature Survey

2.1   Feature Extraction

2.1.1   Local Binary Pattern

Local Binary Pattern (LBP), first proposed by Ojala *et al.* [20], can be described as one of the widely utilised image descriptor feature extraction methods in the field. Through utilisation of a sliding window with a fixed radius, LBP method involves scanning an image pixel-by-pixel, starting from the top-left corner to the bottom-right corner. Each central pixel in the sliding window is assigned a value based on a binary comparison with its adjacent pixels. Using these values, a histogram that represents the texture of the overall image is constructed.

LBP can be classified into two patterns, uniform and non-uniform, which can be categorised based on the nature of the bitwise transitions. The former (uniform) consists of at most two bitwise transitions (either from 0-to-1 or 1-to-0), whilst the latter (non-uniform) may contain more than two such transitions. For example, the pattern (00011110) is considered uniform, whereas (01011100) can be classified as non-uniform. Furthermore, the classification of the pattern has an impact on the resulting vector, this vector may be reduced if it stems from an LBP pattern. To elaborate, the size of the vector can be represented as 2^b where b is the number of adjacent pixels, but can be reduced to $b(b - 1) + 3$ if the pattern is uniform.

2.1.2   GP-Criptor

In [19], GP is utilised to automatically evolve a feature extractor using the raw pixel values, known as GP-criptor. Although GP-criptor operates in a somewhat similar manner to LBP, it automatically evolves a set of pixel formulas, thus replacing the expert designed ones. For this reason, the evolved individual consists of three stages: pixel value extraction, arithmetic operators, and encoding. Pixel extraction involves the position of the sliding window, wherein the values of pixels adjacent to the center are extracted, and then used as the terminal set. From there, arithmetic operators are applied to these extracted pixels, adhering to the evolved individual's function set and GP tree structure. The final step involves encoding the results into a binary string, which is then converted to a decimal number. This number is subsequently tallied in a histogram-like data structure. Essentially, this image descriptor allows the conversion of an image to a histogram like feature vector, which can be fed into a machine learning classifier such as a support vector machine or k-nearest neighbour for classification purposes.

2.2   Related Work

The swift advancement of AI in CADS has led to the emergence of innovative technologies once considered purely fictional. A groundbreaking demonstration of this shift occurred in 2017 by Esteva et al., wherein they introduced a computational model that outperformed board-certified dermatologists in diagnosing

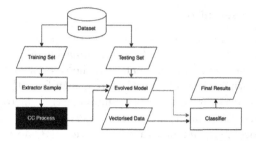

Fig. 1. The overall algorithm of the proposed method.

skin cancer in terms of both accuracy and speed [12]. This achievement has sparked a surge in research on the application of deep learning in skin cancer detection, focusing on the development and refinement of FE algorithms.

Deep learning (DL), deeply rooted in neuroscience, is a subset of AI that utilizes complex ML techniques to replicate the brain's intricate neural networks [21]. Codella et al. Were among the pioneers using DL for skin cancer diagnosis, applying the Caffe DL framework to extract features from skin cancer images that are essential for distinguishing between classes [22]. Convolutional Neural Networks (CNNs) are especially prominent in this field, renowned for their ability to identify high-level feature maps with strong discriminatory power due to their convolutional structure [23]. The prevalence of DL in skin cancer research was evident in 2017, with 22 of the 23 studies using a CNN variant [24].

The development of research is not only confined within DL, however, with researchers exploring other avenues to ascertain an effective AI-based CADS system. Ain et al. juxtapose four GP-based classification methods in one of their related works, [25]. After evaluation on two publicly-available datasets, the authors demonstrate the efficacious nature of GP in the task of feature selection and construction. Upon further analysis of the interpretable models, authors also discovered that asymmetry was crucial in the task of skin cancer detection.

3 Co-operative Co-evolution Image Descriptor

3.1 Algorithm Overview

Illustrated by Fig. 1 is a diagrammatic overview of the proposed algorithm: Co-operative Co-evolution Image Descriptor (CC-Criptor). Firstly, the contents of the dataset is divided into a training and test split. During the training process, the *extractor sample* is randomly chosen, in which 20 samples from each class are chosen. These extractor samples are fed to the CC evolutionary process, wherein a feature extraction model is obtained. This newly evolved model is then utilised to *vectorise* both the training and the test data. This *vectorised training data* is then leveraged to train a classifier, which is then tested using the *vectorised test data*, to obtain the final performance of the feature extraction model.

3.2   Model Representation

The complete evolved model consists of three sub-trees, wherein each tree represents the best performing individual of a sub-population. The representation of a sub-population individual is in alignment with the tree-based GP principles put forth by Koza [13], wherein terminal nodes are drawn from the terminal set and non-terminal nodes are sourced from the function set. Numerous studies underscore the importance of colour in the realm of melanoma detection [26,27]. Consequently, it was deemed prudent to partition each data instance (image) into three discrete sub-populations, corresponding respectively to the Red, Green, and Blue colour channels of the data instances, wherein these sub-populations in adherence to the co-operative co-evolution principles. Each sub-population individual represents an evolved mathematical formula that is utilised to obtain the feature vector. It should be noted that the decomposition into colour channels constitutes the fundamental augmentation of CC-criptor compared to GP-criptor. Unlike GP-criptor, which generates a single individual corresponding to the grayscale representation of the image, CC-criptor produces multiple individuals, each corresponding to a different colour channel of the image.

3.2.1   Terminal Set

The terminal set for each GP tree is composed of the raw pixel values extracted by each sliding window position. For this reason, the size of the terminal set is determined by the size of the sliding window (w). Furthermore, it is contingent upon the associate sub-population, as each tree encapsulates a distinct subset of the feature space, i.e., colour channel. For example, if $w = 3$ (i.e., the sliding window is 3×3), the terminal set for the three colour channels would consist of R_a, G_b, and B_c, where a, b, and c each independently range over $\{1, 2, \ldots, 9\}$.

3.2.2   Function Set

The function set utilised in this algorithm adheres to that which was proposed in [19]. Specifically, this consists of four binary arithmetic operators, and one *encode* operator. The binary operators, including addition, subtraction, multiplication, and protected division, take two values as inputs, perform the specified operation, and return a single value. Importantly, the modulation of the division function is designed to address the *division by zero* scenario, returning a value of 1 in such instances. The *encode* node serves as a unique function node consistently located at the root of each tree. Specifically, the objective of this code node is to synthesise a binary number at each position of the sliding window. It should be noted that the number of children, denoted as h, in the code node determines the length of the resulting binary number. This length defines the range of values that can be represented, specifically 2^h.

3.3   Feature Vector Synthesis

The primary objective of CC-criptor is to generate three GP-trees which in turn can be utilised to convert a colour image into a one dimensional vector

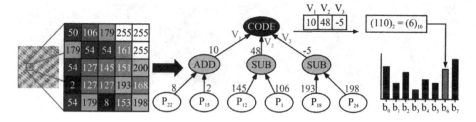

Fig. 2. Diagrammatic explanation of vectorisation process [19].

(or histogram). A sliding window used by the system which traverses the whole image horizontally and vertically, beginning at the top left, in order to extract pixel values at each position. In order to clearly elucidate this process, the method can be broken down into four distinct steps, which is demonstrated by Fig. 2.

1. The pixel values of the current window are extracted and fed to the terminal set of the GP program.
2. The GP program is utilised to evaluate this terminal set, wherein the *encode* node is left with h number of children (i.e. h number of integer values).
3. The integer values of the children of the *encode* node are evaluated against a threshold (*thresh*), in which 1 and 0 are returned for each child, if they are greater than or less than the threshold, respectively. This creates a binary code of length h.
4. The generated binary code is converted to decimal, and the corresponding bin value of the histogram is incremented by one.

It must be noted that this process is repeated for each colour channel, thereby producing a feature vector of length 3×2^h. Consequently, the generated feature vector directly represent a histogram with the frequencies of each binary code in each window position, which can be represented as **x**.

3.4   Individual Fitness Evaluation

A hybrid fitness evaluation is proposed in this work, wherein two objectives are performed simultaneously, in a manner analogous to that proposed in [28]. To elaborate, this fitness function was designed with the objective of having a high accuracy on the training set, whilst concurrently maximising and minimising the distances between instances of different and the same classes, respectively. Specifically, this can be mathematically defined as

$$Fitness(\mathbf{x}) = \left(\frac{1}{1 + e^{-5(D_w - D_b)}} \right) + (1 - Accuracy) \tag{1}$$

$$Accuracy = \frac{Correct}{Total} \tag{2}$$

where D_w is the mean distance of instances that are of the same class, and D_b is the mean distance of instances that are of a different class. It must be noted

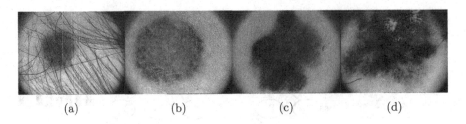

(a)   (b)   (c)   (d)

Fig. 3. Skin lesion image samples from PH2 dataset.

that the specified distance measures are in complete adherence to the work of Al-Sahaf et al. in [29].

In addition to the distance based component of the proposed fitness function, a *wrapper* based function is implemented in order to ascertain the discriminatory capability of the evolved feature extractor. In particular, the k-Nearest Neighbour (k-NN) algorithm is leveraged, wherein the sample set of extracted features are utilised to evaluate efficacy of the feature extractor as it pertains to selecting discriminatory features. Equation 2 delineates the accuracy evaluation of the *wrapper* based function. To reduce bias arising from random train/test splits, stratified k-fold cross-validation is employed, addressing the imbalanced distribution of class samples. It must be noted that the inverse of the mean accuracy provided by the evaluation is utilised, as fitness minimisation is the objective of the fitness evaluation.

As elucidated [16], the fitness assessment of an individual is determined not only on its intrinsic performance but also in conjunction with the performance of leading individuals, termed "representatives", from distinct species. In adherence with this paradigm, every individual tree subject to a fitness evaluation is concatenated with representatives from other species. It is of importance to highlight that the fitness evaluation of an individual tree is contingent upon the collective assessment derived from the entire context vector (CV), as opposed to its isolated performance. Therefore, within a given CV, a tree exhibiting sub-optimal individual performance may surpass the performance of a tree that exhibits superior individual performance.

4   Experimental Design

Outlined in this section is the execution of the proposed approach, aimed at facilitating the replication of the study's outcomes.

4.1   Dataset

In this work, the Pedro Hispano Hospital (PH2) dataset was utilised. Established at the Pedro Hispano Hospital in Mastosinhos, Portugal [30], the PH2 dataset was made publicly available in 2013. This compilation of images encompasses a total 200 dermoscopic images captured with a Tubinger Mole Analyser

Table 1. PH2 Dataset Characteristics

Classes	Num of Instances	Image Size
Melanoma	40	754×576–768×675
Common Nevi	80	763×552–769×577
Atypical Nevi	80	764×575–768×576

system at a magnification factor of 20, each with a pixel resolution of 765 × 560 pixels. The PH2 dataset consists of 8-bit RGB skin lesion images, binary segmentation masks, and clinical diagnoses. A sample of the images utilised in this study is exhibited in Fig. 3, wherein Figs. 3(a) and 3(b) demonstrate benign instances, Figs. 3(c) and 3(d) illustrate malignant instances, respectively. More specifically, a dermatologist evaluated each instance in the dataset, from which clinical evaluations, manual segmentation of skin lesions, clinical diagnoses, and features from the *7-point checklist* were derived and provided. Table 1 elucidates the distribution of classes present in the PH2 dataset.

In the context of this study, both Atypical Nevi and Common Nevi instances are considered as the negative class, while Melanoma instances are designated as the positive class. This classification is consistent with the literature [31], thereby transforming a multi-class problem into a binary classification task.

4.2 Data Pre-processing

In the dataset, the proportion of each image occupied by the Region of Interest (ROI) varied across instances. For instance, while the ROI might encompass half of one image, it could cover only a quarter of another. To address this variability and eliminate potential extraneous details, each image was cropped using a *bounding-box* delineated around the ROI. Given that the focus of this task is skin lesion classification, only the information within the ROI holds relevance for the algorithm. Consequently, the binary mask was superimposed on the original image, ensuring that the proposed algorithm would exclusively process the ROI.

4.3 Methods for Benchmark Comparison

In the present study, a descriptor analogous to the LBP method is proposed, through the integration of CC and GP. LBP for feature extraction has been leveraged in multi-stage GP-based classification models, exhibiting state-of-the-art results [25]. As such, it would be judicious to evaluate the discriminatory ability of features solely-extracted by LBP, and compare it to that which has been extracted by CC-criptor. Additionally, the CC-criptor represents an advancement over the original GP-criptor algorithm. Given this context, it is appropriate to elucidate the performance disparities between these two algorithms. The primary objective is to ascertain whether the incorporation of CC has introduced a notable enhancement in performance. Consequently, the GP-criptor is integrated into this study for comparative analysis.

Furthermore, it was considered prudent to contrast the outcomes of the images derived from the CC-criptor with those from a recently published study in the domain. In [32], Ain et al. introduced a novel technique capable of automatically extracting global features from skin cancer images. This method, termed Feature Learning approach using GP (FLGP$_{SCD}$), leverages GP and combines six commonly used image descriptors to extract these high-level features. The researchers demonstrated the superiority of performance of their method over both the baseline approach and the six feature descriptors on real-world datasets.

Given the pervasive use of DL in computer vision tasks, it was deemed suitable to benchmark the performance of our proposed method against a standard CNN. Consequently, we adopted the LeNet-5 framework, as outlined in [33]. Where relevant, the same methodology was applied, including the utilisation of a 5-fold cross-validation and identical pre-processing techniques.

4.4   Parameters

In this research, the k tournament selection method with $k = 7$ was utilised for parent selection, combined with a one-point crossover mechanism to generate unique offspring. Uniform mutation was the chosen mutation technique, involving random sub-tree generation and replacement in a GP tree, with the evolutionary process concluding after 50 generations. Aligning with [19], the experimental parameters were meticulously chosen to facilitate a fair comparison. The species population size (θ) was set at 50, and the Ramped Half-and-Half method, incorporating grow and full initialisation techniques, was adopted for population initiation. The maximum depth size was limited to 10, with initialisation sizes ranging from 2 to 5. Genetic operators were configured at 80% for crossover and 19% for mutation, with an additional 1% dedicated to elitism. This approach ensures the retention of the fittest individuals in the population, preserving their fitness through subsequent generations.

4.5   Experimental Setup

The imbalanced structure of the dataset, combined with a limited number of positive instances, prompted the decision to employ stratified 5-fold cross-validation. This method aims to minimise the influence of random data splits on fluctuations in algorithm performance. In order to furthermore mitigate the possibility of stochasticity playing a determining role in deviations of the performance of the algorithms, 30 distinct experiments were conducted using different random seeds for stochastic methods, specifically GP-criptor, CC-criptor, and LeNet-5. It should be noted that GP-criptor was evolved with one grayscale colour channel per instance, whilst CC-criptor utilised the red, green, and blue colour channels during evolution. While the primary focus of this study is FE, the efficacy of the extracted features of the GP-criptor and CC-criptor were ascertained by their performance in a foundational classifier. To this end, k-NN, Gaussian Naïve Bayes (GNB), Support Vector Machine (SVM), Linear Discriminant Analysis (LDA) and Multi-layer Perceptron (MLP) classifiers were employed. It must be noted that LeNet-5 instead utilised fully connected neural network layers for classification purposes, as in adherence to [33].

Table 2. Classification Performance of LBP, GP-Criptor, and CC-criptor (%). Values represent the mean and standard deviation (mean ± std. dev.)

	Classifier	F1-Score	Balanced Accuracy
LBP	k-NN	66.39	68.63
	GNB	72.96	67.37
	SVM	47.45	50.00
	LDA	59.82	68.22
	MLP	47.45	50.00
GP-criptor	k-NN	64.03 ± 4.15	59.58 ± 3.76
	GNB	71.58 ± 3.01	62.55 ± 2.24
	SVM	65.78 ± 3.81	61.80 ± 3.68
	LDA	70.05 ± 3.57	66.35 ± 4.00
	MLP	70.33 ± 3.57	63.02 ± 4.02
CC-criptor	k-NN	78.28 ± 2.42	65.43 ± 2.89=
	GNB	76.40 ± 2.84	65.25 ± 2.91=
	SVM	75.10 ± 3.57	65.25 ± 2.81=
	LDA	79.58 ± 3.13	**74.83 ± 3.72**↑
	MLP	80.28 ± 2.27	68.87 ± 2.54=

4.6   Performance Evaluation Metrics

Owing to the imbalanced nature of both datasets, employing balanced accuracy as the primary statistical metric was considered appropriate. Balanced accuracy is defined as:

$$Balanced\ Accuracy = \frac{1}{2}\left(\frac{TP}{TP+FN} + \frac{TN}{TN+FP}\right), \qquad (3)$$

TP, TN, FN, and FP denote true positive, true negative, false negative, and false positive values, respectively. The use of balanced accuracy in this study, rather than conventional raw classification accuracy, helps to offset potential performance disparities in the proposed method that might arise from the predominance of specific classes, as discussed earlier in this chapter. In addition to balanced accuracy, the *F1-score* was utilised to provide a more comprehensive analysis of the model's classification capability.

5   Results and Discussions

To determine the statistical significance between each variant in the experiments, a two-sided, paired t-test at a 95% significance level was used for comparisons between models. The symbols =, ↑, and ↓ indicate whether the average metrics of the CC-criptor models are not significantly different, significantly better, or significantly worse than those of the other models, respectively.

5.1   Binary Classification Performance of GP-Criptor and CC-Criptor

The binary classification results exhibited by CC-criptor and GP-criptor are shown in Table 2. Specifically, these are decomposed into algorithmic blocks vertically, wherein each block demonstrates the efficacy of the features on the conventional classifiers delineated in the previous section. Upon inspection of the results, it becomes clear that LDA provides the best performance in comparison with the other typical ML algorithms, demonstrating a mean accuracy of 74.83 ± 3.72 and 66.35 ± 4.00, for the CC-criptor and GP-criptor algorithms, respectively. Moreover, the results indicate that when using the CC-criptor extracted features with an LDA classifier, the performance is statistically significantly superior than all other permutations.

Table 3. Comparison against Benchmark Methods (%)

	Classifier with Best Performance	Mean Balanced Accuracy	Best Model Performance
LBP	GNB	67.37	73.67
FLGP$_{SCD}$	SVM	66.58 ± 5.40	73.14
LeNet-5	CNN	**74.30 ± 4.12=**	82.03
GP-criptor	LDA	66.35 ± 4.00	77.19
CC-criptor	LDA	**74.83 ± 3.72=**	80.63

5.2   Performance Comparison with Benchmark Methods

Delineated by Table 3, it becomes apparent that although exhibiting similar performance for k-NN, GNB, and SVM, CC-criptor significantly outperforms LBP in effective feature extraction when coupled with the LDA classifier. Additionally, CC-criptor has demonstrated statically significant superiority over the FLGP$_{SCD}$ method when coupled with the LDA classifier. When comparing the performance of CC-criptor with LeNet-5, both exhibit similar results. However, a unique feature of CC-criptor is its ability to automatically evolve three interpretable GP trees. These trees offer insights into the algorithm's decision-making, particularly in relation to the interactions among colour channels. Additionally, CC-criptor extracts model-agnostic features suitable for various classifiers.

6   Further Analysis

6.1   Algorithm Convergence

To evaluate the evolutionary efficacy of both GP-criptor and CC-criptor, it is appropriate to examine the convergence graphs that display the average fitness of the context vector for the GP and CC methodologies, respectively, across each generation. This evaluation includes the average from the 5-folds over all 50 generations, spanning the 30 independent runs, as depicted in Figs. 4a and b. The y-axis represents the average CV fitness for the CC-criptor and GP-criptor algorithms, respectively. It should be noted that the fitness assessment varies between the two methods, leading to a substantially different range of output values for each algorithm. To address this, the y-axis has been scaled to facilitate detailed examination of the fitness convergence. In contrast, the x-axis displays the count of generations for both algorithms. Elucidated by the GP-criptor plot, evolution begins at a starting fitness of 0.90 ± 0.089, then drastically decreases to 0.75 ± 0.072 by generation 25, and then smoothly decreases to 0.72 ± 0.063 by the last generation. In contrast, the CC-criptor plot demonstrates commences evolution with a starting fitness of 0.993 ± 0.007, however follows a more linear-like convergence to 0.986 ± 0.009 at the halfway point of evolution, and finally terminates evolution with a fitness value of 0.982 ± 0.010.

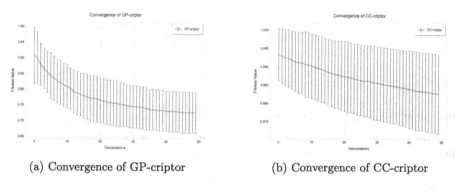

(a) Convergence of GP-criptor   (b) Convergence of CC-criptor

Fig. 4. Comparison of convergence between CC-criptor and GP-criptor.

From this dichotomy, it is immediately apparent that the convergence of GP-criptor is substantially more drastic than that of CC-criptor. However, it must be noted that GP-criptor only relies on the learning of one tree, as opposed to the CV of three trees as in CC-criptor. Furthermore, the linear convergence of CC-criptor may indicate that the algorithm has not reached full convergence, as the last 10 generations do not demonstrate a stagnation in fitness as it does with GP-criptor. Additionally, the smaller relative scale of change in fitness over the evolutionary process of CC-criptor must be taken into account. Since the population size of each method is kept consistent for performance comparison purposes, where GP-criptor has 150 individuals in the whole population whereas CC-criptor has 50 individuals in each subpopulation, it is possible that the lack of genetic diversity in each CC-criptor sub-population is not conducive to more productive convergence.

Table 4. Performance of models on different colour decompositions (%).

	Model	F1-Score	Balanced Accuracy
Red Colour Channel	KNN	80.96 ± 3.15	58.75 ± 5.94
	GNB	79.54 ± 8.07	60.02 ± 8.29
	SVM	80.35 ± 0.86	51.01 ± 2.25
	LDA	80.19 ± 7.38	63.53 ± 8.54
	MLP	80.12 ± 0.52	50.29 ± 1.31
Green Colour Channel	KNN	82.19 ± 2.83	62.26 ± 8.21
	GNB	81.00 ± 4.74	60.79 ± 5.68
	SVM	80.04 ± 0.31	50.10 ± 0.77
	LDA	77.38 ± 7.86	59.90 ± 9.69
	MLP	80.00 ± 0.00	50.00 ± 0.00
Blue Colour Channel	KNN	84.08 ± 3.71	62.36 ± 8.10
	GNB	84.50 ± 4.23	64.93 ± 6.23
	SVM	80.65 ± 1.34	51.63 ± 3.34
	LDA	74.38 ± 10.99	59.11 ± 9.61
	MLP	80.12 ± 0.52	50.29 ± 1.31

6.2 Colour Decomposition Analysis

For the purpose of ascertaining the role of each colour channel on the overall performance of the CC-criptor algorithm, each colour channel was vectorised through utilisation of the corresponding best performing individual, with the results of this being illustrated in Table 4. Upon observation of these results, it becomes apparent that the results of each individual component of the model are considerably worse than that of the model as a whole.

The juxtaposition of the performance of the whole system and each of the individual performance is likely a reflection of the co-operative nature of the evolutionary process. To elaborate, when evaluating each of sub-population individuals, their performance is gauged based upon its collaborative performance with the other members of the CV. Thus, this arises the possibility that although the individual possesses good collaborative performance, it lacks individual discriminatory ability. With none of the trees demonstrating statistically significantly superior performance over any of the others, it cannot be claimed that any of the colour channels contribute substantially more than the others in this proposed method.

7   Conclusions and Future Work

Inspired by the effectiveness of GP in FE, this paper proposes an automatically-evolving feature extraction model for skin cancer. The comparative analysis reveals that the CC-criptor model, especially when combined with LDA, demonstrates superior performance over GP-criptor in binary classification tasks. A salient aspect of CC-criptor is its unique ability to automatically evolve three distinct GP trees, thus shedding light on the interactions of colour channels and asserting its adaptability across a spectrum of classifiers. The integrated model of CC-criptor, leveraging the strengths of individual colour channels, exhibits enhanced effectiveness, suggesting synergistic contributions rather than dependence on a single channel for its efficacy in feature extraction.

In future research, investigation into a modified elitism function is to be conducted. This approach will focus on selecting permutations of individuals that exhibit the highest collective fitness. Furthermore, the exploration a two-stage GP process may yield promising results. Features extracted by CC-criptor will form the terminal set for evolving a GP tree, offering two benefits: enabling the creation of complex features and offering insights into colour channel interactions. Understanding these interactions may have significant clinical implications, by enhancing the efficacy of skin cancer diagnosis and treatment strategies.

References

1. Sung,H., et al.: Global cancer statistics 2020: GLOBOCAN estimates of incidence and mortality worldwide for 36 cancers in 185 countries. CA Cancer Clin. **71**, 209–49 (2021). This report provides the latest global cancer statistics of incidence and mortality worldwide, 2022
2. Pehamberger, H., Steiner, A., Wolff, K.: In vivo epiluminescence microscopy of pigmented skin lesions. I. Pattern analysis of pigmented skin lesions. J. Am. Acad. Dermatol. **17**(4), 571–583 (1987)
3. Argenziano, G., Fabbrocini, G., Carli, P., De Giorgi, V., Sammarco, E., Delfino, M.: Epiluminescence microscopy for the diagnosis of doubtful melanocytic skin lesions: comparison of the ABCD rule of dermatoscopy and a new 7-point checklist based on pattern analysis. Arch. Dermatol. **134**(12), 1563–1570 (1998)
4. Menzies, S.W., Crotty, K., Ingvar, C., McCarthy, W.: Dermoscopy: An Atlas, 3rd edn. McGraw-Hill Education, Australia (2009)
5. Henning, J.S., et al.: The CASH (color, architecture, symmetry, and homogeneity) algorithm for dermoscopy. J. Am. Acad. Dermatol. **56**(1), 45–52 (2007)
6. Stolz, W.: ABCD rule of dermatoscopy: a new practical method for early recognition of malignant melanoma. Eur. J. Dermatol. **4**, 521–527 (1994)
7. Loescher, L.J., Janda, M., Soyer, H.P., Shea, K., Curiel-Lewandrowski, C.: Advances in skin cancer early detection and diagnosis. In: Proceedings of Seminars in Oncology Nursing, vol. 29, pp. 170–181. Elsevier (2013)
8. Carrera, C., et al.: Validity and reliability of dermoscopic criteria used to differentiate nevi from melanoma: a web-based international dermoscopy society study. JAMA Dermatol. **152**(7), 798–806 (2016)

9. Resneck, J., Jr., Pletcher, M.J., Lozano, N.: Medicare, medicaid, and access to dermatologists: the effect of patient insurance on appointment access and wait times. J. Am. Acad. Dermatol. **50**(1), 85–92 (2004)
10. Bichakjian, C.K., et al.: Guidelines of care for the management of primary cutaneous melanoma. J. Am. Acad. Dermatol. **65**(5), 1032–1047 (2011)
11. Schadendorf, D., et al.: Melanoma. The Lancet **392**(10151), 971–984 (2018)
12. Esteva, A., et al.: Dermatologist-level classification of skin cancer with deep neural networks. Nature **542**(7639), 115–118 (2017)
13. Koza, J.R.: Genetic Programming: On the Programming of Computers by Means of Natural Selection. MIT Press (1992)
14. Liang, J., Wen, J., Wang, Z., Wang, J.: Evolving semantic object segmentation methods automatically by genetic programming from images and image processing operators. Soft. Comput. **24**, 12887–12900 (2020)
15. Cano, A., Ventura, S., Cios, K.J.: Multi-objective genetic programming for feature extraction and data visualization. Soft. Comput. **21**, 2069–2089 (2017)
16. Potter, M.A., De Jong, K.A.: A cooperative coevolutionary approach to function optimization. In: Davidor, Y., Schwefel, H.-P., Männer, R. (eds.) PPSN 1994. LNCS, vol. 866, pp. 249–257. Springer, Heidelberg (1994). https://doi.org/10.1007/3-540-58484-6_269
17. Ain, Q.U., Xue, B., Al-Sahaf, H., Zhang, M.: Genetic programming for skin cancer detection in dermoscopic images. In: Proceedings of the 2017 IEEE Congress on Evolutionary Computation, pp. 2420–2427. IEEE (2017)
18. Ain, Q.U., Al-Sahaf, H., Xue, B., Zhang, M.: Automatically diagnosing skin cancers from multimodality images using two-stage genetic programming. IEEE Trans. Cybern. **53**(5), 2727–2740 (2022)
19. Al-Sahaf, H., Zhang, M., Johnston, M., Verma, B.: Image descriptor: a genetic programming approach to multiclass texture classification. In: Proceedings of the 2015 IEEE Congress on Evolutionary Computation, pp. 2460–2467. IEEE (2015)
20. Ojala, T., Pietikainen, M., Maenpaa, T.: Multiresolution gray-scale and rotation invariant texture classification with local binary patterns. IEEE Trans. Pattern Anal. Mach. Intell. **24**(7), 971–987 (2002)
21. Goodfellow, I., Bengio, Y., Courville, A.: Deep Learning. MIT Press (2016)
22. Codella, N., Cai, J., Abedini, M., Garnavi, R., Halpern, A., Smith, J.R.: Deep learning, sparse coding, and SVM for melanoma recognition in dermoscopy images. In: Zhou, L., Wang, L., Wang, Q., Shi, Y. (eds.) MLMI 2015. LNCS, vol. 9352, pp. 118–126. Springer, Cham (2015). https://doi.org/10.1007/978-3-319-24888-2_15
23. Barata, C., Celebi, M.E., Marques, J.S.: A survey of feature extraction in dermoscopy image analysis of skin cancer. IEEE J. Biomed. Health Inf. **23**(3), 1096–1109 (2018)
24. Codella, N.C., et al.: Skin lesion analysis toward melanoma detection: a challenge at the 2017 international symposium on biomedical imaging (ISBI), hosted by the international skin imaging collaboration (ISIC). In: Proceedings of the 15th International Symposium on Biomedical Imaging, pp. 168–172. IEEE (2018)
25. Ain, Q.U., Al-Sahaf, H., Xue, B., Zhang, M.: Automatically diagnosing skin cancers from multimodality images using two-stage genetic programming. IEEE Trans. Cybern. **53**, 2727–2740 (2022)
26. Barata, C., Ruela, M., Francisco, M., Mendonça, T., Marques, J.S.: Two systems for the detection of melanomas in dermoscopy images using texture and color features. IEEE Syst. J. **8**(3), 965–979 (2013)
27. Barata, C., Celebi, M.E., Marques, J.S.: Improving dermoscopy image classification using color constancy. IEEE J. Biomed. Health Inform. **19**(3), 1146–1152 (2014)

28. Al-Sahaf, H., Zhang, M., Johnston, M.: Genetic programming for multiclass texture classification using a small number of instances. In: Dick, G., et al. (eds.) SEAL 2014. LNCS, vol. 8886, pp. 335–346. Springer, Cham (2014). https://doi.org/10.1007/978-3-319-13563-2_29

29. Al-Sahaf, H., Al-Sahaf, A., Xue, B., Johnston, M., Zhang, M.: Automatically evolving rotation-invariant texture image descriptors by genetic programming. IEEE Trans. Evol. Comput. **21**(1), 83–101 (2017)

30. Mendonça, T., Ferreira, P.M., Marques, J.S., Marcal, A.R., Rozeira, J.: PH^2 - a dermoscopic image database for research and benchmarking. In: Proceedings of the 35th Annual International Conference of the IEEE Engineering in Medicine and Biology Society, pp. 5437–5440. IEEE (2013)

31. Ain, Q.U., Al-Sahaf, H., Xue, B., Zhang, M.: Generating knowledge-guided discriminative features using genetic programming for melanoma detection. IEEE Trans. Emerg. Top. Computat. Intell. **5**(4), 554–569 (2020)

32. Ain, Q.U., Al-Sahaf, H., Xue, B., Zhang, M.: A new genetic programming representation for feature learning in skin cancer detection. In: Proceedings of the Companion Conference on Genetic and Evolutionary Computation, pp. 707–710 (2023)

33. LeCun, Y., Bottou, L., Bengio, Y., Haffner, P.: Gradient-based learning applied to document recognition. Proc. IEEE **86**(11), 2278–2324 (1998)

3D Motion Analysis in MRI Using a Multi-objective Evolutionary k-means Clustering

Conor Spann[1]($\boxtimes$)(iD), Evelyne Lutton[2](iD), François Boué[3](iD), and Franck Vidal[4](iD)

[1] School of Computer Science and Engineering, Bangor University, Bangor, UK
c.spann@bangor.ac.uk
[2] INRAE-UMR518, AgroParisTech, Univ Paris-Saclay, 91120 Palaiseau, France
[3] LLB-CEA, CNRS-UMR12, Univ Paris-Saclay, 91190 Gif-sur-Yvette, France
[4] Scientific Computing Department, Science Technology Facilities Council, UK Research and Innovation, Daresbury, UK
franck.vidal@stfc.ac.uk

Abstract. Many studies focused on gastric motility require the use of synthetic tracers to map the motion of content. Our study instead takes advantage of an unusual MRI acquisition protocol, combined with multi-objective optimised clustering to map the motion of food (peas, a natural 'tracer') in a human stomach. We chose NSGA-II to optimise the starting positions for a modified k-means to create optimum clusters. We compared our optimisation approach with a purely random approach that took an equal amount of processing time. Since we have no ground truth available, we have created alternative measures to evaluate our solutions: if the resulting pea velocities are within an expected range, and if each pea's motion is correlated with neighbouring peas. We found that the optimised version has a significant improvement over the purely random search. Furthermore, we found many interesting food motion behaviours, such as correlated pea motion and more complex motion dynamics such as collision. Overall we found that the combined optimisation and clustering approach produced interesting findings relating to food dynamics in a human stomach.

Keywords: NSGA-II · k-means · MRI

1 Introduction

Magnetic resonance imaging (MRI) of the gastrointestinal tract (GIT) is a non invasive technique providing good spatial resolution and tissue contrast which has been used since the late 90s to diagnose various diseases of the GIT and more recently to evaluate functional disorders [17]. Research on digestion (human and animal) are also based on MRI [5], but the imaging of stomach, as a soft tissue that moves and expands, is particularly challenging. MRI is currently used to measure gastric volume, emptying, and contractile activity in various conditions

© The Author(s), under exclusive license to Springer Nature Switzerland AG 2024
S. Smith et al. (Eds.): EvoApplications 2024, LNCS 14634, pp. 430–445, 2024.
https://doi.org/10.1007/978-3-031-56852-7_27

(empty stomach, liquid or solid meals) [7]. These measurements rely on the design of appropriate acquisition and image reconstruction protocols. But image analysis still relies on human intervention, for instance to identify the limits of the organ by manually placing reference points, slowing down the analysis of large volumes of data. The analysis of gastric motility is another difficult and challenging issue [13]. Various experimental protocols have been developed [10], often based on the ingestion of tracers [26], aiming at providing a rapid and high resolution acquisition [17]. Semi-automated analysis of such datasets have been recently proposed [20], mainly focused on the external shape of the organ.

In this paper we propose an automatic image analysis method based on an experimental protocol developed in [5], focused on the visualisation of the stomach content instead of its limits, which is rather novel. The idea is to identify the trajectories of a set of harmless tracers (frozen peas) in the MRI of a human stomach to gain information on the distribution of food inside the stomach during the digestion process. A major limit for the analysis of dynamic stomach content is the MRI acquisition protocol itself, that, to limit the interaction between slices during the measurement, interleaves the acquisition of the slices (see Sect. 3 for details). As the stomach content constantly moves, objects observed in a 2D slice may also appear in another slice: it puzzles the interpretation of the 3D geometry. We propose a 4D reconstruction approach (3D + time) that takes this acquisition protocol into account. By doing this, we are able to group duplicated peas into clusters captured at different time steps. To do this, we have created a multi-objective optimised clustering method using NSGA-II and k-means to cluster these duplicated peas. Then, subsequent analysis of each cluster enables estimating a local trajectory of it, turning what might have been considered as a defect into valuable information about the local movements. In addition, simultaneous imaging of a set of tracers makes it possible to visualise where local movements are correlated or not, providing further valuable information for understanding stomach mechanisms.

The rest of this paper is structured as follows: Sect. 2 reports related research in stomach mechanics and clustering; Sect. 3 describes the MRI datasets used; Sect. 4 expresses our problem in more detail; Sect. 5 describes our optimised clustering approach, Sect. 6 describes our methods of evaluation and finally Sect. 7 reviews whether our solution has achieved its aims and makes suggestions for future work.

2   Related Work

The human stomach can be split in four regions: the fundus, stomach body, antrum and pylorus [8]. The fundus and stomach body serves as a highly-flexible content storage area. The antrum serves as a soft mixer, the pylorus initial aim is a tap and as a second effect, it may generate some higher shear. After a meal is ingested, the stomach wall contracts which causes peristaltic waves to move food through the stomach and aid digestion [11,14]. These peristaltic waves start from the stomach wall and proceed in the direction of the antrum, where the food contents are mixed and any large components are dispersed in smaller parts. The contents are pushed into the pylorus, which itself contracts and pushes the contents back into the stomach [11]. Most of the large particles are reduced into smaller pieces (1–3 mm size), which make it easier to pass through the pylorus [8,11]. The frequency of peristaltic waves is approximately 3 cycles per minute [14,19] and the average range of peristaltic speed is 1.5–5.0 mm/s [14]. Furthermore, in the gastric emptying phase, these contractile waves significantly increase in amplitude and velocity [14]. The speed of stomach-emptying will depend on the food consumed, since solid foods must be ground to a sufficient size first before proceeding [14]. For example, over 50% of a calorific liquid meal will be emptied in 1 h [8,22], and over 2 h for a solid meal [8]. Zero-calorie liquids such as water are emptied almost immediately [14]. There have been several studies focused on quantifying motion in medical imaging. [21] used cine-MRI to compare stomach motion in fasting and post-prandial states. They quantified gastric motility based on the minimum and maximum antrum diameters. An alternative method of quantifying motion in medical imaging is by using the Hausdorff Distance, which has been applied in [28] to show differences in the inhalation states of the human diaphragm.

Our work relies on k-means clustering algorithm, which involves partitioning a multi dimensional dataset into 'k' clusters [18]. The algorithm involves defining a set number of centroids, relating each point to its closest centroid, and then moving centroids to the mean position of its cluster. The initial placement of the centroids can have an impact on the final result of the algorithm [9,23]. Some methods of initialising centroids include the Forgy, MacQueen, Kaufman and random methods [23]. Another approach is the use of wrapper methods, which repeatedly execute k-means with different initial centroids and then picking the best result [9,15]. Alternatively, the initial starting points could be initialised using an evolutionary algorithm to produce optimal clusters [12]. A modified k-means has been used to detect object motion [27]. In the study presented here, we will also used a modified k-means to analyse the motion in the MRI datasets presented next.

3   Datasets

This paper builds upon a previous experiment using MRI to examine digestion. Volunteers swallowed whole frozen garden peas as well as carbohydrates (bread or pasta) with water or lemon juice [5]. They then had MRI scans of their stomachs. Frozen peas have been chosen as they have a simple shape (sphere) and they keep their shape through the initial steps of the digestion. The original experiment focused on the carbohydrate digestion and gastric emptying, whereas here we are interested in the information that can be obtained by following the peas. Each dataset referred to in this study refers to a single patient's stomach scanned in MRI. A numerical identifier is used to allow for anonymity (see Table 1). For consistency we use the same identifiers as in previous articles [1,6,25]. Each scan consists of several 2D cross sectional 'slices'. The key principle to understand is

Table 1. Dataset description

Dataset #	Type of peas	Pixel spacing (in mm)	Pea diameter (in mm)
1	Garden peas	[0.5, 0.5, 2.0]	∼8
2	None	[0.98, 0.98, 2.0]	N/A
3 to 7	Petits pois	[0.9, 0.9, 2.0]	∼5
8 to 10	Petits pois	[0.83, 0.83, 3.0]	∼5

that slices were not scanned in a sequential order. To reduce interference between contiguous slices in the 3D space, atoms of hydrogen must be given time to 'relax'. For this reason, the slices were scanned in groups with spatial gaps in between, and there is a time delay between the scanning of each slice group (see Fig. 1). Due to the time delay, we consider each dataset to be 4D (3D + time). Each cell of Fig. 1 corresponds to the Z index of the corresponding 2D slice in the 3D space. Light blue cells show the cells being acquired at a given timestep; dark blue cells are already acquired at a given timestep. Each row in the figure shows a particular timestep (T = 1, T = 2, and T = 3) from the same scanned volume. All the slices of "Group slice 1", namely 1, 10, 19 and 28, correspond to the timestep T = 1, etc. and were scanned almost simultaneously. Ten datasets were produced in this way.

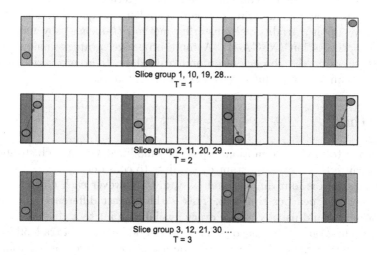

Fig. 1. Demonstrating the unusual scanning order for each of our datasets. Slices are scanned in groups with a time delay in between, meaning that peas can be duplicated in multiple slices.

There are three additional concerns to be raised. Firstly, Dataset 1, a different size of peas were used that were much larger in diameter than the peas used in other datasets (see Table 1). Secondly, Dataset 2 contains no peas, and was used as a control in the original experiments. Thirdly, some of the Datasets when analysed visually had clearer defined peas, in particular Datasets 3–5. In this study, we are focusing on Dataset 3 for the comparison between the optimised and un-optimised k-means. We also use Dataset 4 and Dataset 5 for visualising and analysing the dataset differences between pea correlations.

4   Problem Statement

Due to the nature of the MRI acquisition method, we believe some of the peas have moved during scanning, and have been scanned more than once. From the experimental protocol, we know that there are at most 20 true peas per dataset, and during the acquisition of Dataset 3 for example there are 9 timesteps. Therefore, for this dataset there is a theoretical maximum of 180 possible "duplicated" peas. The focus of this study is whether or not we can identify duplicated peas that are most likely be the same pea but observed at 9 different timesteps, and then recreate each pea's motion through the stomach. We saw this as a clustering problem. The goal is to identify 20 possible clusters, each containing up to 9 peas, with a max of 1 pea per timestep. Each cluster represents 1 true pea, but at different timesteps. We are taking what could have been an inconvenient detail, and exploiting it. By assigning detected peas into clusters, it becomes possible to model the motion of each clustered pea. This provides some important information about stomach content movement patterns.

Due to this being a clustering problem, a challenging question is whether we can improve upon a traditional clustering method (k-means), by using multi-objective optimisation (NSGA-II). A multi-objective strategy makes sense in this context, as, in the absence of ground truth data, there are various ways to evaluate the quality of a cluster (see details in Sect. 5.3).

5   Method

Our solution to model the motion of peas is a modified k-means clustering approach, because there are important constraints to consider:

(i) Our pea data is four dimensional (x, y, z, t), however each cluster can only have a max of 1 pea per timestep. This is the biggest difference to vanilla k-means, which would usually assign every available point to a cluster.

(ii) The evaluation of what is considered as a good cluster is a tradeoff between at least two criteria, this is why we rely on a multi-objective optimisation.

5.1   Definitions

A pea in terms of this paper is a point in 4D space (x, y, z, t) (see Fig. 2). We call a physical pea in the stomach a **true pea**. Each **detected pea** is assigned a unique id. Due to various causes, among which false positives due to the image detection algorithm and the above mentioned acquisition process, there are many more detected peas than actual true peas. These pea positions have already been identified in previous work [1, 6, 25].

A **cluster** is a collection of peas, identified as corresponding to the same true pea, but at different timesteps. There are 9 timesteps in Dataset 3 for example, meaning that each cluster can contain up to 9 peas. We know that there are 20 true peas swallowed during the experiments, therefore there are 20 clusters. Not all peas will be clustered. This is because there are more detected peas than possible duplicates.

A **pea vector** is defined as the vector between a pea in cluster c at a timestep t, and another pea in the same cluster c at timestep $t+1$. We define the velocity of a pea at timestep t to be equal to the magnitude of its vector.

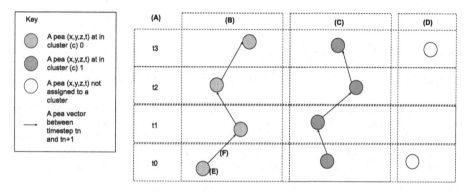

Fig. 2. Our definitions. (A) lists a selection of timesteps. (B) and (C) show two distinct clusters of peas. (D) shows some peas not assigned to a cluster. (E) shows an example pea in a cluster, and (F) is a pea vector between two peas in the same cluster but at different timesteps.

5.2   Evolutionary Algorithm

As part of our solution, we will use NSGA-II [3] for optimisation, using the multi-objective Python framework Pymoo [2]. We have chosen to run it with a population size of 100, and for 100 generations (see Table 2) because it enabled us to obtain good results while keeping computing time rather short. An individual in this study represents the starting positions for the centroids (see Fig. 3) for the k-mean algorithm. In our pipeline, the starting positions for centroids must equal a pea position in the data. Therefore, each individual is defined as a list of size equal to the number of clusters (20), with each value equal to a unique pea id. There cannot be repeated pea ids per individual. Due to changes when mutating individuals, an invalid pea id may be assigned to an individual. Therefore, we include a repair function to change any invalid or duplicate pea ids to a valid id from the left over available pool of pea ids.

Table 2. NSGA-II parameters

Parameter	Value
Individuals	100
Generations	100
Crossover	Simulated Binary Crossover (SBX) [4]
Mutation	Polynomial Mutation (PM) [4]

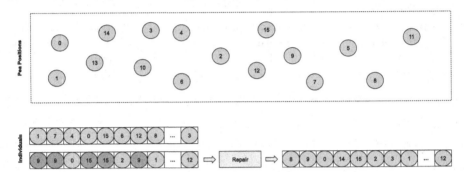

Fig. 3. The individual creation process. It is important to note that here only a small set of peas is shown for simplicity. Each pea has a temporal position and a unique id. Each individual is made up of 20 unique ids. If an invalid id occurs during the mutation process, a repair function is used to replace invalid ids with left over valid ones. The corresponding pea positions from the pea ids are used as the centroid initial positions.

5.3   Assessment of the Clustering Results

Since we have no ground truth available to determine whether the produced clusters are correct, we have to produce alternate measures to assess the quality of our solutions. One measure is to check that each cluster pea velocity is within an expected range. We know a range of food velocity is 1.5–5.0 mm/s [14]. However, there is a degree of uncertainty so we apply a 20 percent tolerance to the upper and lower bounds of the range. To create a metric from this range, we count for each pea, how many velocities are inside the range versus outside, which gives us a percentage out of one hundred.

$$rs = 100 \times \frac{nv}{tv} \tag{1}$$

where: nv = number of velocities in range
tv = total number of velocities

Also, we make the presumption that if the pea motion is accurate, each pea's motion might be correlated with neighbouring peas due to the peristaltic motion caused by the stomach wall. To measure this, we identify the pea vectors for each timestep. We calculate the dot product of each pea vector with each other pea vector at time step t, for all timesteps and calculate the mean of all the dot products:

$$md = \frac{1}{N} \sum_{t=1}^{N} \sum_{j=1}^{n} a_{t,i} \times b_{t,i} \qquad (2)$$

where: $N =$ the number of timesteps

$n =$ the number of components in each vector

$a_{t,i} =$ the i-th of component of vector a at timestep t

$b_{t,i} =$ the i-th of component of vector b at timestep t

A result of 1.0 means that the two vectors a and b are pointing in the same direction, and a result of -1.0 means that the two vectors a and b are pointing in opposite directions. A satisfying clustering result is then a tradeoff between two objectives: a set of smooth pea trajectories but having plausible velocities (in the above specified ranges). We use NSGA-II to supply k-means with potential centroid starting positions. NSGA-II initially starts with a population of random but bounded starting positions and supplies them to k-means. The resulting clusters are then evaluated according to the percentage of pea velocities inside the acceptable stomach velocity range (f_1), and the mean dot product of each cluster compared to every other cluster (f_2) (see Fig. 4).

$$f_1 = rs \qquad (3)$$
$$f_2 = md \qquad (4)$$

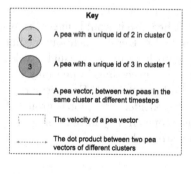
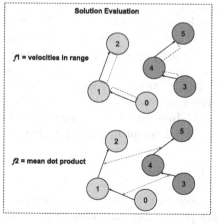

Fig. 4. Each of our metrics for evaluating clustering solutions shown here with only two clusters for simplicity (orange and purple). (Color figure online)

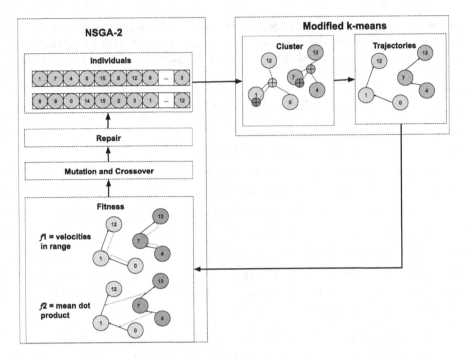

Fig. 5. The pipeline for our optimised k-means algorithm.

A satisfactory $f1$ objective would imply that the euclidean distance between two peas at different timesteps does not exceed a specific range. This is important because there is a physical limit to how fast content can move in the stomach [14]. Furthermore, due to the many different motions the stomach performs (e.g. mixing, grinding), peas should not be able to 'skip' stages by moving too quickly. A satisfactory $f2$ objective would imply that each pea trajectory is similar to many other pea trajectories. This is important because it would suggest that the motion of the peas is correlated as they are mixed in the stomach. Succeeding in both objectives may suggest that the peas are both within a reasonable velocity and are correlated with the other peas, however succeeding in one may have a trade-off with the other, which is why we employ NSGA-II. Furthermore, it is important to note that the $f2$ objective only considers the global group of 20 true peas. In reality, peas in one region of the stomach may have their own local trajectories different to another region of the stomach.

5.4   Pipeline

The customised k-means algorithm runs as follows. First, NSGA-II supplies 20 unique pea ids per individual, which are used to identify the initial centroid starting positions. Next, for each timestep t, find the closest pea to each centroid in the current timestep. Once all timesteps are completed, we compute the mean of each cluster's point set. If none of the means are different to the current centroid's position, end the algorithm. Otherwise, the centroids position becomes the mean and the algorithm iterates again. To prevent an infinite loop, an iteration limit is set to 20 iterations. Once the peas are clustered, each cluster trajectory is calculated and supplied back to NSGA-II to evaluate fitness (see Fig. 5). A tool was implemented to visualise the results of the optimised k-means written in Python with VTK [24]. It shows a semi-transparent boundary of the stomach volume rendered using Marching Cubes [16] with the peas inside rendered using the Glyph Filter from VTK. Each cluster of peas is given a unique cluster colour to differentiate, and arrows showing the previous pea positions at each timestep are also shown. Furthermore, to verify if the optimisation approach is beneficial, it is compared to an approach without optimisation. For the optimisation version, NSGA-II is run with a population of 100 individuals for 100 generations, and for the without version, the modified k-means is run with 10000 random seeds.

6   Results

We have used the MRI Dataset 3 for testing the varying performance of optimised versus non-optimised. Due to the nature of multi objective optimisation, a set of non-dominated solutions which are all non-compromising in either objective are produced (see Fig. 6). The highest range accuracy encountered is 59.17% which has a corresponding dot product score of 0.03. The highest mean dot product is 0.19 which has a corresponding range score of 43.33. To compare the performance of the optimised versus non-optimised, we compare the non dominated solutions with the average results from the non-optimised run. Both methods were run on a PC with an Intel Core i5 CPU with 8 GB of RAM. In terms of computation time, they are both very similar and take around 40 min to complete each (see Table 3). In terms of comparing the k-means to the optimised pipeline, the optimised version appears to perform better. The un-optimised version has a mean of 42.85% for the range metric (see Fig. 7a), and a mean of 0.047 for the mean dot product metric (see Fig. 7b). Compared to the optimised version which has a mean of 54.46% for the range metric and 0.107 for the mean dot product, the optimised version clearly has a performance benefit over the non-optimised pipeline. The optimised version appears to perform better for the first objective (range score), with its minimum (Q0) scoring better than the without optimisation's maximum (Q4), excluding outliers. For the dot product, the optimised version has a much broader range between Q1 and Q3, and it's mean is only slightly higher than the un-optimised versions maximum (Q4). From the pareto front generated by NSGA-II, two solutions were selected. One with the highest range score, and one with the highest mean dot product score. These two solutions provided varying trajectory outcomes depending on the dataset.

A single pea cluster may be more strongly correlated with one other cluster, but have much weaker correlation than others. For this reason, we retrieve the individual mean dot products between each clusters, so these correlation

Table 3. Computation times for the un-optimised versus the optimised.

Method	Running Time (in minutes)
Seeds	41.17
NSGA-II	40.15

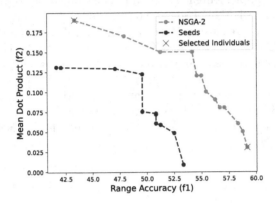

Fig. 6. Pareto front for the optimised k-means, and pareto front from the 10000 seeds approach. The two solutions picked from NSGA-II are marked.

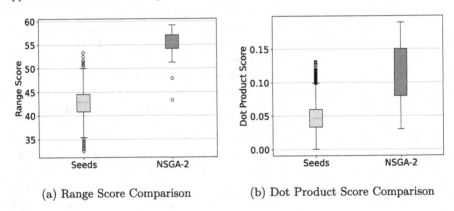

(a) Range Score Comparison   (b) Dot Product Score Comparison

Fig. 7. Comparison between the two k-means centroid initialisation methods: no optimisation (run with 10000 seeds) versus NSGA-II optimisation (non dominated solutions). Figure a shows the comparison in pea velocity range metric, and Figure b shows the comparison in the mean dot product score.

relationships can be observed (see Fig. 8). For example, Cluster 8 is poorly correlated with Cluster 16, but strongly correlated with Cluster 6. Cluster 4 appears to be broadly correlated with many other clusters, whereas Cluster 2 is poorly correlated with most other clusters.

Further analysis was done using the custom visualisation tool we created. Using this tool, we can visualise the pea motion, as well as analyse the various

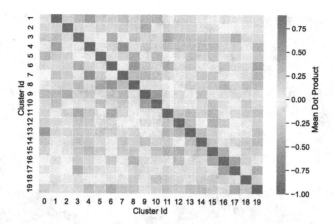

Fig. 8. Heatmap showing how each pea cluster is correlated with each other pea cluster. The solution picked here was the highest range scoring individual so that a more distinct variance of correlation can be shown. Dark blue cells show clusters moving in the same directions at each timestep; dark red cells in opposite directions; and white cells in unrelated directions. (Color figure online)

correlations that we have calculated (see Fig. 9). In Dataset 3, we identified several peas that have correlated movements. One such correlation is located near the antrum of the stomach, and the final pea trajectory appears like the peas are about to leave the stomach (see Fig. 10a). This implies that the peas here may be at a later stage of digestion and are about to be passed to the small intestine. Elsewhere in the stomach, we have found some peas that are correlated for a set amount of timesteps, but depart on later timesteps (see Fig. 10b). This could be due to reconstruction errors, or other factors such as other stomach content changing the pea trajectories. We also performed the optimised k-means on other datasets. Dataset 4 in particular had some interesting findings. Many of the peas had poor correlation, but instead there were what appear to be 'collisions' between peas, resulting in many altered trajectories (see Fig. 11). This could be due to the scan taking place at a different stage of digestion, such as a grinding or mixing stages, and due to a large number of detected peas confined in a smaller area, causing many collisions. Furthermore, its important to note that due to different combinations of bread, pasta, lemon juice or water used in the MRI protocol, the different contents could also have an impact on the pea trajectories.

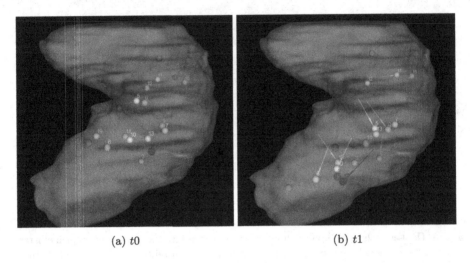

(a) $t0$ (b) $t1$

Fig. 9. The 3D stomach model and peas, showing the highest correlation scoring individual. Each cluster is represented with its own unique colour and ID. Figure a shows the initial pea positions at $t0$, and Figure b shows the trajectories of each pea as they move to $t1$.

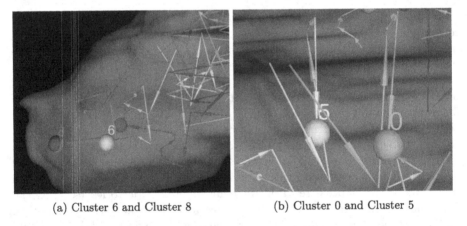

(a) Cluster 6 and Cluster 8   (b) Cluster 0 and Cluster 5

Fig. 10. Two examples of correlated peas from Dataset 3 using the highest correlation scoring individual. Figure a shows an example of two correlated pea clusters: Cluster 6 and Cluster 8 from Dataset 3. Figure b shows an example of two partially correlated pea clusters: Cluster 0 and Cluster 5.

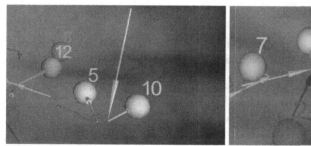 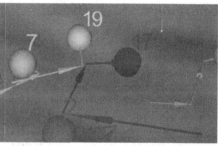

(a) Cluster 5 and Cluster 10   (b) Cluster 17 and Cluster 19

Fig. 11. Using the higher range scoring individual for Dataset 4 had fewer correlated peas compared to Dataset 3, instead featuring many 'collisions' between peas. Figure a shows a collision between Cluster 5 and Cluster 10, and Figure b shows a collision between Cluster 17 and Cluster 19.

7   Conclusion and Future Work

In this paper, we have shown that optimising a clustering algorithm using multi-objective optimisation is highly beneficial. Our optimised version overall out-performed the un-optimised method in each of the measures we have defined. Furthermore, we have found many interesting findings as a result of the clustering, such as groups of peas with similar trajectories and what appear to be peas having altered trajectories due to collisions, varying depending on the dataset. Further work could be done to analyse the trajectories of the peas after they have exited the stomach and entered the small intestine, and beyond. In terms of the evolutionary approach, further parameters could be tuned. For example, since the peas are moving and may after a certain time leave the stomach, the genetic approach could be modified to account for a variable individual size. Furthermore, in these experiments a population of 100 was used, in future much larger runs could be completed to see if it improves the results.

Acknowledgements. This work was partially supported by a STSM Grant from COST Action CA15118 (FoodMC) and the MRI data used in this study were collected at CEA-SHFJ with the support of IR4M CNRS/Orsay University (Xavier Maître and Luc Darrasse) in the framework of the IDI/Paris Saclay PhD of Daniela Freitas.

Ethical Approval. The study protocol has been approved by the Ethics Committee Lyon Sud-Est IV, and it has been registered in the Clinical Trial Registry (clinical-trials.gov; NCT03265392) (see https://clinicaltrials.gov/ct2/show/NCT03265392). All volunteers gave their written informed consent after being provided with oral and written information about the aims and protocol of the study.

References

1. Al-Maliki, S.F., Vidal, F.P.: Visualisation, optimisation and Machine Learning: application in PET Reconstruction and Pea segmentation in MRI Images. Ph.D. thesis, Bangor University (2020)
2. Blank, J., Deb, K.: pymoo: multi-Objective Optimization in Python. IEEE Access **8**, 89497–89509 (2020). https://doi.org/10.1109/ACCESS.2020.2990567
3. Deb, K., Pratap, A., Agarwal, S., Meyarivan, T.: A fast and elitist multiobjective genetic algorithm: NSGA-II. IEEE Trans. Evol. Comput. **6**(2), 182–197 (2002). https://doi.org/10.1109/4235.996017
4. Deb, K., Sindhya, K., Okabe, T.: Self-adaptive simulated binary crossover for real-parameter optimization. In: Proceedings of the 9th Annual Conference on Genetic and Evolutionary Computation, GECCO 2007, pp. 1187–1194. Association for Computing Machinery, New York (2007). https://doi.org/10.1145/1276958. 1277190
5. Freitas, D.: Novel insights into starch digestion and the glycaemic response: from in vitro digestions to a human study using magnetic resonance imaging (MRI). Ph.D. thesis, Université Paris-Saclay (2018)
6. Gardner, J., Al-Maliki, S., Lutton, E., Boué, F., Vidal, F.: Recognising specific foods in MRI scans using CNN and visualisation. In: Ritsos, P.D., Xu, K. (eds.) Computer Graphics and Visual Computing (CGVC). The Eurographics Association (2020). https://doi.org/10.2312/cgvc.20201145
7. Goetze, O., et al.: The effect of macronutrients on gastric volume responses and gastric emptying in humans: a magnetic resonance imaging study. Am. J. Physiol. Gastrointest. Liver Physiol. **292**(1), G11–G17 (2007). https://doi.org/10.1152/ajpgi.00498.2005
8. Goyal, R.K., Guo, Y., Mashimo, H.: Advances in the physiology of gastric emptying. Neurogastroenterol. Motil. **31**(4), e13546 (2019). https://doi.org/10.1111/nmo.13546
9. Hamerly, G., Elkan, C.: Alternatives to the k-means algorithm that find better clusterings. In: Proceedings of the Eleventh International Conference on Information and Knowledge Management, CIKM 2002, pp. 600–607. Association for Computing Machinery, New York (2002). https://doi.org/10.1145/584792.584890
10. Heissam, K., et al.: Measurement of fasted state gastric antral motility before and after a standard bioavailability and bioequivalence 240 mL drink of water: validation of MRI method against concomitant perfused manometry in healthy participants. PLOS ONE **15**(11), e0241441 (2020). https://doi.org/10.1371/journal. pone.0241441. https://dx.plos.org/10.1371/journal.pone.0241441
11. Kong, F., Singh, R.: A model stomach system to investigate disintegration kinetics of solid foods during gastric digestion. J. Food Sci. **73**(5), E202–E210 (2008). https://doi.org/10.1111/j.1750-3841.2008.00745.x
12. Krishna, K., Narasimha Murty, M.: Genetic k-means algorithm. IEEE Trans. Syst. Man Cybern. Part B (Cybern.) **29**(3), 433–439 (1999). https://doi.org/10.1109/3477.764879
13. Kunz, P., Feinle, C., Schwizer, W., Fried, M., Boesiger, P.: Assessment of gastric motor function during the emptying of solid and liquid meals in humans by MRI. J. Magn. Reson. Imaging **9**(1), 75–80 (1999). https://doi.org/10.1002/(SICI)1522-2586(199901)9:1<75::AID-JMRI10>3.0.CO;2-I. https://onlinelibrary.wiley.com/doi/10.1002/(SICI)1522-2586(199901)9:1<75::AID-JMRI10>3.0.CO;2-I

14. Li, Y., Kong, F.: Simulating human gastrointestinal motility in dynamic in vitro models. Compr. Rev. Food Sci. Food Saf. **21**(5), 3804–3833 (2022). https://doi.org/10.1111/1541-4337.13007

15. Likas, A., Vlassis, N., J. Verbeek, J.: The global k-means clustering algorithm. Pattern Recogn. **36**(2), 451–461 (2003). https://doi.org/10.1016/S0031-3203(02)00060-2

16. Lorensen, W., Cline, H.: Marching cubes: a high resolution 3D surface construction algorithm. Comput. Graph. **21**(4), 163–169 (1987)

17. Maccioni, F., Busato, L., Valenti, A., Cardaccio, S., Longhi, A., Catalano, C.: Magnetic resonance imaging of the gastrointestinal tract: current role, recent advancements and future prospectives. Diagnostics **13**(14), 2410 (2023). https://doi.org/10.3390/diagnostics13142410. https://www.mdpi.com/2075-4418/13/14/2410

18. MacQueen, J.: Some methods for classification and analysis of multivariate observations. In: Proceedings of the 5th Berkeley Symposium on Mathematical Statistics and Probability (1967). https://api.semanticscholar.org/CorpusID:6278891

19. Marciani, L., et al.: Antral motility measurements by magnetic resonance imaging. Neurogastroenterol. Motil. **13**(5), 511–518 (2001). https://doi.org/10.1046/j.1365-2982.2001.00285.x

20. Menys, A., et al.: Spatio-temporal motility MRI analysis of the stomach and colon. Neurogastroenterol. Motil. **31**(5), e13557 (2019). https://doi.org/10.1111/nmo.13557. https://onlinelibrary.wiley.com/doi/10.1111/nmo.13557

21. Nonaka, H., Onishi, H., Watanabe, M., Nam, V.H.: Assessment of abdominal organ motion using cine magnetic resonance imaging in different gastric motilities: a comparison between fasting and postprandial states. J. Radiat. Res. **60**(6), 837–843 (2019)

22. Parker, H.L., et al.: Clinical assessment of gastric emptying and sensory function utilizing gamma scintigraphy: establishment of reference intervals for the liquid and solid components of the nottingham test meal in healthy subjects. Neurogastroenterol. Motil. **29**(11), e13122 (2017). https://doi.org/10.1111/nmo.13122

23. Peña, J., Lozano, J., Larrañaga, P.: An empirical comparison of four initialization methods for the k-means algorithm. Pattern Recogn. Lett. **20**(10), 1027–1040 (1999). https://doi.org/10.1016/S0167-8655(99)00069-0

24. Schroeder, W., Martin, K., Lorensen, B.: The Visualization Toolkit, 4th edn. Kitware (2006)

25. Spann, C., Al-Maliki, S., Boué, F., Lutton, E., Vidal, F.P.: Interactive visualisation of the food content of a human stomach in MRI. In: Computer Graphics and Visual Computing (CGVC), pp. 47–54. The Eurographics Association (2022). https://doi.org/10.2312/cgvc.20221171

26. Steingoetter, A., et al.: Magnetic resonance imaging for the in vivo evaluation of gastric-retentive tablets. Pharm. Res. **20**(12), 2001–2007 (2003). https://doi.org/10.1023/B:PHAM.0000008049.40370.5a. http://link.springer.com/10.1023/B:PHAM.0000008049.40370.5a

27. Tao, F., Lin-sheng, L., Qi-chuan, T.: A novel adaptive motion detection based on k-means clustering. In: 2010 3rd International Conference on Computer Science and Information Technology, vol. 3, pp. 136–140 (2010). https://doi.org/10.1109/ICCSIT.2010.5564529

28. Vidal, F.P., Villard, P.F., Lutton, É.: Tuning of patient-specific deformable models using an adaptive evolutionary optimization strategy. IEEE Trans. Biomed. Eng. **59**, 2942–2949 (2012). https://doi.org/10.1109/TBME.2012.2213251

Author Index

Printed in the United States
by Baker & Taylor Publisher Services